Advanced Reinforced Concrete Design

(IS:456-2000)
(BSEN:1992-1-1-2004)
(ACI:318M-2011)

Third Edition

Advanced Reinforced Concrete Design

(IS:456-2000)
(BSEN:1992-1-1-2004)
(ACI:318M-2011)

Third Edition

N Krishna Raju
BE, MSc (Engg), PhD, MIE, MI (Struct E)

Emeritus Professor of Civil Engineering
MS Ramaiah Institute of Technology
Bengaluru

CBS

CBS Publishers & Distributors Pvt Ltd

New Delhi • Bengaluru • Chennai • Kochi • Kolkata • Mumbai
Bhopal • Bhubaneswar • Hyderabad • Jharkhand • Nagpur • Patna • Pune
• Uttarakhand • Dhaka (Bangladesh) • Kathmandu (Nepal)

ISBN: 978-81-239-2960-6

Copyright © Author and Publisher

Third Edition: 2016
  Reprint: 2020

First Edition: 1986
  Reprint: 1988, 1998, 1999, 2001, 2003
Second Edition: 2005
  Reprint: 2007, 2008, 2009, 2010, 2012, 2013, 2014

Published by Satish Kumar Jain and produced by Varun Jain for

CBS Publishers & Distributors Pvt Ltd

4819/XI Prahlad Street, 24 Ansari Road, Daryaganj, New Delhi 110 002, India.
Ph: 23289259, 23266861, 23266867   Website: www.cbspd.com
Fax: 011-23243014   e-mail: delhi@cbspd.com; cbspubs@airtelmail.in.
Corporate Office: 204 FIE, Industrial Area, Patparganj, Delhi 110 092

Ph: 011-4934 4934   Fax: 011-4934 4935   e-mail: publishing@cbspd.com; publicity@cbspd.com

Branches

- **Bengaluru:** Seema House 2975, 17th Cross, K.R. Road,
  Banasankari 2nd Stage, Bengaluru 560 070, Karnataka
  Ph: +91-80-26771678/79   Fax: +91-80-26771680   e-mail: bangalore@cbspd.com
- **Chennai:** 7, Subbaraya Street, Shenoy Nagar, Chennai 600 030, Tamil Nadu
  Ph: +91-44-26680620, 26681266   Fax: +91-44-42032115   e-mail: chennai@cbspd.com
- **Kochi:** 68/1534, 35, 36, Power House Road, Opp. KSEB, Kochi 682018, Kerala
  Ph: +91-484-4059061-65   Fax: +91-484-4059065   e-mail: kochi@cbspd.com
- **Kolkata:** 6/B, Ground Floor, Rameswar Shaw Road, Kolkata-700 014, West Bengal
  Ph: +91-33-22891126, 22891127, 22891128   e-mail: kolkata@cbspd.com
- **Mumbai:** 83-C, Dr E Moses Road, Worli, Mumbai-400018, Maharashtra
  Ph: +91-22-24902340/41   Fax: +91-22-24902342   e-mail: mumbai@cbspd.com

Representatives

• **Bhopal** 0-8319310552	• **Bhubaneswar** 0-9911037372	• **Hyderabad** 0-9885175004	• **Jharkhand** 0-9811541605
• **Nagpur** 0-9421945513	• **Patna** 0-9334159340	• **Pune** 0-9623451994	• **Uttarakhand** 0-9716462459
• **Dhaka (Bangladesh)** 01912-003485	• **Kathmandu (Nepal)** 977-9818742655		

Printed at: Rashtriya Printers Delhi-110095

Preface to the Third Edition

During the last nine years, the second edition of *Advanced Reinforced Concrete Design* was reprinted seven times indicating the ever increasing popularity of the book among engineering students, teachers and practising structural engineers. The book has served as a useful text for students, reference guide for research workers and examiners, and as a ready reckoner in the design offices of structural engineering firms. The second edition has served its purpose by clearly establishing itself as the leading title on the subject of advanced reinforced concrete structural design among the various books presently available in the Indian market.

The third edition of this popular title incorporates the latest developments in the field of structural design of various types of reinforced concrete structural elements conforming to the latest Indian (IS:456-2000), British (BS EN:1992-1-1-2004) and American (ACI:318M-2011) codes of practice. A brief historical review along with the basic principles and philosophy of elastic and limit state designs are included. The new edition updates the contents in each chapter and new chapters are added to emphasize the latest developments in the field of reinforced concrete structures.

Major improvements incorporated in this edition include the design and constructional aspects of large capacity storage structures like bins using dome technology and detailed design procedure for reinforced concrete column brackets or corbels. The introduction to each of the chapters with copious references along with additional design examples will immensely help the students. The assignment compiled at the end of each chapter will be very useful to the students as well as teachers and examiners. The review and objective type questions are intended to test the deeper understanding of the subject by the readers and will also help them to prepare for competitive examinations.

The author gratefully acknowledges the help rendered by his wife Pramila and many of his colleagues and practising structural engineers for their periodical constructive suggestions in the revision of this monograph. Finally, the author welcomes constructive criticism and useful suggestions which will help in updating the contents of the book.

N Krishna Raju

Preface to the First Edition

The extensive use of reinforced concrete for a variety of structural members has necessitated a proper understanding of the design in structural concrete members by the structural engineers. The widespread use of reinforced concrete is the natural outcome of the rapid development in the theory and design procedures with the introduction of the philosophy of limit state design.

The book presents the design of a variety of reinforced concrete structures like continuous beams, portal frames, silos, buners, chimneys, shells, overhead water tanks, Vierendeel girders, trusses, deep beams, box culverts, folded plates, hyperbolic cooling towers, curved girders, poles, pipes and bridge deep systems. The designs of these advanced reinforced concrete structures conform to the revised Indian standard code IS:456-1978 and SI units have been adopted for all the design examples. The topics covered are intended to meet the requirements of graduate and postgraduate curricula of most of the engineering institutions in India. The book is primarily design oriented with more emphasis on types of design with minimum extent of theory, presented wherever required for application in design. The various design steps are identified and provided in a logical sequence.

The book is extensively illustrated with working drawings showing the reinforcement details. The example for practice provided at the end of each chapter is intended to help the students preparing for university examination.

The references provided at the end of the book have been extensively used in the preparation of the text and are gratefully acknowledged. The author is grateful to his wife Pramila and daughters Sarvamangala and Amrutha for extending their fullest cooperation in the preparation of the typescript. Finally, the author welcomes constructive criticisms and suggestions which will immensely help in updating the contents of the book.

N Krishna Raju

Contents

Notations

A	Cross-sectional area
A_c	Concrete cross-sectional area
a	Lever arm
b	Breadth of beam or shorter dimension of a rectangular column
b_{ef}	Effective width of slab
b_f	Effective width of flange
b_w	Breadth of web or rib
D	Overall depth of beam or slab or diameter of column; dimension of a rectangular column in the direction under consideration
D_f	Thickness of flange
DL	Dead load
d	Effective depth
d'	Depth of compression reinforcement from the highly compressed face
E_c	Modulus of elasticity of concrete
E_{ce}	Effective modulus of elasticity of concrete
EL	Earthquake load
E_s	Modulus of elasticity of steel
e	Eccentricity
F	Resisting force
f_{ck}	Characteristic cube compressive strength of concrete
f_c'	Cylinder compressive strength
f_{ctm}	Tensile strength of concrete
f_{ct}	Split tensile strength of concrete
f_{cr}	Modulus of rupture of concrete (flexural strength of concrete)
f_d	Design strength
f_y	Characteristic tensile strength of steel
g	Gravity load or dead load
h	Overall height of retaining wall
h_s	Height of stem
I	Second moment of area or moment of inertia
I_{ef}	Effective moment of inertia
I_{gr}	Moment of inertia of gross-section excluding reinforcement
I_r	Moment of inertia of cracked section
j	Lever arm factor
K	Stiffness of member
k	Constant or coefficient or factor
Ld	Development length
LL	Live load
L	Length of a beam or column between adequate lateral restraints or the unsupported length of a column
L_{ef}	Effective span of beam or slab
L_x	Length of shorter side of slab
L_y	Length of longer side of slab
L_{ex}	Effective span length along XX axis
L_{ey}	Effective span length along YY axis
L_n	Clear span face to face of supports
L_1	Span in the direction in which moments are determined, c/c of supports

L_2	Span transverse to L_1, centre to centre of supports
L_0	Distance between points to zero moments in a beam
M	Bending moment
M_r	Moment of resistance
m	Modular ratio
n_a	Actual neutral axis depth
n_c	Critical neutral axis depth
P	Axial load on a compression member (wind intensity)
p	Safe bearing capacity of soil or intensity of pressure
p_t	Percentage reinforcement in tension
p_c	Percentage reinforcement in compression
q	Shear stress
Q	Design coefficient
r	Radius
S_v	Spacings of stirrups
T	Torsional moment
V	Shear force
w	Distributed load per unit area
W	Total load or concentrated load
WL	Wind load
x_u	Neutral axis depth
Z	Modulus of section
δ	Displacement
γ_f	Partial safety factor for load
γ_m	Partial safety factor for material
μ	Coefficient of friction or coefficient of orthotropy
σ_{cbc}	Permissible stress in concrete in bending compression
σ_{cc}	Permissible stress in concrete in direct compression
σ_{sc}	Permissible stress in steel in compression
σ_{st}	Permissible stress in steel in tension
σ_{sv}	Permissible tensile stress in shear reinforcement
τ_{bd}	Design bond stress
τ_c	Shear stress in concrete
$\tau_{c,max}$	Maximum shear stress in concrete with shear reinforcement
τ_v	Nominal shear stress
ϕ	Diameter of bar
ψ_{cs}	Shrinkage curvature
ε	Support of span ratio
ε_{cc}	Strain in concrete
ε_{cd}	Drying shrinkage strain
ε_{ca}	Autogenous shrinkage strain
ε_{cs}	Total shrinkage strain
ε_{sc}	Strain in steel
φ	Creep coefficient
ϑ	Poisson's ratio
α, β	Angles or ratio
λ	Multiplying factor

1

Basic Principles of Reinforced Concrete Design

1.1　INTRODUCTION

Reinforced concrete is a well established construction material often preferred to steel mainly due to its universal adaptability, versatility, coupled with resistance to fire and corrosion resulting in negligible maintenance costs. Reinforced concrete is a universally suitable composite material ideally suited for structural members like slabs, beams, columns, walls, footings, water tanks, retaining walls, staircases, electric poles, pipes, piles and dams, pavements, marine structures, swimming pools, cooling towers, bunkers, silos, chimneys, tunnels, shells and folded plates. The ideal combination of concrete and steel in a composite form serves as an integrated material mainly due to the similarity of coefficients of thermal expansion of the individual materials in the range $7–12 \times 10^{-6}/°C$. The composite action of this material in structural elements is attributed to the excellent bond between concrete and steel reinforcements which ensures strain compatibility so that the external loads on the structural elements is shared by both steel and concrete without disruption of the composite material.

The development of reliable design and construction procedures during 20th century has paved the way for extensive use of reinforced concrete in the construction industry throughout the world. The amenability of concrete to be cast in various shapes and with attractive surface characteristics is a salient feature for preference of this material by architects and engineers in comparison with other materials. Reinforced concrete is a structural material with desirable properties like mouldability, strength, elasticity, durability and impermeability, good resistance to static, fatigue and dynamic loads. Reinforced concrete has established itself as an universally economical material mainly due to the sustaining efforts of various research workers during the last two centuries resulting in its acceptance as a standard material codified for widespread use by the leading countries of the world.

1.2　HISTORICAL BACKGROUND

The present development in the field of reinforced concrete is attributed to the continuous research work done during the last 150 years. The invention of Portland cement by Joseph Aspdin in 1845 paved the way for early pioneers like Joseph Monier[1], Francois Coignet[2] and Joseph Lambot[3] to embed steel rods in concrete heralding the birth of reinforced concrete in 1849. The first authentic publication of a book on *Reinforced Concrete Construction* is attributed to Turneaure and Maurer[4] and the cover page of this book is shown in Fig. 1.1. During the first few decades of 20th century several books were authored by engineering scientists like Turneaure and Maurer, George Hool[5], Faber & Bowie[6] and Whitney *et al*[7] providing valuable information on reinforced concrete. Phenomenal developments in the field of concrete technology has paved the way for the production of high strength, ultra high strength, high performance[8] and nano concretes[9] with a compressive strength in the range 120–200 N/mm².

The development of prestressed concrete by Freyssinet[10] in 1950 paved the way for production of high quality and high strength cements coupled with vibration

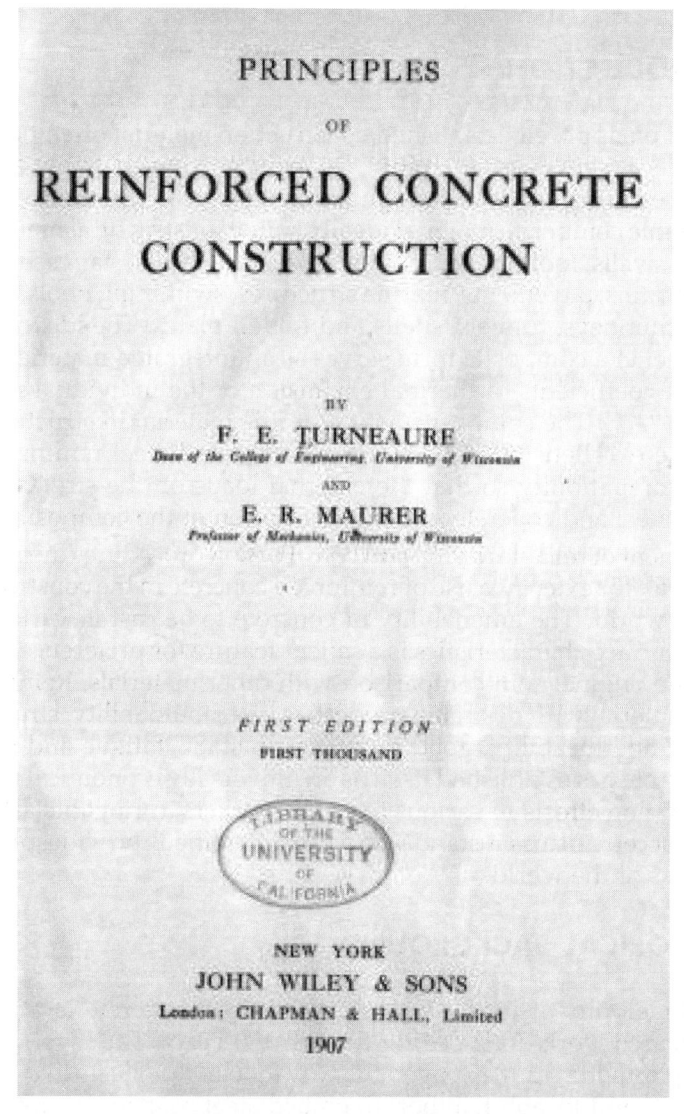

Fig. 1.1: First book on reinforced concrete

techniques for compacting concrete resulting in high strength and durability of hardened concrete. The quality of steel also improved with the production of different types of steel like mild and high yield strength deformed bars of different grades suitable for different types of reinforced concrete structures.

In 1906, the French commission on reinforced concrete formulated the design rules while the Prussian regulations comprising the complete set of design rules of reinforced concrete appeared in 1907. Professional societies like the American Concrete Institute (ACI) and the American Society of Civil Engineers (ASCE) introduced the first joint code on reinforced concrete in 1909.

The first major application of reinforced concrete was in bridges mainly due to the economy in comparison with steel bridges. The *elastic method of design* was firmly established and widely used during this period. Rebuilding of bridges and buildings during the post war periods resulted in establishing reinforced concrete as an economical structural material for use in different types of structures. However, the inadequacy of the elastic or working load design in predicting the ultimate loads of a structure paved the way for the ultimate load theories and designs based on ultimate loads computed by applying load factors to the working loads.

Various investigators like Emperger (1936), Whitney (1937), Jenson (1943), Chambaud (1949), Hognestad[11](1951) and Evans[12] developed the *ultimate load design* based on different types of stress blocks. Reinforced concrete structures designed solely on the basis of ultimate load theory resulted in slender structural elements and their serviceability characteristics (deflections and cracks) under working loads were not within the codified acceptable limits.

The ultimate load method of design ensures the safety of the structures against the collapse limit state only and as such does not give any information about the behaviour of the structure at service loads and the range between service and collapse loads. The inadequacy of the ultimate load method in not ensuring the serviceability of the structure resulted in the development of *limit state design*[13].

The deficiencies in elastic and ultimate load design resulted in the evolution of *limit state design philosophy*[14,15], first incorporated in the Russian code in 1955. Basically, limit state design is a method of designing structures based on a statistical concept of safety and the associated statistical probability of failure. Limit state design is based on the concept of probability and comprises the application of the method of statistics to the variations that occur in practice in the loads acting on the structure and the strength of the materials.

The limit state design overcomes the inadequacies of the working stress and ultimate load methods and ensures the safety of the structure against excessive deflections and cracking under service loads and also provides the desirable load factor against failure. Hence, the British, American, Australian, German, Canadian and the Indian standard codes have adopted the limit state design concepts.

1.3 STRUCTURAL DESIGN PHILOSOPHY

The philosophy of structural design incorporated in the various national and international codes specifies that any design should comply with the following essential requirements:

i. Structures designed should satisfy the criterion of desirable ultimate strength, in flexure, shear, compression, tension and torsion developed under a given system of loads, and their combinations. In addition, the stresses developed in the structure under the given system of loads should be within the safe permissible limits under service loads.

ii. The structure designed should satisfy the criterion of serviceability, which limits the deflections and cracking to be within acceptable limits. The structure should also have adequate durability and impermeability, resistance to acids, corrosion, frost, etc.

iii. The structure should have adequate stability against overturning, sliding, buckling, and vibration under the action of loads.

A satisfactory structural design should ensure the three basic criteria of strength, serviceability and stability. In addition, the structural designer should also consider aesthetics and economy. The structural designer and the architect should coordinate so that the structures designed are not only aesthetically superior, but strong enough to safely sustain the designed loads without any distress during the lifetime of the structure.

According to Moseley et al[16], the structural design process may be considered as a series of inter-related and overlapping stages, in their simplest forms consisting of:

- Conceptual design in which a range of potential structural forms and materials are considered.
- Preliminary design which will typically involve simple and approximate hand computations to assess the viability of a range of alternative conceptual solutions.
- Detailed design to include comprehensive analysis and calculations for the selected scheme of design using suitable computer software.

1.4 REINFORCED CONCRETE APPLICATIONS

The universal adaptability, versatility, moldability, durability and fire resistance of reinforced concrete is established. Consequently, the material is widely adopted for innumerable number of structural elements due to its low cost in comparison with steel. The following list illustrates the salient applications of reinforced concrete in the field of construction industry:

1. Building elements comprising floor and roof slabs, beams, columns
2. Flat slabs, grid or coffered floors suited for conference halls and auditoriums
3. Folded plates, trusses and shell structures for aircraft hangers and workshops
4. Liquid retaining structures like water tanks, cement silos, chimneys and cooling towers
5. Tall TV transmission towers, multistorey buildings
6. Radiation absorption rooms for cancer treatment in hospitals
7. Pressure vessel construction in atomic power generation structures
8. Pipes for water supply and sewage systems
9. Poles for electric transmission lines
10. Sleepers for laying railway tracks
11. Soil retaining structures like embankments, retaining walls

12. Hydraulic structures like dams, canals, barrages and aqueducts

13. Foundation structures like piles, wells and caissons

14. Marine structures like wharfs, quay walls, watch towers and lighthouses

15. Highway and railway bridge structures

1.5 NATIONAL CODES AND HANDBOOKS

Objectives of National Codes

Standard national codes documenting the design philosophy of reinforced concrete structures have been released by most of the countries periodically based on extensive research and practical knowledge. These serve as guidelines for the design of reinforced concrete structures. The principal objectives of the code of practice can be summarized as:

1. Provision of adequate safety by ensuring strength, serviceability and durability.

2. Protection of structural engineers from any liability due to failure of structures caused because of faculty design and improper materials and lack of proper supervision during construction.

3. To provide a uniform set of design guidelines to be followed by various structural designers and engineers in the country.

4. To provide simple design procedures, design tables and formulae for easy computations.

The national building codes are periodically revised to reflect the improvements in the quality of materials and design practices evolved as a result of comprehensive research investigations conducted at various institutions in the country and abroad.

Design Codes and Handbooks

Codified procedures have been made mandatory for the design of a reinforced concrete structures by most of the countries. The various national codes of the prominent countries are listed below:

1. Indian standard code of practice: IS:456-2000 for plain and reinforced concrete[17].

2. British code: BS EN:1992-1-1, Euro code 2-Design of Concrete Structures, General Rules and Rules for Buildings, British Standards Institution, 2004[18].

3. American code: ACI:318M-11, (metric), Building Code Requirements for Structural Concrete and Commentary, American Concrete Institute, 2005[19].

4. AS:3800-1988, Concrete Structures, Standards Association of Australia, 1988[20].

5. DIN:1045-1988, Structural Use of Concrete, Design & Construction, Din Deutsches Institute Fir Normung EV, 1988[21].

6. CSA Standard A23.1-00/A23.2-00, Concrete Materials and Methods of Concrete Construction/Methods of Test for Concrete, Canadian Standards Associations, Toronto, 2000[22].

The following special publications and handbooks will be useful for structural engineers in design offices for rapid computations involved in the design of structural concrete elements.

1. SP:16-1980[23] → *Design Aids for Reinforced Concrete (IS:456)*

2. SP:24-1983[24] → *Explanatory Handbook on IS:456-1978*

3. SP:34-1987[25] → *Handbook on Concrete Reinforcement and Detailing*
4. IS:10262-1982[26] → *Recommended Guidelines for Concrete Mix design*
5. *Reynold's Reinforced Concrete Designers Handbook*[27]
6. *Handbook of Concrete Engineering* by Mark Fintel[28]

1.6 LOADING STANDARDS

Reinforced concrete structural design requires knowledge of various types of loads acting on the member. Generally, there are five types of loads which can act on any member. These may be grouped as: (a) dead loads, (b) live loads, (c) wind loads, (d) earthquake loads and (e) dynamic loads.

a. *Static or dead loads*: The dead or static loads constitute generally the dead weight of various types of building materials as specified in the Indian standard code, IS:875-1987 (part 1)[29], which are shown in Table 1.1.

b. *Live or imposed loads*: Live or imposed loads changing with time are due to people or vehicles using the floor. These are prescribed in IS:875-1987 (part 2)[30] and are shown in Table 1.2.

c. *Wind loads*: In case of tall structures like cooling towers, poles, chimneys and multistorey buildings, wind loads have to be considered in the design. Wind loads depend upon the intensity of wind in the locality and the height and shape of the structure to be designed. The Indian standard code prescribes basic wind speeds in various zones by dividing the country into 6 zones. The design wind pressure is computed as,

$$p_x = V_x^2$$

where,

p_x = design wind pressure in N/mm² at a height Z
V_x = design wind velocity in m/s at a height Z.

Wind load F, acting in a direction normal to the individual structural element or cladding unit is computed using the relation,

$$F = (C_{pe} - C_{pi}) A p_d$$

where,

C_{pe} = external pressure coefficient
C_{pi} = internal pressure coefficient
A = surface area of structural element or cladding unit
p_d = design wind pressure.

Table 1.1: Unit weights of building materials (IS:875-1987) (part 1)

Type of material	Unit weight (kN/m³)
Brick masonry	18.85–22.00
Stone masonry	20.40–26.50
Cement concrete	22.00–23.50
Reinforced concrete	22.75–26.50

(Contd.)

Table 1.1: Unit weights of building materials (IS:875-1987) (part 1) *(Contd.)*

Type of material	Unit weight (kN/m^3)
Cement mortar	20.40
Terrazzo (10 mm thick)	0.24
Mastic asphalt (10 mm thick)	0.215
Brick wall (100 mm thick)	1.91
Brick wall (200 mm thick)	3.84
Laterite	20.40–23.55
Granite	25.90–27.45
Marble	26.70
Asbestos cement sheeting	0.118–0.130
Common burnt clay bricks	15.70–18.85
Hollow concrete blocks	1.41
Solid concrete blocks	17.65
Plain concrete (IS:456-2000)	24
Reinforced concrete (IS:456-2000)	25
Rubble masonry	20.8
Concrete tile flooring (25 mm thick)	0.5
Dry soil	13.85–18.05
Moist soil	15.70–19.60
Fine sand (dry)	15.10–15.70
Aggregate (stone dry)	15.70–18.35
Teakwood	0.28
Steel	78.5
Cement plaster (10 mm thick)	0.20
Floor finishes	0.6–1.2
Roof finishes	0.2–1.2

Table 1.2: Live or imposed loads (IS:875-1987) (part 2)

Loading class (kN/m^2)	Types of floors	Minimum load (kN/m^2)
2	Floors in dwelling houses, tenements, hospital wards, bed rooms and private sitting rooms in hostels and dormitories	2
2.5	Office floors other than entrance hall floors of light work rooms	2.5 to 4.0
3.0	Floors in banking halls, office entrance halls and reading rooms	3.0
4.0	Shop floors used for display and sale of merchandise, floors of work rooms, floors of classrooms, restaurants, machinery halls, power, stations, etc., where not occupied by plant or equipment	4.0

(Contd.)

Table 1.2: Live or imposed loads (IS:875-1987) (part 2) *(Contd.)*

Loading class (kN/m²)	Types of floors	Minimum load (kN/m²)
5.0	Floors of warehouses, workshops, factories and other buildings or parts of building or similar category for light weight loads, office floors for storage and filling purposes. Assembly floor space without fixed seating, public rooms in hotels, dance halls and waiting halls.	5.0
7.5	Floors of warehouses, workshops, factories and other buildings or part of buildings of similar category for medium weight loads	7.5
10.0	Floors of warehouses, workshops, factories and other buildings or parts of buildings of similar category for heavy weight loads, floors of book stores and libraries.	10.0
	Garages (light)	
	Floors used for garages for vehicles not exceeding 25 kN gross weight	
	Slabs	4.0
	Beams	2.5
	Garages (heavy)	
	Floors used for garages for vehicles not exceeding 40 kN gross weight	7.5
	Staircases	
	Stairs, landings and corridors for class 2, but not liable for overcrowding	3.0
	Stairs, landings and corridors for class 2, but liable for overcrowding	5.0
	Balcony	
	Balconies not liable to overcrowding for class 2 loading	3.0
	Loading for other classes	5.0
	Balconies liable to overcrowding	5.0
	Roofs	
	Flat, sloping or curved roof with slopes up to and including 10° a. Access provided	1.5
	b. Access not provided, except for maintenance	0.75
	c. Sloping roof with slope greater than 10°...0.75 kN/m² less 0.001 kN/m² for every increase in slope of over 10° up to and including 20° and 0.002 kN/m² for every degree increase in slope over 20°.	

The values of internal and external pressure coefficients depend upon the type of structure and are compiled in a tabular form in IS:875-1987 (part 3)[31].

d. *Snow loads and local combinations*: In locations where snow is encountered, structures located in such zones have to be designed to resist the snow loads prevailing in the region and also the various load combinations. The snow loads are prescribed in IS:875-1987 (part 4)[32]and the special loads and their combinations are specified in IS:875-1987 (part 5)[33].

e. *Seismic or earthquake loads:* Based on seismic studies, the country has been divided into five zones depending upon the severity of the intensity of earthquake prevalent in the zone. Horizontal seismic forces induced due to the earthquake are specified in IS:1893-1984[34] and these forces should be considered in the design of reinforced concrete structures located in the respective zones. The horizontal seismic force is computed as,

$$F_{eq} = [\alpha\beta\lambda G]$$

where,

α = horizontal seismic coefficient depending on location with values of 0.08, 0.05, 0.04, 0.02 and 0.01 for zones V, IV, III, II and I respectively

β = a coefficient depending on soil-foundation system ranging from 1.0 to 1.5

λ = a coefficient depending upon the importance of the structure varying from 1.5 to 1.0

G = dead load above the section considered.

Structures located in zones III to V are grouped under severe earthquake zones and it is mandatory to design them as specified in IS:4326-1993[35].

In the limit state method, the Indian code specifies that for earthquake resistant structures, the following load combination should be considered in design:

a. At limit state of collapse $\qquad$ [1.5 DL + 1.2 LL + 1.2 EL]

b. At limit state of serviceability $\qquad$ [1.0 DL + 0.8 LL + 0.8 EL]

The American code, ACI:318M-11 code categorises six zones designated as A, B, C, D and E depending upon the severity of the earthquake. Concrete structures resisting seismic forces should be designed for the load combination defined as,

U = [1.2 D + 2 E + 1.0 L + 0.2 S]

where,

U = ultimate load

D = dead load

E = earthquake load

L = live load

S = snow load

The British Euro code BS EN 1998-1:2004[36] specifies detailed rules and recommendations for the design of structures for earthquake resistance. This code is comprehensive and covers the various aspects like the type soil profile, intensity of the seismic activity, type of structure and its behavior under earth quake forces, ductility of the structure and also covers the detailed design and detailing of reinforced concrete structures prone to earthquakes.

The British Euro code is comprehensive since it covers in six parts the various aspects of seismic design of structures like, buildings, bridges, strengthening and repairs of buildings (retro fitting), silos, liquid retaining structures, pipelines, foundations, retaining structures, towers, masts and chimneys.

References

1. Monier J, *History of Concrete and Cement*, Inventors Library Records, Paris, 1849.
2. Coignet F, First Reinforced Concrete 3 Storeyed Building, Library Archives, Paris, 1853, Acquired a French Patent for Reinforced Concrete in 1855.
3. Lambot J, Concrete Boat with Embedded Iron Bars Exhibited in Paris Exhibition, French Library Archives, 1854.
4. Adams H and Mathews ER, *Reinforced Concrete Construction*, 1911–1920.
5. Hool GA, *Reinforced Concrete Construction–Fundamental Principles*, Vol.1, 1912.
6. Faber O and Bowie PG, *Reinforced Concrete–Theory & Practice*, Vol. 1 & 2, 1912–1920.
7. Whitney CA and Hool GA, Concrete Designers Manual, 1921.
8. Russell HG, What is high performance concrete? *Journal of Concrete Products*, January, 1999.
9. Balaguru PS and Chong K, Nanotechnology & Concrete, Proceedings of the ACI Session on Nanotechnology of Concrete, Recent Developments & Future Prospective, Nov. 7, 2006, Denver, USA, pp. 15–28.
10. Freyssinet E, *The Birth of Prestressing*, Cement and Concrete Association, Translation No. CJ. 59, London, 1956, p. 44.
11. Hognestad *et al*, Concrete stress distribution in ultimate strength design, *J. of the ACI*, Vol. 52, Dec. 1955, pp. 455-479.
12. Evans RH, The plastic theories for the ultimate strength of reinforced concrete beams, *Journal of the Institution of Civil Engineers*, London, Dec. 1943, pp. 98–121.
13. Bate SCC, Why limit state design, *Concrete*, March 1968, pp. 103–108.
14. Rowe RE, Cranston WB, and Best BC, New concepts in the design of structural concrete, *Structural Engineer*, Vol. 43, 1965, pp. 339–403.
15. Krishna Raju N, Limit State Design for Structural Concrete, Proceedings of the Institution of Engineers (India), Vol. 51, Jan. 1971, pp. 138–143.
16. Moseley B, Bungey J and Hulse R, *Reinforced Concrete to Euro Code-2*, 7 edn, Palgrave Macmillan, London, 2012.
17. IS:456-2000, Indian Standard Code of Practice for Plain and Reinforced Concrete (Fourth Revision), Bureau of Indian Standards, New Delhi, July 2000.
18. BS EN:1992-1-1, Euro Code-2, Design of Concrete Structures, General Rules & Rules for Buildings, British Standards Institution, London, 2004.
19. ACI:318M-11, Building Code Requirements for Structural Concrete, American Concrete Institute, Farmington Hills, Michigan, 2005, p. 443.
20. AS:3800-1988, Concrete Structures, Standards Association of Australia, 1988.
21. DIN:1045-1988, Structural Use of Concrete, Design & Construction, Din Deutsches Institute Fir Normung EV, 1988.
22. CSA Standard A23.1-00/A23.2-00, Concrete Materials and Methods of Concrete Construction/Methods of Test for Concrete, Canadian Standards Associations, Toronto, 2000.
23. SP:16-1980, Design Aids for Reinforced Concrete to IS:456, Bureau of Indian Standards, New Delhi, 1980
24. SP:24-1983, Explanatory Handbook on IS:456, Bureau of Indian Standards, New Delhi, 1983.
25. SP:34-1987, Handbook of Concrete Reinforcement and Detailing, Bureau of Indian Standards, New Delhi, 1987.

26. IS:10262-1982 (reaffirmed in 1999), Recommended Guidelines for Concrete Mix Design, Bureau of Indian Standards, New Delhi.

27. Reynolds CE, Steedman JC and Thvelfall AJ, *Reinforced Concrete Designers Handbook*, 11 edn, Psychology Press, London, 2008, p. 401.

28. Fintel M, *Handbook of Concrete Engineering*, 2 edn, CBS Publishers, New Delhi, 1986, p. 892.

29. IS:875-1987 (part 1), Code of Practice for Design Loads (other than earthquake) for Buildings and Structures, part 1, Dead Loads (Second Revision), Bureau of Indian Standards, New Delhi. 1989.

30. IS:875-1987 (part 2), Code of Practice for Design Loads (other than earthquake) for Buildings and Structures, part 2, Imposed Loads (Second Revision), Bureau of Indian Standards, New Delhi, 1989.

31. IS:875-1987 (part 3), Code of Practice for Design Loads (other than earthquake) for Buildings and Structures, part 3, Wind Loads (Second Revision), Bureau of Indian Standards, New Delhi, 1989.

32. IS:875-1987 (part 4), Code of Practice for Design Loads (other than earthquake) for Buildings and Structures, part 4, Snow Loads (Second Revision), Bureau of Indian Standards, New Delhi, 1989.

33. IS:875-1987 (part 5), Code of Practice for Design Loads (other than earthquake) for Buildings and Structures, part 5, Special Loads and Combinations (Second Revision), Bureau of Indian Standards, New Delhi, 1989.

34. IS:1893-2002, Criteria for Earthquake Design of Structures (part 1), General Provisions and Buildings (Fifth Revision), Bureau of Indian Standards, New Delhi, 2005.

35. IS:4326-1993, Code of Practice for Earthquake Resistant Design and Construction of Buildings, Bureau of Indian Standards, New Delhi, 1993 (reaffirmed in 1998).

36. BS EN 1998-1:2004, British Standard/Euro Code-8: Design of Structures for Earthquake Resistance, part 1 to part 6, British Standards Institution, London, 2004-2006.

Review Questions

1. What makes reinforced concrete to behave like a composite material?

2. List the pioneers who developed the design rules for reinforced concrete.

3. What desirable properties make reinforced concrete a successful structural material?

4. What type of reinforced concrete structural system would you recommend for covering large column free space used in office floors, assembly halls and auditoriums?

5. List the various applications of reinforced concrete in construction industry.

6. Outline the basic design philosophy of reinforced concrete structural design.

7. What are the different types of loads to be considered in the design of reinforced concrete structures?

8. Briefly outline the importance of national codes in the design of reinforced concrete structures.

9. Briefly mention the advantages of using the handbooks in structural design offices.

10. List the various factors influencing the seismic forces on a building.

Objective Type Questions

1. Reinforced concrete is a successful structural material because
   a. concrete is strong in compression
   b. steel is strong in tension
   c. composite action of the combined material

2. Reinforced concrete is preferred to steel in the construction of bridges due to
   a. high strength of reinforced concrete
   b. higher durability and lower maintenance cost
   c. faster construction using reinforced concrete

3. Reinforced concrete is a viable structural material because
   a. steel is liable to rust
   b. concrete is weak in tension
   c. superior durability of reinforced concrete

4. Reinforced concrete is preferred to steel in the construction of marine structures due to
   a. high strength of reinforcements
   b. high resistance to destructive action of marine atmosphere
   c. faster construction using reinforced concrete

5. In coastal areas the ideal material for construction is
   a. plain concrete
   b. reinforced concrete
   c. steel

6. The strain compatibility of reinforced concrete is attributed to
   a. high strength of steel
   b. high durability of concrete
   c. high bond strength between steel and concrete

7. In multistoried structures, overall costs can be reduced by using
   a. steel
   b. reinforced concrete
   c. plain concrete

8. Reinforced concrete structures in seismic zones are designed to resist mainly
   a. gravity loads
   b. wind loads
   c. earthquake loads

9. Structural concrete members should meet the criterion of
   a. strength
   b. strength and serviceability
   c. serviceability

10. In 21st century, structural concrete designs should conform to the philosophy of
   a. elastic design
   b. ultimate load design
   c. limit state design

2

Limit State Design

2.1 PRINCIPLES OF LIMIT STATE DESIGN

In the beginning of 20th century, the elastic or working stress method of design, developed by German professor Morsch, was firmly established and widely used in the design of reinforced concrete structures like bridges and buildings replacing steel. In 1960, the ultimate load method was developed in Europe and America and was popular for a short period. Due to the inadequacies in the elastic and ultimate load methods, limit state method of design[1] was developed combining the salient features of the earlier methods. Limit state philosophy[2] incorporates proper safeguards against the failure of the structure both at service and ultimate failure states. It is an integrated method of designing structures based on a statistical concept of safety[3] and the associated statistical probability of failure.

The limit state method of design ensures the safety of the structure against excessive deflections and cracking under service loads and also provides for the desirable load factor against failure. Hence, the British[4], American[5], Australian[6], German[7], Canadian[8] and the Indian standard[9] codes have adopted the limit state design concepts in their latest revised versions.

According to Bennett[10], design is essentially a creative rather than a routine analytical activity in which the structural behaviour is only one of the number of functional, constructional, aesthetic and economic considerations. A successful design should not only satisfy the requirements of safety against collapse of the structure due to various causes, but ensure that the serviceability of the structure is not impaired while resisting normal working loads.

The basic purpose of a design is defined as *the provision of a structure complying with the client's requirements and the codes in vogue in the country*. The structural engineer is often required to assist the client in defining his requirements more precisely. Once the basic requirements have been defined, the structural engineer becomes a member of an integrated team. The primary object of a structural design is to obtain a structural solution which can result in the greatest overall economy by providing the maximum assistance in satisfying the various requirements of the structure.

The limit state design incorporates the principles of statistical probability of failure at various limit states leading to collapse of the structure and also ensures the serviceability of the structure at working loads by preventing excessive deflections and local distress in the form of cracks. The evolution of limit state method of design over the years is presented in Fig. 2.1.

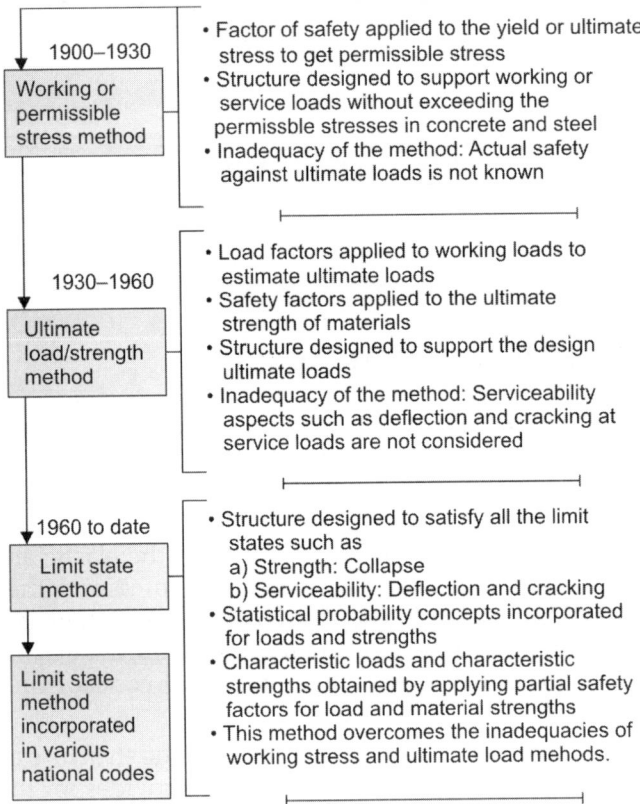

Fig. 2.1: Evolution of limit state method of design

2.2 ░ LIMIT STATE DESIGN AND CLASSICAL RELIABILITY THEORY

In limit state design, probabilistic concepts are explicitly incorporated for the first time. Applications of classical reliability theory[11,12] to structural design require comprehensive statistical data regarding loads and strengths and their exact shapes of normal distribution curves. At present the probabilities of failure that are socially acceptable must be kept very low (1 in a million). At such low levels, the probability of failure is very sensitive to the exact shape of the normal distribution curves. To determine exact shapes of normal distribution curves, we require very large numbers of statistical data and such comprehensive data is not yet available. In particular, sufficient numbers of extreme values of the strengths of complete structures (to define accurately the shapes of the tails of the normal distribution curves) may never be available.

In a simple example, only one type of load and one-strength variable is used. For a real structure, there will, in general, be many types of loads and many modes of failure, normally with complex correlations between them making it very difficult to calculate the probability of failure[13]. Hence in the limit state design, our engineering experience and judgement have been used to modify and to remedy the inadequacies of earlier design methods and partly use the probabilistic concepts. Hence, it is appropriate to designate the limit state design method currently practiced as *semi probabilistic approach*[14] to structural design.

The interaction between load effects and strength is shown in Fig. 2.2 where the normal distribution curves for loads and material strengths are superposed along with the characteristic loads and strengths.

For good design the characteristic loads and strengths are expressed in terms of the standard deviation, mean strength and the probability factor as,

$$F = [F_m + 1.65\sigma]$$
$$f_{ck} = [f_m - 1.65\sigma]$$

where,

F = characteristic load
F_m = mean load
f_{ck} = characteristic strength
f_m = mean strength
σ = standard deviation

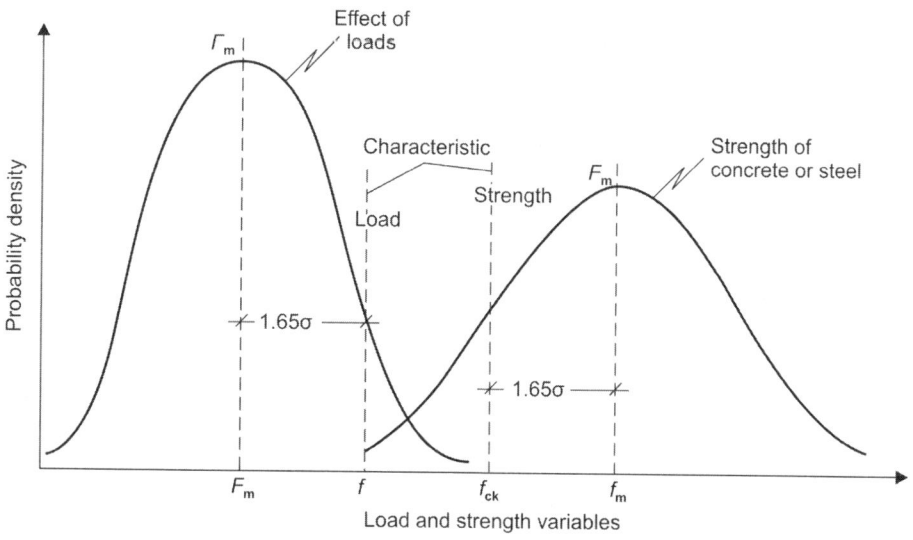

Fig. 2.2: Interaction between load effects and strength

2.3 ▌ CRITERIA FOR LIMIT STATES

In general, a satisfactory design must ensure the achievement of an acceptable probability that the specified life of a structure is not curtailed prematurely due to the attainment of an unsatisfactory condition or limit state, which covers the various forms of failure. There are several limit states at which the structure ceases to function; the most important among them being the limit state of collapse, excessive deflection and cracking. Each of these limit states may be attained due to different types of loading configurations; however, in practice, only one or two of these are of primary significance in design. Some of the important criteria[15] concerning reinforced concrete for the ultimate limit state are given below:

   i. Failure of one or more critical sections in flexure, shear, torsion, or due to their combinations

   ii. Bond and anchorage failure of reinforcement

   iii. Failure of connections between precast and cast *in situ* elements

   iv. Failure due to elastic instability of members.

   The limit state of collapse may also be attained due to fatigue, vibrations, corrosive environment, impact as a consequence of explosions or earthquakes and disintegration due to fire or frost. The structure may be rendered unfit for its intended purpose due to various serviceability limit states being reached, such as:

   i. Excessive deflection or displacement, adversely affecting the finishes and causing discomfort to the users of the structure

   ii. Excessive local damage resulting in cracking or spalling of concrete, which impairs the efficiency or appearance of the structure.

2.4   SAFETY FACTORS

Safety is expressed in terms of the probability that the structure will not become unfit for its intended function during its useful life, i.e. the structure will not reach a limit state. The initial idea of referring to a single failure criterion has been replaced by the comprehensive concept of multiple limit states. With this concept the local or the overall behavior in all stages, *viz.* elastic, cracked, inelastic and ultimate are considered. In the limit state approach, a structure is considered as well designed if it could be shown that the probability of any limit state being attained is substantially constant for all the component members and for the structure as a whole and that consequently, the latter possesses adequate and uniform structural safety. Due to the number of variables involved, a rational determination of safety of a structure, based on probability theory is not yet practical in the design office. Partial safety factors are therefore introduced for each limit state and these consist of γ_m, reduction factor for characteristic strength of materials and γ_f, enhancement factor for characteristic loads on the structure.

Safety may be defined as an acceptable degree of security against complete collapse or failure, which in concrete structures can occur by various modes such as compression, tension, flexure, shear, torsion, fatigue or their combinations.

Serviceability requirement means that the member or structure should not in its intended lifetime deteriorate to such an extent that it fails to fulfill its function for which it is designed. In concrete structures, this state may be reached due to excessive deflection, cracking, vibration, corrosion of reinforcement, etc.

2.5   CHARACTERISTIC AND DESIGN LOADS AND PARTIAL SAFETY FACTORS

The characteristic load F which is independent of the limit state considered and is seldom exceeded in service is defined as

   Characteristic load = [(mean load) + ($k \times$ standard deviation)]

where, k is a factor so chosen as to ensure that the probability of the characteristic load being exceeded is small. A value of 1.64 for k ensures the probability that the characteristic load is exceeded by only 5% during the intended life of the structure.

The statistical data required to define the characteristic loads for different types of occupancy is not readily available, since loading statistics are invariably difficult to compile as they need systematic observations and recording of data over a long period of time. In the absence of statistical observations and recording of data on loading, the nominal imposed loads provided in various national codes, such as IS:875–1987[16], BS: 6399[17], BS:8110-2005[18] and standards ANSI-A-58.1[19] may be treated as characteristic loads.

The characteristic values of the loads take account of expected variations but do not allow for the following:

i. Possible unusual increases in load beyond those considered in deriving the characteristic

ii. Inaccurate assessment of effects of loading and unforeseen stress distribution within the structure

iii. Variations in dimensional accuracy achieved in construction.

The design loads are obtained by enhancing the characteristic loads by suitable partial safety factors for the various limit states. The values of partial safety factors for loads given in the Indian standard code recommendations are compiled in Table 2.1.

Design load = [(characteristic load) × (partial safety factor)]

Table 2.1: Partial safety factor for loads (γ_f) (IS:456-2000)

Load combination			Limit state of collapse			Limit state of serviceability		
DL	IL	WL	DL	IL	WL			
	(1)		(2)	(3)	(4)	(5)	(6)	(7)
(DL + IL)			1.5	1.5	–	1.0	1.0	–
(DL + WL)			1.5	–	1.5	1.0	–	1.0
(DL + IL + WL)			1.2	1.2	1.2	1.0	0.8	0.8
0.9*								

Notes:

1. While considering earthquake effects, substitute *EL* for *WL*.
2. For the limit states of serviceability, the values of γ_f given in this table are applicable for short term effects. While assessing the long term effects due to creep, the dead load and that part of the live load likely to be permanent may only be considered.

* This value is to be considered when stability against overturning or stress reversal is critical.

2.6   CHARACTERISTIC AND DESIGN STRENGTHS AND PARTIAL SAFETY FACTORS

The variation in the properties of concrete and steel are expressed as characteristic values related to the mean values and standard variation,

Characteristic strength (f) = [mean strength – k × standard deviation]

where, k is a factor chosen to ensure that the probability of the characteristic strength not being exceeded is small. Many of the national codes including the Indian standard code IS:456-2000 have recommended a value of 1.65 for k so that only 5% of the test results could have a strength less than the characteristic strength. In the absence of statistical data, the characteristic strength of concrete and steel may be taken as the works cube strength and minimum proof or yield strength respectively as recommended in the current codes.

Since the materials in the structure are likely to differ in quality from those tested, design strengths are obtained by dividing the characteristic strength by γ_m, the appropriate partial safety factor for the limit state being considered. The proposed values for the partial safety factors for materials are given in Table 2.2.

Design strength = [characteristic strength/partial safety factor]
$$f_d = [f/\gamma_m]$$
where, γ_m = partial safety factor appropriate to the material and the limit state being considered.

Table 2.2: Partial safety factors for material strengths (γ_m) (IS:456-2000)

Material	Collapse	Deflection	Local damage
Steel	1.15	1.00	1.00
Concrete	1.50	1.00	1.00 or 1.30

The ACI code specifies slightly different partial safety factors for loads for the limit states of strength and serviceability given as,
$$U = 1.2D = 1.6L + 0.5 \ (L_r \text{ or } S \text{ or } R)$$
$$U = 1.2D + 1.6 \ (L_r \text{ or } S \text{ or } R) + (1.0L \text{ or } 0.5W)$$
$$U = 1.2D + 1.0W + 1.0L + 0.5 \ (L_r \text{ or } S \text{ or } R)$$
$$U = 1.2D + 1.0E + 1.0L + 0.2S$$
$$U = 0.9D + 1.0W$$
$$U = 0.9D + 1.0E$$
where,

U = ultimate load

D = dead load

L = live load

S = snow load

R = rain load or related internal moments and forces

W = wind load

E = earthquake load

L_r = roof live load

The American code values are marginally less in comparison with the Indian code IS:456 in the estimation of ultimate loads for design.

In contrast to the fixed values of partial safety factor for material strengths specified in the Indian code, the ACI code provides for these deficiencies in a more comprehensive way in the form of capacity reduction factors to compute the design strengths. These capacity reduction factors having values varying from 0.60 to 0.90 and depending upon members subjected to various types of forces are similar to the partial safety factors of the Indian code but cover varying types of stresses.

The capacity reduction factors recommended in ACI:318M-11 are compiled in Table 2.3.

Table 2.3: Capacity reduction factors (ACI:318M-11)

Sl. No.	Type force	ϕ
1.	Tension controlled sections	0.90
2.	Compression controlled sections	
	a. Members with spiral reinforcement	0.75
	b. Other reinforced concrete members	0.65
3.	Shear and torsion	0.75
4.	Bearing on concrete	0.65
5.	Post tensioned anchorage zones	0.85
6.	Strut and tie models	0.75
7.	Flexural sections in pretensioned members	0.75
	a. Free end of member to end of transfer length	0.75 to
	b. Free end of member to development length	0.9
8.	Structural members resisting earthquakes (shear)	0.60
9.	For joint and diagonally reinforced coupling beams (shear)	0.85
10.	For structural plain concrete (flexure, compression, shear)	0.60

The British Euro code-2 specifies partial safety factors for loads at the ultimate limit state in a comprehensive manner by categorizing load actions as permanent, variable and for favourable and unfavorable situations as shown in Table 2.4. However, the partial safety factors at the serviceability limit states for all design situations are specified as 1.0 both for permanent and variable actions. The terms favorable and unfavorable refer to the effect of the action(s) on the design situation under consi-deration. For example, if a continuous beam is to be designed for the largest sagging bending moment, it will have to sustain any action that has the effect of increasing the bending moment which will be considered unfavourable whilst any action that reduces the bending moment will be considered as favourable.

The partial safety factors for materials are prescribed for various limit states under the categories persistent/transient and accidental for concrete and steel as compiled in Table 2.5.

2.7 GLOBAL FACTOR OF SAFETY

According to Bill Moseley et al[20], the global factor of safety against a particular type of failure can be computed by multiplying the appropriate partial safety factors for loads and materials. The use of partial safety factors on materials and load actions provides for considerable flexibility and this can be used for special situations such as very high standards of control and construction, e.g. construction of nuclear reactors or in structural elements where failure without warning may be very serious.

Table 2.4: Partial safety factors at the ultimate limit state for loads (BS EN:1992-1-1-Euro code-2)

Persistent or transient design situation	Permanent actions (G_k)		Leading variable actions ($Q_{k,1}$)		Accompanying variable action(Q_{k1})	
	Unfavourable	Favourable	Unfavourable	Favourable	Unfavourable	Favourable
a. For checking the static equilibrium of a building structure	1.10	0.90	1.50	0	1.50	0
b. For the design of structural members (excluding geotechnical actions)	1.35	1.50	1.50	0	1.50	0
c. As an alternative to (a) and (b) to design for both situations with one set of calculations	1.35	1.15	1.50	0	1.50	0

where, G_k = characteristic permanent load and Q_k = characteristic variable load

Table 2.5: Partial safety factors for materials (BS EN:1992-1-1-Euro code-2)

Limit state	Persistent and transient		Accidental	
	Concrete	Reinforcing and prestressing steel	Concrete	Reinforcing and prestressing steel
Ultimate				
Flexure	1.50	1.15	1.20	1.00
Shear	1.50	1.15	1.20	1.00
Bond	1.50	1.15	1.20	1.00
Serviceability	1.00	1.00	–	–

The global factor of safety, for example, in the case of a beam failure caused by yielding of tensile reinforcement is computed as:

$$\gamma_f \times \gamma_m = (1.50 \times 1.15) = 1.725 \text{ for permanent (dead) loads as per IS:456 code}$$
$$= (1.35 \times 1.15) = 1.552 \text{ for permanent loads as per BS EN:1992-1-1 code}$$

Alternatively, failure by crushing of concrete in the compression zone may have a global factor of safety of $(1.5 \times 1.5) = 2.25$ due to variable actions only emphasizing the fact that such failure is generally explosive without warning and may result in serious consequences.

References

1. Rowe RE, Cranston WB and Best BC, New concepts in the design of concrete, *Structural Engineer*, Vol. 43, 1965, pp. 339–403.
2. Bate SCC, Why limit state design, *Concrete*, March 1968, pp. 103–108.
3. CEB recommendations for International Code of Practice for Reinforced Concrete, American Concrete Institute and Cement & Concrete Association, London, 1964.

4. BS EN 1992-1-1-2004, Design of Concrete Structures, General Rules & Rules for Buildings, British Standards Institution, 2004.

5. ACI:318M-11, Building Code requirements for Structural Concrete, American Concrete Institute, Farmington Hills, Michigan, 2005.

6. AS: 3800-1988, Concrete Structures, Standards Association of Australia, 1988.

7. DIN: 1045-1988, Structural Use of Concrete, Design & Construction, Din Deutsches Institute Fir Normung EV, 1988.

8. CSA Standard A23.3-94, Design of Concrete Structures, Canadian Standards Association, Rexdale, Ontario, 1994.

9. IS:456-2000, Indian Standard Code of Practice for Plain and Reinforced Concrete (Fourth Revision), Bureau of Indian Standards, 2000, p. 100.

10. Bennett EW, *Structural Concrete Elements*, Chapman & Hall Ltd, London, 1973, pp. 149–170.

11. Cornell CA, A probability based structural code, *J. of the ACI*, Vol. 66, 1969, pp. 974–985.

12. Ranganathan R, *Reliability Analysis & Design of Structures*, Tata McGraw-Hill, New Delhi, 1990.

13. Ellingwood B, Reliability Basis for Load and Resistance Factors for RC Design, NBS Building Science Series 110, National Bureau of Standards, Washington DC, 1978.

14. Cornell CA, A probability based structural code, *J. of the ACI*, Vol. 66, Dec. 1969, pp. 974–985.

15. Verghese PC, *Limit State Design of Reinforced Concrete*, Prentice Hall of India Pvt Ltd, New Delhi, 1994.

16. IS:875-1987 (part 1), Code of Practice for Design Loads (other than earthquake) for Buildings and Structures, part 2, Dead Loads, Second Revision, Bureau of Indian Standards, New Delhi. 1989.

17. BS 6399-1996, *Code of Practice for Building Loads*, BSI, London, 1996

18. BS 8110-1 and 2, Code of Practice for Structural Use of Concrete, BSI, London, 2005.

19. ANSI-A-58.1-1982, Minimum Design Loads for Buildings of Concrete and Other Structures, American Standards Institute, New York, 1982.

20. Mosley B, Bungey J and Hulse R, *Reinforced Concrete Design to Euro Code-2*, Palgrave Macmillan, London, 2012.

Review Questions

1. Briefly outline the basic concepts of limit state design.

2. Outline the differences between deterministic design and probabilistic design with reference to structural concrete members.

3. What are the various limit states to be considered in the design of structural concrete members?

4. Differentiate between safety and serviceability with respect to structural concrete members.

5. What are the various serviceability states and why they should be considered in design?

6. Explain the terms: (a) characteristic load and (b) characteristic strength.

7. Distinguish between the terms partial safety factors and capacity reduction factors.

8. Explain clearly the reasons for assigning different safety factors for different types of loads.

9. Briefly outline the reasons for applying different partial safety factors for the ultimate limit state in the Indian and British standard codes.

10. What is global factor of safety? How do you compute the global factor of safety for flexural concrete members?

Objective Type Questions

1. Probabilistic concepts are incorporated in
   a. ultimate load design
   b. limit state design
   c. working stress design

2. The partial safety factor specified in the Indian standard code for the combination of live, dead and wind loads is
   a. 1.5
   b. 1.6
   c. 1.2

3. Excessive deflection in a reinforced concrete beam leads to
   a. sudden collapse
   b. damage to partitions
   c. instability of the structure

4. In limit state design process, design loads are obtained by
   a. equating the characteristic loads
   b. enhancing the characteristic loads
   c. decreasing the the characteristic loads

5. The first method to be used in the design of structural concrete members is
   a. the ultimate load method
   b. elastic method
   c. limit state method

6. The partial safety factor used for material strength of steel at the limit state of collapse is
   a. 1.5
   b. 1.0
   c. 1.15

7. The design strength of material is
   a. directly proportional to the partial safety factor
   b. nearly equal to the characteristic strength
   c. inversely proportional to the partial safety factor

8. In the computation of characteristic strength, most of the national codes recommend the value of standard deviation as

   a. 1.5
   b. 2.0
   c. 1.65

9. The partial safety factor specified in the Indian standard code for evaluating the design strength of concrete for the limit state of deflection is
   a. 1.3
   b. 1.5
   c. 1.0

10. The capacity reduction factor specified in the American code for designing structural concrete members subjected to earthquake is
   a. 0.90
   b. 0.75
   c. 0.60

3

Continuous Beams

3.1 INTRODUCTION

Continuous beams are generally adopted in the construction of multistoreyed buildings and long span bridges[1]. In the case of buildings, the floor slabs are generally cast integral with secondary beams framing into main beams and columns. Continuous beams are designed for moments and shear forces developed due to dead and imposed loads. Analysis of the continuous beam indicates that the maximum positive span moments develop when alternate spans are loaded with live load and maximum negative support moments develop when adjacent spans are loaded[2,3]. The design of continuous beams involves the assumption of the size of the member based on the recommendations of the Indian standard code IS:456-2000[4] followed by the determination of moments and shear forces for which suitable reinforcements are designed.

3.2 EFFECTIVE SPAN

According to the Indian standard code IS:456-1978, for a continuous beam having a support width less than 1/12th the clear span, the effective span shall be as per freely supported beams, i.e. clear span plus the effective depth of beam or centre to centre of supports whichever is less.

If the supports are wider than 1/12th the clear span or 600 mm whichever is less, the effective span shall be as given below:

a. For end span with one end fixed and the other continuous or for intermediate spans, the effective span shall be the clear span between the supports.
b. For end span with one end free and the other continuous, the effective span shall be equal to the clear span plus half the effective depth of beam or the clear span plus half the width of the discontinuous support, whichever is less.
c. In the case of spans with roller and rocker bearings, the effective span shall always be the distance between the centre of bearings.
d. In the case of continuous monolithic frames, the effective span of continuous beams are taken as the centre line distance between the members.

3.3 ░ SPAN/DEPTH RATIOS

Basic Span/Depth Ratios

The following factors are considered in the specifications of basic span/depth ratios recommended in IS:456-2000 code:

a. Percentage tension and compression reinforcement in the section

b. Type of beam (rectangular or flanged)

c. Type of supports (cantilever, simply supported, fixed or continuous).

Table 3.1 gives the basic span to effective depth ratios to be used for beams and slabs with spans up to 10 m. For spans greater than 10 m, the ratios have to be multiplied by a factor $F = (10/\text{span})$ in metre. A graphical representation of the basic span/depth ratios is shown in Fig. 3.1.

Table 3.1: Basic span/effective depth ratios for beams and slabs (clause 23.2.1 of IS:456-2000)

Type of support	Rectangular sections	Flanged sections
Cantilever	7	Multiply values for rectangular sections by factor K_f (Refer to Fig. 3.4)
Simply supported	20	
Continuous	26	

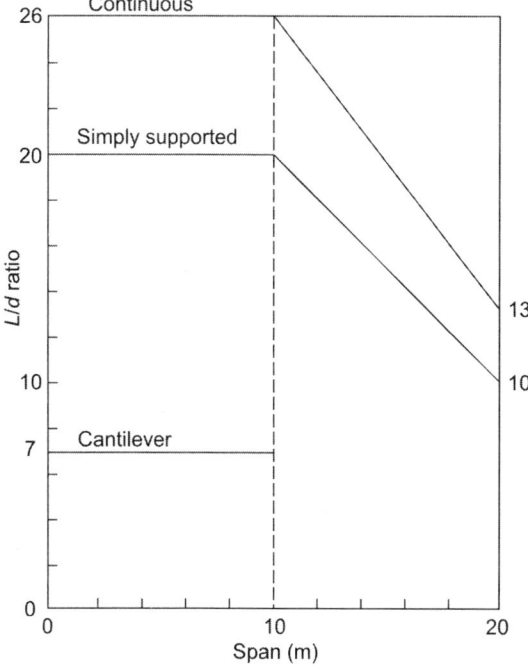

Fig. 3.1: Basic span/depth ratios for beams and slabs

Modification Factors for Basic Span/Depth Ratios

Modification factors have to be applied to the basic span/depth ratio to account for the percentage of tension and compression reinforcements in the section and also the type of section such as rectangular or flanged. The final expression of the span/effective depth ratio can be expressed as

$$(L/d) = [(L/d)_{basic} \times K_t \times K_c \times K_f],$$

where,

$(L/d)_{basic}$ is as given in Table 3.1

K_t = modification factor for tension reinforcement (Refer to Fig. 3.2)

K_c = modification factor for compression reinforcement (Refer to Fig. 3.3)

K_f = modification or reduction factor for flanged sections (Refer to Fig. 3.4)

In general, higher percentages of tension reinforcement are associated with lower values of K_t and higher values of K_c. For Fe 415 grade HYSD bars, the value of K_t is unity corresponding to the unit percentage of tension reinforcement.

For spans above 10 m, the $(L/d)_{basic}$ values have to be multiplied by (10/span) in metre, except for cantilevers in which case deflection computations are required to satisfy the limit state of deflection.

In case of flanged beams (T and L beams), the modification factors K_t and K_c should be based on an area of section $b_f d$ (flange width × effective depth) and the calculated (L/d) ratio is further modified by a reduction factor which depends on the ratio (b_w/b_f) as shown in Fig. 3.4 (Fig. 6 of IS:456-2000 code). The codal procedure yields anomalous results in the case of flanged beams as outlined in the explanatory handbook to the

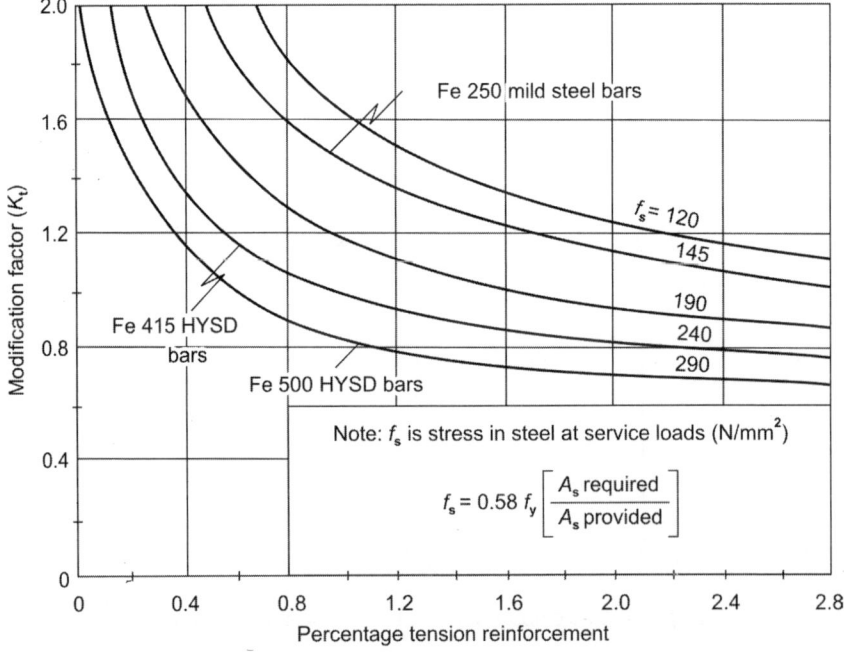

Fig. 3.2: Modification factor for tension reinforcement (IS:456-2000)

code[5]. Hence, it is preferable to consider the width of the web b_w in place of b_f in computations and this procedure yields conservative results.

The empirical procedure recommended for control of deflection in slabs is the same as in beams, i.e. to limit the span/depth ratios and the use of same modification factors.

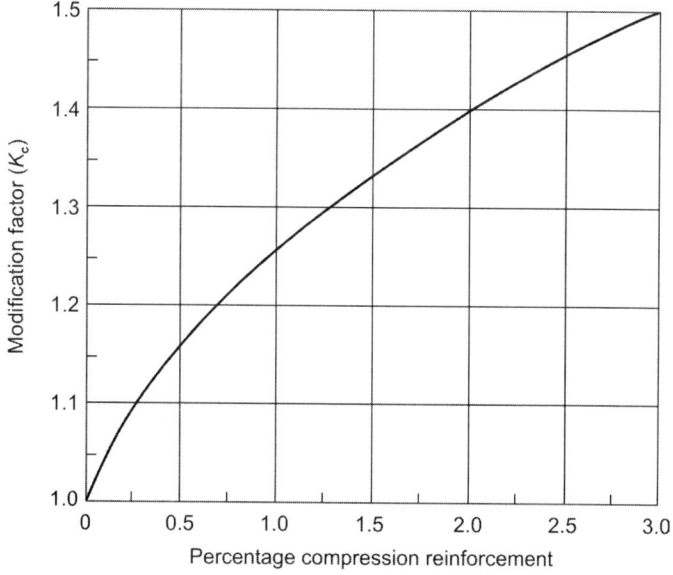

Fig. 3.3: Modification factor for compression reinforcement (IS:456-2000)

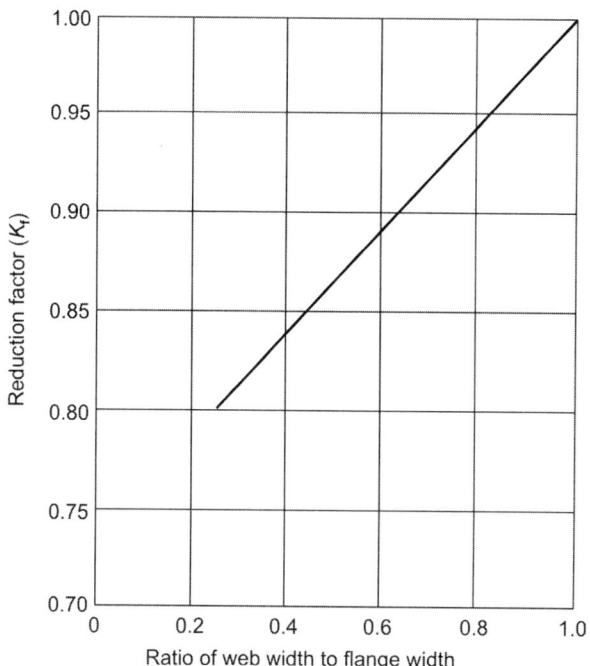

Fig. 3.4: Reduction factor for ratios of span to effective depth for flanged beams (IS:456-2000)

For preliminary proportioning, the thickness of slabs using Fe 415 HYSD bars, it is recommended to assume 0.4% value for p_t which gives a value of K_t of about 1.25 and the corresponding (L/d) ratio being 25. In the design of beams which carry heavy loading, it is preferable to assume span/effective depth ratio in the range 10–12 for practical considerations.

The British Euro code BS EN:1992-1-1-2004[6] provides comprehensive safeguards for the limit state of deflection by limiting the final deflection to a maximum of span/250. Since the value of allowable span/effective depth ratio is influenced by both reinforcement ratio and concrete strength, it recommends the use of a chart to select the span/effective depth ratio as a function of percentage reinforcement and concrete grades varying from M25 to M60.

The ACI code 318M-11[7] specifies minimum thickness for one way slabs when deflection computations are not made. The minimum overall thickness varies from span/28 to span/8 depending upon the type of member and support conditions. Also the maximum permissible computed deflection varies from span/180 to span/240 depending upon the type of member. Separate recommendations are made for the minimum thickness of flat slabs with or without drop panels.

3.4　BENDING MOMENTS AND SHEAR FORCES

The following arrangement of superimposed loading is generally considered for computation of maximum positive and negative moments at the cross-section of a continuous member:

a. Design dead load on all spans with full design live load on alternate spans

b. Design dead load on all spans with full design live load on adjacent spans

The position of incidental or live loads for maximum span and support moments are shown in Fig. 3.5. Tables 3.2 and 3.3 contain the moment and shear coefficients recommended in IS:456-2000 for slabs and beams continuous over three or more approximately equal spans. However, redistribution is not permitted when using these coefficients.

Appendix 1 shows the bending moment and shear force coefficients for continuous beams of two to five equal spans with uniformly distributed and central point loads. These bending moment coefficients are useful in designing beams allowing for

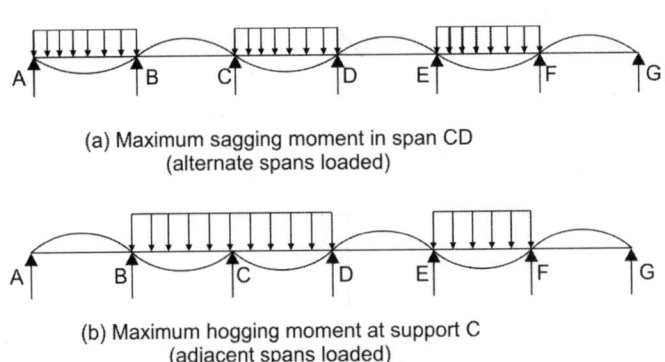

(a) Maximum sagging moment in span CD
(alternate spans loaded)

(b) Maximum hogging moment at support C
(adjacent spans loaded)

Fig. 3.5: Position of live loads for maximum moments in continuous beams

redistribution of moments. According to the IS:456-2000 code, it is permitted to allow 15% redistribution in the working stress method and 30% redistribution in the limit state method of design.

3.5 ELASTIC OR WORKING STRESS AND LIMIT STATE METHODS OF DESIGN

Elastic or Working Stress Method

The bending moment and shear force coefficients given in Tables 3.2 and 3.3 are applicable only in the case of working stress method of design. The coefficients are based on elastic analysis[8] and they are limited to continuous beams supporting substantially uniformly distributed loads over three or more spans which do not differ by more than 15% of the longest span. Designs based on working stress method generally ensure safety against collapse and also satisfy the limit state of serviceability[9].

Table 3.2: Bending moment coefficients (IS:456-2000)

Type of load	Span moments		Support moments	
	Near middle of end span	At middle of interior span	At support next to the end support	At other interior supports
Dead load and imposed load (fixed)	+1/12	+1/16	–1/10	–1/12
Imposed load (not fixed)	+1/10	+1/12	–1/9	–1/9

Note: For obtaining the bending moment, the coefficient shall be multiplied by the total design load and effective span.

Table 3.3: Shear force coefficients

Type of load	At end support	At support next to the end support		At all other interior supports
		Outer side	Inner side	
Dead load and imposed load (fixed)	0.4	0.6	0.55	0.5
Imposed load (not fixed)	0.45	0.6	0.6	0.6

Note: For obtaining the shear force, the coefficient shall be multiplied by the total design load.

Limit State Method

In the limit state method of design, the structure should not only satisfy the limit state of collapse, but also comply with the limit state of serviceability implying that at working loads, the deflections and cracks are within permissible limits together with the availability of prescribed load factors at the limit state of collapse[10]. In indeterminate structures like continuous beams, the collapse of the whole structure is preceded by the formation of plastic hinges depending upon the degree of indeterminacy of the structure. The formation of hinges facilitates the redistribution of moments thus reducing the peak moments at the critical sections.

The Indian standard code in its recommendations in clause 37.1.1 permits redistribution of up to 30% of the numerically largest moment computed by the elastic analysis covering all appropriate combination of loads. However, in structures in which the structural frame provides lateral stability and also over 4 storeys in height, the magnitude of redistribution is restricted to 10%. This limitation has been imposed mainly to safeguard the performance of the structures at the limit state of serviceability.

The following design examples will illustrate the design aspects of continuous beams using various methods.

3.6 ■ DESIGN EXAMPLES

1. A three span continuous beam of equal spans of 8 m, supports uniformly distributed dead and live loads of equal magnitude 10 kN/m. Compute the maximum service load bending moments in the beam using the following methods:

   a. Elastic analysis with total load on all spans
   b. Elastic analysis considering pattern loading
   c. Elastic analysis using IS:456 code recommendations of bending moment coefficients

   Data:

   Span length L = 8 m
   Dead load g = 10 kN/m
   Live load q = 10 kN/m
   Total load $w = (g + q)$ = 20 kN/m

   Figure 3.5 shows the three span continuous beam supporting uniformly distributed load of equal magnitude on all the spans.

   a. *Elastic analysis with total load on all spans*: Referring to Appendix 1 containing the table of bending moment coefficients, the maximum positive and negative moments are computed as,

   Maximum positive span moment $M_1 = 0.080\ wL^2 = (0.08 \times 20 \times 8^2) = 102.4$ kN·m

   Maximum negative support moment $M_2 = 0.100\ wL^2 = (0.1 \times 20 \times 8^2) = 128$ kN·m

   b. *Elastic analysis considering pattern loading*: Referring to Appendix 1 and reading out the moment coefficients corresponding to dead and live loads separately, the maximum positive and negative moments are computed as,

   Maximum positive span moment due to dead load M_{d1}
   $$= 0.080gL^2 = (0.08 \times 10 \times 8^2) = 51.2 \text{ kN·m}$$

   Maximum positive span moment due to live load M_{q1}
   $$= 0.101qL^2 = (0.101 \times 10 \times 8^2) = 64.64 \text{ kN·m}$$

   Hence, maximum positive span moment $= (M_{d1} + M_{q1})$
   $$= (51.2 + 64.64) = 115.84 \text{ kN·m}$$

   Maximum negative support moment due to dead load M_{d2}
   $$= 0.100gL^2 = (0.1 \times 10 \times 8^2) = 64 \text{ kN·m}$$

   Maximum negative support moment due to live load M_{q2}
   $$= 0.117qL^2 = (0.117 \times 10 \times 8^2) = 74.88 \text{ kN·m}$$

Hence, maximum negative support moment $= (M_{d2} + M_{q2})$
$$= (64 + 74.88) = 138.88 \text{ kN·m}$$

c. *Elastic analysis using IS:456 code recommendations of bending moment coefficients:*

Maximum positive span moment due to dead load M_{d1}
$$= (1/12) \, gL^2 = [(1/12) \times 10 \times 8^2] = 53.33 \text{ kN·m}$$

Maximum positive span moment due to live load M_{q1}
$$= (1/10) \, qL^2 = [(1/10) \times 10 \times 8^2] = 64.00 \text{ kN·m}$$

Hence, maximum positive span moment $= (M_{d1} + M_{q1})$
$$= (53.33 + 64.00) = 117.33 \text{ kN·m}$$

Maximum negative support moment due to dead load M_{d2}
$$= (1/10) \, gL^2 = [(1/10) \times 10 \times 8^2] = 64 \text{ kN·m}$$

Maximum negative support moment due to live load M_{q2}
$$= (1/9) \, qL^2 = [(1/9) \times 10 \times 8^2] = 71.11 \text{ kN·m}$$

Hence, maximum negative support moment $= (M_{d2} + M_{q2})$
$$= (64 + 71.11) = 135.11 \text{ kN·m}$$

A comparative analysis of the moments evaluated using three different methods is presented in Table 3.4.

Table 3.4: Maximum moments by various methods

Method	Span moment (M_1) (kN·m)	Support moment (M_2) (kN·m)
1. Total load on all spans	102.40	128.00
2. Pattern loading	115.84	138.88
3. Code coefficients	117.33	135.11

The analysis indicates that the simplified assumption of total load on all spans results in the lowest values of both positive and negative moments. The exact analysis of pattern loading yields 10% higher values compared to the first method. The values obtained by the code coefficient method are comparable to that of the exact method and can be conveniently used for design purposes.

2. A continuous beam of a multistorey frame has three spans each of 8 m. The characteristic dead load is 10 kN/m and the characteristic live load is 15 kN/m. Design the critical sections of the beam and sketch the details of reinforcements using the limit state method. Adopt M20 grade concrete and Fe 415 HYSD bars.

Method 1 (using IS:456-2000 formulae)

   i. *Data:*

Effective span $L = 8$ m $f_{ck} = 20$ N/mm^2

Dead load $= 10$ kN/m $f_y = 415$ N/mm^2

Live load $= 15$ kN/m

Concrete: M20 grade

Steel: Fe 415 HYSD bars

ii. *Cross-sectional dimensions*:

As the continuous beam supports heavy loads, a span/depth ratio of 10 is provided to satisfy the limit state of serviceability.

$$\therefore \text{Effective depth} = \left(\frac{\text{span}}{10}\right) = \left(\frac{8000}{10}\right) = 800 \text{ mm}$$

Adopt overall depth $D = 900$ mm

Effective depth $d = 850$ mm

Width $b = 300$ mm

Cover to compression steel = 50 mm

iii. *Effective span*:

According to clause 22.2 of IS:456-2000, the effective span $L = 8$ m

iv. *Loads*:

Self weight of beam = $(0.3 \times 0.9 \times 25)$ = 6.75 kN/m

Dead load on beam   = 10.00 kN/m

Total dead load on beam g = 16.75 kN/m

Live load on beam q = 15.00 kN/m

$\therefore$ Design ultimate dead load g_u = (1.5×16.75) = 25.125 kN/m

Design ultimate live load q_u = (1.5×15.00) = 22.5 kN/m

v. *Design bending moments and shear forces*:

Referring to the bending moment and shear force coefficients (Tables 3.2 and 3.3):

Negative bending moment at interior support is computed as

$M_u(-ve) = [(g_u L^2/10) + (q_u L^2/9)]$

$= [(25.125 \times 8^2)/10 + (22.5 \times 8^2)/9] = 320.8$ kN·m

Positive bending moment at centre of span is computed as

$M_u(+ve) = [(g_u L^2/12) + (q_u L^2/10)]$

$= [(25.125 \times 8^2)/12 + (22.5 \times 8^2)/10] = 278$ kN·m

Maximum shear force at support next to the end support is computed as

$V_u = [0.6L (g_u + q_u)] = [(0.6 \times 8)(25.125 + 22.5)] = 228.6$ kN

vi. *Limiting moment of resistance*:

$M_{u, \text{lim}} = (0.138 f_{ck} bd^2) = (0.138 \times 20 \times 300 \times 850^2) \times 10^{-6} = 598$ kN·m

Since $M_u < M_{u, \text{lim}}$, the section is under-reinforced.

vii. *Main reinforcement*:

a. For negative bending moment:

$$M_u = (0.87 f_y A_{st} d) \left[1 - \left(\frac{A_{st} f_y}{bd f_{ck}}\right)\right]$$

$$(320.8 \times 10^6) = (0.87 \times 415 A_{st} \times 850) \left[1 - \left(\frac{415 A_{st}}{300 \times 850 \times 20}\right)\right]$$

Solving, $A_{st} = 1160$ mm^2

b. For positive bending moment:

$$(278.8 \times 10^6) = (0.87 \times 415\, A_{st} \times 850) \left[1 - \left(\frac{415\, A_{st}}{300 \times 850 \times 20}\right)\right]$$

Solving, $A_{st} = 990$ mm^2

Provide 4 bars of 20 mm diameter on the tension side over supports ($A_{st} = 1256$ mm^2) and 4 bars of 18 mm diameter on the tension face at mid span ($A_{st} = 1017$ mm^2).

viii. *Check for shear stress*:

$$\tau_v = \left(\frac{V_u}{bd}\right) = \left(\frac{228.6 \times 10^3}{300 \times 850}\right) = 0.398 \text{ N/mm}^2$$

$$p_t = \left(\frac{100\, A_{st}}{bd}\right) = \left(\frac{100 \times 1017}{300 \times 850}\right) = 0.398$$

Refer to Table 19 of IS:456-2000 and readout the permissible shear stress in concrete as $\tau_c = 0.43$ N/mm^2.

Since $\tau_v < \tau_c$, only nominal shear reinforcements are provided.

Adopting 8 mm diameter 2 legged stirrups, the spacing S_v is computed as

$$S_v = \left(\frac{0.87\, f_y\, A_{sv}\, d}{0.4b}\right) = \left(\frac{0.87 \times 415 \times 2 \times 50}{0.4 \times 300}\right) = 300 \text{ mm}$$

Provide 8 mm diameter two-legged stirrups at 300 mm centres throughout the length of the span.

ix. *Check for deflection control*:

Reinforcement percentage $p_t = 0.398$

Refer to Fig. 4 of IS:456-2000 and readout the modification factor $K_t = 1.3$

Neglecting hanger bars $K_c = 1.0$ and $K_f = 1.0$

$$\therefore \quad \left(\frac{L}{d}\right)_{max} = \left(\frac{L}{d}\right)_{basic} \times K_t \times K_c \times K_f$$

$$= (26 \times 1.3 \times 1 \times 1)$$

$$= 33.8$$

$$\left(\frac{L}{d}\right)_{actual} = \left(\frac{8000}{850}\right) = 9.4 < 33.8$$

Hence, the deflections are within safe permissible limits.

x. *Details of reinforcements*:

The details of reinforcements in the continuous beam are shown in Fig. 3.6.

Method 2 (using SP:16 design charts)

i. *Main reinforcements*:

Compute the parameter $= \left(\frac{M_u}{bd^2}\right) = \left(\frac{320.8 \times 10^6}{300 \times 850^2}\right) = 147$

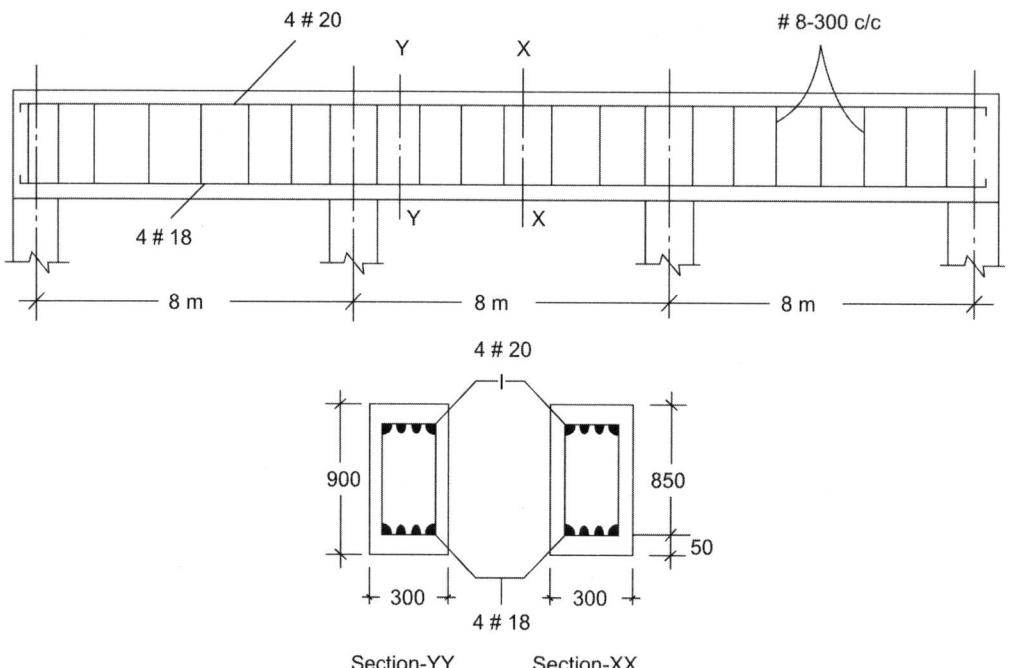

Fig. 3.6: Reinforcement details in continuous beam

Refer to Table 2 of SP:16 and readout the percentage of reinforcement as

$$p_t = \left(\frac{100 A_{st}}{bd}\right) = 0.45$$

$$\therefore \quad A_{st} = \left[\frac{0.45 \times 300 \times 850}{100}\right] = 1150 \text{ mm}^2$$

ii. *Shear reinforcements*:

Shear carried by concrete is computed as

$$V_{uc} = (\tau_c bd) = (0.43 \times 300 \times 850)10^{-3} = 109.6 \text{ kN}$$

Shear to be carried by stirrups is computed as

$$V_{us} = (V_u - V_{uc}) = [228.6 - 109.6] = 119 \text{ kN}$$

Compute the parameter $(V_{us}/d) = (119/85) = 1.4 \text{ kN/cm}$

Refer to Table 6.17 of SP:16 and readout the spacing of 8 mm diameter two-legged stirrups as 260 mm.

The reinforcement details are more or less similar to that computed by the theoretical method.

3. Design a continuous beam of two spans supported on stone masonry walls using the limit state method and allowing for 15% redistribution of moments. The following data may be used.

i. *Data*:

Clear span between supports = 6 m

Width of masonry supports = 330 mm

Thickness of reinforced concrete slab = 150 mm
Spacings of continuous beams = 3 m c/c
Self weight of floor finish = 0.4 kN/m^2
Live load on office floor = 4 kN/m^2
Characteristic cube strength of concrete f_{ck} = 20 N/mm^2
Characteristic strength of steel f_y = 415 N/mm^2

ii. *Characteristic strength*:

$$f_y = 415 \text{ N/mm}^2$$
$$f_{ck} = 20 \text{ N/mm}^2$$

iii. *Depth of beam*:

As the continuous beam carried heavy load, a span/effective depth ratio of 12 is adopted.

$$\text{Effective depth} = \left(\frac{\text{span}}{12} \right) = \left(\frac{6000}{12} \right) = 500 \text{ mm}$$

Overall depth = 600 mm
Cover = 50 mm
Effective depth = 550 mm
Width of beam = 300 mm

iv. *Effective span*:

Width of support = 300 mm

$$1/12 \text{ clear span} = \left(\frac{1}{2} \times 6000 \right) = 500 \text{ mm}$$

Support width is less than 1/12 clear span therefore, effective span is the least of:

a. Centre to centre of supports = (6000 + 300) = 6300 mm = 6.3 m
b. Clear span + effective depth = (6000 + 550) = 6550 mm = 6.55 m
   Effective span, L = 6.3 m

v. *Loads*:

Self weight of RC slab = (0.15 × 24) = 3.6 kN/m^2

Floor finish $= \dfrac{0.4}{4.0} \text{ kN/m}^2$

Dead load on beam = (4.0 × 3) = 12 kN/m
Self weight of beam = (0.3 × 0.6 × 24) = 4.32 kN/m
Total dead load on beam (g) = 16.32 kN/m
Live load on beam (q)   = (4 × 3) = 12 kN/m
The factored ultimate loads on beam are

$$g_u = (1.5 \times 16.32) = 24.48 \text{ kN/m}$$
$$q_u = (1.5 \times 12) = 18.00 \text{ kN/m}$$

vi. *Bending moments and shear forces*:

Referring to Appendix 1, the maximum negative and positive moments at support and mid span sections are obtained as:

$$\text{Negative maximum moment} = 0.125(g_u + q_u)L^2$$
$$= 0.125(24.48 + 18.0)6.3^2$$
$$= 211 \text{ kN·m}$$

$$\text{Positive maximum moment} = (0.071g_u + 0.096q_u)L^2$$
$$= (0.071 \times 24.48 + 0.096 \times 18.0)6.3^2 = 138 \text{ kN·m}$$

$$\text{Maximum support shear force } V_u = 0.62(g_u + q_u)L$$
$$= 0.62(24.48 + 18.0)6.3$$
$$= 166 \text{ kN}$$

vii. *Redistribution of moments*:

The design maximum moment is obtained by decreasing the negative moment at support by 15% and increasing the positive span moment by the same magnitude.

15% of $M_{max} = (0.15 \times 211) = 32 \text{ kN/m}$

Design moment at support section = $(211 - 32) = 179 \text{ kN/m}$

Design moment at centre of span section = $(138 + 32) = 170 \text{ kN/m}$

Since the design ultimate moment, positive and negative are almost equal in magnitude, the sections are designed for an ultimate moment,

∴ $M_u = 179 \text{ kN/m}$

viii. *Moment of resistance*:

Moment of resistance of the balanced section M_u
$$= 0.138 f_{ck} bd^2 \text{ for } f_y = 415 \text{ N/mm}^2$$
$$= (0.138 \times 20 \times 300 \times 550^2)10^{-6}$$
$$= 250.47 \text{ kN·m} > 179 \text{ kN·m}$$

Hence, the section is under-reinforced.

ix. *Main reinforcements in section*:

If A_{st} = area of tensile steel in the section

$$M_u = (0.87 f_y A_{st} d)\left[1 - \left(\frac{A_{st} f_y}{bd f_{ck}}\right)\right]$$

$$179 \times 10^6 = 0.87 \times 415 \times A_{st} \times 550 \left[1 - \frac{A_{st} \times 415}{300 \times 550 \times 20}\right]$$

$$A_{st}^2 - 5966.8 A_{st} + 5.37 \times 10^6 = 0$$

Solving, $A_{st} = 1109 \text{ mm}^2$

Provide 4 bars of 20 mm diameter ($A_{st} = 1256 \text{ mm}^2$)

x. *Shear stresses*:

Ultimate shear force $V_u = 166 \text{ kN}$

$$\tau_v = \left(\frac{V_u}{bd}\right) = \left(\frac{166 \times 10^3}{300 \times 550}\right) = 1.00 \text{ N/mm}^2$$

$$\left(\frac{100 A_{st}}{bd}\right) = \left(\frac{100 \times 1256}{300 \times 550}\right) = 0.76$$

From Table 19 of IS:456, $\tau_c = 0.54 \text{ N/mm}^2$

If two bars are bent up at quarter span points, shear force taken by bent bars V_{us}

$$= 0.87 f_y A_{sv} \sin \alpha$$

$$= \left(\frac{0.87 \times 415 \times 2 \times 314 \times 0.707}{1000} \right) = 160 \text{ kN}$$

Shear taken by concrete $= \tau_c bd$

$$= \left(\frac{0.54 \times 300 \times 550}{1000} \right) = 160 \text{ kN}$$

Total shear resisted $= (160 + 89) = 249 \text{ kN} > 166 \text{ kN}$

Provide nominal shear reinforcements. Using 6 mm two-legged stirrups, the spacing is given by

$$S_v = \left(\frac{0.87 f_y A_{sv}}{0.4b} \right) = \left(\frac{0.87 \times 415 \times 2 \times 50}{0.4 \times 300} \right) = 168 \text{ mm}$$

Adopt 6 mm diameter two-legged stirrups at 160 mm centres, throughout the length of the beam.

4. A continuous beam has a width of 300 mm and an overall depth of 660 mm with three equal spans of 5 m. In the transverse direction, the beams are spaced at 4 m intervals with a slab thickness of 180 mm. The uniformly distributed ultimate design load is 190 kN/m. The ultimate design bending moments and shear forces are shown in the table given below:

Location of section	Bending moment (kN·m)	Shear force (kN)
Mid span	430	0
Support	530	570

The characteristic strength of the concrete and steel are 30 and 500 N/mm² respectively. Design the reinforcements for the mid span section of the end span and interior support section of the continuous beam.

   i. *Data:*

   Width of beam $b_w = 300$ mm

   Overall depth $D = 660$ mm

   Effective depth $d = 600$ mm

   Thickness of slab $D_f = 180$ mm

   Effective span $L = 5$ m

   Characteristic strength of concrete $f_{ck} = 30 \text{ N/mm}^2$

   Characteristic strength of steel $f_y = 500 \text{ N/mm}^2$

   ii. *Design moments and shear forces:*

   Maximum positive span moment $M_{pu} = 430 \text{ kN·m}$

Maximum negative support moment M_{nu} = 530 kN·m
Maximum shear force V_u = 570 kN

iii. *Design of mid span section*:

The mid span section is designed as T-section.

Effective width of flange $b_f = [(L/6) + b_w + 6D_f]$

$$= [(5000/6) + 300 + (6 \times 180)]$$
$$= 2213 \text{ mm}$$

Moment capacity of the flanged section is given by

$$M_{uf} = 0.36 f_{ck} b_f D_f (d - 0.42 D_f)$$
$$= [(0.36 \times 30 \times 2213 \times 180)(600 \times 0.42 \times 180)]$$
$$= (2256 \times 10^6) \text{ N·mm}$$
$$= 2256 \text{ kN·m}$$

Since $M_{pu} < M_{uf}$, the neutral axis depth is less than the thickness of the flange. Hence, the section is designed as a rectangular section with $b = b_f$ for designing reinforcements in the section.

$$M_u = (0.87 f_y A_{st} d) \left[1 - \left(\frac{A_{st} f_y}{b d f_{ck}} \right) \right]$$

$$(430 \times 10^6) = (0.87 \times 500 \times A_{st} \times 600) \left[1 - \left(\frac{A_{st} \times 500}{2256 \times 600 \times 30} \right) \right]$$

Solving, A_{st} = 1700 mm²

Provide 4 bars of 25 mm diameter (A_{st} = 1963 mm²)

Nominal area of shear reinforcements $A_{sv} = \left[\dfrac{0.4 \times b \times S_v}{0.87 f_y} \right] = \left[\dfrac{0.4 \times 300 \times 300}{0.87 \times 500} \right]$

$$= 83 \text{ mm}^2$$

Provide two-legged stirrups of 8 mm bars (A_{sv} = 100 mm²)

iv. *Design of support section*:

The support section is designed as a rectangular section of width $b = b_w$ = 300 mm

Design moment M_u = 530 kN·m

Design shear force V_u = 570 kN

Limiting moment of resistance of singly reinforced section is computed as

$$M_{u, \text{lim}} = 0.133 f_{ck} b d^2$$
$$= (0.133 \times 30 \times 300 \times 600^2) = 430 \times 10^6 \text{ N·mm} = 430 \text{ kN·m}$$

Since $M_u > M_{u, \text{lim}}$, the section has to be designed as doubly reinforced.

$(M_u - M_{u, \text{lim}}) = (530 - 430) = 100$ kN·m

For the grade of steel $(x_{u, \text{max}}) f_y$ = 500 N/mm², the limiting ratio of d = 0.46

Stress in compression steel $f_{sc} = \left[\dfrac{0.0035 (x_{u, \text{max}} - d')}{x_{u, \text{max}}} \right] E_s$

$$f_{sc} = \left[\frac{0.0035(0.46 \times 600 - 50)}{(0.46 \times 600)}\right](2 \times 10^5) = 548 \text{ N/mm}^2$$

But f_{sc} is limited to a value of $0.87 f_y = (0.87 \times 500) = 435 \text{ N/mm}^2$

Hence $A_{sc} = \left[\frac{M_u - M_{u,\text{lim}}}{f_{sc}(d - d')}\right] = \left[\frac{100 \times 10^6}{435 \times 550}\right] = 418 \text{ mm}^2$

Two bars of 25 mm diameter from mid span can be continued at the supports as compression reinforcement at the support section.

$$A_{st2} = \left(\frac{A_{sc} f_{sc}}{0.87 f_y}\right) = \left(\frac{418 \times 435}{0.87 \times 500}\right) = 347 \text{ mm}^2$$

$$A_{st1} = \left(\frac{0.36 f_{ck}(x_{u,\text{lim}})}{0.87 f_y}\right) = \left(\frac{0.36 \times 20 \times 0.46 \times 600}{0.87 \times 500}\right) = 2055 \text{ mm}^2$$

$[A_{st1} + A_{st2}] = [2055 + 347] = 2402 \text{ mm}^2$

Provide 5 bars of 25 mm diameter ($A_{st} = 2455 \text{ mm}^2$)

Design shear force at support $V_u = 570$ kN

$$\tau_v = \left(\frac{V_u}{bd}\right) = \left(\frac{570 \times 10^3}{300 \times 600}\right) = 3.16 \text{ N/mm}^2$$

$$\left(\frac{100 A_{st}}{bd}\right) = \left(\frac{100 \times 2455}{300 \times 600}\right) = 1.36$$

Refer to Table 19 of IS:456-2000 and readout $\tau_e = 0.73 \text{ N/mm}^2$

Since $\tau_v > \tau_c$, shear reinforcements are required.

Shear at a distance d from the face of the support $V_u = [570 - 190(0.6)]$
$$= 446 \text{ kN}$$

$V_{us} = [V_u - (\tau_c bd)] = [446 - (0.73 \times 300 \times 600) 10^{-3}] = 315 \text{ kN}$

Using 10 mm diameter two-legged stirrups, the spacing is obtained as

$$S_v = \left[\frac{0.87 f_y A_{sv} d}{V_{us}}\right] = \left[\frac{0.87 \times 500 \times 2 \times 79 \times 600}{315 \times 1000}\right] = 130 \text{ mm}$$

Provide 10 mm diameter two-legged stirrups at 130 mm centers near supports.

References

1. Continuity in Concrete Building Frames, 4 edn, Portland Cement Association, Chicago, Illinois, 1959.

2. The Applications of Moment Distribution, The Concrete Association of India, Publication of the Associated Cement Companies Limited, Bombay, 1978.

3. Wang K, *Indeterminate Structural Analysis*, McGraw-Hill International Edition, 1983.

4. IS:456-2000, Indian Standard Code of Practice for Plain and Reinforced Concrete (Fourth Revision), Bureau of Indian Standards, New Delhi, July 2000.

5. SP:24-1983, Explanatory Handbook on IS:456, Bureau of Indian Standards, New Delhi, 1983.

6. BS EN:1992-1-1-2004, Design of Concrete Structures, General Rules & Rules for Buildings, British Standards Institution, London, 2004.

7. ACI:318M-11, Building Code Requirements for Structural Concrete (ACI standard), American Concrete Institute, Farmington Hills, Michigan, USA, 2011.

8. Reynolds CE and Steedman J, *Reinforced Concrete Designer's Handbook*, Concrete Publications Ltd, London, 1974.

9. Bate SCC, Why limit state design, *Concrete*, March 1968, pp. 103–108.

10. Verghese PC, *Limit State Design of Reinforced Concrete*, Prentice Hall of India Pvt Ltd, New Delhi, 1994.

Assignment

1. Design a continuous beam of three spans supported on stone masonry piers 300 mm × 300 mm.

   Clear span between the supports = 8 m

   Thickness of reinforced concrete slab = 120 mm

   Spacing of continuous beams = 2.75 m c/c

   Self weight of floor finish = 0.6 kN/m^2

   Live load on office floor = 4 kN/m^2

   Concrete of M20 grade and Fe 415 steel are available for use.

   Design the reinforcements for the critical sections and sketch the details using working stress method.

2. A continuous beam with simple supports has two spans each of 6 m, from centre to centre of supports. The characteristic dead load is 15 kN/m and characteristic live load is 20 kN/m. Design the critical section of the beam and sketch the details of reinforcements using the limit state method. Given, characteristic strength of concrete $f_{ck} = 20$ N/mm^2 and characteristic strength of steel $f_y = 415$ N/mm^2. Sketch the details in the continuous beam.

3. A two span simply supported continuous beam is to be designed to suit the following data:

   Distance between centre to centre of supports = 6 m

   The beam supports dead and live loads of 100 and 150 kN respectively at the centre of spans. Using M20 grade concrete and ribbed for steel, design the continuous beam using working stress method.

4. A continuous beam of two spans is fixed at the end supports and continuous over the central support.

   Distance between the centre of supports = 8 m

   The beam supports a reinforced concrete slab 150 mm thick with live load of 2 kN/m^2 on floor.

   Spacings of continuous beams = 4 m

   Characteristic strength of concrete = 20 N/mm^2

   Characteristic strength of steel = 415 N/mm^2

Design the continuous beam using limit state method and allowing for 15% redistribution of moments.

5. Design the interior span of a continuous T-beam of effective span 6 m. The beams are placed at 4 m centres and carry a slab of thickness 100 mm. Dead load on slab inclusive of self weight may be taken as 3 kN/m² and live load as 6 kN/m². Use M20 grade concrete and Fe 415 HYSD bars.

6. Design the interior span of a rectangular beam supported on 500 mm square column and beam continuous over 8 m span. The beams support a slab 120 mm thick and are placed 4 m centres and are cast monolithic with the slab. The loading on the slab is due to self weight, floor finish and plastering and live load of magnitude 5 kN/m². Adopt M20 grade concrete and Fe 415 HYSD bars.

7. A reinforced concrete beam is continuous over two equal spans of length 6 m each. It is simply supported at its ends. It carries a uniformly distributed load of intensity 15 kN/m inclusive of its own weight over its entire length of 12 m.

Design for flexure and shear a suitable rectangular section for the beam. Adopt M20 grade concrete and Fe 415 HYSD bars.

8. A reinforced concrete continuous beam of three equal spans AB = BC = CD = 7 m, is simply supported at the ends A and D.

If the beam supports a uniformly distributed live load of 10 kN/m in addition to its own weight, design the beam for flexure and shear and sketch the details of reinforcements in the beam.

9. A reinforced concrete beam, continuous over two spans AB = BC = 5 m and supported over masonry walls at ends A and C is required to support a uniformly distributed live load of 20 kN/m. Design a suitable rectangular section for the beam and sketch the details of reinforcements. Adopt M20 grade concrete and Fe 415 HYSD bars.

10. A three span continuous beam with equal spans of 8 m supports carries uniformly distributed dead and live loads of each 15 kN/m over the spans. Adopt M20 grade concrete and Fe 415 HYSD bars, design a suitable rectangular section and reinforcements in the beam using limit state design method conforming to the Indian standard code IS:456-2000.

Review Questions

1. Explain the reasons for adopting pattern loading to evaluate the design moments and shear forces in continuous beams.

2. In continuous beams of several spans, mention the locations where maximum design bending moments and shear forces develop.

3. Explain the reasons for disallowing the redistribution of moments in continuous beams when bending moment coefficients specified in the code are used.

4. Among the various methods of loading in continuous beams, which method yields the maximum support bending moment?

5. What are the various factors to be considered in checking the limit state of deflection in continuous beams?

6. Mention the maximum spacing of the stirrups permitted by the IS:456-2000 code in the case of continuous beams.

7. What limitations are imposed in the Indian standard code regarding the maximum area of compression reinforcement in continuous beams?

8. Outline the various methods by which one can increase the shear resistance in the design of continuous beams.

9. Explain the significance of stipulating the ultimate moment of resistance at any section of a continuous beam to be not less than 70% of the factored moment at that section in the IS:456-2000 code specifications.

10. Outline the advantages of redistribution of bending moments in the design of continuous beams.

Objective Type Questions

1. Reinforced concrete continuous beams under uniformly distributed loads exhibit maximum moments at
   a. centre of interior spans
   b. at end supports
   c. at penultimate support

2. The moments developed in the case of loaded continuous beams are comparatively
   a. higher than in simply supported beams
   b. less than in simply supported beams
   c. similar to that in simply supported beams

3. Maximum support moments in loaded continuous beam develop when
   a. all the spans are loaded simultaneously
   b. alternate spans are loaded
   c. adjacent spans on either side of the support is loaded

4. Bending moment coefficients specified in the Indian standard code IS:456-2000 is applicable only when
   a. the beam has two spans
   b. the beam is fully loaded on all spans
   c. the beam should have at least three spans

5. For multi-span reinforced concrete bridge constructions, it is economical to adopt
   a. simply supported spans
   b. continuous spans of variable cross-section
   c. continuous spans of constant cross-section

6. The bending moment coefficients recommended in the Indian standard code for the design of continuous beams are based on
   a. limit state analysis
   b. elastic analysis
   c. inelastic analysis

7. In the case of reinforced concrete continuous beams, it is more economical to use beams having larger depth at
   a. centre of spans
   b. support sections
   c. end supports

8. Reinforced concrete continuous beams fail suddenly when
   a. the centre of span section yields due to plastic hinge formation
   b. the support section yields due to plastic hinge formation
   c. sufficient number of hinges are formed in the continuous beam

9. Compression reinforcement will be required in case of continuous beams
   a. with variable cross-section
   b. with constant cross-section
   c. with restricted depth at supports sections

10. In the design of continuous beams, the principle of moment redistribution will help to
    a. reduce the critical moments at all sections
    b. even out the moments at span and supports
    c. reduce the moment at end supports

4

Bunkers and Silos

4.1 INTRODUCTION

Bunkers and silos are storage structures in the domain of reinforced concrete structures. Rapid industrialisation during 20th century, resulted in the production of large quantities of materials like cement, coal, aggregates, fertilizers, food grains, and other chemical materials in the granular form. To store these materials, steel bunkers were used before the development of cement storage structures. Reinforced concrete bunkers and silos have almost replaced the steel storage structures used in 18th century. RCC bunkers and silos are easy to maintain and they exhibit excellent resistance to adverse environmental conditions in contrast to steel storage structures. The present day cement factories invariably opt for single or battery of silos to store the manufactured cement. Coal is generally stored in bunkers in various collieries spread over the country. The development of slip form method of casting tall cylindrical reinforced concrete structures has resulted in rapid construction of silos.

A large number of reference books, dealing with the fundamentals of theory and design of bunkers and silos are available to date on this subject. The earliest monograph by Gray[1] details in a lucid manner the theories of Janssen and Airy, widely used by the various national and international codes and recommendations. The handbooks by Reynolds[2] and Martens[3] are very useful in the design of storage structures. The design and construction of silos and bins are covered in detail in monographs by Brown & Nielsen[4] and Sargis Safarin[5].

The technology of precast concrete has facilitated the construction of assembling bunkers and silos using modular precast segments to suit the required capacity of the bulk storage structures. Precast concrete bunkers and silos are widely used in USA, Europe and China. Changli Machinery company[6] is famous for the production of precast modular bunkers and silos which are either bolted or welded having storage capacities in the range of 30 to 500 tons of cement.

At present, a new and novel type of *dome technology* developed by Barry South of USA[7] has been used to construct bulk coal storage structure of size 95 m diameter and 53 m height using thin shell concrete units at Idaho Falls, Idaho; Mississippi, USA in 2014. Dome technology has been recently used in several countries for industrial bulk storage of materials like coal, fly ash, fertilizers, grain, etc. Dome technology is the world's leading dome builder with more than 550 dome structures worldwide for the storage of various types of materials.

At present, dome technology is very popular and the construction of several novel storage structures in USA, China, Brazil, Argentina, Iraq, and Mexico have been reported by Emily Wilson[8]. The main advantage of dome technology is the reduction in cost up to 40–50%, in comparison with the traditional methods used earlier for the construction of bunkers and silos. Fox of Farmers Cooperative, Elevator and Mercantile organization of Dighton, Kansas, USA reports the construction of a large bin to store one million bushels of wheat using dome technology with significant savings in cost and construction time.

The new grain import terminal of Commercializadora La Junta (CLJ), located in Manzanillo, Colima, Mexico, is one of the rare grain storage facility built using dome technology. The terminal opened in June 1999 with all grain storage in two 25,000 ton concrete storage domes. At its peak operations, the terminal handles 2 million tons of grain annually.

David South, President of Monolithic Dome Institute and Monolithic Constructions, Inc, Italy; Texas, USA has reported that concrete storage structures built using dome technology can withstand up to F-5 tornadoes (winds of 400 kmph), hurricanes and earthquakes. Dome technology uses modern architectural fabrics welded into dome shapes, which are inflated and used as the form to which polyurethane foam and concrete are sprayed. The spray applied concrete in conjunction with steel rebar and polyurethane foam renders the dome strong, durable, economical and energy efficient.

4.2 DIFFERENCES BETWEEN BUNKERS AND SILOS

Bunkers

Bunkers are shallow structures in which the plane of rupture of the material stored meets the top horizontal surface of the material before meeting the opposite sides of the structure, as shown in Fig. 4.1.

The angle of rupture is $\left(\dfrac{90+\phi}{2}\right)^{\circ}$ from the horizontal, where ϕ is the angle of repose of the material.

The side walls resist the lateral pressure and the total load of the material is supported by the floor of the bunker. The intensity of lateral pressure on the sides is determined by Rankine's theory.

Silos

In a silo, the vertical walls are considerably taller than the lateral dimensions resulting in a tall structure. Consequently, the plane of rupture of the material stored meets the opposite sides of the structure before meeting the top horizontal surface of the material as shown in Fig. 4.2.

Due to the high ratio of height to the lateral dimension, a significant portion of the load is resisted by friction between the material and the wall. Only a fraction of the total weight of the material acts on the floor of the structure.

Let b = breadth

h = height of the structure

ϕ = angle of repose

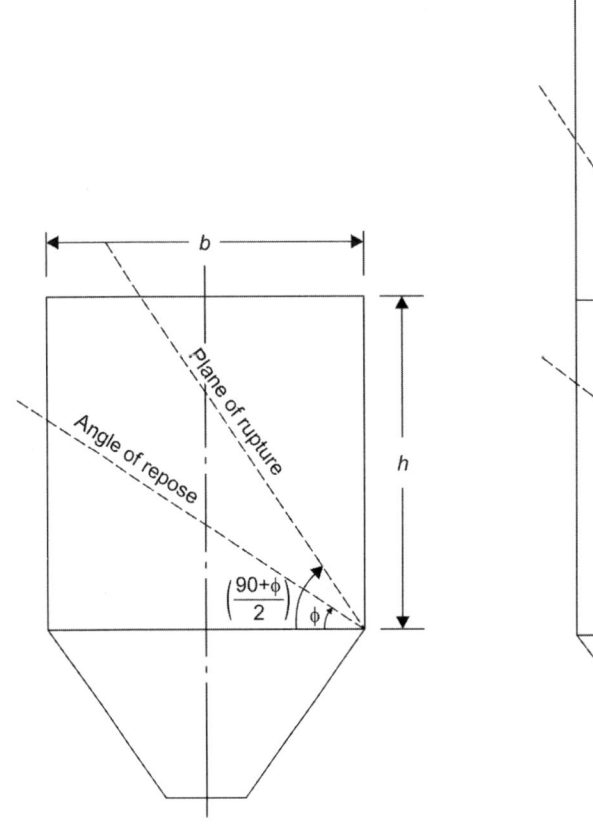

Fig. 4.1: Bunker or shallow bin

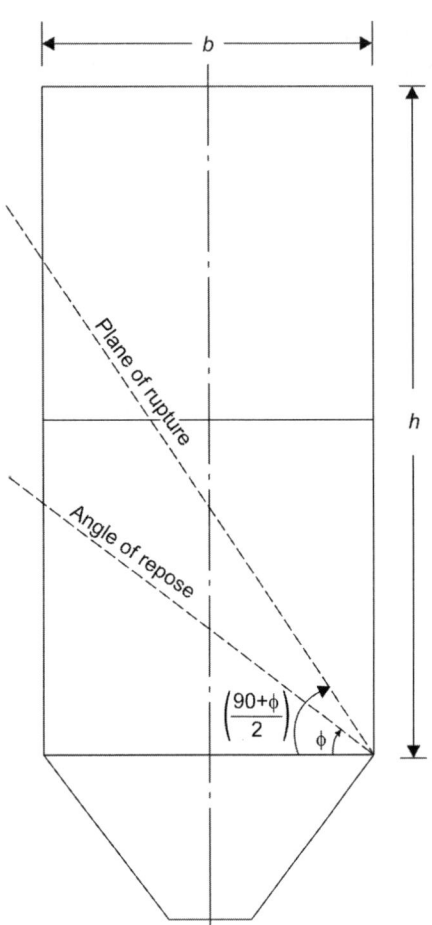

Fig. 4.2: Silo or deep bin

For a structure to be classified as a silo,

$$h > b \tan\left(\frac{90+\phi}{2}\right)^{\circ}$$

If $\phi = 30°$, $h > 1.732b$

4.3 DESIGN OF SQUARE OR RECTANGULAR BUNKERS

The structural elements of a bunker are shown in Fig. 4.3.

The various parts are:

   i. Vertical side walls
  ii. Hopper bottom
 iii. Edge beams
 iv. Columns

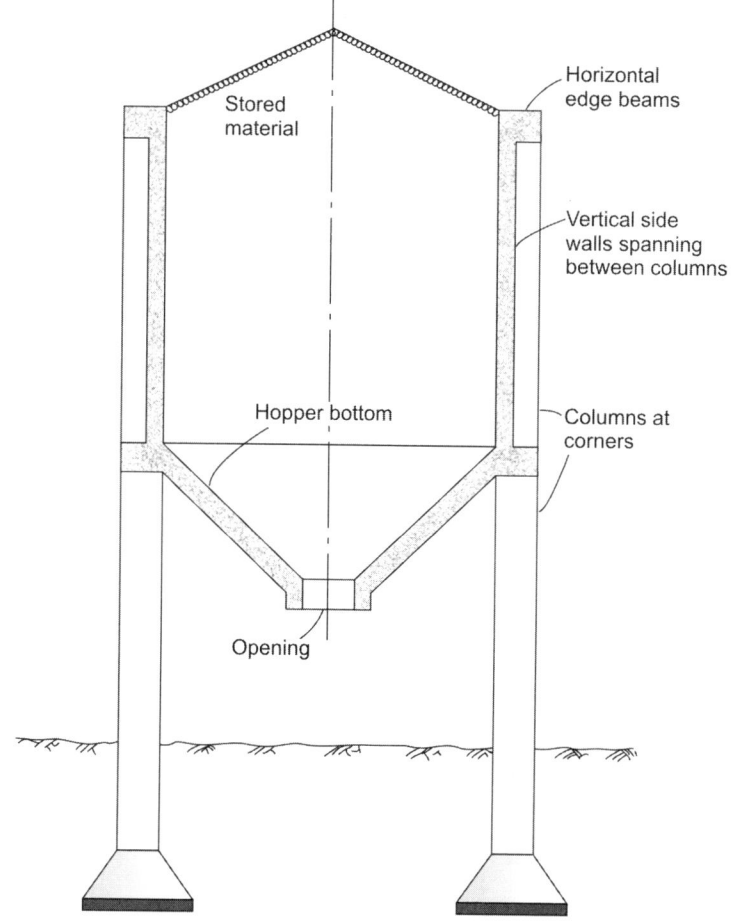

Fig. 4.3: Structural elements of a bunker

a. *Design of vertical side walls:*

Figure 4.4 shows the cross-sectional plan and elevation of a rectangular bunker of length L, breadth B, and height h. If p_u is the intensity of lateral pressure at a depth h, then according to Rankine's theory

$$p_\alpha = wh \cos \alpha \left[\frac{\cos \alpha - \sqrt{\cos^2 \alpha - \cos^2 \phi}}{\cos \alpha + \sqrt{\cos^2 \alpha - \cos^2 \phi}} \right]$$

where,

  α = angle of surcharge

  ϕ = angle of repose

  w = density of material stored

The pressure acts in a direction parallel to the surface of the retained material.

Horizontal component $p = p_\alpha \cos \alpha$

If $\alpha = \phi$, then $p_\alpha = wh \cos \phi$

∴   Service load pressure $p = wh \cos^2 \phi$

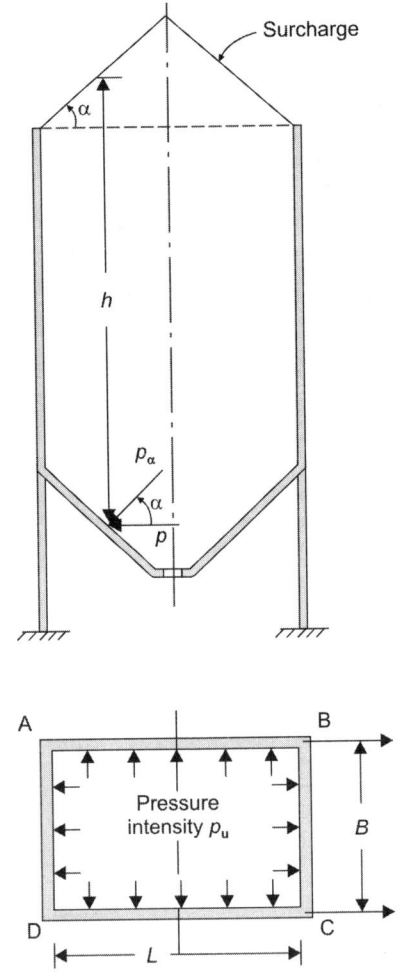

Fig. 4.4: Pressure intensity on walls of a bunker

Design ultimate pressure $p_u = 1.5wh \cos^2 \phi$

The design moments at the supports and centre of long and short walls are given as:

$$M_A = M_B = M_C = M_D = -\frac{p_u}{12}[L^2 + B^2 - BL]$$

Positive moment at centre of AB or CD $= \left[\frac{p_u L^2}{8} - \frac{p_u}{8}(L^2 + B^2 - BL)\right]$

Positive moment at centre of BC or AD $= \left[\frac{p_u L^2}{8} - \frac{p_u}{8}(L^2 + B^2 - BL)\right]$

Direct tension in long walls $= \left(\frac{p_u B}{2}\right)$

Direct tension in long walls $= \left(\dfrac{p_u L}{2}\right)$

The thickness of side walls is designed for maximum bending moment. The reinforcement in the walls is designed for bending moment and direct tension.

If
M_u = design ultimate bending moment
T_u = design ultimate tension
x = distance between the centre of section and reinforcement position
d = effective depth of side walls
Q = design constant = 0.138 for f_y = 415 N/mm^2
b = width of section
A_{st} = area of tension reinforcement
f_{ck} = characteristic compressive strength of concrete
f_y = characteristic tensile strength of reinforcement

then the effective depth is expressed as

$$d = \sqrt{\frac{M_u - T_u x}{Q f_{ck} b}}$$

Let A_{st1} = area of reinforcement required for the net moment, then

$$(M_u - T_u x) = 0.87 f_y A_{st1} d \left[1 - \frac{A_{st1} f_y}{b d f_{ck}} \right]$$

A_{st2} = area of reinforcement required to resist direct tension $= \left(\dfrac{T_u}{0.87 f_y}\right)$

Total area of reinforcement $A_{st} = [A_{st1} + A_{st2}]$

These reinforcements are arranged in the horizontal direction. Distribution steel is provided in the vertical direction. At the top and bottom of the vertical walls, edge beams of 300 mm × 300 mm section are provided to allow for attachment of conveyor supports.

b. *Design of hopper bottom:*

The hopper bottom which is a sloping slab is designed for direct tension developed due to the weight of material and the self weight of the sloping slab.

Referring to Fig. 4.5, if w_t = weight of the material, sloping bottom, etc.,

then direct tension $= w_t \operatorname{cosec} \theta$

where, θ = angle between the horizontal and the sloping slab.

The sloping slab is considered to span horizontally between the intersections of the adjacent sloping faces. The section of the slab at the centre of the slope is designed.

If
w = density of the material stored
h = average height at centre of slope
L = effective span at centre of slope

normal pressure intensity at a depth h is given by

$p_n = wh \cos^2 \theta + p_h \sin^2 \theta$

but $p_h = wh \cos^2 \phi$

$\therefore \quad p_n = wh[\cos^2 \theta + \cos^2 \phi \sin^2 \theta]$

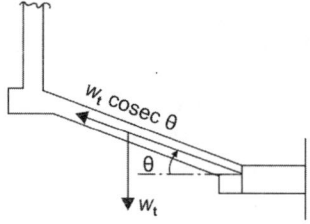

(a) Direct tension in sloping slab

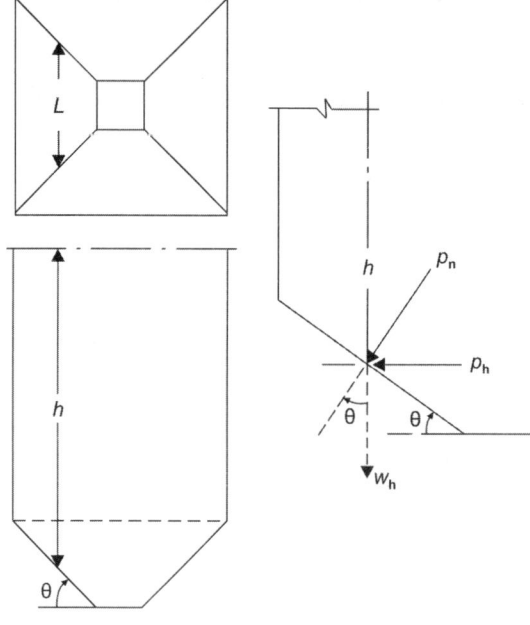

(b) Bending of sloping slab

Fig. 4.5: Forces acting on hopper bottom

If self weight of slab per unit length is w_d

then   normal component $= w_d \cos \theta$

total normal working pressure $p = (p_n + w_d \cos \theta)$

design ultimate pressure $p_u = 1.5(p_n + w_d \cos \theta)$

maximum negative bending moment at supports $= [p_u L^2 / 12]$

maximum positive bending moment at centre of span $= [p_u L^2 / 24]$

4.4   DESIGN OF CIRCULAR BUNKERS

In case of circular bunkers, the vertical walls are subjected to hoop tension.

If D = diameter of the bunker

p_h = horizontal component of pressure at a depth h from the top

then   hoop tension $= 0.5 p_h D$

The reinforcements in the walls are designed to resist the hoop tension.

A minimum thickness of 120 mm is recommended from practical considerations for the vertical walls. Distribution reinforcement of 0.15% of the gross cross-section is provided in the vertical direction.

The hopper bottom is designed for direct tension and hoop tension developed due to the normal pressure on the slopping slab.

The columns of bunkers are designed for compression and bending developed due to the vertical loads such as the stored material and self weight of members and horizontal loads such as wind loads.

4.5 DESIGN EXAMPLES

1. Design the side walls and hopper bottom of a 3 m × 3 m square bunker to store 30 tonnes of coal. Density of coal is 9 kN/m³ and angle of repose is 30°. Adopt M20 grade concrete and Fe 415 HYSD bars. Sketch the details of reinforcements in the bunker.

   i. *Data*:

   Total weight of coal = 300 kN, $f_{ck} = 20 \, \text{N/mm}^2$
   Density of coal = 9 kN/m³, $f_y = 415 \, \text{N/mm}^2$
   Size of bunker in plan = 3 m × 3 m
   Angle of repose = 30°
   Grade of concrete = M20
   Type of reinforcements = Fe 415 HYSD bars

   ii. *Dimensions of bunker*:

   $$\text{Volume of bunker} = \left(\frac{300}{9}\right) = 33.33 \, \text{m}^3$$

   Adopt a bunker of size 3 m × 3 m × 3 m with the depth of the hopper bottom 1.2 m as shown in Fig. 4.6.

   iii. *Design of side walls*:

   Horizontal working pressure $p = wh\cos^2 \phi$
   $= (9 \times 3 \times 0.866^2) = 20.25 \, \text{kN/m}^2$
   Design ultimate pressure $= p_u$
   $= (1.5 \times 20.25) = 30.375 \, \text{kN/m}^2$
   Assuming 180 mm thick side walls,
   effective span $= (3.00 + 0.18) = 3.18 \, \text{m}$
   Maximum bending moment at corners is computed as

   $$M_u = \left(\frac{p_u L^2}{12}\right) = \left(\frac{30.375 \times 3.18^2}{12}\right)$$
   $$= 25.59 \, \text{kN·m}$$

   Direct tension in wall $T_u = \left(\frac{30.375 \times 3}{2}\right)$
   $$= 45.56 \, \text{kN}$$

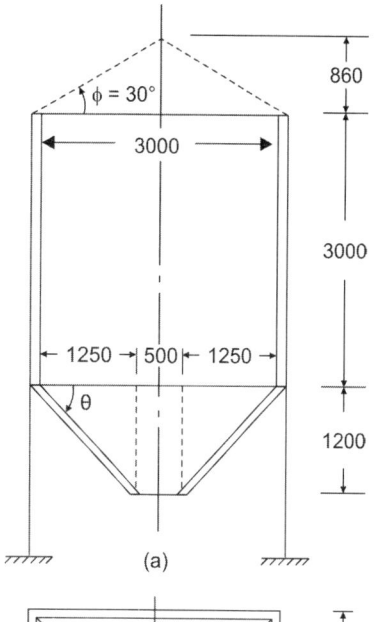

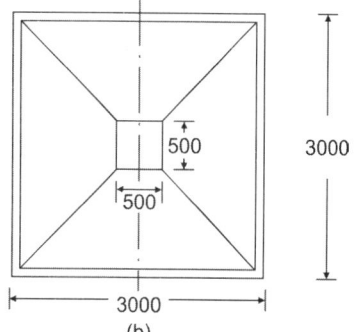

Fig. 4.6: Dimensions of bunker

Providing a cover of 30 mm, effective depth d = 150 mm

Distance between reinforcement and centre line of slab x = 60 mm

Net design moment = $(M_u - T_u x)$ = (25.59 – 45.56 × 0.06) = 22.86 kN·m

Based on limiting moment of resistance, effective depth required is given by

$$d = \sqrt{\frac{M_u - T_u x}{Q f_{ck} b}}$$

For f_{ck} = 20 N/mm², f_y = 415 N/mm², Q = 0.138

$$d = \sqrt{\frac{22.86 \times 10^6}{0.138 \times 20 \times 10^3}} = 91 \text{ mm}$$

Since the depth provided is more, the section is under-reinforced.
Hence, the area of steel required is computed from the relation:

$$\text{Net } M_u = (0.87 f_y A_{st} d)\left[1 - \frac{A_{st} f_y}{b d f_{ck}}\right]$$

$$(22.86 \times 10^6) = (0.87 \times 415 \times A_{st} \times 150)\left[1 - \frac{A_{st} \times 415}{(103 \times 150 \times 20)}\right]$$

Solving, A_{st} = 450 mm²

Provide 12 mm diameter bars at 200 mm centres (A_{st} = 565 mm²)

Positive BM at centre of span = $\left(\dfrac{p_u L^2}{24}\right) = \left(\dfrac{30.375 \times 3.18^2}{24}\right)$ = 12.80 kN·m

∴ $[M_u - T_u x] = [12.80 - (45.56 \times 0.06)] = 10.07$ kN·m

Hence, steel required to resist the moment is computed from the relation:

$$(10.07 \times 10^6) = (0.87 \times 415 \times A_{st} \times 150)\left[1 - \frac{A_{st} \times 415}{(103 \times 150 \times 20)}\right]$$

Solving, A_{st} = 192 mm²

iv. *Design of hopper bottom*

Provide 12 mm diameter bars at 300 mm centres (A_{st} = 192 mm²)

Distribution reinforcement = (0.0012 × 1000 × 180) = 216 mm²

Use 8 mm diameter bars at 200 mm centres (A_{st} = 251 mm²)

Weight of coal = 300 kN

Weight of sloping hopper bottom (180 mm thick) is computed as

$$W_h = \left[\frac{3 + 0.5}{2}\right]\left(\sqrt{1.25^2 + 1.2^2}\right)(4 \times 0.18 \times 25) = 55 \text{ kN}$$

Total load on 4 walls = (300 + 55) = 355 kN

Load on one wall w_t = (0.25 × 355) = 88.75 kN

$$\tan\theta = \left(\frac{1.20}{1.25}\right) = 1.00$$

∴ $\theta = 45°$ and $\csc\theta = 1.44$

Direct tension in sloping wall = $w_t \cosec \theta$ = (88.75 × 1.44) = 128 kN

Working tension per metre run = (128/3) = 42.66 kN/m

Design ultimate tension = (1.5 × 42.66) = 63.99 kN/m

Area of reinforcement for resisting direct tension is

$$A_{st} = \left[\frac{63.99 \times 10^3}{0.87 \times 415} \right] = 177 \text{ mm}^2$$

Main reinforcement = (0.0012 × 180 × 1000) = 216 mm²

Provide 10 mm diameter bars at 300 mm centres (A_{st} = 262 mm²) in the direction of sloping faces.

Normal component of coal pressure at centre of sloping slab is computed as

$$p_n = wh \,(\cos^2 \theta + \cos^2 \phi \, \sin^2 \theta)$$

where,

$w = 9 \text{ kN/m}^2$

$h = [3 + (0.5 \times 1.20) + (0.5 \times 0.86)] = 4.03 \text{ m}$

$\theta = 45°$

$\phi = 30°$

Working pressure p_n = (9 × 4.03) [cos² 45° + cos² 30° × sin² 45°]

= 31.67 kN/m²

Normal component due to weight of sloping slab = $w_d \cos\theta$

= (0.18 × 25 × cos 45°)

= 3.18 kN/m²

Total normal pressure $p = (p_n + w_d \cos\theta) = (31.67 + 3.18) = 34.85$ kN/m²

Design ultimate normal pressure p_u = (1.5 × 34.85) = 52.3 kN/m²

Effective span $L = \left(\dfrac{3+0.5}{2} \right) + 0.18 = 1.93$ m

Maximum negative BM, $M_u = \left(\dfrac{p_u L^2}{12} \right) = \left[\dfrac{52.3 \times 1.93^2}{12} \right] = 16.23$ kN·m/m

Effective depth available = (180 – 30) mm = 150 mm

Limiting moment of resistance $M_{u, \text{lim}}$ = 0.138 $f_{ck} bd^2$

= (0.138 × 20 × 1000 × 150²) × 10⁻⁶

= 62.1 kN·m

Since $M_u < M_{u, \text{lim}}$, the section is under-reinforced

∴ $M_u = (0.87 f_y A_{st} d) \left[1 - \dfrac{A_{st} f_y}{bd f_{ck}} \right]$

$(16.23 \times 10^6) = (0.87 \times 415 \times A_{st} \times 150) \left[1 - \dfrac{A_{st} \times 415}{(103 \times 150 \times 20)} \right]$

Solving, $A_{st} = 313 \text{ m}^2$

Use 10 mm diameter bars at 250 mm centres

Positive maximum BM at centre, $M_u = \left(\dfrac{p_u L^2}{24} \right) = 8.12$ kN·m

Reinforcement area is obtained from the relation:

$$(8.12 \times 10^6) = (0.87 \times 415 \times A_{st} \times 150) \left[1 - \dfrac{A_{st} \times 415}{(103 \times 150 \times 20)} \right]$$

Solving, $A_{st} = 157$ mm²

Minimum reinforcement = $(0.0012 \times 180 \times 1000) = 216$ mm²

Use 8 mm diameter bars at 200 mm centres ($A_{st} = 251$ mm²).

v. *Edge beams*:

Provide edge beams of 300 mm × 300 mm connecting the corner columns at the top and the junction of vertical walls and sloping slab. The details of reinforcements in the side walls, hopper bottom and edge beams are shown in Fig. 4.7.

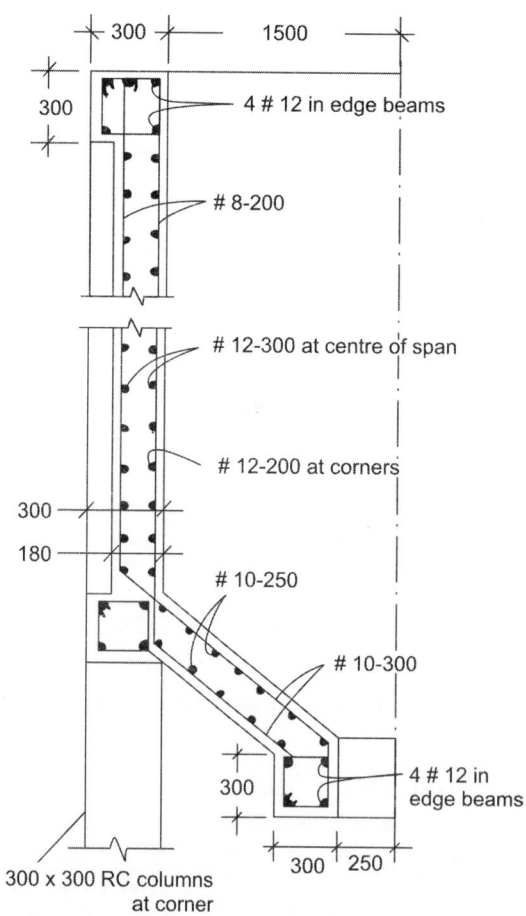

Fig. 4.7: Reinforcement details in a bunker

2. Design a circular bunker to store 20 tonnes of coal. Density of coal is 9 kN/m³ and angle of repose is 30°. Use limit state method of design and adopt characteristic strength of concrete and steel as 20 and 415 N/mm² respectively. Sketch the details of reinforcements in the bunker.

i. *Data*:

Total weight of coal = (20 × 10) = 200 kN

Density of coal = 9 kN/m³

Angle of repose = 30°

Grade of concrete = M20

Type of reinforcements = Fe 415 HYSD bars.

ii. *Characteristic strength and partial safety factors*:

Characteristic strength of concrete f_{ck}
$$= 20 \text{ N/mm}^2$$

Characteristic strength of steel f_y
$$= 415 \text{ N/mm}^2$$

Partial safety factor for live and dead loads = 1.5

Partial safety factor for concrete = 1.5

Partial safety factor for steel = 1.15

iii. *Dimensions of bunker*:

Volume of bunker $= \left(\dfrac{200}{9}\right) = 22.2 \text{ m}^3$

Adopt a bunker of diameter = 2.6 m, height of cylindrical portion = 3 m, depth of hopper bottom = 1.2 m

The overall dimensions of the bunker are shown in Fig. 4.8.

The volume of the bunker is computed as given below:

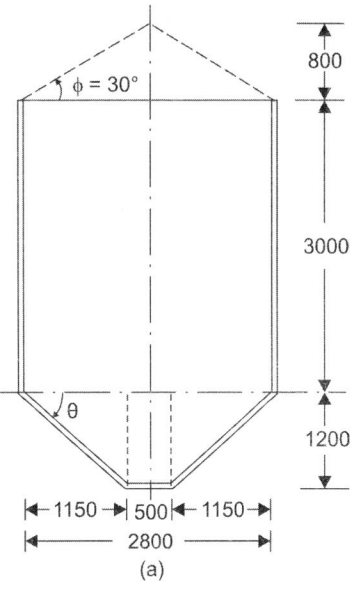

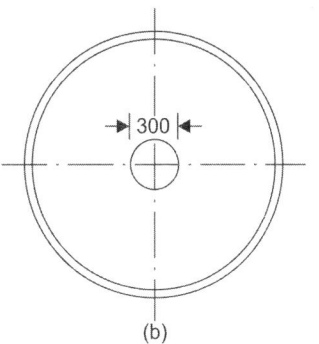

Fig. 4.8: Dimensions of circular bunker

Volume of surcharge
$$= (1/3 \times \pi \times 1.4^2 \times 1.4 \tan 30°) = 1.64 \text{ m}^3$$

Volume of cylindrical portion $= \left(\dfrac{\pi \times 2.8^2}{4} \times 3.00\right) = 18.46 \text{ m}^3$

Volume of frustum of cone $= \dfrac{\pi h}{12}\left(d_1^2 + d_1 d_2 + d_2^2\right)$

$$= \dfrac{\pi \times 1.2}{12}(2.8^2 + 2.8 \times 0.5 + 0.5^2)$$

$$= 2.90 \text{ m}^3$$

Total volume = 23.00 m³ > 22.22 m³

iv. *Design of cylindrical walls*:

Horizontal pressure $p = wh\cos^2\phi$

$$= 9 \times 3 \times (0.886)^2$$

$$= 20.25 \text{ kN/m}^2$$

Hoop tension in cylindrical wall per metre height $= 0.5\, pD$

$$= (0.5 \times 20.25 \times 2.8)$$

$$= 28.35 \text{ kN}$$

Ultimate hoop tension $= (1.5 \times 28.35) = 42.53 \text{ kN}$

Area of reinforcement, $A_{st} = \left(\dfrac{42.53 \times 10^3}{0.87 \times 415}\right) = 118 \text{ mm}^2$

Using 150 mm thick walls,

minimum reinforcement $= (0.0012 \times 150 \times 1000) - 180 \text{ mm}^?$

Use 8 mm diameter hoops at 250 mm centres, provide vertical reinforcement of 8 mm diameter bars at 250 mm centres.

v. *Design of hopper bottom*:

Provide a sloping slab 150 mm thick with 30 mm lining.

Total thickness = 180 mm

Weight of coal $= (23 \times 9) = 207 \text{ kN}$

Weight of sloping bottom $= \pi\left[\dfrac{2.8+0.5}{2}+0.18\sqrt{2}\,\right]0.18 \times \sqrt{2} \times 24 = 42 \text{ kN}$

$\therefore$ Total load $= (207 + 42) = 249 \text{ kN}$

If mean diameter at centre of sloping slab = 1.85 m

Tension per metre run, $T = \left(\dfrac{249}{\pi \times 1.85}\right) \text{cosec } 45° = 60.6 \text{ kN}$

Ultimate tension $= (1.5 \times 60.6) = 90.9 \text{ kN}$

$\therefore$ Steel reinforcement for direct tension $= \left(\dfrac{90.9 \times 10^3}{0.87 \times 415}\right) = 250 \text{ mm}^2$

Use 8 mm diameter bars at 200 mm centres in the direction of the sloping slab.

Normal component of coal pressure at centre of sloping slab is given by

$$p_n = wh\,(\cos^2\theta + \cos^2\phi\sin^2\theta)$$

where,

$w = 9 \text{ kN/m}^3$

$h = 3 + (0.5 \times 1.2) + (0.5 \times 0.80) = 4.00 \text{ m}$

$\theta = 45°$

$\phi = 30°$

$\therefore$ $p_n = 9 \times 4\,(\cos^2 45° + \cos^2 30° \times \sin^2 45°) = 31.5 \text{ kN/m}^2$

Normal component due to weight of the sloping slab $= w_d\cos\theta$

$$= (0.18 \times 24 \times \cos 45°)$$

$$= 3 \text{ kN/m}^2$$

Total normal pressure $p = (p_n + w_d \cos\theta)$
$$= (31.5 + 3) = 34.5 \text{ kN/m}^2$$

Mean diameter at centre of sloping slab $= \left(\dfrac{2.8 + 0.5}{2}\right) + 0.18\sqrt{2} = 1.90$ m

Hoop tension/metre $= (0.5 \times 34.5 \times 1.90) = 32.8$ kN
Ultimate hoop tension $= (1.5 \times 32.8) = 49.2$ kN

Area of hoop reinforcement $= \left(\dfrac{49.2 \times 10^3}{0.87 \times 415}\right) = 137 \text{ mm}^2$

But minimum area of steel $= (0.0012 \times 180 \times 1000) = 216 \text{ mm}^2$

Use 8 mm diameter hoop reinforcement at 220 mm centres in the sloping slab. The details of reinforcement in the circular bunker are shown in Fig. 4.9.

vi. *Edge beams*:

At the junction of the cylindrical wall and hopper bottom and at the top of bunker, edge beams of 300 mm × 300 mm are provided to increase the rigidity of the structure.

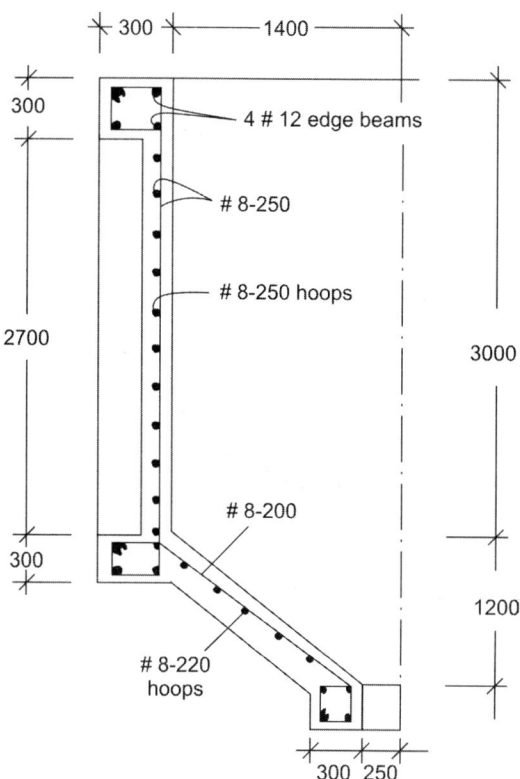

Fig. 4.9: Reinforcement details in circular bunker

4.6 ░ DESIGN OF A BATTERY OF BUNKERS

A battery of bunkers consists of more than one compartment interconnected to form a series of cells, generally square or rectangular with different ratios of side length of walls to the height of the bunker. The design moments to be considered are influenced by the ratio of side length to height and the loading pattern of the compartments.

The following two cases are generally considered for obtaining the design positive and negative moments in the side walls of a battery of three bunkers.

Case I: Battery of Bunkers with High Side Walls

When the side of the bunker is less than 0.75 times the height, the slabs span horizontally and moments are induced at the junction of side walls and partition walls. The typical direction of moments developed in a battery of bunkers is shown in Fig. 4.10. By analysis, the maximum moments developed in each of the walls are determined.

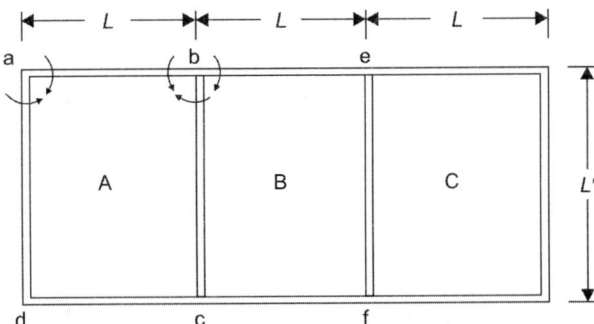

Fig. 4.10: Moments in a battery of bunkers

The moment coefficients compiled in Table 4.1 may be used for such a battery of bunkers.

$$M = \text{coefficient} \times pL^2$$

p = maximum pressure at the bottom due to the stored material

$$R = \left(\frac{L}{L'}\right) = \text{ratio of shorter side to longer side}$$

Moment is positive when it causes tension on the inside face of walls.

Table 4.1 gives the support moments developed when each of the bunkers are loaded. The maximum span moments are obtained as the difference between the values of $(pL^2/8)$ and the support moments.

Table 4.1: Moment coefficients for battery of bunkers (with high side walls)

Moment	A-loaded	B-loaded	C-loaded
$M_{ab} = M_{ad}$	$0.0362 + 0.470R^2$	$0.0076(R^2 - 1)$	–
M_{ba}	$0.0695 + 0.0138R^2$	$-0.038(R^2 - 1)$	$0.0089(R^2 - 1)$
M_{bc}	$0.0250 + 0.0585R^2$	$-(0.0228 + 0.0605R^2)$	$-0.00535(R^2 - 1)$
M_{be}	$0.445(1 - R^2)$	$0.0606 + 0.0227R^2$	$+0.0143(R^2 - 1)$

Case II: Battery of Bunkers with Low Side Walls

When the side walls are low, the slabs behave as two way slabs in the vertical as well as horizontal direction. The approximate moment coefficients for practical purposes are given in Table 4.2.

These coefficients are applicable under the following conditions:
a. The bunker is nearly square in plan.
b. The side of bunker is more than 0.75 and less than 1.25 times the height.
c. The bunker may be an isolated one or any one in the battery.
d. In case of battery of bunkers, all walls, top and bottom slabs should be reinforced on both faces for positive and negative moments.

Table 4.2: Moment coefficients for battery of bunkers (with low side walls) (Refer to figure alongside for notations)

Position of moment	Values of coefficients when L_2/L_1 is		
	1.25	1.20	0.75
Vertical span L_1			
Top end A	1/30	1/45	1/90
Bottom B	1/20	1/30	1/60
Mid span C	1/40	1/60	1/120
Horizontal span L_2			
End E	1/80	1/45	1/30
Mid span D	1/160	1/90	1/60

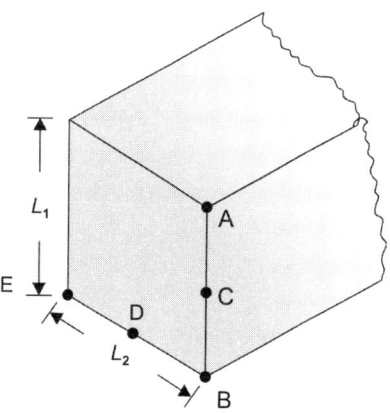

4.7　DESIGN OF SILOS

In deep bins (silos), the weight of the material stored is not completely supported by the bottom of the silo. A part of the load is resisted by friction between material and side walls of silo, resulting in the reduction of lateral pressure. Hence, Rankine's theory cannot be used for determining the pressure intensity. The vertical weight carried by walls causes direct compression in the walls.

The pressure intensity in silos where friction exists between material and wall surface can be determined by H Janssen's and W Airy's theories.

Janssen's Theory

The following assumptions are made in the design of silos according to Janssen's theory:
　i. The material has uniform texture.
　ii. The material has a definite angle of repose.
　iii. The coefficient of friction between material and side walls has a constant value.

The following notations are adopted in the analysis by Janssen's theory:
p_h = horizontal component of intensity of pressure at a depth h (kN/m²)
p_v = vertical component of intensity of pressure at a depth h (kN/m²)
r = radius of the silo (m)

μ' = coefficient of friction between wall and material

R = hydraulic mean radius

$$= \left(\frac{\pi r^2}{2\pi r}\right) = \left(\frac{r}{2}\right)$$

$$n = \left(\frac{p_h}{p_v}\right)$$

w = density of material (kN/m³)

For the equilibrium of vertical forces in the disc (Fig. 4.11), we have

$$\pi r^2 \cdot dh \cdot w = \pi r^2 \cdot dp_v + \mu' p_h \cdot 2\pi r \cdot dh$$

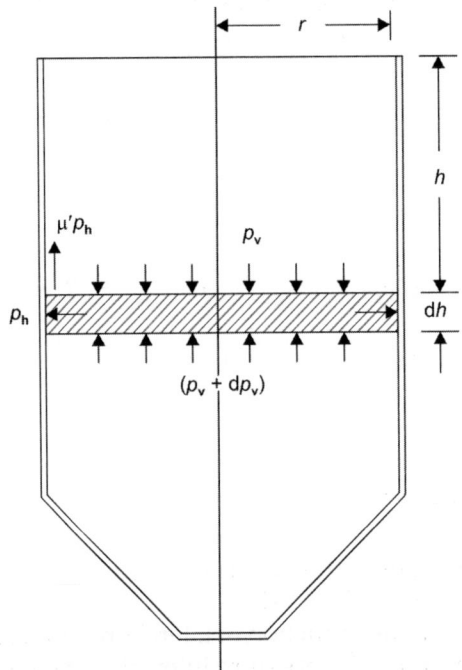

Fig. 4.11: Equilibrium of forces in a silo

Dividing by πr and putting $p_h = np_v$ and rearranging the terms, we get

$$dh = \left(\frac{r}{wr - 2\mu' np_v}\right) dp_v \qquad \qquad ...(4.1)$$

Integrating, we get

$$h = \frac{r}{2\mu' n} \log_e (wr - 2\mu' np_v) + c_1$$

But at

$$h = 0, \ p_v = 0$$

$\therefore$

$$c_1 = -\left(\frac{r}{2\mu' n} \log_e wr\right)$$

The solution of the differential Eq. (4.1) is given by

$$h = -\frac{r}{2\mu'n}\log_e(wr - 2\mu'np_v) + \frac{r}{2\mu'n}\log_e wr$$

$$h = -\frac{r}{2\mu'n}\log_e\left(\frac{wr - 2\mu'np_v}{wr}\right)$$

$$\therefore \qquad -\frac{2\mu'nh}{r} = \log_e\left(\frac{wr - 2\mu'np_v}{wr}\right)$$

$$\exp\left(-\frac{2\mu'nh}{r}\right) = \left(1 - \frac{2\mu'np_v}{wr}\right)$$

$$\therefore \qquad \left(\frac{2\mu'np_v}{wr}\right) = 1 - \exp\left(-\frac{2\mu'nh}{r}\right)$$

$$\therefore \qquad p_v = \frac{wr}{2\mu'n}\left(1 - \exp\left(\frac{2\mu'nh}{r}\right)\right)$$

But hydraulic mean radius $R = \left(\dfrac{r}{2}\right)$

$$\therefore \qquad p_v = \frac{wR}{\mu'n}\left[1 - \exp\left(-\frac{\mu'nh}{R}\right)\right]$$

Also $\qquad p_h = np_v$

$$\therefore \qquad p_h = \frac{wR}{\mu'}\left[1 - \exp\left(-\frac{\mu'nh}{R}\right)\right]$$

For large values of h;

$$\exp\left(-\frac{\mu'nh}{R}\right) = \left[\frac{1}{\exp\left(\dfrac{\mu'nh}{R}\right)}\right] = \text{very small quantity}$$

Hence, $\qquad p_h = \dfrac{wR}{\mu'}$

The vertical walls are designed for hoop tension of $p_h r$ and also for the vertical load supported by the wall.

The load taken by wall due to friction = $\mu' p_h$

Total vertical load taken by wall for a depth h

$$= \mu'\int_0^h \frac{wR}{\mu'}\left[1 - \exp\left(-\frac{\mu'nh}{R}\right)\right]2\pi r dh$$

$$= wA\int_0^h\left[1 - \exp\left(-\frac{\mu'nh}{R}\right)\right]dh$$

$$= wA\left[h + \frac{R}{\mu'n}\exp\left(-\frac{\mu'nh}{R}\right)\right]_0^h$$

$$= wA\left[h - \frac{R}{\mu'n}\left\{1 - \exp\left(-\frac{\mu'nh}{R}\right)\right\}\right]$$

$$= wAh - Ap_v = A(wh - p_v)$$

where, A = cross-sectional area of silo.

Airy's Theory

Airy's theory of design of silos is based on Coulomb's wedge theory of earth pressure. The results obtained from this theory also fairly agree with the experimental results although the basis of the theory is different from that of Janssen's theory. Using Airy's formula, horizontal pressure per unit length of periphery and position of plane of rupture can be determined. Knowing the horizontal pressure, vertical pressure and also vertical load taken by walls can be evaluated. Depending upon the plane of rupture, following two cases are considered.

Case 1: *Plane of rupture cuts the top horizontal surface (shallow bin or silo)*

From Fig. 4.12, the forces acting on the wedge ADE of grain are given below:

W = weight of wedge

R_1 = total reaction on side AC

R_2 = total reaction on side AE

P = reaction from wall

R = reaction from material

b = diameter of the silo

h = height of the silo

μ = coefficient of friction between the particles of material stored

μ' = coefficient of friction between material and concrete wall

θ = angle between plane of rupture and horizontal base

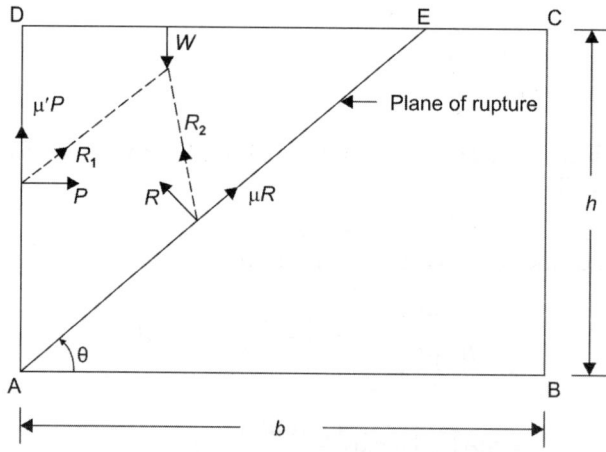

Fig. 4.12: Equilibrium of forces (shallow bin)

Resolving the forces in equilibrium along the plane of AE and perpendicular to AE,
$$\mu R + P\cos\theta = (W - \mu'P)\sin\theta$$
$$R - A\sin\theta = (W - \mu'P)\cos\theta$$

Solving,
$$P = \frac{W(\tan\theta - \mu)}{\tan\theta(\mu + \mu') + (1 - \mu\mu')}$$

Since
$$DE = h\cot\theta$$

$$W = (1/2 \times h\cot\theta \times h \times w) = \left(\frac{wh^2}{2}\right)\cot\theta$$

$\therefore$
$$P = \left[\frac{(\tan\theta - \mu)}{(1 - \mu\mu') + \tan\theta(\mu + \mu')}\right]\frac{wh^2}{2}\cot\theta \qquad \ldots(4.2)$$

For maximum value of P,

$$\left(\frac{dP}{d\theta}\right) = 0$$

Differentiating and equating to zero, it can be shown that

$$\tan\theta = \mu + \sqrt{\frac{\mu(1+\mu^2)}{(\mu+\mu')}}$$

Substituting this value in Eq. (4.2), we get

$$P = \frac{wh^2}{2}\left[\frac{1}{\left\{\sqrt{\mu(\mu+\mu')} + \sqrt{(1+\mu^2)}\right\}^2}\right]$$

$$p_h = \left(\frac{dP}{dh}\right) = wh\left[\frac{1}{\left\{\sqrt{\mu(\mu+\mu')} + \sqrt{(1+\mu^2)}\right\}^2}\right]$$

Total lateral pressure $= \pi b p_h$
Vertical load taken by walls $= \pi b p_h \mu'$
The depth up to which the silo will act as a shallow bin is given by

$$\tan\theta = \left(\frac{h}{b}\right) = \mu + \sqrt{\frac{\mu(1+\mu^2)}{(\mu+\mu')}}$$

$$h = b\left[\mu + \sqrt{\frac{\mu(1+\mu^2)}{(\mu+\mu')}}\right]$$

Case 2: *Plane of rupture cuts opposite side (deep bin or silo)*

Referring to Fig. 4.13, the forces acting on the wedge ADCE of grain is given below:
$$W = \text{weight of wedge ADCE}$$
$$CE = (h - b\tan\theta)$$

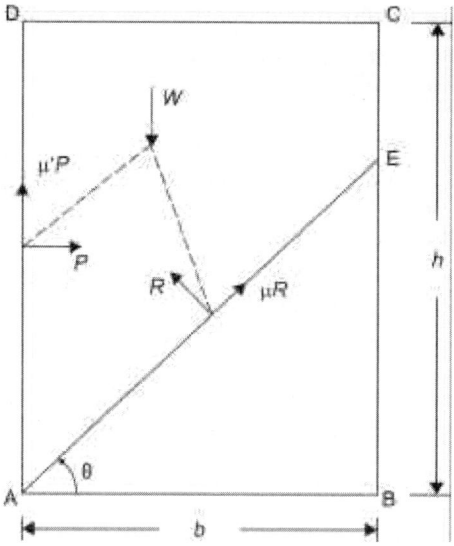

Fig. 4.13: Equilibrium of forces (deep bin)

$$W = wbh - \frac{wb}{2}b\tan\theta$$

$$W = \frac{wb}{2}(2h - b\tan\theta)$$

As in case 1, by studying the equilibrium of forces on the wedge,

$$P = \frac{W\tan\theta}{\tan\theta(\mu+\mu')+(1-\mu\mu')}$$

Substituting for W and differentiating for maximum value of P, we have

$$P = \frac{wb\left(\dfrac{2h-b\tan\theta}{2}\right)(\tan\theta-\mu)}{(\mu+\mu')\tan\theta+(1-\mu\mu')}$$

$$= \frac{wb}{2}\left[\frac{(2h-b\tan\theta)(\tan\theta-\mu)}{(\mu+\mu')\tan\theta+(1-\mu\mu')}\right]$$

$$= \frac{wb^2}{2(\mu+\mu')^2}\left[\sqrt{\frac{2h}{b}(\mu+\mu')+(1-\mu\mu')}-\sqrt{1+\mu^2}\right]^2$$

Differentiating and equating to zero, i.e. $\dfrac{dP}{dh} = 0$, yields

$$\tan\theta = -\frac{(1-\mu\mu')}{(\mu+\mu')}+\sqrt{\frac{2h}{b}\frac{(1+\mu^2)}{(\mu+\mu')}+\frac{(1+\mu^2)(1-\mu\mu')}{(\mu+\mu')^2}}$$

Substituting this value for $\tan\theta$,

$$P = \frac{wb}{2}(2h - b\tan\theta)\left[\frac{(\tan\theta - \mu)}{(\mu + \mu')\tan\theta + (1 - \mu\mu')}\right]$$

$$p_{\mathrm{h}} = \frac{\mathrm{d}p}{\mathrm{d}h} = \frac{wb}{(\mu + \mu')}\left[1 - \frac{\sqrt{(1 + \mu^2)}}{\sqrt{\frac{2h}{b}(\mu + \mu') + (1 - \mu\mu')}}\right]$$

Vertical load taken by walls $= \pi b P \mu'$

For the design of conical hopper bottom, the surcharge pressure is given by the expression:

$$\text{Surcharge pressure} = \left[\frac{\left(\frac{\pi b^2}{4}wh\right) - \pi b p_{\mathrm{h}}\mu'}{\left(\frac{\pi b^2}{4}\right)}\right]$$

$$= wh - \frac{4p_{\mathrm{h}}\mu'}{b}$$

The values of the coefficient of friction μ and μ' are compiled in Table 4.3.

Table 4.3: Values of friction coefficients

Material	Density kN/m³	Coefficient of friction Filling on filling (μ)	Filling on concrete (μ')	$K = \left(\dfrac{1 - \sin\phi}{1 + \sin\phi}\right)$
Cement	14.00	0.316	0.554	0.5371
Coal	8.00	0.700	0.700	0.2709
Anthracite	8.35	0.510	0.510	0.3753
Coke	4.50	0.839	0.839	0.2174
Sand	16.00	0.674	0.577	0.2830
Wheat	8.50	0.466	0.444	0.4062

4.8　SILOS FOR STORAGE OF CEMENT

The cement manufactured in a factory is generally stored in a battery of silos which are circular in shape with diameters of cylindrical containers up to 10 m and height 25 to 30 m. A typical battery of silos to store 130,000 kN of cement is shown in Fig. 4.14.

The horizontal pressure distribution at various depths in cement silos predicted by Janssen's and Airy's theories are compared with the experimental values reported by Faber and Mead in Fig. 4.15. Truly frictional noncohesive materials behave in accordance with Janssen's theory. In materials like cement, due to the erratic cohesive nature of the material, the pressures obtained using Janssen's theory varied widely with the actual experimental observations.

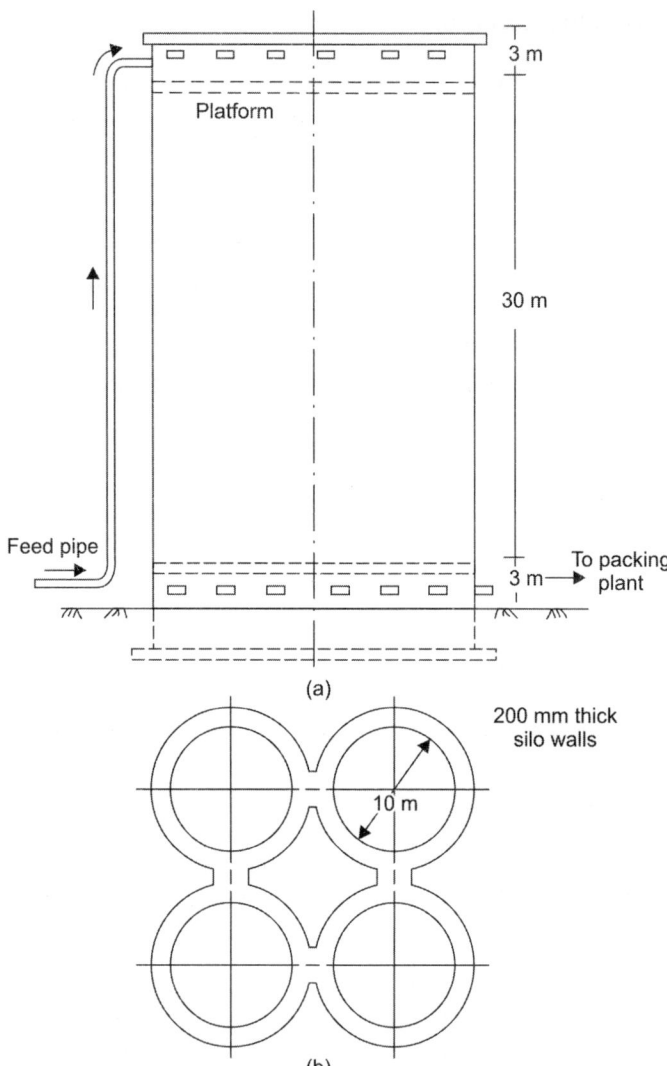

Fig. 4.14: Battery of cement silos

Full scale experiments of cement silos indicate that the lateral pressures are very much underestimated towards the upper half of the silo by the Janssen's theory. In contrast, the lateral pressures are very much overestimated towards the lower half of the silo by the Airy's theory. Towards the bottom of the silo, the lateral pressures reduce due to the arching action.

The depth of cement occurring below any particular level under consideration influences the lateral pressure. The cement below suffers a cushioning effect packing tighter as the head of cement overburden is increased until it acquires its state of maximum density. At depths where maximum density is reached, the maximum angle of internal friction is also developed. Compacted cement can stand for considerable heights with a vertical face, whereas shallow lightly sprinkled cement exhibits properties not far removed from that of a liquid. Maximum pressures indicated in Fig. 4.15 are useful in design of silos.

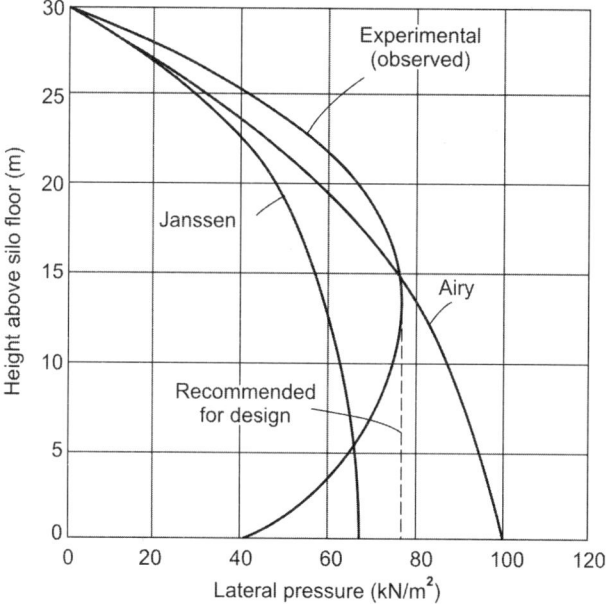

Fig. 4.15: Horizontal pressure distribution in cement silos

4.9 DESIGN EXAMPLES

1. A cylindrical silo has an internal diameter of 6 m and 20 m deep (cylindrical portion) with a conical hopper bottom. The material stored is wheat with density of 8 kN/m³. The coefficient of friction between wall and material is 0.444. The ratio of horizontal to vertical pressure is 0.40. Angle of repose is 25°. Design the reinforcements in the silo walls. Adopt M20 grade concrete and Fe 415 HYSD bars. Adopt Janssen's theory for pressure calculations.

   i. *Data*:

     Diameter of silo = 6 m

     Depth of cylindrical portion = 20 m

     Density of wheat = 8 kN/m³

     Coefficient of friction between wall and material = 0.444

     Ratio of horizontal to vertical pressure intensity = 0.40

     Angle of repose = 25°

     Type of reinforcements = Fe 415 HYSD bars

     Grade of concrete = M20

  ii. *Characteristic strength*:

     f_{ck} = 20 N/mm²

     Permissible tensile stress in concrete = 2.8 N/mm²

     f_y = 415 N/mm² (according to clause B-2.1.1 of IS:456-2000)

     m = 13

 iii. *Dimensions of silo*:

     Diameter of silo = 6 m

Height of cylindrical portion = 20 m
Depth of hopper bottom = 2.5 m
Diameter of opening in hopper bottom = 1 m

iv. *Design of cylindrical walls*:
Using Janssen's theory:

Horizontal pressure $p_h = \dfrac{wR}{\mu'}\left[1-\exp\left(-\dfrac{\mu' nh}{R}\right)\right]$

Now $n = \left(\dfrac{1-\sin\phi}{1+\sin\phi}\right) = \left(\dfrac{1-\sin 25°}{1+\sin 25°}\right) = 0.40$

R = hydraulic mean radius $= \left(\dfrac{D}{4}\right) = \left(\dfrac{6}{4}\right) = 1.5$ m

$\therefore \quad p_h = \left(\dfrac{8.00\times 1.5}{0.444}\right)\left[1-\exp\left(-\dfrac{0.444\times 0.40\times h}{1.5}\right)\right]$

$= 27.03(1 - e^{-x})$

For different values of h = 4, 8, 12, 16, 20 and 22.5 m from the top, the value of e^{-x} is calculated and the value of p_h is compiled in Table 4.4. Appendix 2 gives the values of e^{-x} for different values of x.

Maximum horizontal pressure in the cylindrical portion (20 m from top) $p_h = 24.58$ kN/m²

Hoop tension in cylindrical wall per metre height $= 0.5\, p_h D$

$= (0.5 \times 24.58 \times 6)$

$= 73.74$ kN

Design ultimate hoop tension $= (1.5 \times 73.74) = 110.61$ kN

Area of hoop reinforcement $A_{st} = \left[\dfrac{110.61\times 1000}{0.87\times 415}\right] = 306.3$ mm²

Adopt 8 mm diameter hoops at 140 mm centres (A_{st} = 359 mm²).

Assuming 150 mm thick cylindrical walls, the tensile stresses developed in concrete under working hoop tension should be limited to the values specified in clause B-2.1.1 of IS:456-2000.

Table 4.4: Horizontal pressure in silo walls

Depth from top, h (metre)	Horizontal pressure, p_h (kN/m²)
4	10.54
8	16.67
12	20.62
16	23.05
20	24.58
22.5	25.13

Tensile stress in concrete $= \left[\dfrac{F_t}{A_c + mA_{st}} \right]$

$$= \left[\frac{73.74 \times 1000}{(150 \times 1000) + (13 \times 359)} \right]$$

$$= 0.47 \text{ N/mm}^2 < 2.8 \text{ N/mm}^2$$

Minimum area of hoop reinforcement $= (0.0012 \times 150 \times 1000) = 180 \text{ mm}^2$

Adopt 8 mm diameter hoops at 200 mm centres ($A_{st} = 251 \text{ mm}^2$) towards the top of silo walls. The details of reinforcements to be provided at different depths of the cylindrical silo walls are compiled in Table 4.5.

Table 4.5: Reinforcement details in silo walls

Depth from top (metre)	Spacings of 8 mm diameter hoops	Vertical distribution reinforcements
0–12	200 mm	8 mm diameter bars at
12–16	160 mm	270 mm centres throughout
16–20	140 mm	the whole depth

Provide vertical reinforcements (temperature reinforcements) of 8 mm diameter bars at 200 mm centres throughout the silo from top to bottom. Refer to Fig. 4.9 for reinforcement details.

v. *Design of hopper bottom*:

Provide a sloping slab 150 mm thick with 30 mm lining.

Total thickness = 180 mm

Surcharge load on hopper bottom/metre $= \left(wh - \dfrac{4p_h \mu'}{b} \right)$

$$= \left[(8 \times 20) - \frac{4 \times 24.58 \times 0.44}{6} \right]$$

$$= 153 \text{ kN}$$

Weight of sloping bottom $= \pi \left[\left(\dfrac{6+1}{2} \right) + 0.18\sqrt{2} \right] \times 0.18 \times \sqrt{2} \times 24 = 72 \text{ kN}$

Total load = (153 + 72) = 225 kN

If mean diameter at centre of sloping slab = 3.5 m

Ultimate tension per metre run, $T_u = \left[\dfrac{1.5 \times 225}{\pi \times 3.5} \right] \text{cosec } 45° = 45 \text{ kN}$

Reinforcement for direct tension $A_{st} = \left[\dfrac{30 \times 10^3}{0.87 \times 3.5} \right] = 83.1 \text{ mm}^2$

Minimum reinforcement = $(0.0012 \times 150 \times 1000) = 180 \text{ mm}^2/\text{m}$

Provide 8 mm diameter bars at 200 mm centres, in the direction of the sloping slab $(A_{st} = 251 \text{ mm}^2)$.

Surcharge pressure on hopper bottom $= \left(\dfrac{153}{\pi \times 6^2} \right) = 1.35 \text{ kN/m}^2$

Maximum horizontal pressure in hopper bottom $p_h = 25.13 \text{ kN/m}^2$

Normal pressure intensity $= p_n$

$\qquad = (1.35 \cos^2 \theta + p_h \sin^2 \theta)$

$\qquad = (1.35 \times \cos^2 45° + 25.13 \times \sin^2 45°)$

$\qquad = 13.24 \text{ kN/m}^2$

Normal component due to self weight of sloping slab $= w_d \cos \theta$

$\qquad\qquad\qquad\qquad\qquad\qquad = (0.180 \times 24 \times \cos 45°)$

$\qquad\qquad\qquad\qquad\qquad\qquad = 3.00 \text{ kN/m}^2$

Total normal pressure $P = (p_n + w_d \cos \theta)$

$\qquad\qquad\qquad\quad = (13.24 + 3.00)$

$\qquad\qquad\qquad\quad = 16.24 \text{ kN/m}^2$

Mean diameter of sloping slab $= \left(\dfrac{6+1}{2} \right) + 0.18\sqrt{2} = 3.75 \text{ m}$

Hoop tension per metre = $(0.5 \times 16.24 \times 3.75) = 30.45 \text{ kN}$

Ultimate hoop tension per metre = $(1.5 \times 30.45) = 45.675 \text{ kN}$

Area of hoop reinforcement $= \left(\dfrac{45.675 \times 10^3}{0.87 \times 415} \right) = 126.5 \text{ mm}^2$

Minimum reinforcement = $(0.0012 \times 150 \times 1000) = 180 \text{ mm}^2/\text{m}$

Provide 8 mm diameter bars at 200 mm centres as hoop reinforcement $(A_{st} = 251 \text{ mm}^2)$

vi. *Edge beams*:

At the junction of the cylindrical wall and hopper bottom at the top of the silo, edge beams of size 300 mm × 300 mm with 4 bars of 12 mm diameter are provided to increase the rigidity of the structure. Refer to Fig. 4.9 for the details of reinforcements.

2. A cement silo has an internal diameter of 10 m with the height of cylindrical portion being 30 m. The density of cement is 15.2 kN/m³. Coefficient of friction between concrete and material is 0.70. The angle of repose of the material is 17.5°. Adopting M20 grade concrete and Fe 415 HYSD bars, design the thickness and the reinforcement required at the bottom of the cylindrical portion of the silo using Janssen's theory.

i. *Data*:

Diameter of silo = 10 m

Depth of cylindrical portion = 30 m

Density of cement = 15.2 kN/m³

Coefficient of friction between wall and material $\mu' = 0.70$

Angle of repose of the material $\phi = 17.5°$

Ratio of horizontal to vertical pressure intensity $n = \left(\dfrac{1-\sin\phi}{1+\sin\phi}\right)$

$$= \left(\dfrac{1-0.3}{1+0.3}\right) = 0.54$$

ii. *Characteristic strength:*

$f_{ck} = 20$ N/mm^2 Permissible tensile stress in concrete = 2.8 N/mm^2

$f_y = 415$ N/mm^2 Modular ratio $m = 13$

iii. *Design of cylindrical walls:*

Using Janssen's theory,

horizontal pressure $p_h = \dfrac{wR}{\mu'}\left[1-\exp\left(-\dfrac{\mu'nh}{R}\right)\right]$

$$R = \left(\dfrac{D}{4}\right) = \left(\dfrac{10}{4}\right) = 2.5$$

$n = 0.54$ $w = 15.2$ kN/m^3

$h = 30$ m $\mu' = 0.70$

At the bottom of the cylindrical portion ($h = 30$ m), the horizontal pressure is

given by $p_h = \left(\dfrac{15.2\times2.5}{0.70}\right)\left[1-\exp\left(-\dfrac{0.70\times0.54\times30}{2.5}\right)\right]$

$$= 54 \text{ kN/m}^2$$

Hoop tension in cylindrical wall per metre height = 0.5 $p_h D$

$$= (0.5 \times 54 \times 10) = 270 \text{ kN}$$

Ultimate hoop tension = $(1.5 \times 270) = 405$ kN

Area of hoop reinforcement $A_{st} = \left(\dfrac{405\times1000}{0.87\times415}\right) = 1122$ mm^2

Adopt 16 mm diameter hoops at 160 mm centres ($A_{st} = 1256$ mm^2)

Using 200 mm thick cylindrical walls, tensile stress developed in concrete is computed as

$$\sigma_{ct} = \left(\dfrac{F_t}{A_c + mA_{st}}\right)$$

$$= \left[\dfrac{270\times1000}{(200\times10^3)+(13\times1256)}\right]$$

$$= 1.248 \text{ N/mm}^2 < 2.8 \text{ N/mm}^2$$

Vertical pressure intensity at a depth $h = 30$ m is computed as

$$p_v = \left(\dfrac{p_h}{n}\right) = \left(\dfrac{54}{0.54}\right) = 100 \text{ kN/mm}^2$$

Equivalent depth of cement $= \left(\dfrac{100}{15.2}\right) = 6.57$ m

Even though there is 30 m depth of cement, the effective vertical load is only due to 6.57 m of cement.

30 m depth of cement corresponds to a pressure $= (15.2 \times 30) = 456$ kN/m^2

Vertical pressure intensity at 30 m depth $= 100$ kN/m^2

and is $\left(\dfrac{100}{456} \times 100\right) = 21.9\%$ of the total weight.

The remaining 78.1% is transferred to the silo walls by friction.

Provide vertical distribution steel of 0.12% of the cross-section equivalent to 8 mm diameter bars at 160 mm centres.

3. Compare the horizontal pressures developed at 5 m intervals in a cement silo of internal diameter 10 m and height 30 m using Janssen's and Airy's theories.

Density of cement $= 15.2$ kN/m^2

Coefficient of friction between concrete and cement $\mu' = 0.554$

Coefficient of friction between filling on filling $\mu = 0.316$

Angle of repose of cement $\phi = 17.5°$

$$n = \left(\dfrac{1-\sin\phi}{1+\sin\phi}\right) = 0.54$$

a. *Janssen's theory*:

$$p_h = \dfrac{wR}{\mu'}\left[1 - \exp\left(-\dfrac{\mu'nh}{R}\right)\right]$$

Now $\qquad R = \left(\dfrac{D}{4}\right) = \left(\dfrac{10}{4}\right) = 2.5$

$\therefore \qquad p_h = \left(\dfrac{15.2\times2.5}{0.554}\right)\left[1 - \exp\left(-\dfrac{0.554\times0.54\times h}{2.5}\right)\right]$

$\qquad\qquad = 68.59\,(1 - e^{-x})$

For different values of $h = 5$, 10, 15, 20, 25 and 30 m from top, the values of horizontal pressure p_h are evaluated.

b. *Airy's theory*:

  i. The depth up to which the silo will act as shallow bin is given by

$$h = b\left[\mu + \sqrt{\dfrac{\mu(1+\mu^2)}{(\mu+\mu')}}\right]$$

$$= 10\left[0.316 + \sqrt{\dfrac{0.316(1+0.316^2)}{(0.316+0.554)}}\right] = 9.48 \text{ m}$$

$$p_h = wh \left[\frac{1}{\sqrt{\mu(\mu + \mu')} + \sqrt{1 + \mu^2}^{-2}} \right]$$

$$= 15.2h \left[\frac{1}{\sqrt{0.316(0.316 + 0.554)} + \sqrt{1 + 0.316^2}^{-2}} \right]$$

$$= 6.14h \text{ kN/m}^2 \text{ valid up to } 9.48 \text{ m.}$$

ii. The horizontal pressure intensity at depths greater than 9.48 m is given by

$$p_h = \frac{wb}{(\mu + \mu')} \left[1 - \frac{\sqrt{1 + \mu^2}}{\sqrt{\frac{2h}{b}(\mu + \mu') + (1 - \mu\mu')}} \right]$$

$$= \frac{15.2 \times 10}{(0.316 + 0.554)} \left[1 - \frac{\sqrt{1 + 0.316^2}}{\sqrt{\frac{2h}{10}(0.316 + 0.554) + (1 - 0.316 \times 0.554)}} \right]$$

$$= 174.7 - 183(0.174h + 0.825)^{-1/2}$$

Using this equation the values of p_h at depths of 10, 15, 20, 25 and 30 m are evaluated and compared in Table 4.6.

Table 4.6: Horizontal pressure in cement silos

Depth from top (metre)	Values of p_h (kN/m²)	
	Janssen's theory	Airy's theory
5	31.00	30.74
10	47.87	60.40
15	57.25	75.88
20	62.41	86.30
25	65.16	94.30
30	66.71	100.00

In deep bins, Airy's theory predicts higher values of horizontal pressure than that resulting from Janssen's theory for increasing depth from top. At greater depths, the pressure computed by Airy's theory is nearly 50% greater than that evaluated by Janssen's theory.

4.10 DESIGN AND CONSTRUCTION OF LARGE CAPACITY BINS USING DOME TECHNOLOGY

The various steps involved in the design and construction of bulk storage structures using dome technology are listed below:

1. *Design aspects of dome silo*: Dome silo comprises a thin concrete shell supported by a rigid reinforced concrete ring beam at the base to resist the hoop tension developed due to self weight of the concrete shell, finishes and any live loads on the shell. The ring beam and the dome floor should have a proper foundation comprising a massive reinforced concrete mat resting on compacted soil at site. The ring beam should be designed to resist the hoop tension developed due to the concrete shell dome. The diameter of the dome silo is based on the volume of storage of the bulk material. Barry South[7] of USA has reported the construction of dome silos having diameters from 30 to 100 m and heights varying from 10 to 50 m.

2. *Geotechnical and subgrade preparation*: In this process, geotechnical investigations of the soil at site are done to determine the design parameters of the soil. The subgrade preparation includes thorough analysis of the existing soil conditions, replacing soils, compaction and other similar requirements determined by the performance parameters of the specific project.

3. *Foundations and tunnels*: After subgrade preparation, the construction of the foundations for the dome silo comprising the reinforced concrete ring beam below the edges of the concrete dome shell and the concrete floor is started. The ring beam should be designed to resist the hoop tension developed due to the thrust transmitted by the shell at the base. Dome technology integrates the construction of tunnels for handling the stored material with the construction of the foundation system as shown in Fig. 4.16.

Fig. 4.16: Formation of tunnels and foundations at the base of dome silo

4. *Inflated air form to support reinforcements for the concrete dome*: After the foundations and tunnel system are completed, the reinforced PVC air form is unfolded over the top of the staged materials and the air form edges are secured to the ring beam foundation system as shown in Fig. 4.17. Temporary openings are positioned to allow access to the interior of the dome structure during construction activities. The air form is checked to ensure integrity prior to inflation.

5. *Inflation of the PVC air form*: Inflation of the dome silo air form is affected by using large fans inside the dome to provide sustained air pressure within the air form. The intensity of pressure depends upon the size of the dome and typically

Fig. 4.17: Ring beam and PVC air form before inflation

range from 1.5 to 2.5 inches of water column. The inflation pressure is maintained throughout the construction process until the reinforced concrete dome is cured.

6. *Formation of penetrations, entrance openings*: Locations for openings, entrances and similar penetrations in the finished dome silo are laid out and marked in the inflated air form. Forming materials may be installed at the boundary of the proposed openings to demarcate them during application of insulation foam and shotcrete.

7. *Primer and foam insulation*: A coat of specialized primer is applied to the interior side of the inflated air form. This primer is followed by several coats of spray applied foam insulation to achieve the designed thickness of the specific project as shown in Fig. 4.18. The foam insulation provides a continuous and uninterrupted thermal barrier between the exterior and interior sides of the dome silo. During application of the insulation, metal sticker rods and depth gauges are embedded in the foam

Fig. 4.18: Application of primer and foam insulation to the interior side of inflated air form

insulation. These stickers allow for attachment of steel reinforcing bars during later stages of construction. Depth gauges provide consistency for the finished thickness of both foam insulation and reinforced concrete dome.

8. *Premat steel mesh:* Premat steel reinforcements are placed over the final thickness of foam insulation and tied to the sticker rods as shown in Fig. 4.19. This premat steel serves as a base for the initial shotcrete application until sufficient designed depth of the reinforced concrete dome is completed.

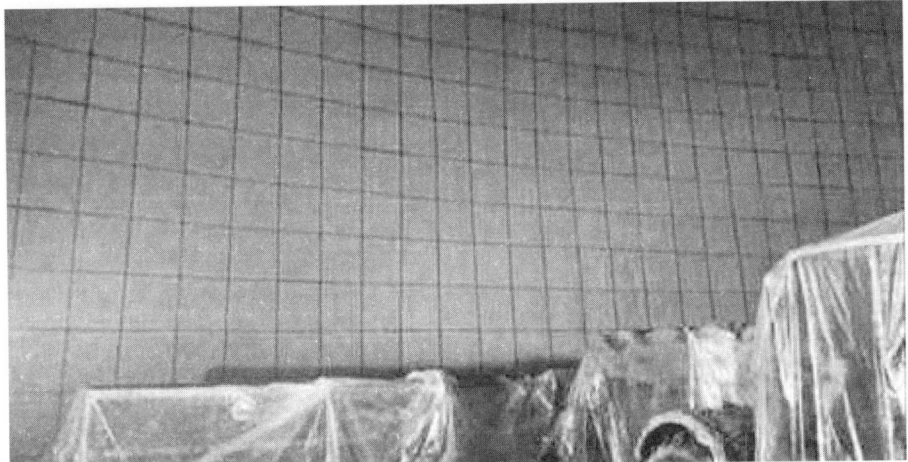

Fig. 4.19: Premat steel reinforcements over the foam insulation

9. *Shotcrete and main reinforcing steel:* The designed quantity of mesh reinforcements is placed before final shotcreting to achieve the designed thickness of dome silo. The thickness depends upon the size of the dome and is designed to support its self weight, finishes and other imposed loads. The shotcrete process is the final primary process in constructing the dome silo and represents completion of the concrete thin shell dome structure as shown in Fig. 4.20.

Fig. 4.20: Shotcreting of dome silo after placing main reinforcements

10. *Entrances, penetrations, overbuilds*: After curing of the concrete in the dome silo, the inflation fans are shut down and removed and the concrete shell structure is self supporting and resting on the main ring beam at the base. The entrance extensions and penetrations are constructed with full integration to the concrete thin shell dome.

11. *Auxiliary equipment installation*: After the concrete shell dome is complete, auxiliary equipment involving mechanical, electrical and other conveyer systems are installed to complete the dome silo project. The completed dome silo with the mechanical conveyer system is shown in Fig. 4.21.

Fig. 4.21: Completed dome silo with the mechanical conveyer system

4.11 DESIGN EXAMPLE

A large capacity dome silo is to be designed to store fertilizers. Volumetric computations indicate that the spherical dome should have diameter of 30 m with a central rise of 10 m. The dome silo should be designed as thin shell concrete spherical structure supported on a rigid ring beam at the ground level. Design the salient dimensions of the thin shell concrete dome and the ring beam at base. The designs should conform to the Indian standard codes IS:4995[9] and IS:456[10]. The dome should be designed to support its own weight together with that of the finishes and nominal live loads. Also design suitable reinforcements in the dome silo and the ring beam using the following data:

i. *Data*:

Diameter of silo = 30 m

Height of silo = 10 m

Density of concrete = 25 kN/m^3

Weight of finishes = 0.75 kN/m^2

Live load on dome = 1.5 kN/m^2

Materials: M20 grade concrete,

Fe 415 HYSD reinforcements

ii. *Permissible stresses*:

$\sigma_{cc} = 5 \text{ N/mm}^2$

$\sigma_{ct} = 2.8 \text{ N/mm}^2$ for ring beam

$\sigma_{st} = 150 \text{ N/mm}^2$

$m = 13$

iii. *Geometrical parameters of dome*:

Diameter of the dome, $D = 30$ m

Central rise, $r = 10$ m

Let R = radius of the spherical dome

then $(R - r)^2 = [R^2 - (D/2)^2]$

Substituting the values, $(R - 10)^2 = [R^2 - (30/2)^2]$

Solving, $R = 16.25$ m and $\cos\theta = (15/16.25) = 0.92$

iv. *Design loads*:

Assume the thickness of the dome as 150 mm

Self weight of the dome = $(0.15 \times 25) = 3.75 \text{ kN/m}^2$

Live load on dome   = 1.50 kN/m^2

Finishes   = 0.75 kN/m^2

Total design load (w)   = 6.00 kN/m^2

v. *Stresses in dome*:

Meridional thrust $T_1 = \left[\dfrac{wR}{1+\cos\theta}\right] = \left[\dfrac{6\times16.25}{1+0.92}\right] = 58.78 \text{ kN/m}$

Meridional compressive stress $= \left[\dfrac{58.78\times10^3}{1000\times150}\right] = 0.39 \text{ N/mm}^2 < 5 \text{ N/mm}^2$

Hoop stress $= \dfrac{wR}{t}\left[\cos\theta - \dfrac{1}{(1+\cos\theta)}\right]$

$= \left(\dfrac{6\times16.25}{0.15}\right)\left[0.92 - \dfrac{1}{(1+0.92)}\right]$

$= 260 \text{ kN/m}^2 = 0.26 \text{ N/mm}^2 < 5 \text{ N/mm}^2$

Hence, the stresses are within safe permissible limits.

vi. *Reinforcements in dome*:

Since the stresses are very low, nominal reinforcements of 0.3% of the gross cross-sectional area is provided.

$A_{st} = (0.003 \times 1000 \times 150) = 450 \text{ mm}^2/\text{m}$

Providing two layers of 8 mm diameter bars,

area of steel in each layer = 225 mm²/m

Spacing of 8 mm diameter bars = $(1000 \times 50)/225 = 222$ mm

Provide 8 mm diameter bars both meridionally and circumferentially at a spacing of 200 mm centres.

vii. *Ring beam at the base of dome*:

Horizontal component of meridional thrust $H = T_1 \cos\theta$

$$= (58.78 \times 0.92) = 54 \text{ kN/m}$$

Hoop tension in ring beam $F_t = (H \times D/2) = (54 \times 30/2) = 810 \text{ kN}$

Area of steel to resist tension $A_{st} = (810 \times 1000)/150 = 5400 \text{ mm}^2$

Provide 12 bars of 25 mm diameter ($A_{st} = 5890 \text{ mm}^2$)

Let A_c = cross-sectional area of the ring beam

Allowing a tensile stress of 2.8 N/mm^2 in concrete, we have the relation from IS:456-2000 as

$$\left[\frac{F_t}{A_c + (m-1)A_{st}}\right] = \left[\frac{810 \times 1000}{A_c + (13-1)5400}\right] = 2.8$$

Solving, $A_c = 224485 \text{ mm}^2$

Provide a ring beam of size 400 mm wide by 600 mm deep ($A_c = 240000 \text{ mm}^2$) reinforced with 12 bars of 25 mm diameter arranged in four rows having three bars in each row.

References

1. Gray WS, *Reinforced Concrete Water Towers, Bunkers, Silos and Gantries*, Concrete Publications, London, 1947.

2. Reynolds CS, Steedman JC and Threlfall AJ, *Reinforced Concrete Designer's Handbook*, 11 edn, Taylor & Francis, London and New York, 2008.

3. Martens P, *Silohandbuch*, Verlag Ernst & Sohn, Berlin, 1988.

4. Brown CJ and Nielson J (Editors), *Silos; Fundamentals of Theory & Design*, CRC Press, Architecture, May 1998, p. 856.

5. Safarin SS, Haris EC, *Design and Construction of Silos and Bunkers*, Von Nostrand Reinhold Co, 1985, p. 468.

6. Changli Machinery, Shangjie District, Zengzhov City, Henanan province, China.

7. Dome Technology for Bins and Silos, Newsletter, Idaho, Mississippi, USA, 2014.

8. FOX, Farmers, Cooperative, Elevator and Mercantile organization, Dighton, Kansas, USA Report by Emily Wilson, Storing in Domes, World-Grain.com, 2000.

9. IS:4995-1974, Criteria for the Design of Reinforced Concrete Bins for the Storage of Granular and Powdery Materials, part I, General Requirements and Assessment of Bin Loads, part II, Design Criteria. New Delhi, 1974.

10. IS:456-2000, Indian Standard Code for Plain and Reinforced Concrete, Bureau of Indian Standards, New Delhi, 2000, pp. 1–100.

Assignment

1. A reinforced concrete circular bunker has 4 m internal diameter. The circular wall is 4.5 m high. It has a 45° conical hopper bottom. The hopper has a concentric circular opening of diameter 0.5 m. The bunker is supported on four columns and is used for storing material weighing 8.5 kN/m^3. Angle of friction is 30°. The coefficient of friction between the concrete surface and the material is 0.4. Design the circular walls of the bunker and the conical hopper bottom. Adopt M20 grade concrete and ribbed tor steel Fe 415 grade.

2. A cylindrical silo having a ratio of height (cylindrical portion) to the diameter as 4, is required to store 2000 kN of wheat having a density of 8 kN/m^3. The coefficient of friction between grain and concrete is 0.444 and the ratio of horizontal to vertical pressure is 0.4. Design the reinforcements in the walls and conical bottom of the silo using Janssen's theory. Adopt M20 grade concrete and Fe 415 HYSD bars.

3. A coal bunker is to be designed to store 300 kN of coal having a unit weight of 8 kN/m^3. The bunker should be a square with sides 3 m. The stored coal is to be surcharged at an angle of repose which is 30° for coal. Adopt M25 grade concrete and Fe 415 steel and design the side walls and hopper bottom, and sketch the details of reinforcements.

4. A rectangular shallow bin 3 m × 2.5 m in plan is required to store 240 kN of foamed slag aggregates, the unit weight of aggregates being 9.5 kN/m^3. The bin is supported on four RC columns at the corners. Using M20 grade concrete and Fe 415 steel, design the side walls, hopper bottom and the column, and sketch the structural details. The angle of repose is 30° and angle of surcharge is 25°.

5. A circular grain silo of 3.6 m diameter is reinforced with horizontal hoop reinforcement of 10 mm diameter. Calculate the maximum pitch for these rods at depth of 24 m below the surface of the grain. The unit weight of grain is 8 kN/m^3 and the coefficient of friction between grain and silo walls is 0.41. The ratio of lateral to vertical pressure at any point is 0.6. Adopt M20 concrete and Fe 415 bars and use Janssen's theory.

   The vertical reinforcement of the 125 mm thick concrete shell consisting of 24 rods is 10 mm diameter equally spaced. Determine the vertical stress in concrete at 24 m depth assuming the silo to be full and neglecting wind forces. Allow for a superimposed load of 11 kN/m^2 at roof level and modular ratio is 19.

6. A cylindrical silo having the ratio of height (cylindrical portion) to the diameter as 4 is required to store 2000 kN of wheat weighing 8 kN/m^3. The coefficient of friction between grain and concrete is 0.444 and that between grains is 0.466. Compare the lateral pressures developed at 4 m intervals using Janssen's and Airy's theories. Design the reinforcements in the wall and the conical bottom for the worst case. Adopt M20 grade concrete and Fe 415 HYSD bars.

7. A spherical dome silo is to be designed to store grain in large quantities. Preliminary computations indicate that the dome should have a diameter at base of 40 m with a central rise of 20 m. Design the dome silo as a thin shell reinforced concrete structure supported on a ring beam at base. Adopt M30 grade concrete and Fe 500 grade HYSD steel bars. Sketch the details of reinforcements in the dome shell and ring beam. Assume suitable values for finishes and nominal live loads on the dome.

8. Design a dome silo with a base diameter of 90 m and a central rise of 50 m to store coal. The silo should be designed as a thin shell concrete structure supported on a massive prestressed concrete ring beam at the base. Adopt M30 grade concrete for the shell dome and M50 grade concrete for the ring beam. Fe 500 grade HYSD bars and high strength steel standard Freyssinet cables are available for use in the ring beam. Assume suitable values for finishes and nominal live loads on the dome. Design the concrete shell dome and the prestressed ring beam, and sketch the details of reinforcements.

Review Questions

1. How do you distinguish between bins and silos? Mention the main criteria identifying the differences between the two types of structures.
2. List the main structural components of a bunker.
3. What are the various forces acting on the hopper bottom of a bunker and how do you design them?
4. What are the various theories used in the analysis of forces in silos? Mention the significant differences between them.
5. What are the various force components to be considered in the design of deep bins?
6. Discuss briefly the horizontal pressure intensity developed in silos using the Janssen's and Airy's theories, and comment on the differences in the values of pressure intensity as a function of the depth of the silo.
7. Sketch the details of reinforcements in the various structural components of a silo.
8. What are dome silos? In what way dome silos are more economical in comparison with the traditional silos?
9. What are the main structural components of a dome silo?
10. List the various steps involved in the construction of a dome silo.

Objective Type Questions

1. In the case of silos, a significant portion of the load is resisted by
   a. the floor portion of the silo
   b. the conical portion of the silo
   c. friction between the material and the side walls of the silo

2. In a shallow bunker, the plane of rupture
   a. meets the side walls of the bunker
   b. coincides with the angle of repose
   c. meets the top horizontal surface

3. The hopper bottom of a coal bunker should be designed to resist
   a. the normal component of coal pressure
   b. the normal component of weight of the sloping slab
   c. both the normal components of coal pressure and self weight

4. The vertical walls of a silo should be designed for
   a. horizontal component of the intensity of pressure
   b. vertical component of the intensity of pressure
   c. both hoop tension and vertical loads

5. The lateral pressure on the walls of a silo is influenced by
   a. the cement stored above the level under consideration
   b. the density of the material stored
   c. the depth of cement stored below the particular level

6. The horizontal component of the intensity of pressure developing at any depth in a silo wall is
   a. directly proportional to the coefficient of friction between wall and the stored material

   b. inversely proportional to the hydraulic mean radius

   c. inversely proportional to the coefficient of friction between wall and stored material

7. In the computation of horizontal pressure in a cement silo, Janssen's theory
   a. overestimates the values over upper half of the silo
   b. overestimates the values over lower half of the silo
   c. underestimates the values towards the upper half of the silo

8. In a dome silo storing large quantities of any material, the load from the concrete shell is resisted by
   a. the shell wall at the bottom
   b. the foundation mat over which the material is stored
   c. the massive ring beam supporting the shell dome

9. In a dome silo, the load due to the stored material is transferred to
   a. curved walls of the shell dome
   b. the ring beam supporting the shell dome
   c. the massive floor mat on the ground

10. The cost of constructing and storing large quantities of any material is the least in
   a. traditional silos with vertical walls
   b. bunkers with sloping bottom
   c. dome silo

5

Chimneys

5.1   GENERAL ASPECTS

Chimneys are an integral part of a hearth in a fire place commonly used in kitchens and industrial structures to fulfill the important function of serving as a conduit to carry and discharge the combustion gases to the atmosphere. Earlier they were known as fire vents and stacks and as they grew larger and taller, they came to be known as chimneys. Human civilization has been in search of an effective system to dispose the undesirable gaseous products of combustion and as the volume of gases increased, small pipe vents were replaced by larger brick chimneys. The size of chimneys gradually increased due to industrial revolution and those larger than 500 m in diameter were classified as *industrial chimneys*. As chimneys grew taller, brick chimneys became uneconomical and steel chimneys were used.

According to Manohar[1] the first concrete chimney was built in Germany in 1876 and reinforced concrete chimneys were introduced in UK and Europe by 1907 and a tapering concrete chimney was built in 1910. Reinforced concrete chimneys were widely used in USA, USSR since 1900 and a 165 m tall concrete chimney built in Japan in 1916 was considered as the tallest chimney in the world for many years. Significant improvements in the design and construction of reinforced concrete chimneys were done between 1930 and 1940, but concrete attained its present predominant status as the most suitable material only after the problem of concrete cracking due to thermal gradient was resolved by 1950. Most of the power projects involve the construction of tall reinforced concrete chimneys to reduce pollution at ground level. Plechik *et al*[2] have reported the design and construction of world's tallest free standing fibre glass stack in 1981.

5.2   HISTORICAL REVIEW

In the early days, the emission of gaseous pollutants from short chimneys continued to be tolerated as a necessary evil until the London smog disaster in 1952 claimed enormous loss of life. Technologists realised that atmosphere cannot absorb an unlimited quantity of pollutants resulting in legislation of Clean Air Act in UK and USA in 1956 and 1963 respectively. Technical guide to air quality was specified in 1967, and an environmental protection agency was established in 1970. In the early days, it was generally supposed that adequate dispersion of pollutants will be achieved if the chimney height is about two and half times the height of buildings around the chimney. This assumption was disapproved and mathematical formulae were proposed to predict the approximate distribution of pollutants from the gases of a chimney based on the investigations by Bosanquet[3] and Cramer[4]. Later many semi-empirical methods were developed to predict the height of the chimney by technologists like Briggs[5], Holland (1953), Priestley (1956), Pasquill and Cramer (1957), and others. These investigations enabled the formulation of mathematical models to reasonably predict the plume behaviour and arrive at an estimate of the ground level concentration of pollutants.

5.3 STRUCTURAL PARTS OF A CHIMNEY

Reinforced concrete chimneys are generally circular in shape to facilitate the expulsion of flue gases without obstructions. At present reinforced concrete is the dominant construction material preferred for the construction of tall and short chimneys, precast concrete modular units are generally used for faster construction. M25 to M30 grade concrete with HYSD reinforcements are generally used to ensure strength and serviceability requirements according to limit state design[6]. A fire brick lining of thickness ranging from 100 to 150 mm is provided inside the concrete shell with an air gap ranging from 80 to 120 mm to reduce the temperature gradient from the interior surface of the fire brick lining to the exterior surface of the concrete shell. Reinforced concrete brackets with holes are provided at regular intervals to support the fire brick lining. At the bottom of the chimney, provision is made for a flue opening. The chimney rests on a reinforced concrete raft foundation. In case of soils with low bearing capacity, pile foundations can be adopted. Figure 5.1 shows the salient parts of a typical reinforced concrete chimney with a raft foundation.

5.4 DESIGN FACTORS

Reinforced concrete chimneys are designed to resist the stresses developed due to the following force components:

a. Self weight of chimney
b. Wind pressure
c. Temperature stresses
d. Seismic forces

a. The self weight of the chimney comprises the dead weight of the concrete shell, fire brick lining, brackets to support the lining and other accessories associated with the concrete chimney.

b. The wind pressure depends upon the location of the chimney such as coastal zones where wind velocity is higher, or the interior places where wind forces are comparatively low. Also the velocity of wind increases with the height of the chimney above the ground. The wind pressure in different zones can be obtained from the Indian standard code IS:875[7] for the purpose of design. The design wind loads on the chimney depend upon the cross-sectional shape of the chimney and the values of the shape factor for different cross-sectional shapes such as circular, octagonal and square. These values are compiled in Table 5.1.

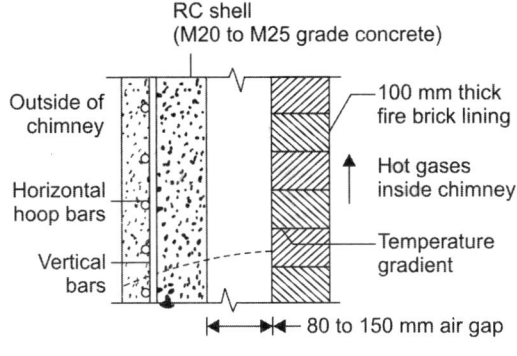

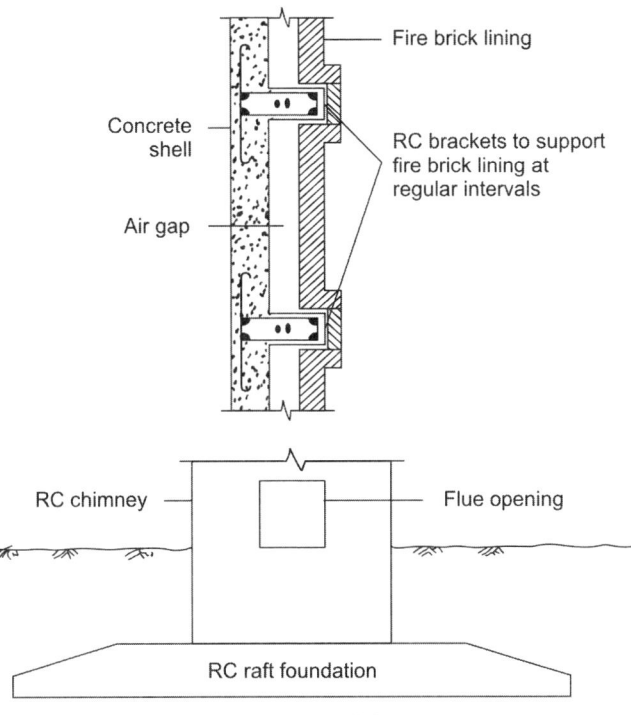

Fig. 5.1: Parts of RCC chimney

Table 5.1: Shape factors for chimneys

Cross-sectional shape of chimney	Ratio of height to base width		
	0 to 4	4 to 8	8 or over
Circular	0.7	0.7	0.7
Octagonal	0.8	0.9	1.0
Square			
a. Wind perpendicular to diagonal	0.8	0.9	1.0
b. Wind perpendicular to face	1.0	1.15	1.3

c. Due to the temperature gradient between the inside and outside faces of the chimney, temperature stresses are induced in the chimney walls both in the vertical and horizontal planes. The inner surface of the shell being at a higher temperature, tends to expand more than the outer surface, which restrains the expansion of the inner fiber to a certain extent. This restrained expansion results in the compression of the inner fibers and tension of the outer fibers. Due to this restrained expansion and contraction, the shell is subjected to a bending moment in the vertical plane as shown in Fig. 5.2.

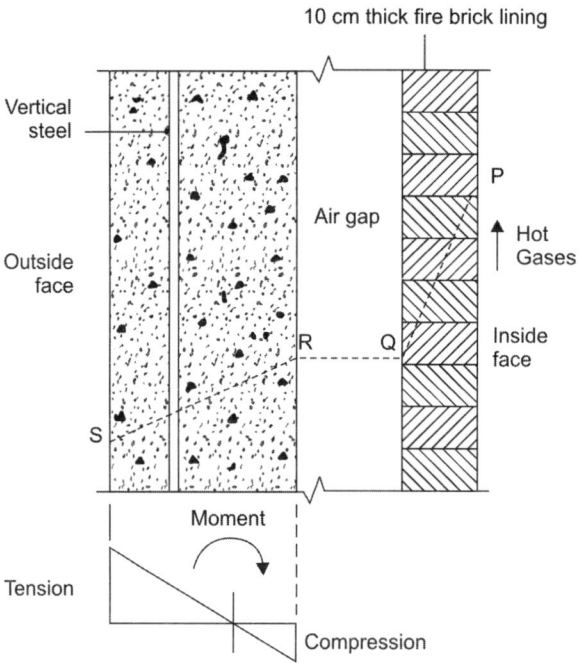

Fig. 5.2: Temperature stresses in RC chimney

In the vertical cross-section, P, Q, R and S represent the temperature gradients across the chimney wall. Experiments have shown that the rate of fall of temperature is steeper in the fire brick lining than in the concrete shell. The slope PQ is nearly 6 to 10 times that of the slope RS. The temperature at P is slightly lower than that of flue gases, and at S, it is slightly higher than the atmospheric temperature. In horizontal sections, the action due to temperature is similar so that the inner fibers are compressed in a horizontal plane while the outer fibers are stretched.

5.5 ▌ STRESSES IN RC SHAFTS DUE TO SELF WEIGHT AND WIND LOADS

The following notations are used in the analysis of stresses in RC shafts:

W = total weight of the shaft above the considered section

P = resultant wind force acting at a distance h from the section

A_s = area of reinforcement assumed to be in the form of a ring at the centre of thickness of shell

t_s = thickness of steel ring = $\left(\dfrac{A_s}{2\pi R}\right)$

R = radius of the centre of thickness

d = outside diameter of the shell

t_c = thickness of concrete shell

n = coefficient of neutral axis depth

σ_c = compressive stress in concrete at the centre of thickness of shell

σ_s = tensile stress in steel

α = angle subtended by the neutral axis at the centre

m = modular ratio

M = bending moment at the section

Consider a strip $R d\theta$ at an angle θ from XX (Fig. 5.3). Stress in concrete at the level of elementary strip is

$$\sigma_c' = \sigma_c \left[\frac{R\cos\theta + R\cos\alpha}{R + R\cos\alpha}\right] = \sigma_c \left[\frac{\cos\theta + \cos\alpha}{1 + \cos\alpha}\right]$$

Area of strip = $(R d\theta) t_c$

Area of steel in strip = $(R d\theta) t_s$

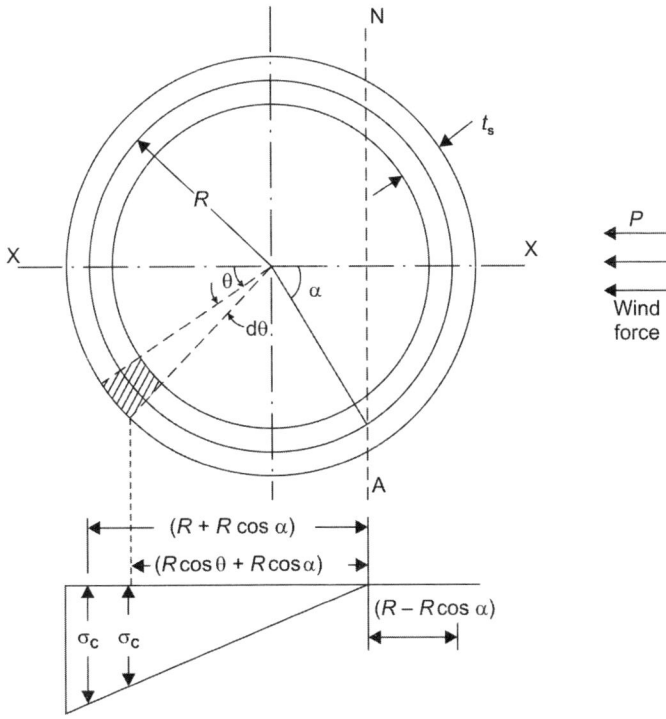

Fig. 5.3: Stresses in chimney shafts

$\therefore$ Total compressive force in concrete and steel

$$C = 2\int_0^{(\pi-\alpha)}(Rd\,\theta)t_c\sigma_c\left[\frac{\cos\theta+\cos\alpha}{1+\cos\alpha}\right]$$

$$+2\int_0^{(\pi-\alpha)}(Rd\,\theta)t_s(m-1)\sigma_c\left[\frac{\cos\theta+\cos\alpha}{1+\cos\alpha}\right]$$

$$C = \left[\frac{2R\sigma_c t_c}{1+\cos\alpha}\right]\int_0^{(\pi-\alpha)}(\cos\theta+\cos\alpha)\,d\theta$$

$$+\left[\frac{2R\sigma_c t_s(m-1)}{1+\cos\alpha}\right]\int_0^{(\pi-\alpha)}(\cos\theta+\cos\alpha)\,d\theta$$

Integrating and simplifying, we get

$$C = \left[\frac{2R\sigma_c}{1+\cos\alpha}\right][\{t_c+(m-1)t_s\}][\sin\alpha+(\pi-\alpha)\cos\alpha]$$

Similarly, total tension in steel is given by

$$T = 2\int_0^\alpha(Rd\,\theta)t_s\sigma_c\left[\frac{R\cos\theta-R\cos\alpha}{R+R\cos\alpha}\right] = \left[\frac{2R\sigma_c mt_s}{1+\cos\alpha}\right][\sin\alpha-\alpha\cos\alpha]$$

Equating the sum of the internal forces to external load W,

$$W = (C-T)$$

$$W = \left[\frac{2R\sigma_c}{1+\cos\alpha}\right][(t_c-t_s)\{\sin\alpha+(\pi-\alpha)\cos\alpha\}+\pi mt_s\cos\alpha] \qquad ...(5.1)$$

Equating the element of external forces to the sum of the moments of the internal forces, we have

$$M = \int_0^{(\pi-\alpha)}CR\cos\theta+\int_0^\pi TR\cos\theta$$

$$M = \left[\frac{2R^2\sigma_c}{1+\cos\alpha}\right][(t_c+(m-1)t_s]\int_0^{(\pi-\alpha)}(\cos^2\theta+\cos\theta\cos\alpha)d\theta$$

$$+\left[\frac{2R^2\sigma_c mt_s}{1+\cos\alpha}\right]\int_0^\pi(\cos^2\theta-\cos\theta\cos\alpha)d\theta$$

Integrating and simplifying, we have the final equation for the moment M as

$$M = \left[\frac{2R^2\sigma_c}{1+\cos\alpha}\right][t_c+(m-1)t_s]\left[\frac{\sin 2\alpha}{4}+\frac{(\pi-\alpha)}{2}\right]$$

$$+\left[\frac{2R^2\sigma_c mt_s}{1+\cos\alpha}\right]\left[\frac{\alpha}{2}-\frac{\sin 2\alpha}{4}\right]$$

$$\therefore \qquad M = \left[\frac{2R^2\sigma_c}{1+\cos\alpha}\right]\left[(t_c - t_s)\left(\frac{\sin 2\alpha}{4} + \frac{(\pi-\alpha)}{2}\right) + \frac{mt_s\pi}{2}\right] \qquad \dots(5.2)$$

Eccentricity $e = (M/W)$,

$$e = R\left(\frac{\left[(t_c - t_s)\left\{\dfrac{\sin 2\alpha}{4} + \dfrac{\pi-\alpha}{2}\right\} + \dfrac{mt_s\pi}{2}\right]}{(t_c - t_s)[\sin\alpha + (\pi-\alpha)\cos\alpha] + \pi m t_s \cos\alpha}\right) \qquad \dots(5.3)$$

The value of α which satisfies Eq. (5.3) is determined by trial and error. Knowing α, the stresses in concrete and steel can be evaluated using Eq. (5.1).

5.6 STRESSES IN HOOP REINFORCEMENTS DUE TO SHEAR FORCE

If $\qquad H$ = horizontal shear force at the section

$\qquad d$ = diameter of the chimney

$\qquad s$ = pitch of hoop bars

$\qquad A_t$ = area of hoop bars in one pitch length

then area of steel resisting shear in one metre height = $\left(\dfrac{2A_t \times 1000}{s}\right)$

If σ_s = stress in steel,

then shear force resisted = $\left(\dfrac{2\times A_t \times 1000 \times \sigma_s}{s}\right)$ $\qquad \dots(5.4)$

If horizontal distance between reinforcement on both sides is assumed as $0.8d$, then

$$\text{shear per metre} = \left(\frac{H\times 1000}{\text{lever arm}}\right) = \left(\frac{1000H}{0.8d}\right) \qquad \dots(5.5)$$

Equating Eqs. (5.4) and (5.5), we get

$$\left(\frac{2A_t \times 1000 \times \sigma_s}{s}\right) = \left(\frac{1000H}{0.8d}\right)$$

$$\therefore \qquad \sigma_s = \left(\frac{Hs}{1.6 A_t d}\right) \qquad \dots(5.6)$$

where, d and s are expressed in mm units and A_t in mm^2 units.

5.7 STRESSES DUE TO TEMPERATURE GRADIENT

In the walls of a reinforced concrete chimney, stresses are developed due to the temperature gradient between the inner and outer surface of the walls. This temperature drop from inside to outside surface tends to expand the inner surface relative to the outer one. Due to the monolithic action of the entire wall, differential expansion is not possible and hence equal expansion takes place so that the shell is compressed on its inside surface and pulled on the outside surface. As a whole, there is an average increase in the length of the shell due to the temperature gradient.

Effect of Temperature Only on Stresses in Chimney Walls

Let T = temperature difference between inside and outside with a linear temperature gradient

α = coefficient of expansion of steel and concrete

e = strain due to temperature difference

m = modular ratio

t_s = area of reinforcement per unit width

t_c = area of concrete per unit width

σ_{ct} = stress in concrete due to temperature

σ_{st} = stress in steel due to temperature

$p = (t_s/t_c)$

k = neutral axis depth constant

Referring to Fig. 5.4 and considering the force equilibrium, we have the following relations:

$$\left(\frac{1}{2}\right)\sigma_{ct}kt_c = t_s\sigma_{st} = pt_c\sigma_{st}$$

$\therefore$
$$\sigma_{st} = \left(\frac{\sigma_{ct}k}{2p}\right) = m\sigma_{ct}\left(\frac{at_c - kt_c}{kt_c}\right) = m\sigma_{ct}\frac{(a-k)}{k}$$

$\Rightarrow$
$$k^2 = 2pm(a-k)$$

$\Rightarrow$
$$k = -mp + \sqrt{2mpa + p^2m^2} \qquad \qquad ...(5.7)$$

Rise in temperature in reinforcement $= (1 - a)T$

Free expansion of steel $= (1 - a)\alpha T$

The tensile stress in steel is due to the difference between strain e and temperature rise $(1 - a)T$.

$\therefore$ Stress in steel, $\sigma_{st} = E_s[e - (1 - a)\alpha T]$

At the neutral axis, there is free expansion due to strain e

$$e = (1 - k)\alpha T$$

$\therefore$ Stress in steel
$$\sigma_{st} = E_s[(1 - k)\alpha T - (1 - a)\alpha T]$$
$$\sigma_{st} = E_s\alpha T(a - k) \qquad \qquad ...(5.8)$$

Stress in concrete
$$\sigma_{ct} = E_c(T\alpha - e) = E_c[T\alpha - (1 - k)\alpha T]$$
$$\sigma_{ct} = E_c\alpha kT \qquad \qquad ...(5.9)$$

Combined Effect of Wind Loads, Self Weight and Temperature on Stresses in Chimney Walls

The stress developed at the neutral axis, compression zone (leeward side) and tension zone (windward side) of the chimney due to the combined effect of wind loads, self weight and temperature will be examined separately.

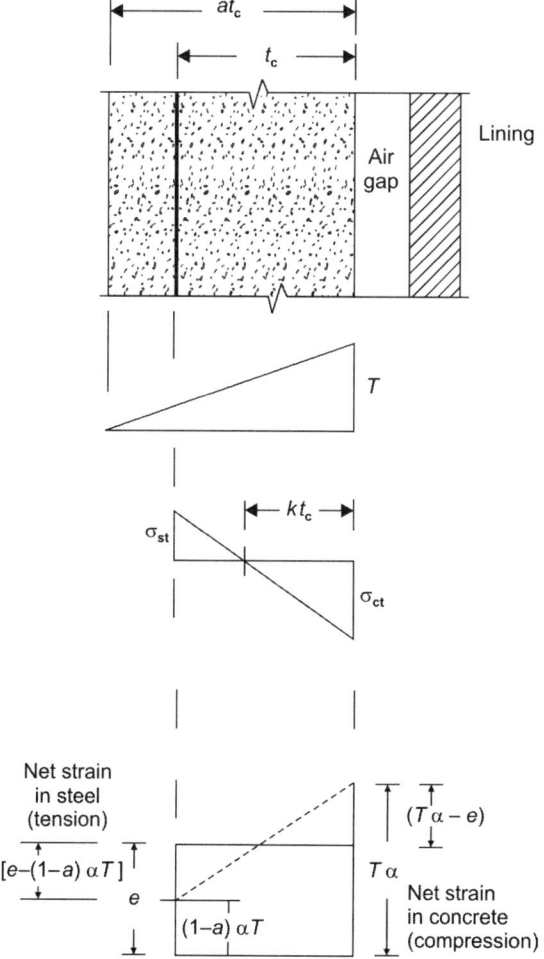

Fig. 5.4: Temperature stresses in chimney walls

Case 1: Stresses at Neutral Axis

There are no stresses at the neutral axis due to the external loads. But stresses are developed due to temperature difference.

∴   Stress in steel $= E_s \alpha T(a - k)$

∴   Stress in concrete $= E_c \alpha k T$

Case 2: Stresses in Compression Zone (Leeward Side)

The leeward side of the chimney is under compression due to the effect of wind loads and self weight. If the temperature stresses are superposed, the final effect will be to increase the compressive stress in concrete and decrease the stress in steel.

Refer to Fig. 5.5 and use the following notations:

    $\alpha_c =$ compressive stress in concrete assumed uniform due to the effect of self weight and wind loads

    $k't_c =$ position of neutral axis

σ'_c = compressive stress in concrete due to the combined effect

σ'_s = stress in steel due to the combined effect

e = strain due to temperature difference of T

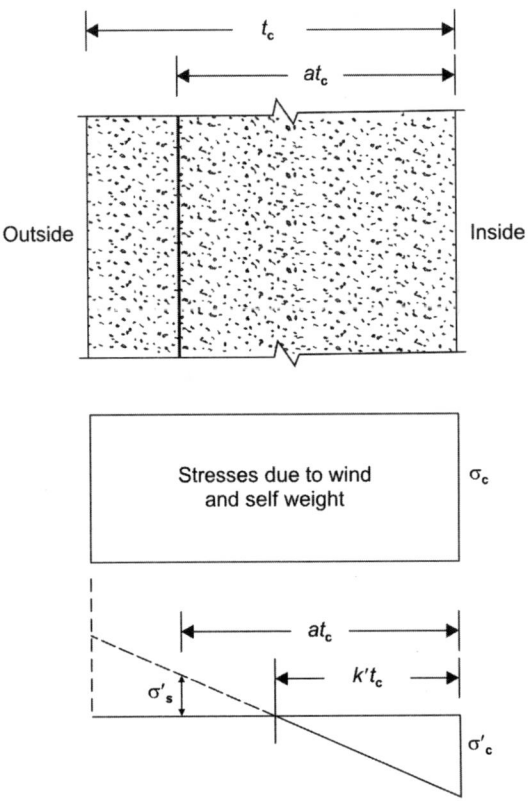

Fig. 5.5: Temperature stresses in compression zone (leeward side)

Total compression will remain unchanged in the section.

Considering the force equilibrium, we have

$$\sigma_c t_c + (m-1)t_s \sigma_c = \left(\frac{1}{2}\sigma'_c k' t_c\right) - t_c \sigma_s$$

If
$$t_s = p t_c$$

then
$$\sigma_c t_c + (m-1)p t_c \sigma_c = \left(\frac{1}{2}\sigma'_c k' t_c\right) - p t_c m_c \sigma'_c \left(\frac{a-k'}{k'}\right)$$

$$\therefore \quad \sigma_c t_c [1 + (m-1)p] = \sigma'_c t_c \left[\frac{k'}{2} - mp\left(\frac{a-k'}{k'}\right)\right]$$

$$\Rightarrow \qquad \sigma'_c = \left[\frac{\sigma_c[1+(m-1)p]}{\dfrac{k'}{2} - mp\left(\dfrac{a-k'}{k'}\right)}\right] \qquad \qquad \ldots(5.10)$$

Change in stress in concrete at inside face $= (\sigma_c' - \sigma_c)$ ...(5.10a)

$\therefore \qquad (\sigma_c' - \sigma_c) = E_c(T\alpha - e)$

Change of stress in steel $= (\sigma_s' + m\sigma_c)$

$\therefore \qquad (\sigma_s' + m\sigma_c) = E_s[e - (1-a)\alpha T]$ ...(5.10b)

But $\qquad \sigma_s' = \sigma_c'm\left(\dfrac{a-k'}{k'}\right)$

$\therefore \qquad \left[\dfrac{\sigma_c'm\left(\dfrac{a-k'}{k'}\right) + m\sigma_c}{E_s}\right] + (1-a)\alpha T = e$ ...(5.10c)

Also from Eq. (5.10a), we have

$$T\alpha - \dfrac{(\sigma_c' - \sigma_c)}{E_c} = e$$ ...(5.10d)

Equating Eqs. (5.10c) and (5.10d), we get

$$\sigma_c' = \left[\dfrac{E_c aT\alpha}{1 + \dfrac{(a-k')}{k'}}\right]$$ ...(5.11)

Equating Eqs. (5.10) and (5.11), we get

$$\dfrac{\sigma_c[1+(m-1)p]}{\dfrac{k'}{2} - mp\left(\dfrac{a-k'}{k'}\right)} = \left[\dfrac{E_c aT\alpha}{1 + \dfrac{(a-k')}{k'}}\right]$$ ...(5.12)

For given values of various variables, we can evaluate k' from Eq. (5.12) and then compute σ_c' [from Eq. (5.11)] and σ_s'.

If k' is more than unity, the whole thickness of concrete t_c will be in compression and the stresses can be analysed using the same procedure.

Case 3: Stresses in Tension Zone (Windward Side)

The chimney section in the windward zone is in tension due to the effect of self weight and wind loads. Concrete is assumed to take negligible tension and hence the whole tension is resisted by steel.

Refer to Fig. 5.6 and use the following notations:

σ_s = tensile stress in steel due to the wind loads and self weight

$k't_c$ = position of neutral axis

σ_c = compressive stress in concrete

σ_s = stress in steel due to the combined effect

The effect of temperature is to develop compressive stress in concrete and increase the tensile stress in steel.

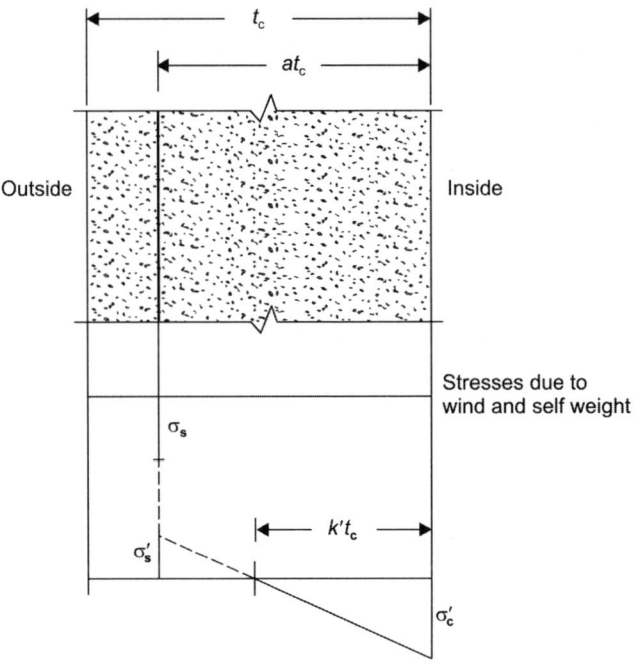

Fig. 5.6: Temperature stresses in tension zone (windward side)

The total force in the section remains unchanged. Considering the equilibrium of forces, we get

$$t_s\sigma_s = \left(t_s\sigma'_s - \frac{1}{2}\sigma'_c k't_c \right)$$

But

$$\sigma'_s = m\sigma'_c\left(\frac{a-k'}{k'}\right) \text{ and } t_s = pt_c$$

$$\therefore \quad pt_c\sigma_s = pt_c m\sigma'_c\left(\frac{a-k'}{k'}\right) - \frac{1}{2}\sigma'_c k't_c$$

$$\sigma'_c = \left[\frac{p\sigma_s}{pm\left(\dfrac{a-k'}{k'}\right) - \dfrac{k'}{2}}\right] \qquad\qquad ...(5.13)$$

Change of strain in concrete at inside face is $\left(\dfrac{\sigma_s}{E_s} + \dfrac{\sigma'_c}{E_c}\right)$

$$\therefore \quad \left(\frac{\sigma_s}{E_s} + \frac{\sigma'_c}{E_c}\right) = (T\alpha - e)$$

$$\therefore \quad e = \left[T\alpha - \frac{\sigma_s}{E_s} - \frac{\sigma'_c}{E_c}\right] \qquad\qquad ...(5.13a)$$

Change of strain in steel is given by

$$\left(\frac{\sigma'_s - \sigma_s}{E_s}\right) = [e - (1 - a)\,\alpha T]$$

$\therefore$
$$e = \left(\frac{\sigma'_s - \sigma_s}{E_s}\right) + (1 - a)\alpha T \qquad \qquad \text{...(5.13b)}$$

Equating Eqs. (5.13a) and (5.13b), we have the following relation:
$$\sigma'_c = [a\alpha TE_c - \sigma'_s/m],$$
substituting for σ'_s and simplifying, we get
$$\sigma'_c = \alpha TE_c k' \qquad \qquad \text{...(5.14)}$$
Equating Eqs. (5.13) and (5.14), we have

$$\left[\frac{p\sigma_s}{pm\left(\dfrac{a - k'}{k'}\right) - \dfrac{k'}{2}}\right] = \alpha TE_c k' \qquad \qquad \text{...(5.15)}$$

The value of k' can be evaluated using the known values of the other variables and then the values of the stress in concrete σ'_c and the stress in steel σ'_s can be computed using the relevant equations.

Stresses in Horizontal Reinforcement Due to Temperature Difference

At high temperatures, the inner surface of the chimney is prevented from expansion and, therefore, gets compressed. The outer surface will expand more than the natural expansion and will be in tension. Due to temperature stresses, generally the hoops try to expand and consequently tensile stresses develop in the hoop reinforcement.

Take unit height of wall, Refer to Fig. 5.7 and use the following notations:

$k't_c$ = position of neutral axis
σ'_c = compressive stress in concrete
σ'_s = tensile stress in steel
A'_s = area of hoop reinforcement/unit height
A_s = cross-sectional area of horizontal steel
S = spacing
$A'_s = (A_s/S) = pt_c$
$\sigma'_s = m\sigma'_c\,(a - k')/k' \qquad \qquad \text{...(5.16)}$

Consider the force equilibrium of the section, compressive force in concrete on the inner side = tensile force in horizontal reinforcement, i.e.

$$\frac{1}{2}\sigma'_c k' t_c = A'_s \sigma'_s = pt_c m\sigma'_c\left(\frac{a - k'}{k'}\right)$$

$\therefore$
$$\frac{k'}{2} = mp\left(\frac{a - k'}{k'}\right)$$

$$k' = \sqrt{2pma + p^2m^2} - pm \qquad \qquad \text{...(5.17)}$$

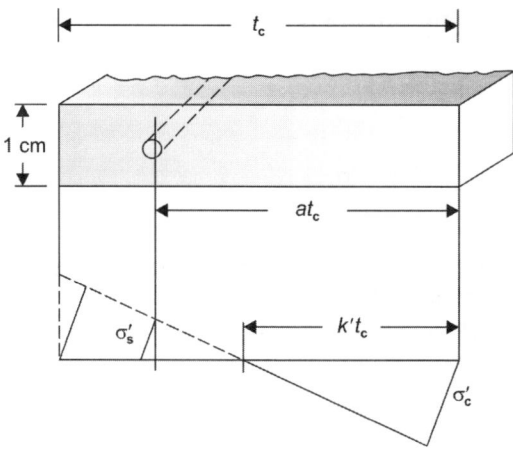

Fig. 5.7: Stresses in horizontal reinforcement due to temperature difference

Using Eq. (5.16), the position of the neutral axis is determined.

Let e be the actual strain

Stress in concrete $\sigma'_c = (\alpha T - e)E_c$

Stress in steel $\sigma'_s = [e - (1 - a)\alpha T]E_s$

From these two relations, we get

$$e = [\alpha T - \sigma'_c/E_c] \text{ and } e = [\sigma'_s/E_s + \alpha T(1 - a)]$$

$$\therefore \qquad [\alpha T - \sigma'_c/E_c] = [\sigma'_s/E_s + \alpha T(1 - a)]$$

$$\therefore \qquad [\sigma'_s + m\sigma'_c] = E_s\alpha Ta \qquad \qquad ...(5.18)$$

Knowing the value of k', the stresses in steel and concrete can be obtained from Eqs. (5.16) and (5.18) respectively.

5.8 ▮ DESIGN OF REINFORCED CONCRETE CHIMNEYS

The design of reinforced concrete chimneys involves several steps of fixing preliminary dimensions of the chimney height, diameter at different heights, sectional dimensions such as thickness of concrete shaft and fire brick lining, expected temperature differences from inside to outside of the shaft. Additional data required are the quality of concrete to be used together with the salient properties of concrete like permissible stresses, modular ratio, modulus of elasticity of concrete, the type of steel reinforcement used and its properties like permissible stresses. The intensity of wind pressure at the locality is also required for computing the effect of wind on the nature of stresses developed in concrete and steel. For designing foundations for the chimney, data regarding the bearing capacity of the soil at site is also required.

The salient steps involved in the design of reinforced concrete chimneys are listed below:

1. *Dimensions of the chimney*: The base dimensions of the chimney are largely governed by structural considerations. The dimensions selected should be such that stresses due to dead load together with those of the wind, temperature, earthquake and other effects are safely resisted by the chimney walls. From flow considerations, a cylindrical chimney is preferred. For concrete chimneys, a taper

of 1:50 to 1:100 is usually provided depending upon the height and geographical location of the chimney. The top dimension of the chimney is fixed such that a given volume of flue gases can be discharged at the design exit velocity.

2. *Height of the chimney*: The following two criteria dictate the physical height of the chimney:
   a. To generate a draft which will cause gases to flow out with the desired exit velocity
   b. To satisfy local regulations in respect of geographic logistic considerations involving pressure of air at the locality and that of the flue gases.
   The physical height is determined such that the draft is adequate to overcome the friction and other head loses as the gases flow up and out of the chimney. Several parameters are involved to design the height of the chimney and the reader may refer to the publications by Manohar[7], Diver[8] and Pinfold[9].

3. *Cross-sectional dimensions*: Generally, thickness of the concrete shell varies from 150 to 200 mm at the top and 300 to 400 mm at the bottom of the chimney. These values are suitable for chimneys of height in the range of 50 to 100 m with base diameter in the range of 4 to 6 m.

4. *Design of reinforcements and analysis of stresses*: Normally reinforcements are first assumed in the range of 1% to 1.5% of the cross-sectional area. Design dead and wind loads are computed and the section is analyzed for the position of neutral axis. The stresses developed in concrete and steel at the base section due to the combined action of dead and wind loads are evaluated and checked with the permissible values. Hoop reinforcements are also designed using the relevant design equations.

5. *Check for temperature stresses*: Temperature stresses are checked in concrete and steel on the leeward and windward sides of the chimney under the combined effect of wind loads, self weight and temperature. The stresses developed should be within the permissible limits for concrete and steel prescribed in the Indian standard code IS:4998-1975[10] dealing with the criteria for the design of reinforced concrete chimneys.

6. *Design of foundations*: Generally for major chimneys, raft foundations are designed by considering the various design loads. Suitable reinforcements are designed for the critical bending moment at the junction of chimney walls and footing. The critical section for shear is checked in the raft and suitable reinforcements are designed.
   The following examples will illustrate the analysis and design of reinforced concrete chimneys.

5.9   ANALYSIS AND DESIGN EXAMPLES

1. A reinforced concrete chimney 50 m high above ground has an outside diameter of 4 m. The thickness of the shell is 20 cm at the top and it is increased to 25 cm and 30 cm at 18 m and 30 m from the top. Vertical steel bars are equla to 1% of the cross-sectional area throughout. The total wind load above the section at 18 m from top may be taken as 93 kN. Find the stresses developed due to wind and dead loads at the section 18 m from the top of the chimney. Assume that modular ratio m is 13.

Weight of concrete shell = $(\pi \times 3.8 \times 0.2 \times 18 \times 24)$

$$W = 1031 \text{ kN}$$

$$A_s = 0.01 \times \pi(2^2 - 1.8^2) \, 10^4 = 240 \text{ cm}^2$$

$$\therefore \quad t_s = \left(\frac{A_s}{2\pi R}\right) = \left(\frac{240}{2 \times \pi \times 190}\right) = 0.20 \text{ cm}$$

Equivalent second moment of area

$$\begin{aligned} I_e &= [\pi/64 \, (D^4 - d^4) + (m-1) \, \pi r^3 t_s] \\ &= [\pi/64 \, (400^4 - 360^4) + (13-1) \, \pi \times 190^3 \times 0.2] \\ &= 480 \times 10^6 \text{ cm}^4 \end{aligned}$$

Equivalent area $A_e = [\pi/4 \, (D^2 - d^2) + (m-1) \, A_s]$

$$= \pi/4 \, (400^2 - 360^2) + (13-1) \, 240 = 26300 \text{ cm}^2$$

Horizontal wind load $P = 93$ kN

Moment due to wind load at a section 18 m from top $M = (93 \times 9) = 837$ kN·m

∴ Eccentricity $e = (M/W) = (837/1031) = 0.81$ m

For no tension to develop, we have

$$e_{max} = \left(\frac{2I_e}{A_e D}\right) = \left(\frac{2 \times 480 \times 10^6}{26300 \times 400}\right) = 91 \text{ cm}$$

Hence, the entire section is in compression.

Maximum compressive stress in concrete is

$$\sigma_c = \left[\frac{W}{A_e} + \frac{My}{I_e}\right] = \left[\frac{1031 \times 10^3}{26300 \times 10^2} + \frac{837 \times 10^6 \times 2000}{480 \times 10^{10}}\right] = 0.74 \text{ N/mm}^2$$

Compressive stress in steel $m\sigma_c = (13 \times 0.74) = 9.62$ N/mm²

2. A concrete chimney of height 80 m with the external diameter of the shaft being 4 m at top and 5 m at bottom is required in a place where the wind intensity is 1.5 kN/m². Thickness of fire brick lining is 10 cm, temperature difference between the inside and outside of shaft is 75 °C and permissible bearing pressure on soil at site is 150 kN/m².

Adopting M25 grade concrete mix and tor steel Fe 415 grade, design base section and foundation for the chimney.

    i. *Permissible stresses*:

        M25 grade concrete   ∴ $\sigma_{cb} = 8.5$ N/mm²

        Fe 415 grade steel $m = 11$

                                   $\sigma_{st} = 230$ N/mm²

    ii. *Load referring to Fig. 5.8*:

        Weight of chimney = $(\pi \times 4.2 \times 0.3)80 \times 24 = 7600$ kN

        Weight of fire brick lining (10 cm thick) = $(\pi \times 3.8 \times 0.1)80 \times 20 = 1920$ kN

        Total wind load above base = $\left[0.7 \times 1.5 \times \dfrac{(4+5)}{2} \times 80\right] = 375$ kN

Acting at a height of 40 m above base,

total dead load above base $(W) = (7600 + 1920) = 9520$ kN

Bending moment at base due to wind loads is

$$M = (375 \times 40) = 15000 \text{ kN·m}$$

Eccentricity $e = \left(\dfrac{M}{W}\right) = \left(\dfrac{15000}{9520}\right) = 1.575$ m

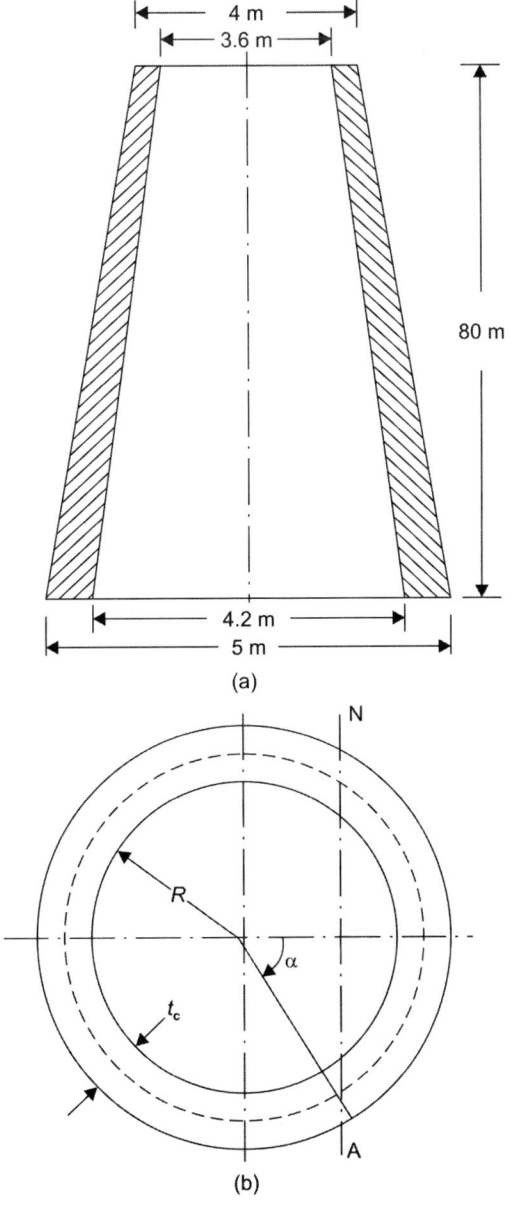

Fig. 5.8: Details of RC chimney

iii. *Reinforcements*:

Providing reinforcements of 1% of the cross-sectional area:

$$A_{st} = 0.01(\pi \times 4600 \times 400) = 57805 \text{ mm}^2$$

Using 25 mm diameter bars,

$$\text{number of bars} = \left(\frac{57805}{491}\right) = 177$$

Provide 120 bars of 25 mm diameter bars.

Equivalent thickness of steel ring is given by

$$t_s = \left(\frac{A_{st}}{\pi d_m}\right) = \left(\frac{58800}{\pi \times 4600}\right) = 4 \text{ mm}$$

iv. *Analysis of stresses at base section*:

Referring to Fig. 5.8, if α is the angle subtended by the neutral axis at the centre, the eccentricity is given by

$$e = R\left(\frac{\left[(t_c - t_s)\left\{\dfrac{\sin 2\alpha}{4} + \dfrac{\pi - \alpha}{2}\right\} + \dfrac{mt_s\pi}{2}\right]}{(t_c - t_s)\{\sin\alpha + (\pi - \alpha)\cos\alpha\} + \pi mt_s \cos\alpha}\right)$$

In this equation:

$$e = 1575 \text{ mm} \qquad t_s = 4 \text{ mm}$$
$$R = 2300 \text{ mm} \qquad t_c = 400 \text{ mm}$$

Trial values of α vary from 70° to 80°.

α is assumed until the value of calculated eccentricity coincides with the actual value of $e = 1575$ mm.

Trial 1: Assume $\alpha = 70°$

The right hand side of equation, $e = 1742$ mm which is greater than 1575 mm. Hence, reduce the value of α for the second trial.

Trial 2: Assume $\alpha = 60°$

$$(\pi m\, t_s \cos\alpha) = (\pi \times 11 \times 4 \times 0.5) = 69.11$$

$$\left(\frac{\sin 2\alpha}{4}\right) = 0.216$$

$$\left(\frac{\pi - \alpha}{2}\right) = \left(\frac{\pi - 1.047}{2}\right) = 1.047$$

$$\sin\alpha = 0.866$$
$$(\pi - \alpha)\cos\alpha = 1.045$$
$$R = 2300 \text{ mm}$$

$$e = \left(\frac{\left[(400 - 4)\left\{\dfrac{\sin 120°}{4} + \dfrac{\pi - 1.047}{2}\right\} + \dfrac{11 \times 4 \times \pi}{2}\right]}{(400 - 4)\{\sin 60° + (\pi - 1.047)\cos 60°\} + \pi \times 11 \times 4 \times \cos 60°}\right)$$

$$= 1586 \text{ mm} \approx 1575 \text{ mm}$$

Hence, the value of angle α is $60°$

Using the value of α in Eq. (5.1) which is given by

$$W = \left[\frac{2R\sigma_c}{1+\cos\alpha}\right][(t_c - t_s)\{\sin\alpha + (\pi - \alpha)\cos\alpha\} + \pi m t_s \cos\alpha]$$

$$9520 \times 10^3 = \left[\frac{2\times2300\sigma_c}{1+\cos60°}\right][(400-4)\{\sin60° + (\pi - 1.047)\cos60°\}$$

$$+\pi\times11\times4\times\cos60°]$$

Solving, $\sigma_c = 3.75$ N/mm²

Stress in steel $\sigma_s = m\sigma_c\left[\dfrac{R(1-\cos\alpha)}{R(1+\cos\alpha)}\right]$

$$= 11\times3.75\left[\frac{1-\cos60°}{1+\cos60°}\right] = 13.73 \text{ N/mm}^2$$

The stresses in concrete and steel are within safe permissible limits.

v. *Design of hoop reinforcement*:

Shear at the base of chimney = 375 kN

Mean diameter at base = 4600 mm

Using 10 mm diameter hoops at 200 mm centres, we get from Eq. (5.6)

$$\sigma_s = \left(\frac{Hs}{1.6A_t d}\right) = \left(\frac{375\times10^3\times200}{1.6\times79\times4600}\right)$$

$$= 129 \text{ N/mm}^2 < 230 \text{ N/mm}^2$$

Hence, stresses are within permissible limits.

vi. *Temperature stresses (combined effect of wind loads, self weight and temperature)*

a. Compression zone (leeward side)

Providing an effective cover of 50 mm to steel

$t_c = 400$ mm $\qquad\qquad\qquad t_s = 4$ mm

$at_c = 350$ mm $\qquad\qquad\qquad \therefore\ a = 0.875$

$p = (t_s/t_c) = (4/400) = 0.01$

$T = 75°C$

$\alpha = 11 \times 10^{-6}/°C \qquad\qquad\qquad m = 11$

$$E_c = (E_s/m) = \left(\frac{210\times10^3}{11}\right)$$

$\sigma_c = 3.75$ N/mm²

Use Eq. (5.12) which is given by

$$\frac{\sigma_c[1+(m-1)p]}{\left[\dfrac{k'}{2} - mp\left(\dfrac{a-k'}{k'}\right)\right]} = \frac{E_c aT\alpha}{\left[1+\left(\dfrac{a-k'}{k'}\right)\right]}$$

$$\frac{3.75[1+10\times0.01]}{0.5k'-11\times0.01\left(\dfrac{0.875-k'}{k'}\right)} = \frac{\left(\dfrac{210\times10^3}{11}\times0.875\times75\times11\times10^{-6}\right)}{\left[1+\left(\dfrac{0.875-k'}{k'}\right)\right]}$$

Solving, $k' = 0.70$

From Eq. (5.11), the stress in concrete σ'_c is given by the following relation:

$$\sigma'_c = \left[\frac{E_c aT\alpha}{1+\dfrac{a-k'}{k'}}\right] = \frac{\left(\dfrac{210\times10^3}{11}\times0.875\times75\times11\times10^{-6}\right)}{\left(1+\dfrac{0.875-0.70}{0.70}\right)} = 11.0\ \text{N/mm}^2$$

The permissible compressive stress for M25 grade concrete when wind loads are also considered is $(1.33 \times 8.5) = 11.33\ \text{N/mm}^2$.

Stress in steel $\quad \sigma'_s = m\sigma'_c\left(\dfrac{a-k'}{k'}\right)$

$$= 11\times11.0\left(\frac{0.875-0.70}{0.70}\right)$$

$$= 30.25\ \text{N/mm}^2 < 230\ \text{N/mm}^2 \text{ permissible stress}$$

b. Tension zone (windward side)

Use Eq. (5.15) which is given by

$$\frac{p\sigma_s}{\left[pm\left(\dfrac{a-k'}{k'}\right)-\dfrac{k'}{2}\right]} = \alpha TE_c k'$$

$$\frac{0.01\times13.73}{\left[0.01\times11\left(\dfrac{0.875-k'}{k'}\right)-\dfrac{k'}{2}\right]} = \left[11\times10^{-6}\times75\times\frac{210\times10^3}{11}\times k'\right]$$

Solving, $k' = 0.55$

Stress in concrete $\sigma_c = \alpha TE_c k'$

$$\sigma_c = \left(11\times10^{-6}\times75\times\frac{210\times10^3}{11}\times0.55\right)$$

$$= 8.66\ \text{N/mm}^2 < 11.33\ \text{N/mm}^2$$

Stress in steel $\quad \sigma_s = m\sigma'_c\left(\dfrac{a-k'}{k'}\right) = 11\times8.66\left(\frac{0.875-0.55}{0.55}\right)$

$$= 56.28\ \text{N/mm}^2 < 230\ \text{N/mm}^2$$

c. Stresses at neutral axis

Use Eq. (5.7) which is given by

$$k = -mp + \sqrt{2mpa + p^2m^2}$$

$$p = 0.01, \qquad m = 11, \qquad a = 0.875$$

$$k = -11 \times 0.01 + \sqrt{2 \times 11 \times 0.01 \times 0.875 + (0.01)^2 (11)^2} = 0.342$$

Stress in concrete $\sigma_c = E_c \alpha kT$

$$\sigma_c = \left[\frac{210 \times 10^3}{11} \times 11 \times 10^{-6} \times 0.342 \times 75 \right] = 5.386 \text{ N/mm}^2$$

Stress in steel $\quad \sigma_s = E_s \alpha T (a - k)$

$$= (210 \times 10^3 \times 11 \times 10^{-6} \times 75)(0.875 - 0.342)$$

$$= 92.34 \text{ N/mm}^2$$

The stresses are within safe permissible limits.

vii. *Stresses in hoop steel due to temperature*:

Hoop steel of 10 mm diameter at 200 mm centres provided at base section.

$$p = \left(\frac{A_s}{st_c} \right) = \left(\frac{79}{200 \times 400} \right) = 0.00098$$

$$a = 0.875 \qquad\qquad m = 11$$

Use Eq. (5.17), which is given by

$$k' = \sqrt{2pma + p^2m^2} - pm$$

$$= \sqrt{2 \times 0.00098 \times 11 \times 0.875 + (0.00098)^2 (11)^2} - (0.00098 \times 11)$$

$$= 0.127$$

Using Eqs. (5.16) and (5.18), we have

$$\sigma_s' = m\sigma_c' \left(\frac{a - k'}{k'} \right) = 11 \times \sigma_c' \left(\frac{0.875 - 0.127}{0.127} \right) = 60.45 \, \sigma_c'$$

Also $\qquad (\sigma_s' + m\sigma_c') = E_s \alpha Ta$

$\therefore \qquad (60.45\sigma_c' + 11\sigma_c') = (210 \times 10^3 \times 11 \times 10^{-6} \times 75 \times 0.875)$

Solving, $\sigma_c' = 2.12 \text{ N/mm}^2$

and $\qquad \sigma_s' = 128.25 \text{ N/mm}^2$

Total stress in hoop steel = (stress due to shear) + (stress due to temperature difference)

$$= (129 + 128.15)$$

$$= 257.25 \text{ N/mm}^2 > 230 \text{ N/mm}^2$$

Hence, the spacing of the hoop reinforcement can be reduced to 150 mm instead of 200 mm centres.

viii. *Design of foundations*:

A circular RC slab foundation is designed for the chimney.

Total vertical load on base = 9520 kN

Bending moment = 15000 kN·m

Allowable bearing pressure = 150 kN/m²

Self weight of footing (assumed at 10%) = 950 kN

Total load on soil = (9520 + 950) = 10470 kN

If D = diameter of the circular footing for no tension to develop, then

$$\left(\frac{W}{A}\right) = \left(\frac{M}{Z}\right)$$

$$\therefore \quad \left(\frac{10470}{\dfrac{\pi D^2}{4}}\right) = \left(\frac{15000}{\dfrac{\pi D^3}{32}}\right)$$

Solving, D = 11.4 m

The loading on the base is taken as annular loading on the mean diameter. The bending moments in the base obtained by superposing the two types of loading are shown in Fig. 5.9.

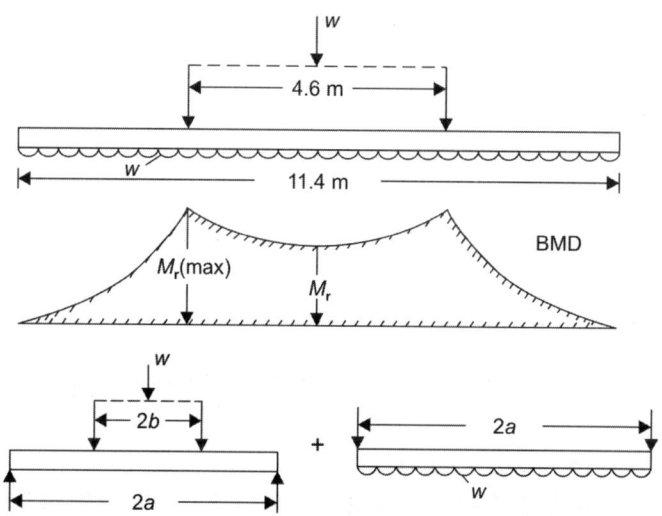

Fig. 5.9: Bending moments in circular foundation footing

Intensity of soil pressure $(w) = \left[\dfrac{10470}{\dfrac{\pi \times 11.4^2}{4}}\right]$ = 103 kN/m²

$2a$ = 11.4 m $2b$ = 4.6 m

Maximum bending moment in the section is governed by the radial moment.

M_t = bending moment at centre of footing

$$= \frac{W}{8\pi}\left[2\log_e(a/b)+1-(b/a)^2\right]-\frac{3}{16}wa^2$$

$$= \frac{10470}{8\pi}\left[2\log_e\left(\frac{5.7}{2.3}\right)+1-\left(\frac{2.3}{5.7}\right)^2\right]-\frac{3}{16}\times103\times5.7^2$$

$$= 476.28 \text{ kN·m/m}$$

$M_{r(max)}$ = moment at junction of footing and chimney walls at a radius of 2.3 m

$$= \frac{W}{8\pi}\left[2\log_e(a/b)+1-(b/a)^2\right]-\frac{3}{16}w(a^2-b^2)$$

$$= \frac{10470}{8\pi}\left[2\log_e\left(\frac{5.7}{2.3}\right)+1-\left(\frac{2.3}{5.7}\right)^2\right]-\frac{3}{16}\times103(5.7^2-2.3^2)$$

$$= 578.38 \text{ kN·m/m}$$

Design ultimate moment M_u = (1.5 × 578.38) = 868 kN·m/m

Using M20 grade concrete and Fe 415 HYSD bars,

effective depth $d = \sqrt{\dfrac{M_u}{0.138\,f_{ck}b}} = \sqrt{\dfrac{868\times10^6}{0.138\times20\times10^3}} = 561$ mm

Since moment is large, adopt larger depth of footing to reduce the quantity of reinforcemets.

Adopt $d = 700$ mm and overall depth = 800 mm.

Using SP:16 design charts, compute the parameter:

$$\left(\frac{M_u}{bd^2}\right) = \left(\frac{868\times10^6}{10^3\times700^2}\right) = 1.77$$

Refer to Table 2 of SP:16 and readout the percentage reinforcement as following:

$$p_t = 0.555 = \left(\frac{100\,A_{st}}{bd}\right)$$

$\therefore \qquad A_{st} = \left(\dfrac{0.555\times10^3\times700}{100}\right) = 3885 \text{ mm}^2/\text{m}$

Provide 28 mm diameter bars at 150 mm centres (A_{st} = 4105 mm²) in perpendicular directions both ways as shown in Fig. 5.10. Also provide 12 mm diameter bars both ways to top of the footing.

ix. *Check for shear*:

Intensity of soil pressure p = 103 kN/m²

Ultimate soil pressure p_u = (1.5 × 103) = 154.5 kN/m²

Cantilever projection = [(0.5 × 11.4) – 2.5] = 3.2 m

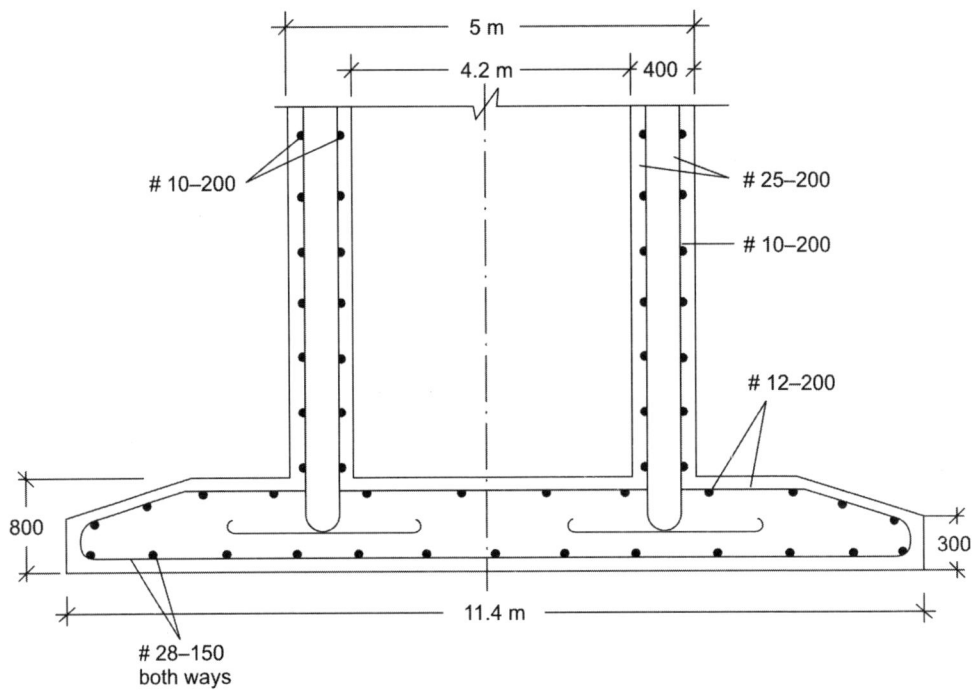

Fig. 5.10: Reinforcement details in chimney and foundations

Maximum shear force at a distance d from the support is given by:

$$V_u = [154.5 \, (3.2 - 0.7)] = 386.25 \text{ kN}$$

$$\tau_v = \left(\frac{V_u}{bd}\right) = \left(\frac{386.25 \times 10^3}{10^3 \times 700}\right) = 0.52 \text{ N/mm}^2$$

$$\left(\frac{100 \, A_{st}}{bd}\right) = \left(\frac{100 \times 4105}{10^3 \times 700}\right) = 0.586$$

Refer to Table 19 of IS:456-2000 and readout the permissible shear stress as following:

$$k\tau_c = (1 \times 0.52) = 0.52 \text{ N/mm}^2$$

Since $\tau_v = k\tau_c$, the slab is safe against shear failure.

3. A circular RC chimney has a constant shell thickness of 300 mm and an external diameter of 4 m. The section is reinforced with 1% steel located at 50 mm from the outer face. The temperature difference between the inside and outside face of concrete is 70°C.

Modulus of elasticity of steel = 210 kN/mm²

Modulus of elasticity of concrete = 19 kN/mm²

Coefficient of expansion of concrete and steel = 11 × 10⁻⁶/°C

Compute the stress developed in concrete and steel due to the temperature gradient.

  i. *Shell properties*:

$$t_c = 300 \text{ mm}, \quad A_{st} = 0.01 \, (\pi \times 3700 \times 300) = 34872 \text{ mm}^2$$

$$t_s = \left(\frac{A_{st}}{\pi d_m}\right) = \left(\frac{34872}{\pi \times 3700}\right) = 3 \text{ mm}$$

$m = (E_s/E_c) = (210/19) = 11$

$p = (t_s/t_c) = (3/300) = 0.01$

$at_c = 250$ mm

$$\therefore \qquad a = \left(\frac{250}{300}\right) = 0.833$$

Neutral axis depth constant $k = -mp + \sqrt{2mpa + p^2 m^2}$

$$\therefore \qquad k = (-11 \times 0.01) + \sqrt{(2 \times 11 \times 0.01 \times 0.833) + (0.01)^2 \times (11)^2}$$

$$= 0.338$$

Stress in steel $\sigma_s = E_s \alpha T(a - k)$

$$= (210 \times 10^3 \times 11 \times 10^{-6} \times 70)(0.833 - 0.338)$$

$$= 80.04 \text{ N/mm}^2$$

Stress in concrete $\sigma_c = E_c \alpha kT$

$$= (19 \times 10^3 \times 11 \times 10^{-6} \times 0.338 \times 70) = 4.944 \text{ N/mm}^2$$

5.10 DESIGN OF REINFORCEMENTS IN CHIMNEYS USING CHARTS

Design the vertical reinforcements at the base section of a chimney using the following data and the design chart shown in Fig. 5.11.

Total vertical load at base section $W = 10000$ kN

Bending moment at base section $M = 17000$ kN·m

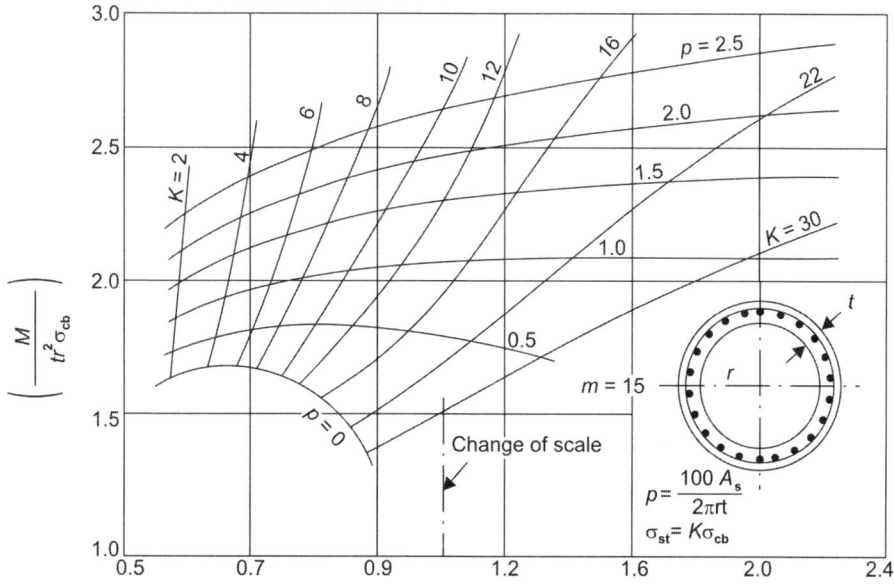

Fig. 5.11: Interaction diagram for circular chimney shell in bending

Thickness of chimney shell t = 300 mm

Mean radius of the shell r = 2 m

Concrete M20 grade and mild steel grade I are used.

The following parameters are computed for using the chart.

$$\text{Eccentricity } e = \left(\frac{M}{W}\right) = \left(\frac{17000}{10000}\right) = 1.7$$

$$(e/r) = \left(\frac{1.7}{2.0}\right) 0.85, \ \sigma_{cb} = 7 \ \text{N/mm}^2 = 7000 \ \text{kN/m}^2$$

$$\left(\frac{M}{tr^2\sigma_{cb}}\right) = \left(\frac{17000}{0.3\times2^2\times7000}\right) = 2.02$$

From the chart, the value of percentage of reinforcement is obtained as

$$p = 1.0\% \text{ and } K = 10$$

Stress in tensile steel $K\sigma_{cb} = (10 \times 7) = 70 \ \text{N/mm}^2$

5.11 ANALYSIS AND DESIGN OF RC CHIMNEYS SUBJECTED TO DYNAMIC LOADS

Tall and slender structures like chimneys are subjected to severe wind loads and earthquake forces. Hence, a wind load analysis of the structure is essential to study the performance of the structure under the dynamic effects of wind forces. The dynamic analysis for wind loads involves the computation of the natural frequencies and fundamental periods of the structure using the several empirical formulas suggested in the Indian standard codes IS:1893-1975[11], IS:4998 (part 1)1975, American Concrete Institute recommendations ACI:307-1998[12]. Readers may refer to the excellent publications by Maugh and Rumman[13] dealing with the dynamic design and earthquake resistant design of chimneys by Housner[14] for detailed information regarding the design of chimneys under dynamic loads.

Due to air pollution control, tall chimneys in the range of 300 to 400 m in height have been constructed. The advent of modern high speed digital computers has made it possible to analyse the tall chimneys subjected to dynamic loads and to compute stresses for various loading combinations in a rigorous and at the same time accurate and economical methods. Typical investigations of the dynamic analysis of a chimney stack of height 180 m and outer diameter varying from 6 m at top to 15 m at bottom and the thickness of concrete shell varying from 220 mm at top to 400 mm at bottom, when subjected to wind loads varying from 2 kN/m² at bottom to 2.88 kN/m² at top, indicate that the design bending moments under dynamic loads are significantly higher than the static load computations. Under dynamic load analysis, the permissible stresses in concrete are restricted to $0.4f_{ck}$ and the permissible stresses in steel are limited to $0.6f_y$. A separate analysis is made to calculate the temperature stresses.

References

1. Manohar SN, *Tall Chimneys*, Tata McGraw-Hill, New Delhi, 1985, pp. 1–6.

2. Plechik JM, Gerwin PH and Pham MG, Design and construction of world's tallest free standing fiber glass stack, J ASCE, Vol. 51, No.1, Jan. 1981, p. 57.

3. Bosanquet CH and Pearson JL,The Spread of Smoke and Gases from Chimneys, Transactions of the Faraday Society, Vol. 32, 1936, London, p. 1249.

4. Cramer JE, A brief survey of the meteorological aspects of atmospheric pollutants, *Bulletin of the American Metallurgical Society*, Vol. 40, 1959, p.165.

5. Briggs RA, Will wind spoilers solve stack vibration problems, *Power Engineering*, Vol. 66, Aug. 1962, p. 49.

6. Krishna Raju N, Limit state design for structural concrete, *Proceedings of the Institution of Engineers (India)*, Vol. 51, Jan. 1971, pp. 138–143.

7. Manohar SN, Design of concrete chimney shaft with multiflue openings, *Journal of the Structural Engineering*, Roorke, India, Vol. 5, July 1977, p. 83.

8. Diver M, Design of reinforced concrete chimneys, *Journal of the American Concrete Institute*, Vol. 67, Oct. 1970, p. 788.

9. Pinfold GM, *Reinforced Concrete Chimneys and Towers*, View Point Publication, Cement and Concrete Association, London, 1975.

10. IS:4998-1975, Indian Standard Code-Criteria for the Design of Reinforced Concrete Chimneys, Bureau of Indian Standards, New Delhi. 1975.

11. IS:1893-1984, Criteria for Earthquake Resistant Design of Structures, (Fourth Revision), Bureau of Indian Standards, New Delhi, 1984.

12. ACI:307-1998, Design and Construction of RC Chimneys, American Concrete Institute Recommendations, 1998.

13. Maugh LC and Rumman WS, Dynamic design of reinforced concrete chimneys, *Journal of the American Concrete Institute*, Sept. 1967, p. 558.

14. Housner GW, Earthquake resistant design base on dynamic properties of earthquakes, *Journal of the American Concrete Institute*, Vol. 28, July 1956, p. 85.

Assignment

1. A reinforced concrete chimney having a wall thickness of 15 cm has a mean diameter of 2.5 m. The section is reinforced with sixty bars of 16 mm diameter. If the effective wind pressure is 1.4 kN/m² on the projected area, evaluate the maximum stresses in concrete and steel at a section 25 m from the top of the chimney. Assume modular ratio to be 15.

2. A reinforced concrete chimney of 60 m height above ground level has an outside diameter of 4 m. Thickness of fire brick lining provided up to 40 m above ground level is 10 cm. Density of fire brick lining is 20 kN/m³. The lining is supported at every 6 m intervals. The thickness of concrete shell is 200 mm at the top and is increased to 250 mm and 300 mm at 24 m and 48 m from the top respectively. Vertical steel bars of 1% of cross-section is provided throughout the chimney. Wind load is 1.65 kN/m² for the top 24 m and 1.5 kN/m² over the rest of the chimney height. Temperature difference between inside and outside of concrete shell is 70 °C. Coefficient of expansion of concrete and steel is 11 × 10⁻⁶/°C and modular ratio is 11.Using M25 grade concrete and Fe 415 HYSD bars, design the chimney and

check for stresses at the base section. Also design suitable foundations for the chimney assuming the safe bearing capacity of the soil as 150 kN/m².

3. A circular RC chimney 50 m high above ground has a constant shell thickness of 300 mm and an external diameter of 4 m. The effective wind pressure on the projected surface is 2 kN/m². The section is reinforced with 1% steel in the vertical direction. If modular ratio is 11, find the position of the section from the top where the resultant stress distribution due to dead and wind loads is such that there is maximum compressive stress in concrete at the leeward side and zero stress at the windward face. Evaluate the maximum stresses in concrete and steel at this section.

4. A reinforced concrete chimney 100 m high above ground has an external diameter 4 m at the top and 5 m at the ground level. The thickness of concrete shell varies from 200 mm at the top to 400 mm at the bottom. The wind pressure at site may be taken as 2 kN/m². Assuming a modular ratio of 15, design suitable reinforcements in the shell walls. Adopt M20 grade concrete and Fe 415 HYSD bars.

5. A concrete chimney of height 60 m with the external diameter of the shaft being 5 m at top and 6 m at bottom is required to withstand an intensity of wind pressure of 1 kN/m². The thickness of concrete shell is 300 mm. The thickness of brick lining is 120 mm. Temperature difference between the inside and outside of chimney is 80 °C. Permissible bearing pressure on soil at site is equal to 200 kN/m². Adopt M20 grade concrete and Fe 415 HYSD bars and design the reinforcements for the chimney at 10 m intervals and a suitable foundation for the chimney.

6. A circular RCC chimney has a constant shell thickness of 270 mm and an external diameter of 5 m. The section is reinforced with 1% reinforcement located at 70 mm from the outer face. The temperature difference between the inside and outside face of concrete is 60 °C. Modulus of elasticity of steel is 210 kN/m² and modulus of elasticity of concrete is 20 kN/mm². Coefficient of expansion of concrete and steel is 11 × 10⁻⁶/°C. Estimate the stresses developed in concrete and steel due to the temperature gradient.

Review Questions

1. Briefly explain the necessity of using reinforced concrete chimneys. In what type of industrial structures, chimneys are used?

2. List the various parts of a reinforced concrete chimney mentioning their function with sketches.

3. What are the salient loads to be considered in the design of reinforced concrete chimneys?

4. Explain the nature of temperature stresses developed in reinforced concrete chimney walls showing the temperature gradient from inside to outside of the chimney.

5. What is shape factor? In what way does it affects the nature of wind forces on RC chimneys?

6. Briefly mention the various factors to be considered in the design of reinforced concrete chimneys.

7. Mention the various design parameters to be considered in the analysis of stresses in the walls of cylindrical chimney shaft due to dead and wind loads.

8. How do you analyse the stresses in the walls of a reinforced concrete chimney due to the combined effect of self weight, wind loads and temperature?

9. Explain the method of analysing the stresses in the hoop reinforcements of reinforced concrete chimney walls due to temperature gradients.
10. What type of foundations are used for reinforced concrete chimneys? Explain the method of designing raft foundations for chimneys.

Objective Type Questions

1. The best cross-sectional shape preferred for the construction of reinforced concrete chimney is
   a. square
   b. circular
   c. octagon

2. The velocity of wind acting on a tall chimney is maximum at the
   a. top
   b. middle
   c. bottom

3. Due to the combined effect of wind and dead loads on a reinforced concrete chimney, maximum compressive stresses in concrete develop at the
   a. windward side
   b. leeward side
   c. mid height

4. In reinforced concrete chimneys under working conditions, the temperature gradient is maximum in the
   a. concrete wall
   b. air gap
   c. fire brick lining

5. The stress in concrete chimney walls due to temperature only depends upon the
   a. bending moment
   b. grade of concrete
   c. modulus of elasticity of concrete

6. In reinforced concrete chimneys, hoop reinforcements are designed to resist the
   a. bending moment
   b. wind loads
   c. shear force

7. In horizontal sections of the chimney, the action due to temperature in a horizontal plane is to
   a. stretch the inner fibres
   b. compress the outer fibres
   c. compress the inner fibres and stretch the outer fibres

8. In seismic zones, chimneys should be designed to resist the action of
   a. dead and wind loads
   b. dead loads and temperature actions
   c. dead, wind, earthquake and temperature effects

9. Tall chimneys located in coastal regions should be analysed for stresses under
   a. dead loads
   b. temperature effects
   c. dynamic analysis under wind loads

10. Tall reinforced concrete chimneys should be provided with foundation system consisting of
    a. spread footings
    b. pile foundations
    c. raft foundations

6

Curved Beams

6.1　GENERAL ASPECTS

Structural concrete elements like curved beams are often used to support the circular cylindrical walls of elevated water tanks, curved balconies, circular domes and ring beams of circular raft foundations. Curved beams forming a quadrant of a circle are generally supported on columns spaced at regular intervals and they are subjected to both bending and torsional moments since the loads or reactions do not lie along the axis at any point of the beam. However, in the case of circular beams supported by symmetrically placed columns, the vertical reactions are provided by the columns and due to symmetry, the torsional moments at the centre of the curved beam between any two consecutive supports will be zero. Also the maximum negative bending moment develops at the support sections and maximum positive bending moment at sections in between the supports.

The theory of torsion[1,2] of prismatic homogeneous members of different types of cross-sections is well established and described in detail in books of mechanics of materials[3,4]. Maximum torsional moments will develop at sections near the supports and where the bending moment is zero. In other words, the maximum torque occurs at points of contra flexure or inflexion. Also the shear force will be maximum at the support sections. It is important to note that the support sections have to be designed for the combined action of maximum negative bending moment and shear force as specified in the Indian standard code IS:456-2000[5]. Sections subjected to maximum torsional moment should also be designed to resist the torque together with the associated shear force at the section. The reader may refer to excellent monographs[6,7] for detailed design of reinforced concrete members subjected to the combined action of flexure, torsion and shear[8,9,10].

6.2　ANALYSIS OF FLEXURAL AND TORSIONAL MOMENTS IN CIRCULAR GIRDERS

The bending and torsional moments developed in curved girders can be analysed by strain energy methods. The magnitude of moments and their location is influenced by the number of supports and the radius of the curved beam. Figure 6.1 shows a typical curved beam circular in plan supported on eight columns. Referring to Fig. 6.2, the maximum positive and negative bending moments and the torsional moments can be expressed in the following form:

Maximum negative bending moment = K_1WR

Maximum positive bending moment = K_2WR

Maximum torsional moment = K_3WR

where,

W = total load on the curved beam = $2\pi Rw$

where,

w = uniformly distributed load per unit length of beam

θ = angle subtended at the centre by the ends of the beam

R = radius of the circular girder

K_1, K_2 and K_3 are moment coefficients, the values of which are compiled in Table 6.1.

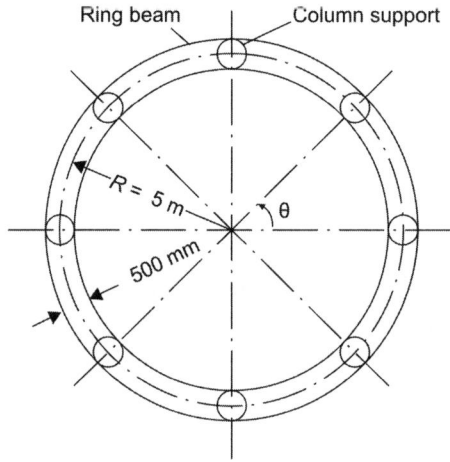

Fig. 6.1: Ring girder supported on eight columns

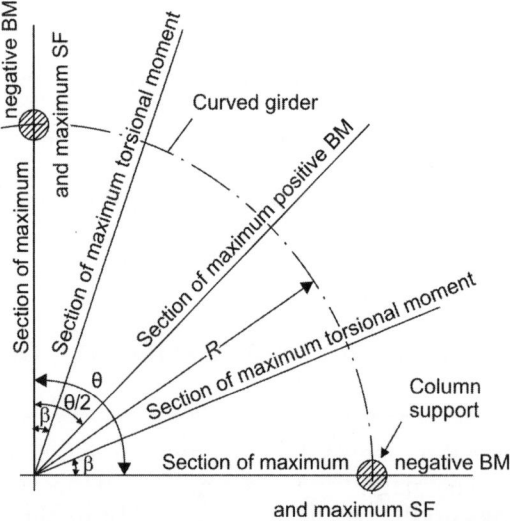

Fig. 6.2: Position of maximum moments in curved girders

Table 6.1: Moment coefficients in circular girders supported on columns moment coefficients

Number of columns	Angle subtended between the columns	Negative bending moment at support	Positive bending moment at centre of spans	Maximum twisting moment or torque	Angular distance for maximum torque
n	θ	K_1	K_2	K_3	β
4	90°	0.0342	0.0176	0.0053	19°12′
5	60°	0.0148	0.0075	0.0015	12°44′
8	45°	0.0083	0.0041	0.0006	9°33′
10	36°	0.0054	0.0023	0.0003	7°30′
12	30°	0.0037	0.0014	0.0017	7°15′

The critical sections to be designed are the support sections subjected to maximum negative and positive bending moments and the section subjected to maximum twisting moment associated with some shear force. At this section, the bending moment is zero. Hence, this section has to be designed for combined torsion and shear.

6.3 MOMENTS IN SEMICIRCULAR BEAMS SUPPORTED ON THREE COLUMNS

The magnitude and position of maximum positive and negative bending moments and the twisting moments in a semicircular beam supported on three equally spaced supports are given by the following reactions:

Maximum positive BM = $0.1520wR^2$ at section 29°44′ from the end columns.

Maximum negative BM over the central support = $-0.4290wR^2$

Maximum twisting moment = $0.103wR^2$ at section 59°29′ from the end columns.

6.4 DESIGN EXAMPLES

1. A circular RC girder for the foundation raft of a water tower has a mean diameter of 10 m. The uniformly distributed load transmitted by eight symmetrically placed columns on the girder being 300 kN/m. The width of the beam is 500 mm and the overall depth is 1000 mm. Using M20 grade concrete and Fe 415 grade steel, design suitable reinforcements in the circular girder and sketch the details of reinforcements at critical sections.

   i. *Data*:
      Radius of girder = R = 5 m
      Width of girder = b = 500 mm
      Depth of girder = D = 1000 mm
      Load on girder = q = 300 kN/m
      Angle θ = 45°

   ii. *Characteristic strength*:
      f_{ck} = 20 N/mm^2
      f_y = 415 N/mm^2

iii. *Loads*:

Self weight of beam = $(0.5 \times 1 \times 24) = 12$ kN/m

Uniformly distributed load = 300 kN/m

Total service load $(w) = 312$ kN/m

Total design ultimate load $(w_u) = (1.5 \times 312) = 468$ kN/m

Total design ultimate load on circular girder $W_u = (2\pi R w_u)$

$$= (2 \times \pi \times 5 \times 468)$$
$$= 14703 \text{ kN}$$

iv. *Design bending moments and shear forces*:

Maximum negative BM at support sections $M_{nu} = (0.0083 \, W'_u R)$

$$= (0.0083 \times 14703 \times 5)$$
$$= 610 \text{ kN·m}$$

Maximum positive BM at centre of spans $M_{pu} = (0.0041 W_u R)$

$$= (0.0041 \times 14703 \times 5)$$
$$= 302 \text{ kN·m}$$

Maximum torsional moment at an angle of 9.5° from support is given by

$$T_u = (0.0006 W_u R) = (0.0006 \times 14703 \times 5) = 45 \text{ kN·m}$$

Shear force at support section is computed as follows:

$$V_u = \left[\frac{468 \times 5 \times (\pi/4)}{2}\right] = 918 \text{ kN}$$

Shear force at section of maximum torsion is computed as follows:

$$V_u = \left[918 - \left(\frac{468 \times 5 \times \pi \times 9.5}{180}\right)\right] = 531 \text{ kN}$$

v. *Design of support section*:

$$M_{nu} = 610 \text{ kN·m}$$
$$V_u = 918 \text{ kN}$$

Assuming a width of beam section, $b = 500$ mm

Effective depth $d = \sqrt{\dfrac{M_{nu}}{0.138 f_{ck} b}} = \sqrt{\dfrac{610 \times 10^6}{(0.138 \times 20 \times 500)}} = 665$ mm

Since shear forces are high, adopt a larger effective depth, $d = 950$ mm and overall depth, $D = 1000$ mm.

The area of steel required is computed by using the following equation:

$$M_u = (0.87 f_y A_{st} d)\left[1 - \left(\frac{A_{st} f_y}{b d f_{ck}}\right)\right]$$

$$(610 \times 10^6) = (0.87 \times 415 \times A_{st} \times 950)\left[1 - \frac{415 A_{st}}{(500 \times 950 \times 20)}\right]$$

Solving, $A_{st} = 1943$ mm^2

Provide 4 bars of 25 mm diameter ($A_s = 1963$ mm^2)

$$\tau_v = \left(\frac{V_u}{bd}\right) = \left(\frac{918 \times 1000}{1000 \times 950}\right) = 0.966 \text{ N/mm}^2$$

From Table 19 of IS:456-2000, readout the permissible shear stress as

$$\tau_c = 0.62 \text{ N/mm}^2 < \tau_v$$

Hence shear reinforcements are required.

Shear force resisted by concrete is computed as

$$V_c = (\tau_c bd) = (0.62 \times 500 \times 950) \ 10^{-3} = 294 \text{ kN}$$

Balance shear force $= V_{us} = (918 - 294) = 624$ kN

Using 10 mm diameter four-legged stirrups, the spacing is given by

$$S_v = \left[\frac{0.87 f_y A_{sv} d}{V_{us}}\right] = \left[\frac{0.87 \times 415 \times 4 \times 79 \times 950}{624 \times 1000}\right] = 173.6 \text{ mm}$$

Adopt 10 mm diameter four-legged stirrups at 170 mm centers.

vi. *Design of mid span section:*

Maximum positive bending moment $= M_{pu} = 302$ kN·m

$$(302 \times 10^6) = (0.87 \times 415 A_{st} \times 950)\left[1 - \frac{415 A_{st}}{(10^3 \times 950 \times 20)}\right]$$

Solving, $A_{st} = 447$ mm^2

But minimum area of reinforcement is given by the following relation:

$$A_{st} = \left(\frac{0.85 bd}{f_y}\right) = \left(\frac{0.85 \times 500 \times 950}{415}\right) = 984 \text{ mm}^2$$

Provide 4 bars of 20 mm diameter ($A_{st} = 1256$ mm^2).

vii. *Design of section subjected to maximum torsion and shear:*

$$T_u = 45 \text{ kN·m} \qquad\qquad D = 1000 \text{ mm}$$
$$V_u = 531 \text{ kN} \qquad\qquad b = 500 \text{ mm}$$
$$M_u = 0$$

$$M_t = T_u\left[\frac{1 + (D/b)}{1.7}\right] = 45\left[\frac{1 + (1000/500)}{1.7}\right] = 80 \text{ kN·m}$$

$$M_{el} = (M_u + M_t) = (0 + 80) = 80 \text{ kN·m}$$

Since the moment is very small, minimum area of reinforcement is computed as

$$A_{st} = \left(\frac{0.85 \times 500 \times 950}{415}\right) = 984 \text{ mm}^2$$

Provide 4 bars of 20 mm diameter ($A_{st} = 1256$ mm^2)

Equivalent shear force $V_e = [V_u + 1.6 \ (T_u/b)]$
$$= [531 + 1.64 \ (45/0.5)] = 680 \ \text{kN}$$

$$\tau_{ve} = \left(\frac{V_e}{bd}\right) = \left(\frac{680 \times 10^3}{500 \times 950}\right) = 1.43 \ \text{N/mm}^2 < \tau_{max} = 2.8 \ \text{N/mm}^2$$

$$\left(\frac{100 \, A_{st}}{bd}\right) = \left(\frac{100 \times 1140}{500 \times 950}\right) = 0.24$$

Refer to Table 19 (IS:456-2000) and readout the permissible shear stress as follows:

$$\tau_c = 0.35 \ \text{N/mm}^2 < 1.43 \ \text{N/mm}^2$$

Hence shear reinforcements are required.

Using 12 mm diameter two-legged stirrups with side covers of 25 mm and bottom and top covers of 50 mm:

$$b_1 = 450 \ \text{mm}, \qquad d_1 = 900 \ \text{mm}, \qquad A_{sv} = (2 \times 113) = 226 \ \text{mm}^2$$

$$S_v = (A_{sv} \times 0.87 f_y) \left[\frac{b_1 d_1}{T_u} + \frac{2.5 d_1}{V_u}\right]$$

$$= (226 \times 0.87 \times 415) \left[\left(\frac{450 \times 900}{45 \times 10^6}\right) + \left(\frac{2.5 \times 900}{531 \times 10^6}\right)\right]$$

$$= 1080 \ \text{mm}$$

But $\qquad S_v = \left[\frac{0.87 A_{sv} f_y}{(\tau_{ve} - \tau_c) b}\right] = \left[\frac{226 \times 0.87 \times 415}{(1.43 - 0.35)500}\right] = 151 \ \text{mm}$

Adopt 12 mm diameter two-legged stirrups at a spacing of 150 mm.

viii. *Reinforcement details*:

The details of reinforcements at the critical sections are shown in Fig. 6.3.

2. Design a semicircular beam supported on 3 columns equally spaced. The centre of the columns are on a curve of diameter 8 m. The superimposed load on the beam is 20 kN. Adopt M20 grade concrete and Fe 415 HYSD bars.

i. *Data*:

Radius of the girder $= R = 4 \ \text{m}$
Assume width of girder $= b = 400 \ \text{mm}$
Depth of girder $= D = 800 \ \text{mm}$
Load on girder $= q = 20 \ \text{kN/m}$

ii. *Characteristic strength*:

$f_{ck} = 20 \ \text{N/mm}^2$ and $f_y = 415 \ \text{N/mm}^2$.

iii. *Loads*:

Self weight of beam $= (0.4 \times 0.8 \times 24) = 7.68 \ \text{kN/m}$
Live load on beam $= 20.00 \ \text{kN/m}$
Finishes etc. $= 2.32 \ \text{kN/m}$

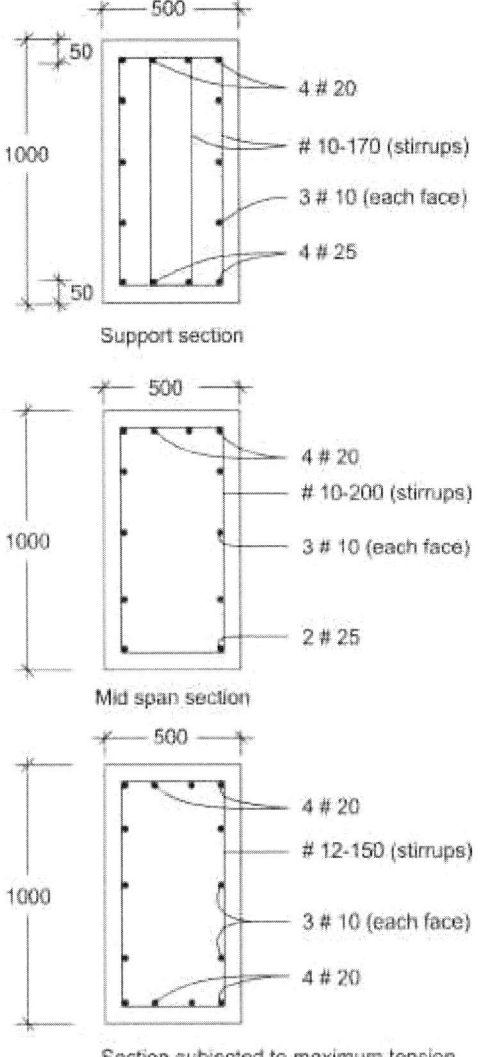

Fig. 6.3: Typical cross-sections of circular RC beam

Total load (w) = 30.00 kN/m

Total design ultimate load (w_u) = (1.5 × 30) = 45 kN/m

iv. *Bending moments and shear forces*:

Maximum negative BM at middle support = $-0.429\,w_u R^2$

$$= -(0.429 \times 45 \times 4^2) = -309 \text{ kN·m}$$

Maximum positive moment = $0.152\,w_u R^2 = (0.152 \times 45 \times 4^2) = 110$ kN·m

Maximum torsional moment = $0.103\,w_u R^2 = (0.103 \times 45 \times 4^2) = 74$ kN·m

Reaction at end support = R_1

Reaction at mid support = R_2

Then, $R_1 = \dfrac{w_u R}{2}(\pi - 2)$ and $R_2 = 2w_u R$

Shear at end support = $R_1 = \dfrac{w_u R}{2}(\pi - 2) = \dfrac{(45 \times 4)}{2}(\pi - 2) = 102.7$

Shear at mid support section = $\left(\dfrac{2w_u R}{2}\right) = w_u R = 45 \times 4 = 180$ kN

Shear at the section of maximum twisting moment = $(R_1 - w_u R\theta)$
where, $\theta = 59°29' = 1.038$ radians.

∴   Shear force = $[102.7 - (45 \times 4 \times 1.038)] = -84$ kN

v. *Design of mid support section*:

Bending moment = $M_u = 309$ kN·m and shear force = $V_u = 180$ kN

$$d = \sqrt{\dfrac{M_u}{0.138\, f_{ck}\, b}} = \sqrt{\dfrac{309 \times 10^6}{0.138 \times 20 \times 400}} = 529 \text{ mm}$$

Since the section is subjected to torsion and shear also, adopt a greater overall depth $D = 800$ mm and effective depth $d = 750$ mm.

The reinforcement area is computed by the following relation:

$$M_u = (0.87 f_y A_{st} d)\left[1 - \left(\dfrac{A_{st} f_y}{bd f_{ck}}\right)\right]$$

$$(309 \times 10^6) = (0.87 \times 415 \times A_{st} \times 750)\left[1 - \dfrac{415\, A_{st}}{(103 \times 750 \times 20)}\right]$$

Solving, $A_{st} = 1248$ mm²

Provide 4 bars of 22 mm diameter at top ($A_s = 1520$ mm²)

$$\tau_v = \left(\dfrac{V_u}{bd}\right) = \left(\dfrac{180 \times 10^3}{400 \times 750}\right) = 0.6 \text{ N/mm}^2$$

$$\left(\dfrac{100\, A_{st}}{bd}\right) = \left(\dfrac{100 \times 1520}{400 \times 750}\right) = 0.50$$

From Table 19 (IS:456-2000), readout the permissible shear stress as follows:

$\tau_v = 0.48$ N/mm² $< \tau_v = 0.6$ N/mm²

Hence shear reinforcements are required.

Shear force resisted by concrete is computed as:

$V_e = (\tau_c bd) = (0.48 \times 400 \times 750)\, 10^{-3} = 144$ kN

Balance shear force = $V_{us} = (V_u - V_e) = (180 - 144) = 36$ kN

Using 10 mm diameter two-legged stirrups, the spacing is given by

$$S_v = \left[\dfrac{0.87 f_y A_{sv} d}{V_{us}}\right] = \left[\dfrac{0.87 \times 415 \times 2 \times 79 \times 750}{36 \times 10^3}\right] = 1188 \text{ mm}$$

According to clause 26.5.17 of IS:456-2000, the spacing of transverse reinforcement should be the least of:

a. $x_1 = (400 - 25 - 25) = 350$ mm

b. $[(x_1 + y_1)/4] = [(350 + 700)/4] = 262.5$ mm

c. 300 mm

where, x_1 and y_1 are respectively the short and long dimensions of the stirrups. Hence adopt a spacing $S_v = 250$ mm.

vi. *Design of maximum positive moment section:*

Maximum positive moment $M_u = 110$ kN·m

$$b = 400 \text{ mm and } d = 750 \text{ mm}$$

$$\left(\frac{M_u}{bd^2}\right) = \left(\frac{110 \times 10^6}{400 \times 750^2}\right) = 0.48$$

Refer to Table 2 of SP:16 and readout the percentage of reinforcement as

$$p_t = \left(\frac{100 A_{st}}{bd}\right) = 0.137$$

$$\therefore \quad A_{st} = \left(\frac{0.137 \times 400 \times 750}{100}\right) = 411 \text{ mm}^2$$

But minimum area of reinforcement should be not less than the value given by

$$A_{st\,min} = \left(\frac{0.85\,bd}{f_y}\right) = \left(\frac{0.85 \times 400 \times 750}{415}\right) = 615 \text{ mm}^2$$

Provide 2 bars of 22 mm diameter as main reinforcement ($A_{st} = 760$ mm²).

vii. *Design of section subjected to maximum torsion and shear:*

$$T_u = 74 \text{ kN·m} \qquad\qquad D = 800 \text{ mm}$$
$$V_u = 84 \text{ kN} \qquad\qquad b = 400 \text{ mm}$$
$$M_u = 0 \qquad\qquad\qquad d = 750 \text{ mm}$$

$$M_t = T_u\left[\frac{1+(D/b)}{1.7}\right] = 74\left[\frac{1+(800/400)}{1.7}\right] = 130.5 \text{ kN·m}$$

$$M_{ct} = (M_u + M_t) = (0 + 130.5) = 130.5 \text{ kN·m}$$

Since the magnitude of moment is small, provide minimum area of tension reinforcement computed as follows:

$$A_s = \left(\frac{0.85\,bd}{f_y}\right) = \left(\frac{0.85 \times 400 \times 750}{415}\right) = 615 \text{ mm}^2$$

Provide 2 bars of 22 mm diameter at top and bottom ($A_{st} = 760$ mm²)

Equivalent shear force $V_e = [V_u + 1.6(T_u/b)]$

$$= [84 + 1.6(74/0.4)] = 380 \text{ kN}$$

$$\tau_{ve} = \left(\frac{V_e}{bd}\right) = \left(\frac{380 \times 10^3}{400 \times 750}\right) = 1.26 \text{ N/mm}^2$$

$$\left(\frac{100 A_{st}}{bd}\right) = \left(\frac{100 \times 760}{400 \times 750}\right) = 0.25$$

Refer to Table 19 (IS:456-2000) and readout the permissible shear stress as

$$\tau_c = 0.36 \text{ N/mm}^2 < \tau_{ve} = 1.43 \text{ N/mm}^2$$

Hence shear reinforcements are required.

Using 10 mm diameter two-legged stirrups with side covers of 25 mm and top covers of 50 mm, we have

$$b_1 = 350 \text{ mm}, \qquad d_1 = 700 \text{ mm}, \qquad A_{sv} = (2 \times 79) = 158 \text{ mm}^2$$

$$S_v = (A_{sv} \times 0.87 f_y) \left[\frac{b_1 d_1}{T_u} + \frac{2.5 d_1}{V_u}\right]$$

$$= (158 \times 0.87 \times 415) \left[\left(\frac{350 \times 700}{74 \times 10^6}\right) + \left(\frac{2.5 \times 700}{84 \times 10^3}\right)\right]$$

$$= 1317 \text{ mm}$$

Also, $\qquad S_v = \left[\dfrac{A_{sv} 0.87 f_y}{(\tau_{ve} - \tau_c)b}\right]$

$$= \left[\frac{2 \times 79 \times 0.87 \times 415}{(1.26 - 0.36)400}\right] = 158 \text{ mm}$$

Adopt 10 mm diameter two-legged stirrups at 150 mm centres.

References

1. Timoshenko S and Goodier JN, *Theory of Elasticity*, 3 edn, McGraw-Hill, New York, 1970.
2. Popov EP, *Mechanics of Materials*, 2 edn, Prentice Hall Englewood Cliffs, New Jersey, 1978.
3. Srinath LS, *Advanced Mechanics of Solids*, Tata McGraw-Hill, New Delhi, 1980, pp. 223–259.
4. Krishna Raju N and Gururaja DR, *Advanced Mechanics of Solids and Structures*, Narosa Publishing House, New Delhi, 1997.
5. IS:456-2000, Indian Standard Code of Practice for Plain and Reinforced Concrete (Fourth Revision), Bureau of Indian Standards, New Delhi, July 2000.
6. Collins MP, The Torque-Twist Characteristics of Reinforced Concrete Beams, Inelasticity and Non Linearity in Structural Concrete, SM Study No. 8, University of Waterloo Press, Waterloo, 1972, pp. 211–232.
7. Verghese PC, *Limit State Design of Reinforced Concrete*, Prentice Hall of India Private Ltd, New Delhi, 1994, pp. 388–419.
8. Krishna Raju N and Pranesh RN, *Reinforced Concrete Design*, (IS:456-2000) (Principles and Practice), New Age International, New Delhi, 2003, pp. 147–172.
9. Purushothaman P, *Reinforced Concrete Structural Elements, Behaviour, Analysis and Design*, Tata McGraw-Hill, New Delhi, 1984.
10. Warner RF, Rangan BV, and Hall AS, *Reinforced Concrete*, Pitman, Australia, 1976.

Assignment

1. A circular girder of a water tank has a mean diameter of 10 m, and it is supported on six symmetrically placed columns. The uniformly distributed load on the girder is 200 kN/m. Design the critical sections of the girder using M20 grade concrete and Fe 415 HYSD bars and sketch the details of reinforcements.

2. The circular girder of an Intze type water tank of mean diameter 10 m and supported on eight symmetrically placed columns, supports an uniformly distributed load of 500 kN/m exclusive of its self weight. Design the girder using M20 grade concrete and Fe 415 HYSD bars.

3. Design a semicircular RC beam supported on three columns equally spaced and supporting an uniformly distributed load of 30 kN/m. The radius of the centre line of the beam is 5 m. Adopt M20 grade concrete and Fe 415 HYSD bars.

4. A circular girder of mean diameter 12 m is supported on 12 symmetrically placed columns. The uniformly distributed load transmitted to the girder is 300 kN/m, exclusive of its self weight. Design the reinforcements in the girder assuming a width of 500 m and overall depth of the girder as 1000 mm. Adopt M25 grade concrete and Fe 415 HYSD bars.

5. A ring girder having a mean diameter of 8 m, supported on eight symmetrically placed columns, is required to support a total load of 10000 kN exclusive of self weight of the girder. Assuming a girder size of 400 mm by 800 mm, $f_{ck} = 20$ N/mm^2 and $f_y = 500$ N/mm^2, design the reinforcements in the girder.

6. Design a semicircular ring girder supported on three columns equally spaced supports carrying a uniformly distributed load of 50 kN/m. The mean diameter of the girder is 12 m. Adopt M20 grade concrete and Fe 415 HYSD bars. Design the reinforcements in the girder assuming a width of 400 mm and an overall depth of 800 mm.

Review Questions

1. Under what situations would you recommend the use of curved beams in structural concrete construction?

2. What type of forces develop in curved girders? List the various types of moments and forces in curved girders.

3. Briefly explain the method of analysis of determining the force and moment components in circular beams subjected to loads.

4. Explain the reasons for the development of torsional moments in curved girders.

5. Discuss briefly the variations in flexural and torsional moments developed in circular girders with increasing number of columns from 4 to 12.

6. What are the critical sections to be designed in the case of circular girders supported on columns?

7. Briefly outline the method of designing a rectangular reinforced concrete section, subjected to the combined action of torsion and shear forces.

8. Explain the term equivalent bending moment. Under what situation would you compute the equivalent moment?

9. What is equivalent shear force? How do you compute the equivalent shear in the case of circular girders subjected to loads?

10. Explain with sketches the method of distributing the reinforcements in a concrete section subjected to the combined action of bending, torsion and shear.

Objective Type Questions

1. In the case of curved beams supported on columns and subjected to uniformly distributed loads, the force and moment components developed at the centre span section are
   a. shear force
   b. torsional moment
   c. bending moment

2. Maximum negative bending moment develops in a curved beam supported on columns at
   a. the centre of span section
   b. section nearer to the centre of the beam
   c. support section

3. The magnitude of flexural and torsional moments in a circular girder
   a. increases with the number of supporting columns
   b. is independent of the number of supporting columns
   c. decreases with the number of supporting columns

6. In the case of circular beams supported on columns, sections subjected to maximum torsional moment are
   a. also subjected to maximum shear force
   b. not subjected to shear force
   c. also subjected to some shear force

7. Reinforced concrete sections subjected to bending, torsion and shear should be designed for
   a. the maximum bending moment only
   b. torsional moment only
   c. the equivalent bending moment

8. In designing reinforced concrete circular beams, the term equivalent shear depends on
   a. the bending moment
   b. the torsional moment
   c. the shear force and torsional moment

9. In the limit state design of curved beam sections, the transverse reinforcements are designed for the combined action of
   a. ultimate bending and torsional moments
   b. ultimate bending and shear forces
   c. ultimate torsion and shear at the section

10. Reinforcements in curved beams designed for torsion should be
    a. provided at the tension zone of section
    b. provided at the compression zone of section
    c. distributed around the periphery of the section

7

Towers

7.1 | INTRODUCTION

Reinforced concrete tower frameworks are mostly used to support overhead water tanks, watching platforms, television transmission frameworks, lighthouses to guide ships under navigation and other special type of structures such as tubular towers to carry high tension electrical transmission lines. Elevated water tanks[1,2] are generally supported on a system of reinforced concrete columns arranged symmetrically to suit the shape of the overhead water tank. Generally, the supporting columns have the same cross-section and are arranged symmetrically to suit the shape of the overhead structure. Tower structures are usually taller and to reduce the effective length of the supporting columns, braces are used at intervals of 1.5 to 2 m. The braces reduce the bending moments developed in the columns.

The tower structure is designed to resist the forces developed due to dead, live and wind loads. The columns and braces are designed for direct loads and bending moments developed at critical cross-sections based on an approximate analysis. An accurate procedure for determination of forces and moments in the columns and braces involves a sway analysis[3,4] of wind loads using any of the well established methods[5,6].

A typical tower framework for supporting a water tank is shown in Fig. 7.1. Generally, the braces are spaced at intervals of 1.5 to 2 m.

7.2 | DESIGN PRINCIPLES

Tall towers are analysed by various rigorous[7,8] and approximate methods[9,10] to determine the moments and shear forces developed due to the action of dead, live, wind and seismic forces acting on the structure. However, the approximate methods also yield fairly reliable data of the design force components and hence are used for the analysis.

The assumptions made in the approximate methods of analysis are summarised below:

  a. The braces are assumed to be stiff and integral with the columns so that they are held in position and direction.
  b. Contraflexure points are assumed to develop at mid points of columns and braces.

c. The inner columns are assumed to resist, twice the shear forces taken by the outer columns, since the inner columns are stiffened by double bracing.

d. The end condition of the column at footing is assumed to be hinged, unless the footing is made rigid by means of piles or rafts or by horizontal braces.

The moment at each end of a brace is the sum of the moments in the column above and below the brace. The shearing force in a brace is equal to the change of bending moment from one end of the brace to the other, divided by the length of the brace.

The brace is designed for bending moment and shear developed due to the external loads and self weight. The brace is generally reinforced with equal area of steel at top and bottom so that it is safe to resist wind from one side or the other.

7.3 COMPUTATION OF MOMENTS DUE TO WIND LOADS

Case I: Two Columns Hinged at Base

Referring to Fig. 7.2, let

W = horizontal wind force resisted by one row of columns

V = vertical reaction

Taking moments about B, we have the relation:

$$V \times L = W \times H$$

$$V = \left(\frac{WH}{L} \right)$$

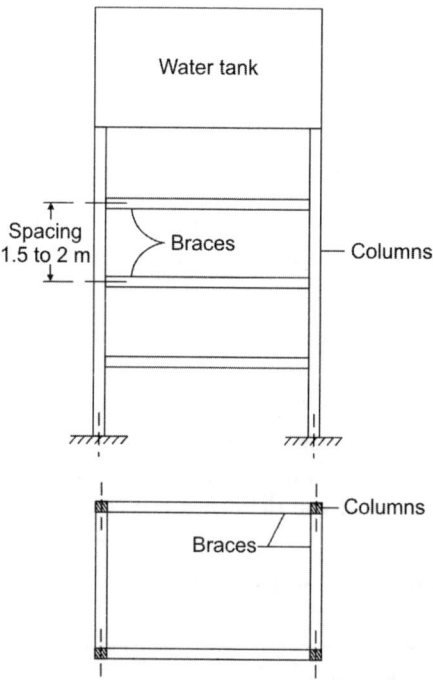

Fig. 7.1: RC supporting tower

BM in each column = $\left(\dfrac{wh}{2}\right)$

SF in each column = $w/2$

Vertical thrust in each column = $\left(\dfrac{wH}{L}\right)$

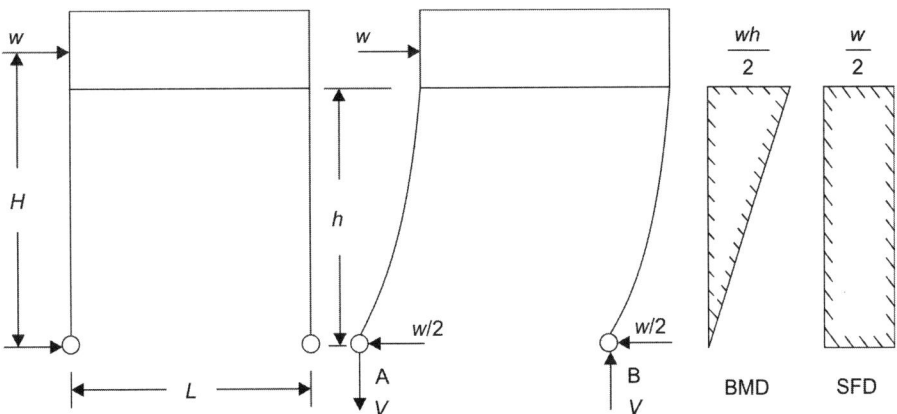

Fig. 7.2: Two columns hinged at base

Case II: Three Columns Hinged at Base

Referring to Fig. 7.3, the central column is assumed to take double the shear resisted by the exterior columns. Hence, by taking moments about A, we have

$$V \cdot 2L = WH \qquad \therefore \quad V = \left(\dfrac{WH}{2L}\right)$$

BM in exterior column = $(wh/4)$

SF in exterior column = $(w/4)$

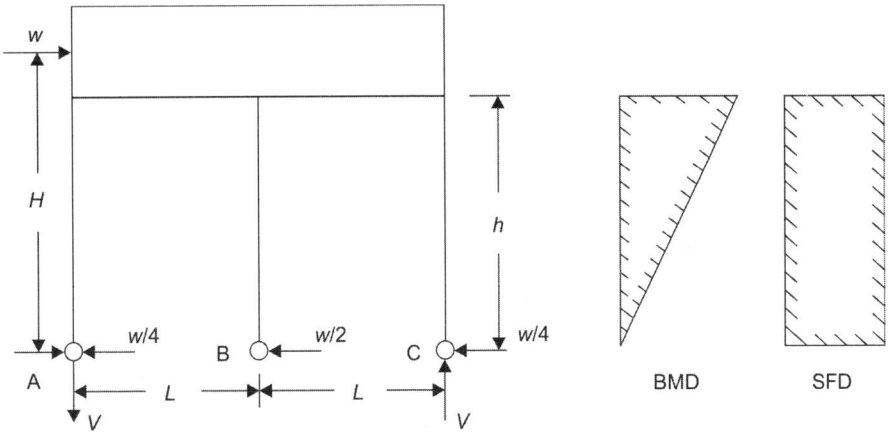

Fig. 7.3: Three columns hinged at base

Vertical thrust in exterior column = $\left(\dfrac{WH}{2L}\right)$

BM in interior column = $(WH/2)$

SF in interior column = $(w/2)$

Vertical thrust in interior column = 0

Case III: Two Columns Fixed at Base

Referring to Fig. 7.4, let

$M_A = M_B$ = moments developed in columns A and B

V = vertical thrust in the columns by taking moments about B

We have

$$WH = VL + M_A + M_B$$
$$\therefore \qquad VL = WH - (M_A + M_B) = WH - (Wh/4 + Wh/4)$$
$$\therefore \qquad V = W/L\,(H - h/2)$$

BM in each column = $\dfrac{Wh}{4}$

SF in each column = $W/2$

Vertical thrust in each column = $W/L\,(H - h/2)$

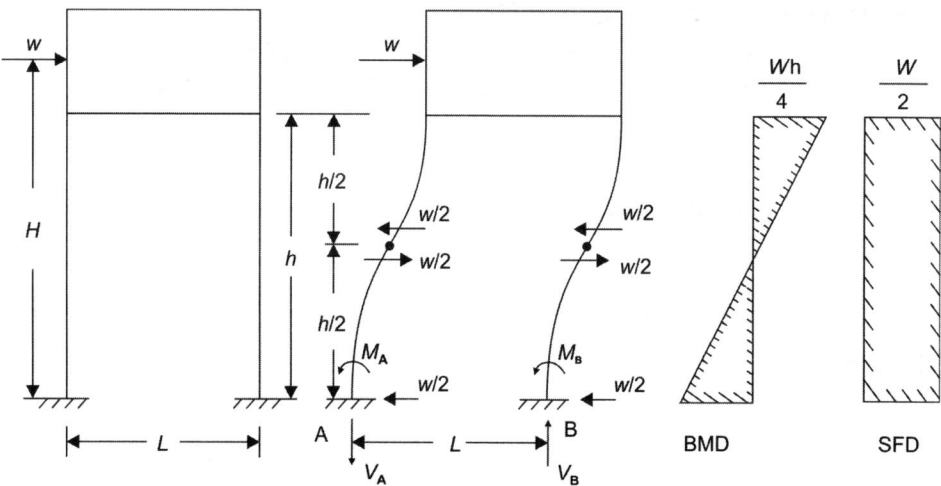

Fig. 7.4: Two columns fixed at base

Case IV: Columns Braced at Intervals with Fixed Base

Referring to Fig. 7.5, let

W = load resisted by one row of columns assuming contraflexure points at centre of panels

$\therefore$ Shear force in each column = $W/2$

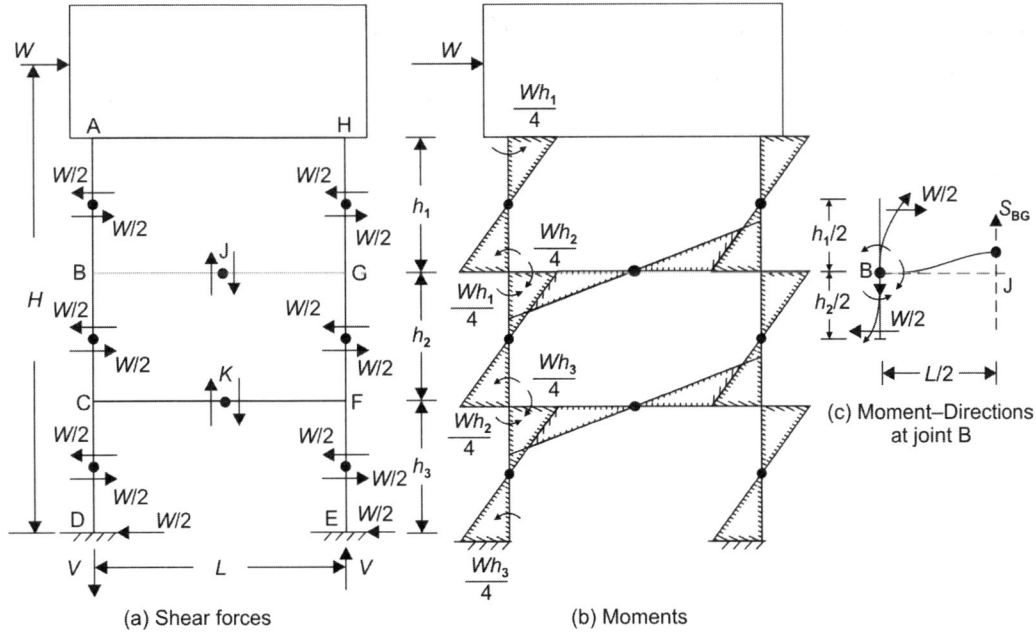

(a) Shear forces

(b) Moments

(c) Moment–Directions at joint B

Fig. 7.5: Columns braced at intervals with fixed base

Moments:

$$M_{AB} = \left(\frac{W}{2}\right)\left(\frac{h_1}{2}\right) = \left(\frac{Wh_1}{4}\right)$$

$$M_{BA} = \left(\frac{Wh_1}{4}\right)$$

$$M_{BC} = M_{CB} = \left(\frac{Wh_2}{4}\right)$$

$$M_{CD} = M_{DC} = \left(\frac{Wh_3}{4}\right)$$

Moments in brace at any junction = sum of the moments in the column above and below the brace

$$\therefore \qquad M_{BG} = M_{GB} = (M_{BA} + M_{BC}) = \frac{W}{4}(h_1 + h_2)$$

$$M_{CF} = M_{FC} = (M_{CB} + M_{CD}) = \frac{W}{4}(h_2 + h_3)$$

Shear force in brace is obtained as

$$S_{BG} = \frac{W}{2L}(h_1 + h_2)$$

$$A_{CF} = \frac{W}{2L}(h_2 + h_3)$$

Case V: Columns Braced at Intervals with Hinged Base

Referring to Fig. 7.6, let

W = load resisted by one row of columns. Assuming contraflexure points at centre of panels, shear force in each column = $W/2$

Moments:

$$M_{AB} = M_{BA} = \left(\frac{W}{2}\right)\left(\frac{h_1}{2}\right) = \left(\frac{Wh_1}{4}\right)$$

$$M_{BC} = M_{CB} = \left(\frac{Wh_2}{4}\right)$$

$$M_{DC} = \left(\frac{Wh_3}{2}\right)$$

$$M_{DC} = 0$$

$$M_{BG} = M_{GB} = (M_{BA} + M_{BC}) = W/4(h_1 + h_2)$$

$$M_{CF} = M_{FC} = (M_{CB} + M_{CD}) = W/4(h_2 + h_3)$$

$$S_{BG} = W/2L(h_1 + h_2)$$

$$S_{CF} = W/2L(h_2 + h_3)$$

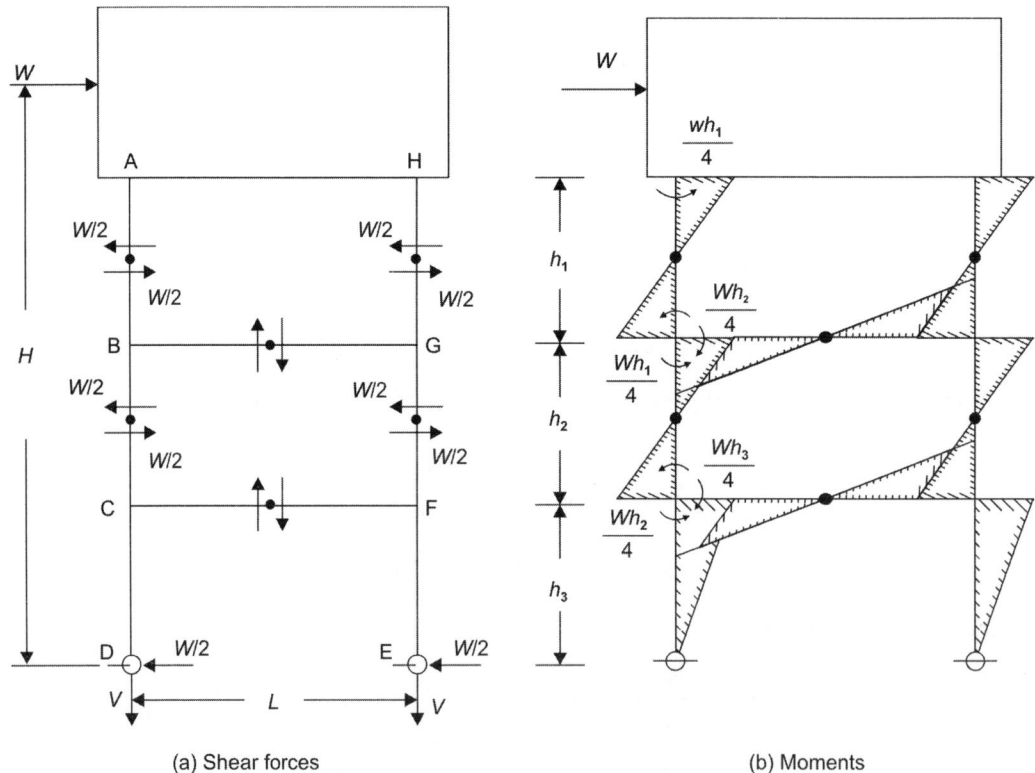

(a) Shear forces   (b) Moments

Fig. 7.6: Columns braced at intervals with hinged base

7.4 WIND LOAD ANALYSIS OF A TOWER WITH CIRCULAR GROUP OF COLUMNS

In the case of overhead circular tanks, the tower consists of a number of columns braced together on the periphery of a circle. The moments and shears in the columns and braces due to wind loads are analysed by assuming contraflexure points at the mid heights of columns between the braces (Fig. 7.7).

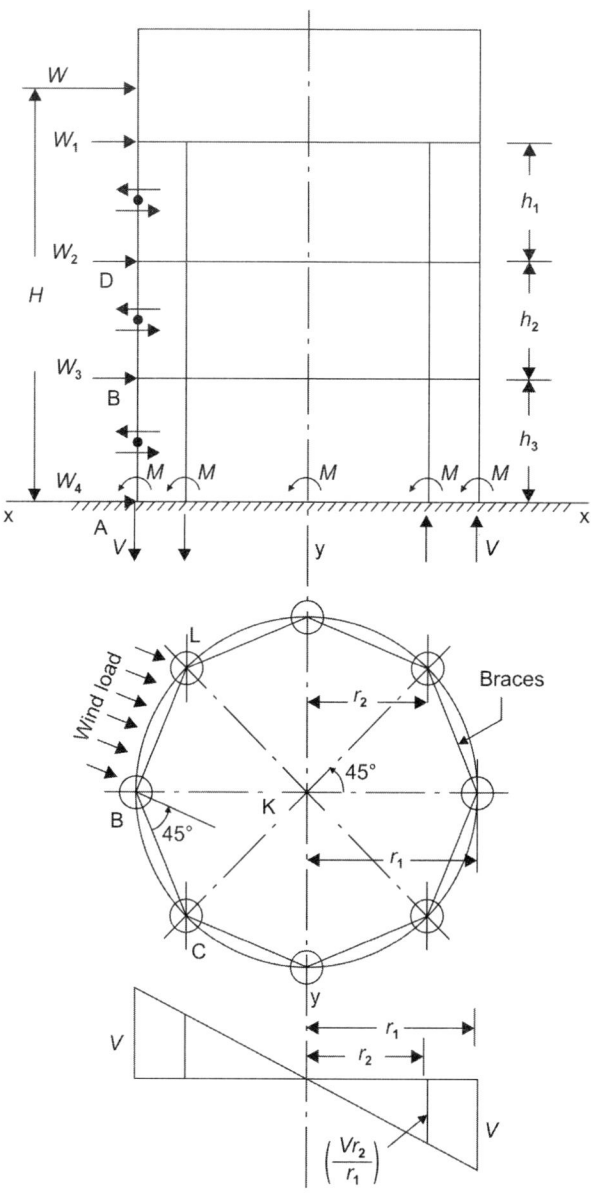

Fig. 7.7: Wind load analysis of circular tower

W, W_1, W_2, W_3 and W_4 are the wind loads acting on the tank and columns at different nodal points. Maximum moments and shears occur in the bottom panel.

Shear in the bottom panel = $(W + W_1 + W_2 + W_3)$

Moment about the base due to wind is calculated as

$$M_1 = WH + W_1(h_1 + h_2 + h_3) + W_2(h_2 + h_3) + W_3h_3$$

Taking moments of all the forces about the axis yy at the base, we have

$$M_1 = \Sigma M + 2Vr_1 + 4\left(V\frac{r_2}{r_1}\right)r_2$$

where,

$$\Sigma M = (W + W_1 + W_2 + W_3)\frac{h_3}{2}$$

V = the vertical reaction developed in the exterior column B, r_1 and r_2 are the distance of the column B and C respectively, measured from the central axis yy.

$\therefore$

$$M_1 = \Sigma M + \frac{V}{r_1}(\Sigma r^2)$$

The maximum moment in brace BC, occurs when wind is blowing normal to brace BL.

Moment at the junction of column in the direction BK = $(M_{BA} + M_{BD})$

Moment in the direction of brace BC is given by

$$M_{BC} = (M_{BA} + M_{BD})\sec 45° = (M_{BA} + M_{BD})\sqrt{2}$$

Shear force in the brace is given by

$$S = \left[\frac{\text{moment in the brace}}{\frac{1}{2}\ (\text{length of brace})}\right]$$

7.5 DESIGN EXAMPLES

1. A rectangular water tank 3 m × 3 m in plan of depth 3 m is supported on a tower of 6 m height. The number of columns is 4. The columns are braced at mid height. The wind pressure on the tank may be taken as 1 kN/m². Assume dead weight of tank is 160 kN. Weight of water in tank is 280 kN. Adopt M20 grade concrete and Fe 415 HYSD bars and design the columns and brace of the supporting tower. The columns are provided with a rigid foundation so that flexity conditions may be assumed at the column base.

   i. *Data*:

   Height of columns = 3 m
   Length of brace = 3 m
   Dead weight of tank = 160 kN
   Weight of water = 280 kN
   Wind pressure = 1 kN/m²
   (M20 grade concrete and Fe 415 HYSD bars, adopted)

ii. *Characteristic strength*:
$$f_{ck} = 20 \text{ N/mm}^2$$
$$f_y = 415 \text{ N/mm}^2$$

iii. *Loads and moments*:

Referring to Fig. 7.8, we have

wind load on tank $= (3 \times 3 \times 1)$
$$= 9 \text{ kN}$$

Loads on columns:

Dead weight of tank = 160 kN

Weight of water = 280 kN

Self weight of columns
$$= (4 \times 0.3 \times 0.3 \times 6 \times 24) = 52 \text{ kN}$$

Self weight of braces
$$= (4 \times 0.3 \times 0.3 \times 3 \times 4) = 26 \text{ kN}$$

Total dead load = 518 kN

Dead load per column $= \left(\dfrac{518}{4}\right)$

$$= 130 \text{ kN}$$

Shear force in each column due to wind $= (9/4) = 2.25$ kN

Bending moment in column
$$= (2.25 \times 1.5) = 3.375 \text{ kN}$$

If V = direct load due to wind, taking moments about B, we have

$(2V \times 3) + (3.375 \times 4) = (9 \times 7.5)$

or $V = 9$ kN

iv. *Design of column section*:

Size of column = 300 mm × 300 mm

Axial service load $P = (130 + 9)$
$$= 139 \text{ kN}$$

Working bending moment $M = 3.375$ kN·m

Design ultimate axial load $P_u = (1.5 \times 139) = 208.5$ kN·m

Design ultimate bending moment $M_u = (1.5 \times 3.375) = 5.0625$ kN·m

Compute the parameters:

$$\left(\frac{P_u}{f_{ck}bD}\right) = \left(\frac{208.5 \times 10^3}{20 \times 300 \times 300}\right) = 0.115$$

$$\left(\frac{M_u}{f_{ck}bD^2}\right) = \left(\frac{5.0625 \times 10^6}{20 \times 300 \times (300)^2}\right) = 0.0093$$

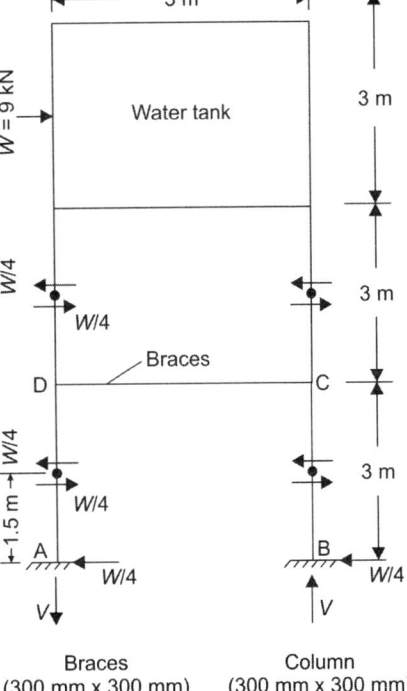

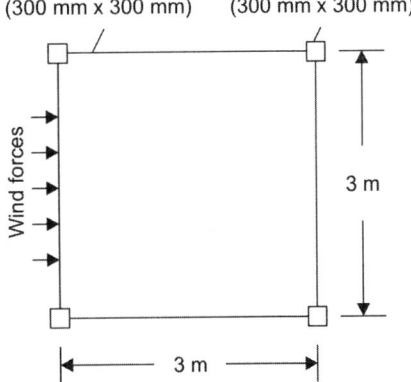

Fig. 7.8: Wind load analysis of water tower

Assuming a cover of 50 mm, the ratio $(d'/D) = (50/300) = 0.16$

Refer to chart 31, SP:16 design charts and readout the percentage reinforcement which is negligible.

Hence, provide minimum reinforcement of 0.8% of the column cross-section as per clause 26.5.3.1(a) of IS:456-2000.

∴ $A_{sc} = (0.008 \times 300 \times 300) = 720$ mm²

Provide 4 bars of 16 mm diameter ($A_{sc} = 804$ mm²)

v. *Lateral reinforcement*:

According to clause 26.5.3.2, IS:456-2000, the diameter and pitch of transverse reinforcement are computed as:

a. Diameter of ties not less than $\left[\dfrac{1}{4} \times 16\right] = 4$ mm and in no case less than 6 mm

b. Pitch shall be the least of the following:
    i. Least lateral dimension of column = 300 mm
    ii. 16 times the diameter of longitudinal bar = $(16 \times 16) = 256$ mm
    iii. 300 mm

Hence, adopt 6 mm diameter ties at 250 mm centres.

vi. *Design of brace*:

Moment in the brace = $(2 \times 2.25 \times 1.5) = 6.75$ kN·m

Ultimate moment $M_u = (1.5 \times 6.75) = 10.125$ kN·m

Shear force in the brace = $\left[\dfrac{\text{moment in brace}}{\text{half length of brace}}\right] = \left(\dfrac{6.75}{1.5}\right) = 4.5$ kN

Ultimate shear force $V_u = (1.5 \times 4.5) = 6.75$ kN

Assuming the cross-section of brace 300 mm × 300 mm, provide minimum reinforcement, since the moment is very small (assuming cover of 30 mm, $d = 270$ mm).

$$A_{st} = \left(\dfrac{0.85\,bd}{f_y}\right) = \left(\dfrac{0.85 \times 300 \times 270}{415}\right) = 166 \text{ mm}^2$$

Provide 2 bars of 12 mm diameter ($A_{st} = 226$ mm²) at top and bottom with a cover of 30 mm.

Nominal shear stress $\tau_v = \left(\dfrac{V_u}{bd}\right) = \left(\dfrac{6.75 \times 10^3}{300 \times 270}\right) = 0.083$ N/mm²

$$\left(\dfrac{100\,A_{st}}{bd}\right) = \left(\dfrac{100 \times 226}{300 \times 270}\right) = 0.279$$

Refer to Table 19, IS:456-2000 and readout the permissible shear stress as

$\tau_c = 0.37$ N/mm² $> \tau_v$

Hence, provide only nominal shear reinforcements. Using 6 mm diameter two-legged stirrups, the spacing is given by the following relation:

$$S_v = \left(\frac{A_{sv}\,0.87\,f_y}{0.4b}\right) = \left(\frac{2 \times 28 \times 0.87 \times 415}{0.4 \times 300}\right) = 168 \text{ mm}$$

Provide 6 mm diameter two-legged stirrups at 160 mm c/c.

2. A circular overhead water tank has a diameter of 10 m and a height of 5 m. The tank is supported on a tower of 13 m height which is braced at intervals of 5 m. The RC columns have a cross-section 300 mm × 500 mm and the braces are 400 mm × 450 mm. The number of columns in the tower is 8. The columns are arranged at intervals of 45°. The total wind loads acting on the tank and tower are given below:

Wind load (kN)	Distance from base (m)
$W = 63$	18
$W_1 = 23$	15
$W_2 = 36$	10
$W_3 = 36$	5

The total dead load due to the self weight of concrete and weight of water acting on each column at the base is equal to 700 kN. If $f_{ck} = 20$ N/mm² and $f_y = 250$ N/mm², design the reinforcements in the column section at base and in the braces assuming raft foundations.

   i. *Data*:

Column section = 300 mm × 500 mm

Brace section = 400 mm × 450 mm

Wind loads: $W = 63$ kN, $W_1 = 23$ kN, $W_2 = W_3 = 36$ kN

$h_1 = h_2 = h_3 = 5$ m $H = 18$ m

Dead load on each column at base = 700 kN

(M20 grade concrete and Fe 250 grade mild steel adopted)

  ii. *Stresses*:

$f_{ck} = 20$ N/mm² $f_y = 250$ N/mm²

 iii. *Loads and moments*:

Referring to Fig. 7.9, we have

M = moment at the base of the columns

$$= (W + W_1 + W_2 + W_3)\frac{h_3}{2}$$

$$= (63 + 23 + 36 + 36)\frac{5}{2} = 395 \text{ kN·m}$$

If M_1 = moment due to wind loads about base

             = $(63 \times 18) + (23 \times 15) + (36 \times 10) + (36 \times 5) = 2019$ kN·m

If V = reaction developed at the base of exterior columns,

$$M_1 = \Sigma M + \frac{V}{r_1}\left[\Sigma r^2\right]$$

$$2019 = 395 + \frac{V}{5}\left[(2 \times 5^2) + 4(5/\sqrt{2})^2\right] \quad \therefore \quad V = 89 \text{ kN}$$

∴   Total load on leeward column at base = $(700 + 82) = 782$ kN

Moment in column at base $= \left(\dfrac{395}{8}\right) = 50$ kN·m

Moments in brace BC

$$= \left[\frac{(63+23+36)}{8} \times \frac{5}{2} + \frac{(63+23+36+36)}{8} \times \frac{5}{2}\right]\sqrt{2} = 124 \text{ kN·m}$$

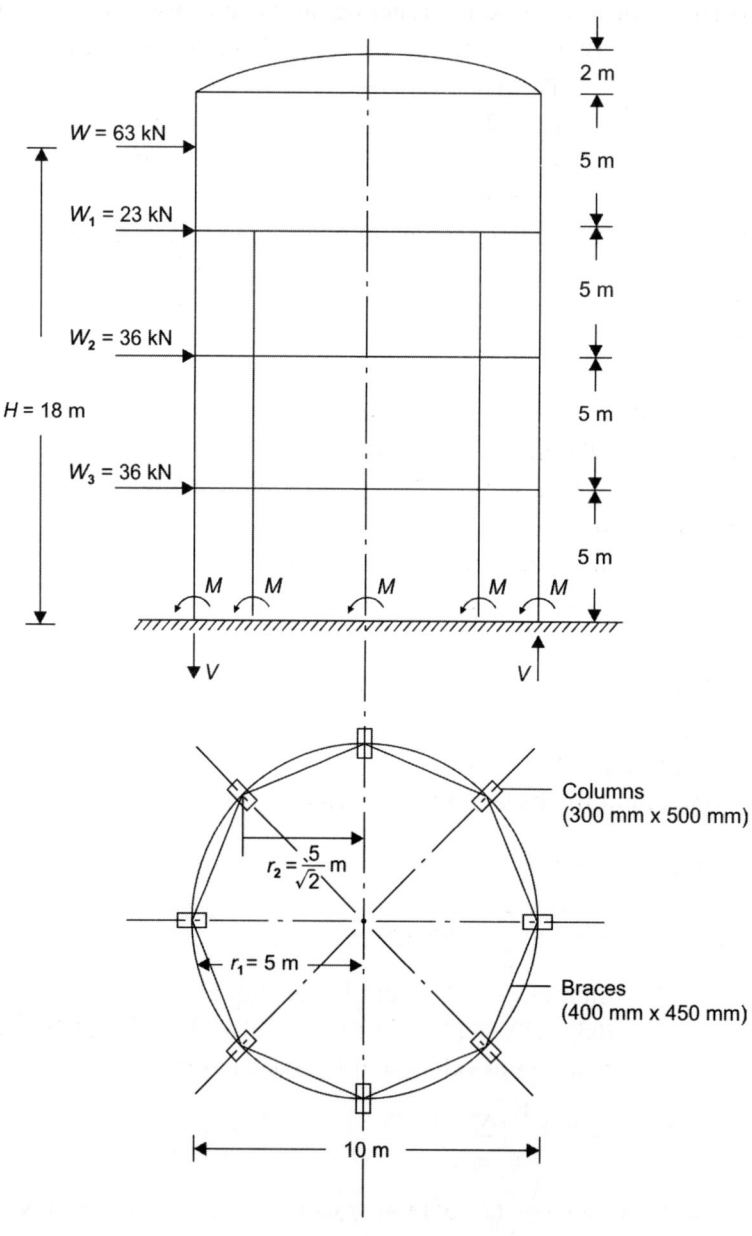

Fig. 7.9: Wind loads on water tank tower

Shear force in brace $= \left(\dfrac{124}{\dfrac{1}{2} \times 3.8}\right) = 65.3$ kN·m

iv. *Design of column section*:

Ultimate load $P_u = (1.7 \times 782) = 1173$ kN

Ultimate moment $M_u = (1.5 \times 50) = 75$ kN·m

$b = 300$ mm $\qquad D = 500 \qquad d' = 50$ mm

Ratio $(d'/D) = 0.10$

$$\left(\frac{P_u}{f_{ck}bD}\right) = \left(\frac{1173 \times 10^3}{20 \times 300 \times 500}\right) = 0.391$$

$$\left(\frac{M_u}{f_{ck}bd^2}\right) = \left(\frac{75 \times 10^6}{20 \times 300 \times 500^2}\right) = 0.05$$

Refer to the interaction diagram, corresponding to

$$f_y = 250 \text{ N/mm}^2 \text{ and } (d'/D) = 0.1$$

(Refer to Chart 28 of SP:16, Design Aids)

$$(p/f_{ck}) = 0.05 \qquad \therefore \qquad p = (0.05 \times 20) = 1\%$$

$$\therefore \qquad A_s = \left(\frac{pbD}{100}\right) = \left(\frac{1 \times 300 \times 500}{100}\right) = 1500 \text{ mm}^2$$

Provide 6 bars of 20 mm diameter distributing three bars on either face.

Transverse reinforcement:

Using 6 mm diameter ties, spacing of the ties is the least of:

a. least lateral dimension = 300 mm

b. $(16 \times 20) = 320$ mm

c. 300 mm

Adopt 6 mm diameter ties at 300 mm centres.

v. *Design of braces*:

$M_u = (1.5 \times 124) = 186$ kN·m

$V_u = (1.5 \times 65.3) = 98$ kN

Section of brace is

$$b = 400 \text{ mm}$$
$$D = 450 \text{ mm}$$
$$d = 400 \text{ mm}$$

$\therefore \qquad M_{u \text{ lim}} = (0.148 f_{ck} bd^2) = (0.148 \times 20 \times 400 \times 400^2)$

$\qquad\qquad = 189.4 \times 10^6$ N·mm $= 189.4$ kN·m

$$\therefore \qquad A_{st} = \left[\frac{0.36 f_{ck} b(0.53 d)}{0.87 f_y}\right] = \left[\frac{0.36 \times 20 \times 400 \times 0.53 \times 400}{0.87 \times 250}\right] = 2807 \text{ mm}^2$$

Use 6 bars of 25 mm diameter on each side ($A_{st} = 2946$ mm^2)

$$\tau_v = \left(\frac{V_u}{bd}\right) = \left(\frac{98 \times 10^3}{400 \times 400}\right) = 0.61 \text{ N/mm}^2$$

$$\left(\frac{100 A_{st}}{bd}\right) = \left(\frac{100 \times 2946}{400 \times 400}\right) = 1.84$$

Refer to Table 19 of IS:456-2000 and readout the permissible shear stress as

$$\tau_c = 0.76 \text{ N/mm}^2 > \tau_v$$

∴ Using 10 mm diameter two-legged stirrups as nominal shear reinforcements, the spacing is given by

$$S_v = \left(\frac{A_{sv} 0.87 f_y}{0.4b}\right) = \left(\frac{2 \times 79 \times 0.87 \times 250}{0.4 \times 400}\right) = 214 \text{ mm}$$

Provide 10 mm diameter two-legged stirrups at 200 mm centres.

References

1. Reynolds CE, Steed Man JC and Threlfall AJ, *Reinforced Concrete Designer's Handbook*, 11 edn, Concrete Publications, London, 2011.

2. Kalani M and Salpekar SA, A comparative study of different methods of analysis of staging of elevated water tanks, *Indian Concrete Journal*, July-August 1978, pp. 210–216.

3. Reddey CS, *Basic Structural Analysis*, Tata McGraw-Hill, New Delhi, 1981, pp. 380–450.

4. Coates RC, Coutie MG and Kong FK, *Structural Analysis*, ELBS, 1975, pp. 180–220.

5. Martin HC, *Introduction to Matrix Methods of Structural Analysis*, McGraw-Hill Book Co, International Student Edition, Kogakusha Co, Japan, 1966, pp. 1–131.

6. Wang CK, *Matrix Methods of Structural Analysis*, 2 edn, (International), Scranton, Pennsylvania, 1970.

7. Argyris JH and Kelsey S, *Energy Theorems and Structural Analysis*, Butterworth Scientific Publications, London, 1960.

8. Meek JL, *Matrix Analysis of Structures*, McGraw-Hill, New York, 1971.

9. Jindall RL, *Indeterminate Structures*, 4 edn, S Chand & Co, New Delhi, 1984, pp. 548–642.

10. Junarkar SB and Shah HJ, *Mechanics of Structures*, Vol. II, 13 edn, Charotar Publishing House, Anand, 1989, pp. 690–724.

Assignment

1. A square water tank 4 m × 4 m × 4 m is to be supported by a four column tower of height 4 m. The columns have independent footings and their base may be considered as hinged. If the dead weight of the tank is 400 kN and weight of water in the tank is 640 kN, design the supporting tower allowing for a wind load of 1.5 kN/m². Adopt M20 grade concrete and Fe 415 grade tor steel.

2. A rectangular water tank 3 m × 4 m with a tank depth of 3 m is supported on a four column tower 6 m in height, braced at the mid height. Assuming the dead weight of tank to be 300 kN and the weight of water as 400 kN, design the columns and braces of the supporting tower. Assume the columns as fixed at the base. Intensity of wind pressure is 1.5 kN/m². Adopt M20 concrete mix and Fe 415 grade tor steel.

3. A circular RC water tank 6 m diameter and 5 m height is supported by a tower consisting of 6 RC columns on a circle of diameter 5 m. The tower height is 8 m with bracing at a height of 4 m from the ground. The tank is designed to hold water up to a depth of 5.5 m. The self weight of the tank is estimated to be 1750 kN. Intensity of wind on tank is 1.5 kN/m². Using M20 grade concrete and Fe 415 grade tor steel, design the supporting tower of the tank.

4. The staging for a water tank of 12 m diameter and 6 m height comprises 12 RC columns arranged concentrically on a circle of 12 m diameter. The height of staging is 12 m with bracing at 3 m intervals. The columns are 450 mm square while the braces are 400 mm square. If the dead weight of the concrete tank is 2250 kN and the weight of water is three times the self weight of tank, design the reinforcements in the column and brace assuming a wind intensity of 1.5 kN/m². The columns may be assumed to be fixed at the base.

5. A reinforced concrete rectangular water tank is 3 m × 6 m with a depth of 3 m. The tank is supported on 6 columns provided with rigid foundation. The dead weight of the tank is 400 kN and the weight of water may be taken as 500 kN. The height of the staging is 6 m with braces at 3 m height above the ground. Assuming a wind pressure of 1.5 kN/m², M20 grade concrete and Fe 415 grade tor steel bars, design the column and brace.

6. A circular overhead water tank has a mean diameter of 12 m and a height of 4 m. The tank is supported on a tower 8 m in height with 8 symmetrically placed columns with a bracing at a height of 4 m from the ground. The tank is designed to hold water up to a depth of 3.5 m. The reinforced concrete columns have a cross-section of 300 mm × 500 mm and the braces are 300 mm × 300 mm. Intensity of wind pressure is 1.5 kN/m². The total dead weight of tank may be assumed as 2000 kN. The columns are fixed at the base of the tank. Adopting M20 grade concrete and Fe 415 HYSD bars, design the columns and braces of the tower.

7. The staging of a circular overhead water tank is 15 m above ground with a diameter of 8 m. The tower is made up of 8 columns symmetrically arranged with bracings arranged at intervals of 3 m. The depth of the tank is 4 m. The intensity of wind at the site is 1.5 kN/m². The total dead weight of tank may be assumed as 2500 kN. The columns may be assumed as fixed at the base resting on a ring beam with a raft foundation. Adopting M25 grade concrete and Fe 500 grade HYSD reinforcements, design the columns and braces for the tank.

8. A circular water tank 12 m in diameter and height 6 m is supported by a tower system comprising eight columns on a circle of diameter 10 m. The height of the tower is 12 m with bracings at intervals of 3 m. The tank is designed to hold water up to depth of 5 m. The self weight of the tank is estimated to be 3000 kN. Intensity of wind acting is 1 kN/m². Adopting M25 grade concrete and Fe 500 HYSD bars, design the tower system of the tank.

Review Questions

1. Explain the necessity of using tower structures with examples. What are the advantages of using towers to support water tanks?

2. Sketch a typical tower structure arrangement for a circular water tank mentioning the various structural components.

3. What is the function of a brace in tower structures? In what way braces help in the stability of the tower skeletal arrangement?

4. List the various forces to be considered in the design of columns in a tower structure.

5. What are the various types of base connections used for the columns? Under what situations would you assume the column base as fixed?

6. Briefly explain the design principles involved in the design of columns and braces in a tower structure.

7. Explain the method of analysing the moments developed in the columns and braces of a circular water tank supported on eight columns.

8. What are the assumptions made in the design of braces in a water tower?

9. What methods would you adopt for designing the columns subjected to direct loads and bending moments?

10. Explain the advantages of using design charts for determining the reinforcements in columns subjected to bending moment and direct loads.

Objective Type Questions

1. Tower structures for water tanks are required to facilitate
   a. easy visibility of the water tank
   b. gravity flow of water
   c. purity of filtered water without pollution

2. Water tank towers are preferably located in higher ground level zones
   a. to cover larger area of distribution by gravity flow
   b. to prevent water logging during failure of staging
   c. to prevent failure during earthquakes

3. To reduce the effective length of the columns in the tower structure
   a. more number of columns are used
   b. larger cross-sections of columns are used
   c. braces are used at suitable intervals

4. At the base of the columns in a water tower structure, moments will develop if the base is
   a. hinged
   b. sliding
   c. fixed

5. The interior columns in a water tower structure resist
   a. the same horizontal forces than that of the exterior columns
   b. less horizontal forces than that of the exterior columns
   c. twice the horizontal force resisted by the exterior columns

6. In the approximate method of analysis, it is assumed that the braces are
   a. hinged to the columns
   b. free to rotate at their supports
   c. stiff and integral with columns

7. Generally the braces in a water tower structure are spaced at intervals of
   a. 4 to 6 m
   b. 1 to 2 m
   c. 2 to 3 m

8. In the approximate analysis of braces, it is assumed that contraflexure points develop at
   a. the end of braces
   b. mid third points
   c. mid points

9. The brace should be designed to resist
   a. bending moment
   b. shear force
   c. bending moment, shear force and self weight

10. The critical section for design in the columns of water tower structure is
    a. just below that of the water tank
    b. at mid height of the tower structure
    c. at the base of the column

8

Elevated Water Tanks

8.1 INTRODUCTION

Overhead water tanks are an essential feature of urban and rural water supply system. Elevated water tanks are generally positioned at higher zones so that the distribution of water is by gravity. In 18th century, steel water tanks were commonly adopted for storage of water. With the development of cement and steel reinforcements and the technology of producing concrete of desired strength and durability, reinforced concrete is widely used for the construction of elevated water tanks. Reinforced concrete elevated water tanks are superior in many respects compared to the steel tanks which have low durability and are liable to deterioration due to rusting. In addition, capital and maintenance costs are higher for steel tanks while it is negligible for reinforced concrete tanks.

At present, reinforced concrete structures have to be designed to conform to the national codes[1,2,3] and most of the codes have incorporated the concepts of limit state design involving that the designed structure should satisfy the limit states of strength and serviceability[4,5,6]. In water tanks, the limit state of cracking is of primary importance to ensure a watertight structure. Advanced computational methods have been evolved in the British code to compute the width of cracks to comply with the serviceability limit states.

8.2 TYPES OF OVERHEAD WATER TANKS

The most common types of overhead water tanks are:
a. Rectangular and circular tanks
b. Intze type tanks[7]
c. Conical or funnel shaped tanks[8]

Circular tanks with a horizontal or flat floor slab are economical for small storage capacity of up to 2 lakh litres. The tank diameter in the range of 4 to 8 m and depth of storage of 3 to 4 m is generally used. Intze type tanks with conical bottom and spherical domes are more economical compared to circular tanks, and hence they are widely used for storing large quantities of water. The conical or funnel shaped tanks, with smaller storage of water are aesthetically superior with the conical tank supported by a hollow cylindrical shaft. In the circular and Intze type tanks, the design of staging or tower is more or less similar. The reader may refer to a separate monograph by the author[9] for the design of circular tanks. Elevated water tanks are usully provided with raft foundations[10].

8.3 INTZE TYPE TANK

In the case of large diameter elevated circular tanks, thicker floor slabs are required resulting in uneconomical designs. In such cases, Intze type tank with conical and bottom spherical domes provides an economical solution. The proportions of the conical and the spherical bottom domes are selected so that the outward thrust from the bottom dome balances the inward thrust due to the conical domed part of the tank floor.

Structural Elements of an Intze Tank

The various structural elements of an Intze type tank comprise the following:
- top spherical dome
- top ring beam
- circular side walls
- bottom ring beam
- conical dome
- bottom spherical dome
- bottom circular girder
- tower with columns and braces
- foundations.

If D is diameter of the tank, the proportions of the various other structural elements in terms of the diameter are shown in Fig. 8.1.

Design Principles of an Intze Tank

Top Spherical Dome
Refer to Fig. 8.2 and use the following notations:

t = thickness of dome generally varies in the range of 75 to 100 mm.

h = rise of dome = $\dfrac{1}{5}$ to $\dfrac{1}{6}D$

R = radius of dome

w = uniformly distributed load per unit area of surface.

The reinforcements in the dome are designed for maximum meridional thrust and circumferential force given by

$$T_1 = \text{meridional thrust} = \left(\frac{wR}{1 + \cos\theta} \right)$$

$$T_2 = \text{circumferential force} = wR\left(\cos\theta - \frac{1}{1 + \cos\theta} \right)$$

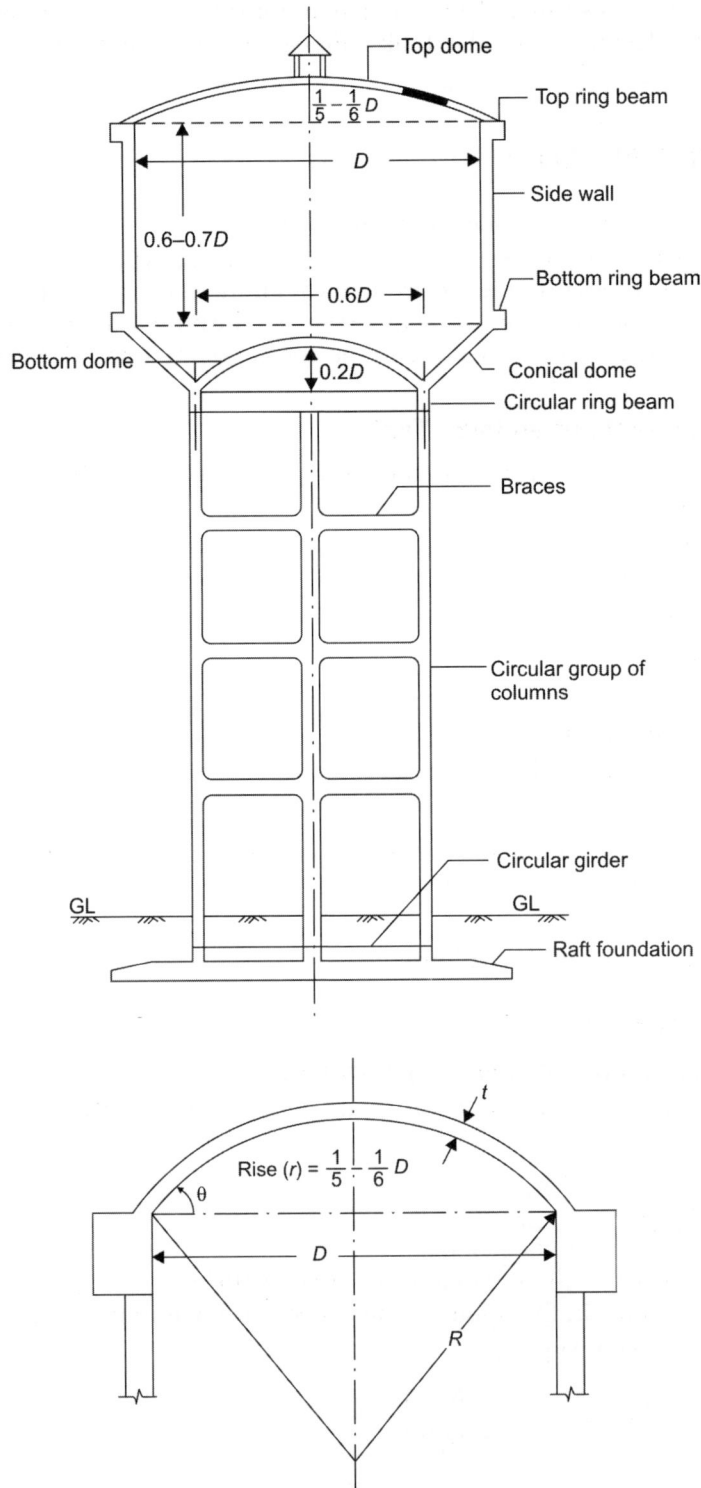

Fig. 8.2: Dimensions of top dome

Top Ring Beam

Hoop tension in ring beam $= \left(\dfrac{T_1 D \cos \theta}{2} \right)$

The cross-sectional area of ring beam is determined by limiting the tensile stress in the ring beam to values specified in IS:456-2000, depending upon the grade of concrete. The tensile stress is calculated by the following equation:

$$f_t = \left(\dfrac{F_t}{A_c + m A_{st}} \right)$$

where,

f_t = direct tensile stress
F_t = direct or hoop tension
A_c = cross-sectional area of concrete
m = modular ratio
A_{st} = cross-sectional area of steel

Side Walls of Tank

The side walls of the tank are designed for hoop tension developed due to the water pressure in the tank.

Maximum hoop tension $= \left(\dfrac{wHD}{2} \right)$

where,

w = density of water (10 kN/m^3)
H = height of vertical walls
D = diameter of tank

A minimum thickness of 150 to 200 mm is provided at the top of the tank and the thickness at the base of the vertical wall is designed by limiting the tensile stress. The spacing of the hoop reinforcement is gradually increased towards the top of the tank. Distribution and temperature reinforcement of 0.3% of the gross-section is provided in the vertical direction.

Bottom Ring Beam

Referring to Fig. 8.3, if

V_1 = weight of roof, side wall and top ring beam per metre run of the ring beam
h = height of water above the ring beam
T = thrust in the conical dome
H = horizontal force developed at the junction
d = depth of ring beam

then for equilibrium of the forces, we have

$$T \sin \theta = V_1$$
$$T \cos \theta = H$$
$$(H/V_1) = \cot \theta$$
$$H = V_1 \cot \theta$$

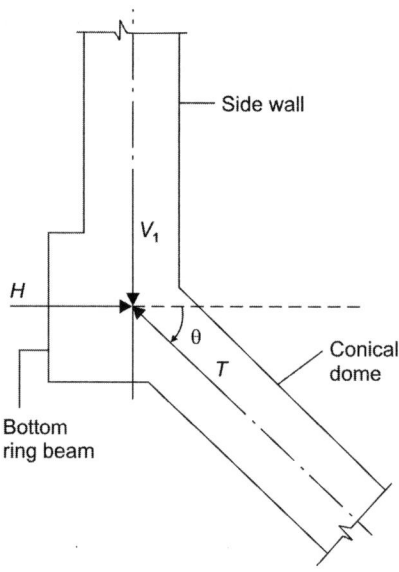

Fig. 8.3: Forces in bottom ring beam

∴ Hoop tension in the beam $= \left(\dfrac{HD}{2} + \dfrac{whdD}{2} \right)$

The reinforcements in the ring beam are designed to resist the hoop tension and the section is designed limiting the tensile stress in concrete.

Conical Dome

Referring to Fig. 8.4, if

V_2 = total load/metre run at the base of the conical dome

T = meridional thrust in the slab due to V_2

H = hoop tension due to water pressure and self weight of conical dome slab

p = intensity of water pressure at a depth h below the water level

q = weight of the conical slab per square metre of the surface area

$θ$ = angle made by the conical slab with the horizontal

D = diameter at a depth h from top

then the meridional thrust and hoop tension in the conical dome is computed using the following equations:

$$T = V_2 \operatorname{cosec} θ$$
$$H = (p \operatorname{cosec} θ + q \cot θ) \, D/2$$

The reinforcements in the conical dome are designed for hoop tension and meridional thrust.

Bottom Spherical Dome

The design of the floor dome is similar to that of the top dome. The design load for the dome includes the self weight of the dome and the weight of water column above the

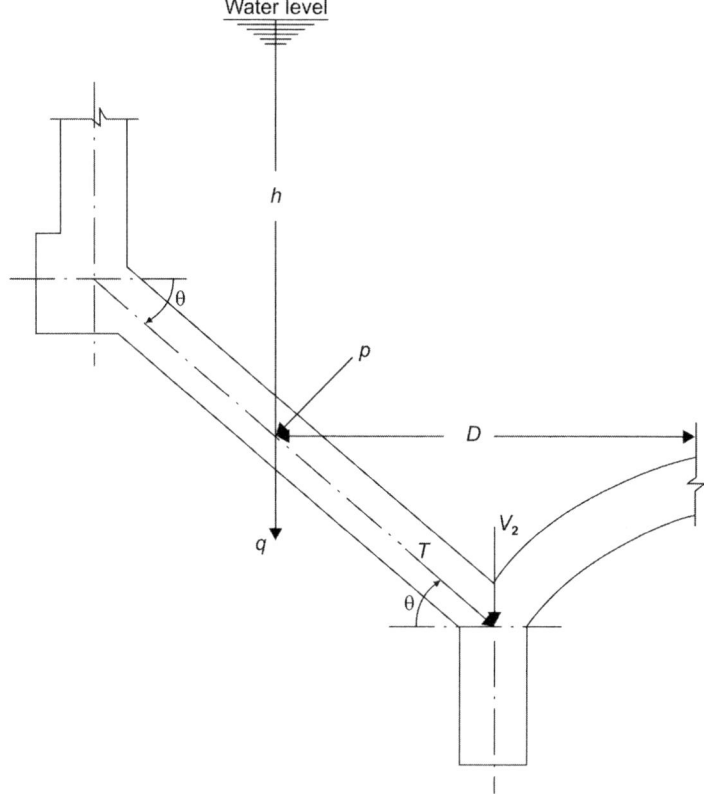

Fig. 8.4: Forces acting on conical dome

dome. The dome reinforcements are designed for meridional thrust and circumferential forces.

Bottom Circular Girder

Refer to Fig. 8.5 and use the following notations:

T_1 = thrust in the direction of the conical dome, acting at an angle α to the horizontal

T_2 = thrust from the bottom spherical dome, acting at an angle β to the horizontal

P = net horizontal force on the ring beam

If $T_1 \cos \alpha > T_2 \cos \beta$, the ring beam is subjected to a compressive force. The magnitude of this compressive force is negligibly small in well proportioned tanks.

The vertical load on the ring beam is obtained by the relation, $(T_1 \sin \alpha + T_2 \sin \alpha)$ or alternatively by dividing the total vertical loads by the perimeter of the bottom ring beam.

The ring beam is supported by a number of columns equally spaced along the periphery of the circle. Depending upon the number of columns, moment coefficients compiled in Table 4.1 (Refer to Chapter 4) are useful in computing the maximum bending and torsional moments in the circular girder. The procedure outlined in Chapter 4 can be used to design the reinforcements in the ring beam.

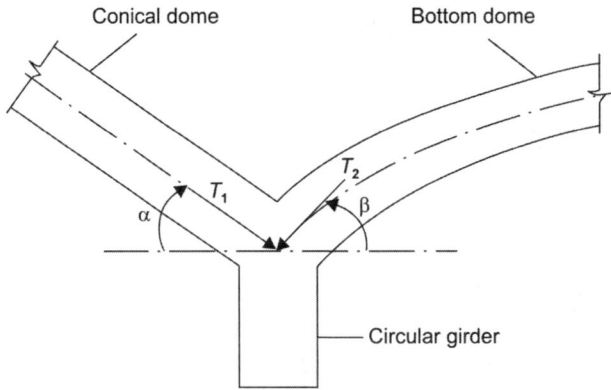

Fig. 8.5: Forces acting on circular girder

Tower with Columns and Braces

The procedure explained in Section 7.4 is useful in the design of a tower with circular group of columns braced together at regular intervals. The columns and braces are designed for the maximum forces and moments developed due to the dead and wind loads acting on the tower.

Foundations

The foundations for a circular group of columns generally comprise a ring beam with raft slab. The ring beam is designed for maximum bending and torsional moments while the annular raft slab is designed for maximum soil pressure from the bottom. The design procedure is illustrated by the following example.

8.4 ▌ DESIGN EXAMPLE OF INTZE TYPE WATER TANK

Design an Intze type water tank of 1 million litres capacity, supported on an elevated tower comprising eight columns. The base of the tank is 16 m above ground level. Depth of foundations is 1 m below ground level. Adopt M20 grade concrete and Fe 415 grade tor steel. The design of the tank should conform to the stresses specificed in IS:3370 and IS:456.

   i. *Data*:

     Capacity of tank = 1 million litres = 1000 m^3

     Height of supporting tower = 16 m

     Number of columns = 8

     Depth of foundations = 1 m below ground level.

  ii. *Permissible stresses*:

     M20 grade concrete and Fe 415 grade tor steel for calculations related to resistance to cracking (IS:3370):

$$\sigma_{ct} = 1.2 \, \text{N/mm}^2 \qquad \sigma_{cb} = 1.7 \, \text{N/mm}^2 \qquad \sigma_{st} = 150 \, \text{N/mm}^2$$

     For strength calculations, the stresses in concrete and steel are same as that recommended in IS:456.

$$\sigma_{cc} = 5 \text{ N/mm}^2 \qquad\qquad m = 13$$
$$\sigma_{cb} = 7 \text{ N/mm}^2 \qquad\qquad Q = 0.897$$
$$j = 0.906$$

iii. *Dimensions of tank*:

Referring to Fig. 8.1, let D be the inside diameter of the tank. Assuming the average depth = $0.75D$, we have

$$\left(\frac{\pi D^2}{4} \times 0.75 D \right) = 1000 \text{ m}^3$$

$\therefore \qquad\qquad D = 12 \text{ m}$

Height of cylindrical portion of tank = 8 m

Depth of conical dome = 2 m

Diameter of the supporting tower = 8 m

Spacings of bracing = 4 m

The salient dimensions of the tank and the staging are shown in Fig. 8.6.

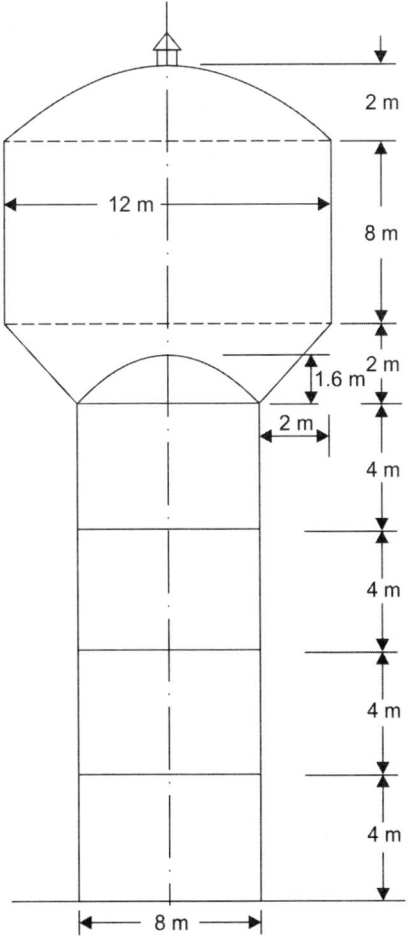

Fig. 8.6: Salient dimensions of Intze tank

iv. *Design of top dome*:

Thickness of dome slab $t = 100$ mm

Live load on dome $= 1.5$ kN/m²

Self weight of dome $= (0.1 \times 24) \quad = 2.4$ kN/m²

Live load $= 1.5$

Finishes $= 0.1$

Total load $w = 4.0$ kN/m²

Diameter at base $D = 12$ m

Central rise $r = (1/6 \times 12) = 2$ m

$$\text{Radius of the dome, } R = \left[\frac{(D/2)^2 + r^2}{2r}\right] = \left[\frac{6^2 + 2^2}{2 \times 2}\right] = 10 \text{ m}$$

$$\cos\theta = (8/10) = 0.8$$
$$\theta = 36°50'$$

$$\text{Meridional thrust } T_1 = \left(\frac{wR}{1 + \cos\theta}\right) = \left(\frac{4 \times 10}{1 + 0.8}\right) = 22.22 \text{ kN/m}$$

$$\text{Circumferential force} = wR\left[\cos\theta - \frac{1}{1 + \cos\theta}\right] = 4 \times 10\left[0.8 - \frac{1}{1.8}\right] = 10 \text{ kN/m}$$

$$\text{Meridional stress} = \left(\frac{22.22 \times 10^3}{1000 \times 100}\right) = 0.22 \text{ N/mm}^2 < 5 \text{ N/mm}^2$$

$$\text{Hoop stress} = \left(\frac{10 \times 10^3}{1000 \times 100}\right) = 0.10 \text{ N/mm}^2 < 5 \text{ N/mm}^2$$

Hence the stresses are within safe limits.

Provide nominal reinforcements of 0.3%

$$A_{st} = \left(\frac{0.3 \times 100 \times 1000}{100}\right) = 300 \text{ mm}^2$$

Provide 8 mm diameter at 160 mm centres both circumferentially and meridionally.

v. *Design of the top ring beam*:

$$\text{Hoop tension } F_t = \left(\frac{T_1 D \cos\theta}{2}\right) = \left(\frac{22.22 \times 0.8 \times 12}{2}\right) = 106.6 \text{ kN}$$

$$A_{st} = \left(\frac{106.6 \times 10^3}{150}\right) = 710 \text{ mm}^2$$

Provide 8 bars of 12 mm diameter ($A_{st} = 904$ mm²)

If A_c = cross-sectional area of ring beam, then

$$\left(\frac{106.6\times10^3}{A_c +13\times904}\right)= 1.2 \qquad \therefore A_c = 77082 \text{ mm}^2$$

Provide 300 mm × 300 mm size top ring beam with 8 bars of 12 mm diameter as main reinforcement and 6 mm diameter stirrups at 200 mm centres.

vi. *Design of cylindrical tank wall:*

Maximum hoop tension at base of wall $F_t =\left(\dfrac{whD}{2}\right)$

where,

   w = density of water
   h = depth of water

$\therefore \qquad\qquad F_t =\left(\dfrac{10\times8\times12}{2}\right)= 480 \text{ kN/m}$

$$A_{st} =\left(\frac{410\times10^3}{150}\right)= 3200 \text{ mm}^2/\text{m height}$$

Provide 20 mm diameter at 180 mm centres on each face (A_{st} = 3492 mm²). Steel area required at 2 m below top is A_{st} = (2/8 × 3200) = 800 mm².
Provide 10 mm diameter at 180 mm centres on each face
If t = thickness of side wall at bottom, then

$$\left[\frac{480\times10^3}{1000t +(13\times3492)}\right]= 1.2$$

$\therefore \qquad\qquad t = 358 \text{ mm}$

Adopt 400 mm thick walls at bottom gradually reducing to 200 mm at top.
Distribution steel:

At bottom, $A_{st} =\left[\dfrac{0.2\times400\times1000}{100}\right]= 800 \text{ mm}^2$

Provide 10 mm diameter at 10 mm centres on both faces.

At top, $A_{st} =\left(\dfrac{0.3\times200\times1000}{100}\right)= 600 \text{ mm}^2$

Provide 10 mm diameter at 250 mm centres on both faces.
The details of reinforcements provided in the cylindrical tank walls at different height are as follows:

Distance from top	Main hoop steel (each face)	Vertical distribution steel (each face)
0–2 m	10–180 mm c/c	10–250 mm c/c
2–4 m	16–200 mm c/c	10–250 mm c/c
4–8 m	20–180 mm c/c	10–180 mm c/c

vii. *Design of bottom ring beam*:

Loads on ring beam are as follows:

a. Load due to top dome = (meridional thrust × sin θ)
$$= (22.22 \times \sin 36°50') = 13.3 \text{ kN/m}$$

b. Load due to top ring beam = (0.3 × 0.3 × 24) = 2.16 kN/m

c. Load due to cylindrical wall = $\left(\dfrac{0.4+0.2}{2}\right) \times 8 \times 24 = 57.6$ kN/m

d. Self weight of ring beam (assuming a section 1.2 m × 0.6 m)
$$= (1.2 \times 0.6 \times 24) = 17.28 \text{ kN/m}$$

∴ Total vertical load $V_1 = 91$ kN/m

Horizontal force $H = V_1 \cot\theta = (91 \times \cot 45°) = 91$ kN

∴ Hoop tension due to vertical loads

$$H_g = \left(\frac{HD}{2}\right) = \left(\frac{91 \times 12}{9}\right) = 546 \text{ kN}$$

Hoop tension due to water pressure

$$H_w = \left(\frac{whdD}{2}\right) = \left(\frac{10 \times 8 \times 0.6 \times 12}{2}\right) = 288 \text{ kN}$$

∴ Total hoop tension = $(H_g + H_w) = (546 + 288) = 843$ kN

∴ $$A_{st} = \left(\frac{834 \times 10^3}{150}\right) = 5560 \text{ mm}^2$$

Provide 18 bars of 20 mm ($A_{st} = 5562$ mm²)

Max. tensile stress = $\left[\dfrac{834 \times 10^3}{(1200 \times 600) + (13 \times 5652)}\right] = 1.05$ N/mm² < 1.2 N/mm²

Provide a ring beam 1200 mm wide by 600 mm deep with 18 bars of 20 mm diameter and distribution bars of 10 mm diameter from cylindrical wall taken round the main bars as stirrups at 180 mm centres.

viii. *Design of conical dome*:

Average diameter of conical dome = $\left(\dfrac{12+8}{2}\right) = 10$ m

Average depth of water = (8 + 2/2) = 9 m

Weight of water above conical dome = (π × 10 × 9 × 2 × 10) = 5655 kN

Assume 600 mm thick slab

Self weight of slab = (π × 10 × 2.83 × 0.6 × 24) = 1280 kN

Load from top dome, top ring beam, cylindrical wall and bottom ring beam
$$= (\pi \times 12 \times 91) = 3430 \text{ kN}$$

∴ Total load at base of conical slab = (5655 + 1280 + 3430) = 10365 kN

Load/unit length $V_2 = \left(\dfrac{10365}{\pi \times 8}\right) = 413$ kN/m

Meridional thrust $T = V_2 \cosec\theta = (413 \times \cosec 45°) = 584$ kN

Meridional stress $= \left(\dfrac{584 \times 10^3}{600 \times 1000}\right) = 0.973$ N/mm² < 5 N/mm² (safe)

Hoop tension in conical dome will be maximum at the top of the conical dome slab since diameter D is maximum at this section.

Water pressure $p = (10 \times 8) = 80$ kN/m²

Weight of conical dome slab per m² is computed as follows:

$$q = (0.6 \times 24) = 14.4 \text{ kN/m}^2$$
$$\theta = 45°$$
$$D = 12 \text{ m}$$

Hoop tension $H = (p\cosec\theta + q\cot\theta)\,D/2$

∴ $\qquad H = (80 \times \cosec 45° + 14.4 \times \cot 45°)\,12/2 = 765$ kN

∴ $\qquad A_{st} = \left(\dfrac{765 \times 10^3}{150}\right) = 5100$ mm²

Provide 25 mm diameter at 180 mm centres ($A_{st} = 5470$ mm²) on both faces of the slab.

Distribution steel $= \left(\dfrac{0.2 \times 600 \times 1000}{100}\right) = 1200$ mm²

Provide 10 mm diameter at 130 mm centres on both faces along the meridions.

Max. tensile stress $= \left(\dfrac{765 \times 10^3}{(600 \times 1000) + (13 \times 5470)}\right) = 1.13$ N/mm² < 1.2 N/mm² (safe)

ix. *Design of bottom spherical dome*:

Thickness of dome slab is assumed as 300 mm.

Diameter at base $D = 8$ m

Central rise $r = (1/5 \times 8) = 1.6$ m

If R = radius of the dome, then

$$(2R - r)r = (D/2)^2$$
$$(2R - 1.6)1.6 = 4^2$$

∴ $\qquad R = 5.8$ m

Self weight of dome slab $= (2 \times \pi \times 5.8 \times 1.6 \times 0.3 \times 24) = 420$ kN

Weight of finishes and fittings etc. $= 300$ kN

Volume of water above the dome

$$= \pi \times 4^2 (8+2) - \left[\frac{2\pi \times 5.8^2 \times 1.6}{3} - \frac{\pi \times 4^2}{3}(5.8 - 1.6)\right] = 460 \text{ mm}^3$$

∴ Weight of water $= (460 \times 10) = 4600$ kN

∴ Total load on dome $= (420 + 300 + 4600) = 5320$ kN

Load/unit area $w = \left(\dfrac{5320}{\pi \times 4^2}\right) = 106 \text{ kN/m}^2$

$\cos\theta = \left(\dfrac{4.2}{106 \times 5.8}\right) = 0.724 \qquad \therefore \quad \theta = 44.5°$

Meridional thrust $T_1 = \left(\dfrac{wR}{1+\cos\theta}\right)$

$\therefore \qquad T_1 = \left(\dfrac{106 \times 5.8}{1+0.724}\right) = 357 \text{ kN/m}$

Meridional stress $= \left(\dfrac{357 \times 10^3}{300 \times 1000}\right) = 1.19 \text{ N/mm}^2 \text{ (safe)}$

Circumferential force $= wR\left(\cos\theta - \dfrac{1}{1+\cos\theta}\right)$

$= (106 \times 5.8)\left(0.724 - \dfrac{1}{1.724}\right) = 88.5 \text{ kN/m}$

$\therefore \qquad$ Hoop stress $= \left(\dfrac{88.5 \times 10^3}{300 \times 1000}\right) = 0.3 \text{ N/mm}^2 \text{ (safe)}$

Provide nominal reinforcement of 0.3%

$A_{st} = \left(\dfrac{0.3 \times 300 \times 1000}{100}\right) = 900 \text{ mm}^2$

Provide 12 mm at 120 mm centres circumferentially and along the meridions.

x. *Design of bottom circular girder*:

Thrust from conical dome $T_1 = 413$ kN/m acting at an angle $\alpha = 45°$ to the horizontal.

Thrust from spherical dome $T_2 = 357$ kN/m acting at an angle $\beta = 44.5°$ to the horizontal.

Net horizontal force on ring beam $= (T_1 \cos\alpha - T_2 \cos\beta)$
$$= [(413 \times 0.707) - (357 \times 0.713)] = 38 \text{ kN/m}$$

Hoop compression in the beam $= \left(\dfrac{38 \times 8}{2}\right) = 152 \text{ kN}$

Assume the size of the ring girder at 600 mm wide by 1200 mm deep.

Hoop stress $= \left(\dfrac{152 \times 10^3}{600 \times 1200}\right) = 0.21 \text{ N/mm}^2 \text{ (safe)}$

Vertical load on ring beam $= (T_1 \sin\alpha + T_2 \sin\beta)$
$$= [(413 \times 0.707) + (357 \times 0.70)] = 542 \text{ kN/m}$$

Self weight of beam = $(0.6 \times 1.2 \times 24) = 17.28$ kN/m

$\therefore$ Total load = $(542 + 17.28) = 560$ kN/m

Total design load on the ring girder = $W = (\pi D w) = (\pi \times 8 \times 560) = 14074$ kN

The circular girder is supported on 8 columns. Using the moment coefficients given in Table 4.1, we have

maximum negative BM at support section $= 0.0083WR$

$$= (0.0083 \times 14074 \times 4) = 467 \text{ kN·m}$$

maximum positive BM at mid span section $= 0.0041WR$

$$= (0.0041 \times 14074 \times 4) = 231 \text{ kN·m}$$

But minimum area of steel in the section $= \left(\dfrac{0.3 \times 600 \times 1200}{100} \right) = 2160 \text{ mm}^2$

Provide 5 bars of 25 mm at mid span section and stirrups 10 mm four-legged at 300 mm centres. Design of section subjected to maximum torsion is as follows:

$T = 34$ kN·m	$D = 1200$ mm
$V = 521$ kN	$b = 600$ mm
$M = 0$	$d = 1150$ mm

$$M_t = T \left[\frac{1 + (D/b)}{1.7} \right] = 34 \left[\frac{1 + (1200/600)}{1.7} \right] = 60 \text{ kN·m}$$

$$M_{el} = (M + M_t) = (0 + 60) = 60 \text{ kN·m}$$

$$A_{st} = \left(\frac{60 \times 10^6}{150 \times 0.9 \times 1150} \right) = 387 \text{ mm}^2$$

But minimum area of tension steel is

$$A_{st} = \left(\frac{0.3 \times 600 \times 1200}{100} \right) = 2160 \text{ mm}^2$$

Provide 5 bars of 25 mm diameter ($A_{st} = 2455$ mm^2)

Equivalent shear force $V_e = [V + 1.6(T/b)] = \left(521 + 1.6 \times \dfrac{34}{0.6} \right) = 612$ kN

$$\tau_{ve} = \left(\frac{V_e}{bd} \right) = \left(\frac{612 \times 10^3}{600 \times 1150} \right) = 0.88 \text{ N/mm}^2$$

$$\left(\frac{100 A_{st}}{bd} \right) = \left(\frac{100 \times 2455}{600 \times 1150} \right) = 0.32$$

$\therefore$ $\tau_c = 0.24$ N/mm^2

Since $\tau_v > \tau_c$, shear reinforcements are required.

Using 12 mm diameter four-legged stirrups with side covers of 25 mm and top and bottom covers of 50 mm, spacing is given as follows:

$$S_v = \left[\frac{A_{sv}\sigma_y}{(\tau_{ve}-\tau_c)b}\right] = \left[\frac{4\times113\times150}{(0.88-0.24)600}\right] = 188 \text{ mm}$$

Adopt 12 mm diameter four-legged stirrups at 180 mm centres.

xi. *Design of columns of supporting tower*:

The supporting tower comprises 8 columns equally spaced on a circle of 8 m diameter.

a. Vertical load on each column = $\left(\dfrac{14074}{8}\right)$ = 1760 kN

b. Self weight of column of height 16 m and diameter 650 mm
$$= (\pi/4 \times 0.65^2 \times 16 \times 24) = 127 \text{ kN}$$

c. Self weight of bracings (three numbers of 4 m intervals, size of bracing is 500 mm × 500 mm) = (3 × 0.5 × 0.5 × π/8 × 24) = 57 kN

Total vertical load on each column = (1760 + 127 + 57) = 1944 kN

Intensity of wind pressure = 1.5 kN/m²

Reduction coefficient for circular shapes = 0.7

Wind forces on columns are as follows:

a. Wind force on top dome and cylindrical wall
$$= (8 + 2/2) \times 0.7 \times 1.5 \times 12 = 114 \text{ kN}$$

b. Wind force on conical dome = (2 × 10 × 0.7 × 1.5) = 21 kN

c. Wind force on bottom ring beam = (1.2 × 8 × 0.7 × 1.5) = 11 kN

d. Wind force on five columns = (5 × 0.65 × 16 × 0.7 × 1.5) = 55 kN

e. Wind force on bracings = (0.5 × 8 × 3 × 1.5) = 18 kN

∴   Total horizontal wind force = (114 + 21 + 11 + 55 + 18) = 219 kN

Assuming contraflexure points at mid height of columns and fixity at the base due to raft foundations, the moment at the base of columns is computed as follows:

$$M = \left(\frac{219\times4}{2}\right) = 438 \text{ kN·m}$$

M_1 = moment at the base of the column due to wind loads
$$= (114 \times 23) + (21 \times 17) + (11 \times 16) + (6 \times 12) + (6 \times 8) + (6 \times 4)$$
$$= 3299 \text{ kN·m}$$

V = reaction developed at the base of exterior columns

∴ $$M_1 = \Sigma M + \frac{V}{r_1}[\Sigma r^2]$$

$$3299 = 438 + \frac{V}{4}\left[(2\times4^2)+4(4/\sqrt{2})^2\right]$$

∴ $V = 179$ kN

∴   Total load on leeward column at base = (1944 + 179) = 2123 kN

Moment in each column at base = $\left(\dfrac{438}{8}\right) = 55$ kN·m

Design ultimate moment in each column $M_u = (1.5 \times 55) = 82.5$ kN·m

Design ultimate axial load $P_u = (1.5 \times 2123) = 3184.5$ kN

Compute the parameters as follows:

$$\left(\frac{P_u}{f_{ck} D^2} \right) = \left(\frac{3184.5 \times 10^3}{20 \times 650^2} \right) = 0.376$$

$$\left(\frac{M_u}{f_{ck} D^3} \right) = \left(\frac{82.5 \times 10^6}{20 \times 650^2} \right) = 0.015$$

Refer to Chart 56 (SP:16) for circular columns with compression and bending and eight longitudinal bars and the ratio $(d'/D) = 0.10$. The corresponding percentage reinforcement is readout as follows:

$$\left(\frac{p}{f_{ck}} \right) = 0.02$$

$\therefore$ $\qquad\qquad p = (20 \times 0.02) = 0.4\%$

But minimum reinforcement in columns = 0.8%

Providing 1% reinforcement, we have

$$A_{sc} = \left(\frac{p\pi D^2}{400} \right) = \left(\frac{1 \times \pi \times 650^2}{400} \right) = 3318 \text{ mm}^2$$

Provide 8 bars of 25 mm diameter ($A_{sc} = 3927$ mm²)

Diameter of lateral ties not less than $[(1/4) \times 25] = 6.25$ mm

Adopt 8 mm diameter lateral ties

Pitch of lateral ties shall be the least of:

    a. Least lateral dimension = 650 mm

    b. $(16 \times 25) = 400$ mm

    c. 300 mm

Hence, adopt 8 mm diameter lateral ties at 300 mm centres.

xii. *Design of bracings*:

Service moment in brace, $M = (2 \times \text{moment in column} \times \sqrt{2})$

$\qquad\qquad\qquad = (2 \times 55 \times \sqrt{2}) = 156$ kN·m

Design ultimate moment $M_u = (1.5 \times 156) = 234$ kN·m

Section of brace = 500 mm × 500 mm

$\therefore$ $b = 500$ mm and effective depth $d = 450$ mm

Limiting moment of resistance of the section is computed as

$\qquad M_{u, lim} = (0.138 f_{ck} bd^2)$

$\qquad\qquad = (0.138 \times 20 \times 500 \times 450^2) \, 10^{-6}$

$\qquad\qquad = 279$ kN·m $< M_u$

Hence the section is under-reinforced.

Compute the parameter as

$$\left(\frac{M_u}{bd^2}\right) = \left(\frac{234\times10^6}{500\times450^2}\right) = 2.3$$

Refer to Table 2 of SP:16 and readout the percentage reinforcement as

$$P_t = \left(\frac{100\,A_{st}}{bd}\right) = 0.757$$

$$\therefore \qquad A_{st} = \left(\frac{0.757\times500\times450}{100}\right) = 1703 \text{ mm}^2$$

Provide 4 bars of 25 mm diameter (A_{st} = 1964 mm²) both at top and bottom since wind direction is reversible.

Length of brace $L = (2 \times 4 \times \sin 22.5°) = 3.06$ m

Maximum service load shear force in brace is computed as

$$V = \left(\frac{\text{moment in brace}}{\text{half length of brace}}\right) = \left(\frac{156}{0.5\times3.06}\right) = 102 \text{ kN}$$

Design ultimate shear force $V_u = (1.5 \times 102) = 153$ kN

$$\therefore \qquad \tau_v = \left(\frac{V_u}{bd}\right) = \left(\frac{153\times10^3}{500\times450}\right) = 0.88 \text{ N/mm}^2$$

$$\left(\frac{100\,A_{st}}{bd}\right) = \left(\frac{100\times1964}{500\times450}\right) = 0.87$$

Refer to Table 19 of IS:456-2000 and readout the permissible shear stress as

$$\tau_c = 0.59 \text{ N/mm}^2$$

Since $\tau_v > \tau_c$, shear reinforcements are required.

Shear force carried by concrete = $(\tau_c bd)$ = $(0.59 \times 500 \times 450) \times 10^{-3}$ = 133 kN

Balance shear force = $(153 - 133)$ = 20 kN

Using 10 mm diameter two-legged stirrups, spacing is computed as

$$S_v = \left[\frac{0.87\times415\times2\times79\times450}{20\times10^3}\right] = 1283 \text{ mm}$$

But S_v is not greater than $0.75d$ or 300 mm whichever is less.

$$\therefore \qquad S_v = (0.75 \times 450) = 337.5 \text{ mm} > 300 \text{ mm}$$

Hence, provide 10 mm diameter two-legged stirrups at 300 mm centres.

xiii. *Design of foundations*:

A circular girder with a raft slab is provided for the tower foundations.

Total load on foundations = (1944×8) = 15552 kN

Self weight of foundations at 10%	= 1555
Total load	= 17107 kN
Safe bearing capacity of soil at site	= 250 kN/m²

$\therefore$ Area of foundation $= \left(\dfrac{17107}{250} \right) = 68.4 \text{ m}^2$

Providing a raft slab with equal projections on either side of a circular ring beam and if b = width of raft slab, then

$$(\pi \times 8 \times b) = 68.4$$

$\therefore$ $\qquad\qquad\qquad\qquad b = 2.72 \text{ m}$

Adopt a raft slab having 5 m inner diameter and 11 m outer diameter (b = 3 m)

Design of circular girder of raft slab:

Total load on circular girder = W = 15552 kN

Load per metre run on girder $= \left(\dfrac{15552}{\pi \times 8} \right) = 618.8 \text{ kN/m}$

Referring to moment coefficients given in Table 4.1, the maximum moments in the circular girder is computed.

Maximum service load negative moment at support

$\qquad\qquad = (0.0083 WR) = (0.0083 \times 15552 \times 4) = 516 \text{ kN·m}$

Design ultimate negative moment $M_u(-ve) = (1.5 \times 516) = 774 \text{ kN·m}$

Maximum service positive moment at mid span

$\qquad\qquad = (0.0041 WR) = (0.0041 \times 15552 \times 4) = 255 \text{ kN·m}$

Design ultimate positive moment $M_u(+ve) = (1.5 \times 255) = 382.5 \text{ kN·m}$

Maximum service torsional moment (at 9.5° from either support)

$\qquad\qquad = (0.0006 WR) = (0.0006 \times 15552 \times 4) = 37.3 \text{ kN·m}$

Design ultimate torsional moment $T_u = (1.5 \times 37.3) = 56 \text{ kN·m}$

Working shear force at support section

$$V = \left[\frac{618.8 \times 4 \times (\pi/4)}{2} \right] = 972 \text{ kN}$$

Design ultimate shear force at support section $V_u = (1.5 \times 972) = 1458 \text{ kN}$

Working shear force at section of maximum torsion

$$V = \left[972 - \left(\frac{618 \times \pi \times 4 \times 9.5}{180} \right) \right] = 562 \text{ kN}$$

Design ultimate shear force = $(1.5 \times 562) = 842 \text{ kN}$

The support section is designed for a maximum negative moment given by

$\qquad\qquad M_u = 774 \text{ kN·m and shear force } V_u = 1458 \text{ kN}$

Assume width of girder section, b = 750 mm to house the column of diameter 650 mm.

Effective depth $= d = \sqrt{\dfrac{M_u}{0.138\, f_{ck} b}} = \sqrt{\dfrac{774 \times 10^6}{(0.138 \times 20 \times 750)}} = 612 \text{ mm}$

The depth required for shear force is nearly 1.5 times that required for moment. Hence, adopt effective depth d = 1000 mm with effective cover of 70 mm.

Compute the parameter as

$$\left(\frac{M_u}{bd^2}\right) = \left(\frac{774 \times 10^6}{750 \times 930^2}\right) = 1.19$$

Refer to Table 2 of SP:16 and readout the percentage of reinforcement as $p_t = 0.356$

$$\therefore \qquad \left(\frac{100 A_{st}}{bd}\right) = 0.356$$

$$A_{st} = \left(\frac{0.356 \times 750 \times 930}{100}\right) = 2483 \text{ mm}^2$$

Provide 6 bars of 25 mm diameter ($A_{st} = 2946$ mm²)

$$\tau_v = \left(\frac{V_u}{bd}\right) = \left(\frac{1458 \times 103}{750 \times 930}\right) = 2.09 \text{ N/mm}^2 < \tau_{c, max} = 2.8 \text{ N/mm}^2$$

$$\left(\frac{100 A_{st}}{bd}\right) = \left(\frac{100 \times 2946}{750 \times 930}\right) = 0.42$$

Refer to Table 19 of IS:456-2000 and readout the permissible shear stress as

$$\tau_c = 0.38 \text{ N/mm}^2$$

Since $\tau_v > \tau_c$, shear reinforcements are required.

Shear force resisted by concrete = $(\tau_c bd) = (0.38 \times 750 \times 930)\, 10^{-3} = 265$ kN

∴ Balance shear force $V_{us} = 1458 - 265 = 1193$ kN

Using 12 mm diameter four-legged stirrups, spacing is computed as

$$S_v = \left[\frac{0.87 \times 415 \times 4 \times 113 \times 930}{1193 \times 10^3}\right] = 127 \text{ mm}$$

Adopt 12 mm diameter stirrups at 120 mm centres.

Reinforcement required for mid span section to resist the bending moment $M_u = 382.5$ kN·m will be less than the minimum quantity. Hence provide minimum reinforcement computed as

$$A_{st(min.)} = \left(\frac{0.85 bd}{f_y}\right) = \left(\frac{0.85 \times 750 \times 930}{415}\right) = 1429 \text{ mm}^2$$

Provide 3 bars of 25 mm diameter at mid span section ($A_{st} = 1473$ mm²).

The section subjected to maximum torsional moment and shear should be designed for the following moments:

$$T_u = 56 \text{ kN·m} \qquad\qquad D = 1000 \text{ mm}$$
$$V_u = 843 \text{ kN} \qquad\qquad b = 750 \text{ mm}$$
$$M_u = 0 \qquad\qquad d = 930 \text{ mm}$$

$$M_t = T_u \left[\frac{1 + (D/b)}{1.7}\right] = 56 \left[\frac{1 + (1000/750)}{1.7}\right] = 77 \text{ kN·m}$$

$$\therefore \qquad M_{el} = (M_u + M_t) = (0 + 77) = 77 \text{ kN·m}$$

The magnitude of moment being very small, provide minimum reinforcement of
$$A_{st(min.)} = 1429 \text{ mm}^2$$
Provide 3 bars of 25 mm diameter (A_{st} = 1473 mm^2).
Equivalent shear force is
$$V_e = [V_u + 1.6(T_u/b)] = [843 + 1.6(56/0.75)] = 963 \text{ kN}$$

$$\therefore \qquad \tau_v = \left(\frac{V_e}{bd}\right) = \left(\frac{963 \times 10^3}{750 \times 930}\right) = 1.38 \text{ N/mm}^2$$

$$\left(\frac{100 A_{st}}{bd}\right) = \left(\frac{100 \times 1473}{750 \times 930}\right) = 0.1$$

Refer to Table 19 of IS:456-2000 and readout the permissible shear stress as
$$\tau_c = 0.28 \text{ N/mm}^2$$
Since $\tau_v > \tau_c$, shear reinforcements are required.
Balance shear = $[843 - (0.38 \times 750 \times 930)10^{-3}] = 648 \text{ kN}$
Using 12 mm diameter four-legged stirrups, spacing is computed as

$$S_v = \left[\frac{A_{sv} \, 0.87 f_y}{(\tau_v - \tau_c)b}\right] = \left[\frac{4 \times 113 \times 0.87 \times 415}{(1.38 - 0.28)750}\right] = 198 \text{ mm}$$

Provide 12 mm diameter four-legged stirrups at 190 mm centres.

xiv. *Design of raft slab*:

Maximum projection of raft slab from face to column = $\left(\dfrac{3 - 0.75}{2}\right) = 1.125 \text{ m}$

Soil pressure = $\dfrac{155.52}{(5.5^2 - 2.5^2)} = 206 \text{ kN/m}^2$

Considerl 1 m width of raft slab along the circular arc.

Maximum BM = $\left(\dfrac{206 \times 1.1^2}{2}\right) = 124.63 \text{ kN·m}$

Design ultimate bending moment $M_u = (1.5 \times 124.63) = 187 \text{ kN·m}$
Effective depth required is computed as

$$d = \sqrt{\frac{M_u}{0.138 f_{ck} b}} = \sqrt{\frac{187 \times 10^6}{(0.138 \times 20 \times 1050)}} = 260 \text{ mm}$$

Provide an overall depth of 500 mm and effective depth d = 450 mm to limit the shear stress within the permissible limits.

Compute $\left(\dfrac{M_u}{bd^2}\right) = \left(\dfrac{187 \times 10^6}{1000 \times 450^2}\right) = 0.92$

Refer to Table 2 of SP:16 and readout the percentage of reinforcement as

$$p_t = \left(\frac{100\,A_{st}}{bd}\right) = 0.27$$

$$\therefore \qquad A_{st} = \left(\frac{0.27\times1000\times450}{100}\right) = 1215 \text{ mm}^2$$

Provide 25 mm diameter bars at 200 mm centres to reduce the shear stress to be within safe permissible limits ($A_{st} = 2454$ mm²).

Distribution reinforcement = $(0.0012 \times 500 \times 1000) = 600$ mm²

Provide 12 mm diameter bars at 180 mm centres.

Ultimate shear force at a section 450 mm from face of column is computed as

$$V_u = 1.5\,(200 \times 0.65 \times 1.00) = 201 \text{ kN}$$

$$\tau_v = \left(\frac{V_u}{bd}\right) = \left(\frac{201\times103}{1000\times450}\right) = 0.44 \text{ N/mm}^2$$

$$\left(\frac{100\,A_{st}}{bd}\right) = \left(\frac{100\times2454}{1000\times450}\right) = 0.545$$

Refer to Table 19 of IS:456-2000 and readout the design shear strength of concrete as

$$\tau_c = 0.48 \text{ N/mm}^2$$

Since $\tau_v > \tau_c$, no shear reinforcements are required.

The thickness of footing is retained at 500 mm up to a distance of 500 mm from the column face and thereafter gradually decreased to 350 mm at the edge. The details of reinforcements in the various structural elements of an Intze type tank are detailed in Figs 8.7–8.9.

8.5 CONICAL OR FUNNEL SHAPED TANK

Conical or funnel shaped overhead water tanks are often preferred to other shapes mainly due to their aesthetic and superior architectural features in comparison with other types of overhead tanks. Basically a conical overhead tank comprises the following structural components (Fig. 8.10):

a. Conical dome covering the tank
b. Top ring beam
c. Conical shell
d. Bottom spherical dome and internal shaft
e. Bottom ring beam
f. Supporting cylindrical shaft
g. Raft foundation

For supporting towers higher than 25 m, reinforced concrete cylindrical shells are economical and can be rapidly constructed using the slip form process of casting.

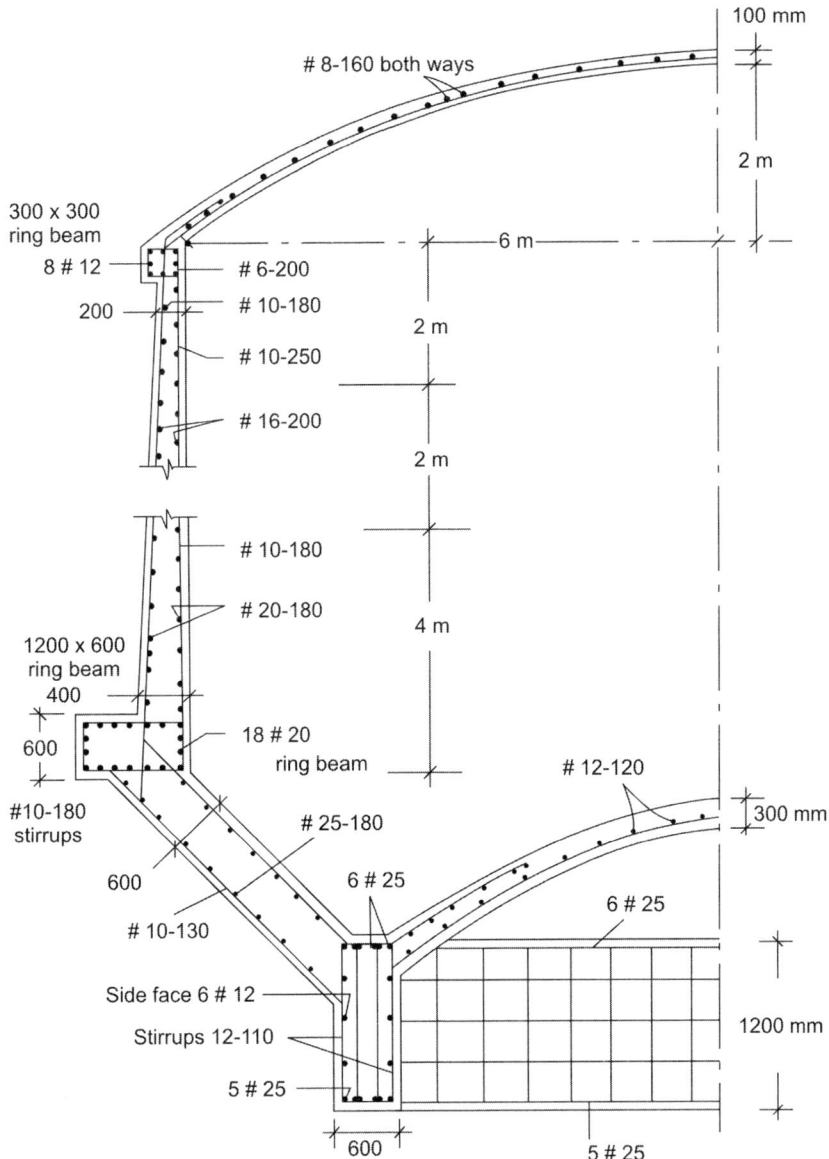

8-160 both ways

100 mm

2 m

300 x 300
ring beam
8 # 12

6-200

10-180

10-250

16-200

6 m

2 m

2 m

10-180

20-180

4 m

1200 x 600
ring beam
400

600

18 # 20
ring beam

12-120

300 mm

#10-180
stirrups

25-180

600

6 # 25

6 # 25

10-130

Side face 6 # 12

Stirrups 12-110

5 # 25

600

1200 mm

5 # 25

Fig. 8.7: Reinforcement details in an Intze tank

They can also be built using precast concrete elements. The conical shell walls are sloping at 45° to the horizontal and the thickness of the walls gradually increased towards the bottom of the tank and designed for hoop tension and meridional thrust. The top and bottom ring beams are designed for hoop tension.

The supporting cylindrical shell tower is designed for combined thrust and bending due to wind and seismic forces. A rigid raft slab foundation is provided to support the shaft. The design of a conical or funnel shaped elevated water tank is illustrated by the following example.

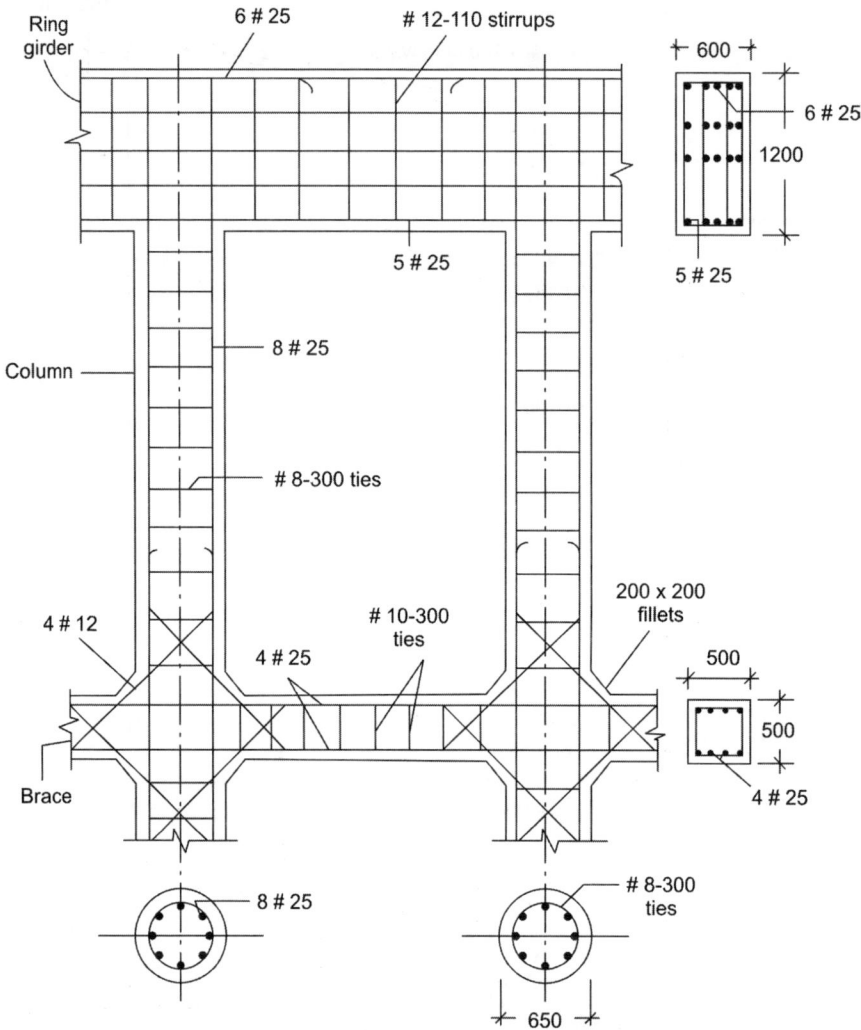

Fig. 8.8: Reinforcement details in ring girder, columns and braces

8.6 ▊ DESIGN EXAMPLE OF FUNNEL SHAPED OVERHEAD TANK

Design a funnel shaped overhead water tank to suit the following data:

Capacity of the tank = 350000 litres

Height of tower shaft = 20 m above GL

Basic wind pressure = 1.5 kN/m²

Adopt M20 grade concrete and Fe 415 grade tor steel for all RCC work

Depth of foundations = 1.5 m

Safe bearing capacity of soil = 200 kN/m²

Design the conical shaped tank, supporting cylindrical shaft and foundations for the tank and sketch the typical reinforcement details in the various structural components.

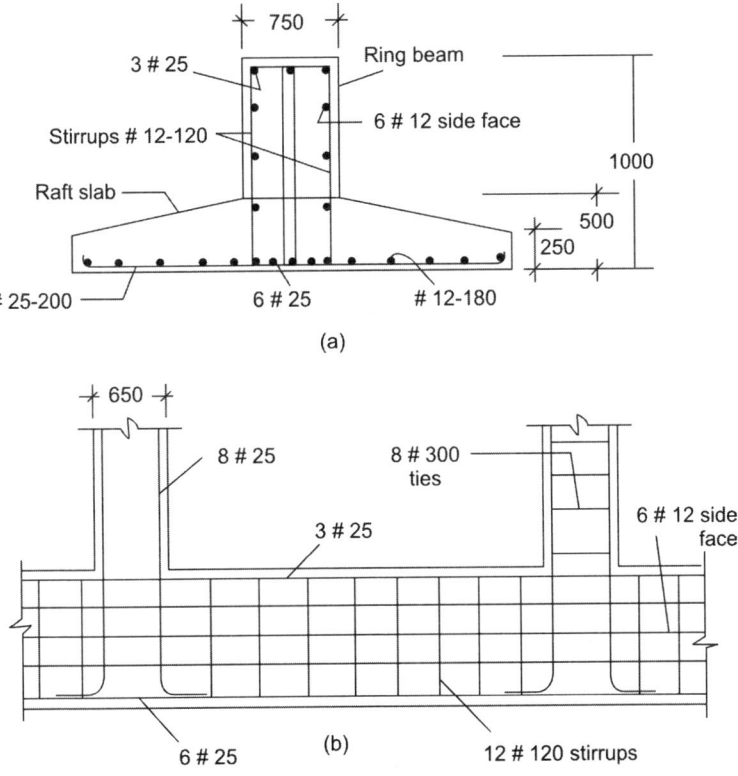

Fig. 8.9: Reinforcement details in ring beam and raft slab

i. *Data*:

Capacity V = 350000 litres

Height of supporting shaft = 20 m

Wind pressure = 1.5 kN/m²

SBC of soil = 200 kN/m²

Depth of foundations = 1.5 m

ii. *Permissible stresses*:

For M20 grade concrete and Fe 415 grade steel

σ_{ct} = 1.2 N/mm² m = 13

σ_{cc} = 5 N/mm² Q = 0.897

σ_{cb} = 7 N/mm² j = 0.906

σ_{st} = 150 N/mm² f_{ck} = 20 N/mm² and f_y = 415 N/mm²

iii. *Dimensions of tank*:

Diameter of the supporting shaft is assumed to be 5 m.

Let R be the radius of the conical shell. The dimensions of the tank are expressed in terms of R as shown in Fig. 8.11.

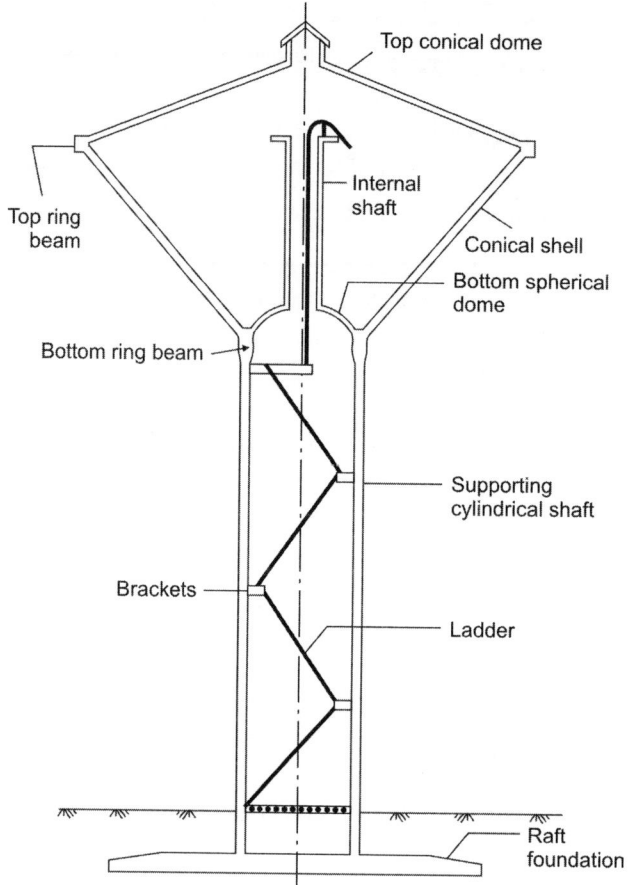

Fig. 8.10: Structural components of funnel shaped tank

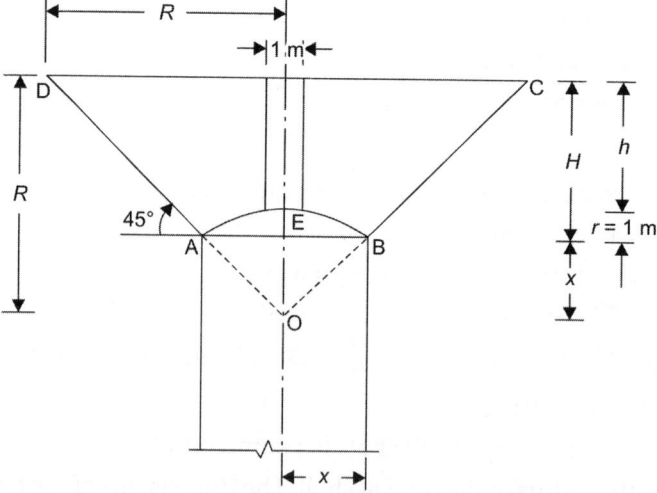

Fig. 8.11: Dimensions of conical shell

x = radius of shaft = 2.5 m

$H = (R - x) = (R - 2.5)$

$h = (H - r) = (R - 2.5 - 1) = (R - 3.5)$

If V = volume of the tank

then V = (volume of cone ODC) – (volume of cone OAB) – (volume of dome AEB) – (volume of inner shaft)

V_1 = volume of cone ODC

$$= \frac{1}{3}\pi RR^2 = \frac{1}{3}\pi R^3$$

V_2 = volume of cone OAB

$$= \frac{1}{3}\pi x^3 = \frac{1}{3}\pi \times 2.5^3$$

V_3 = volume of dome AEB

$$= \frac{\pi r^2}{3}(3R - r) = \frac{\pi \times 1^2}{3}(3R - 1)$$

V_4 = volume of inner shaft = $\pi(0.5)^2 (R - 3.5)$

∴ $V = (V_1 - V_2 - V_3 - V_4)$

$$= \frac{1}{3}\pi R^3 - \frac{1}{3}\pi(2.5)^3 - \frac{\pi}{3}(3R - 1) - \pi(0.5)^2(R - 3.5)$$

But $V = \left(\dfrac{350000}{1000}\right) = 350 \ m^3$

∴ $350 = 1.047 R^3 - 16.36 - 3.14 R + 1.047 - 0.785 R + 2.7489$

⇒ $1.047 R^3 - 3.925 R - 362.57 = 0$

Solving, $R = 7.15$ m

Adopt $R = 7.5$ m to allow for free board. The capacity of the tank with radius $R = 7.5$ m is computed as $V = 3.88 \ m^3$ which is marginally greater than the required volume capacity of 350 m^3.

iv. *Design of top conical dome:*

The geometry of the conical dome is shown in Fig. 8.12.

Assume a central rise of 2 m.

If α is the semi-vertical angle at apex, $\tan\alpha = \left(\dfrac{7.5}{2}\right) = 3.75$

∴ $\alpha = 75°, \qquad \theta = 15°$

t = thickness of slab = 100 mm

Self weight of slab = $(0.1 \times 24) = 2.4 \ kN/m^2$

Live load   = 0.6 kN/m^2

Total load w = 3.0 kN/m^2

Maximum meridional thrust at base is

$$T_1 = \left(\frac{wh}{2\cos^2\alpha}\right) = \left(\frac{3 \times 2}{2\cos^2 75°}\right) = 45 \ kN/m$$

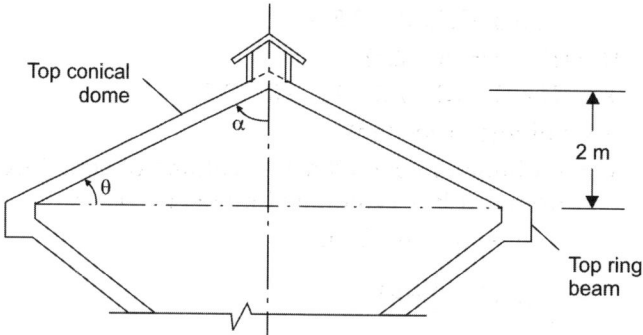

Fig. 8.12: Top conical dome

Maximum circumferential force at base is

$$N_\theta = wh\tan^2\alpha = (3 \times 2 \times \tan^2 75°) = 84.3 \text{ kN/m}$$

$\therefore$ Meridional stress $= \left(\dfrac{45 \times 10^3}{1000 \times 100}\right)$

$$= 0.45 \text{ kN/mm}^2 < 5 \text{ N/mm}^2 \text{ (safe)}$$

Hoop stress $= \left(\dfrac{84.3 \times 10^3}{1000 \times 100}\right) = 0.843 \text{ N/mm}^2 < 5 \text{ N/mm}^2 \text{ (safe)}$

Since stresses are within permissible limits, nominal reinforcements are provided in the conical dome.

$$A_{st} = \left(\frac{0.3 \times 1000 \times 100}{100}\right) = 300 \text{ mm}^2$$

Provide 8 mm diameter at 160 mm centres both meridionally and circumferentially.

v. *Design of top ring beam*:

Hoop tension $F_t = \left(\dfrac{T_1 D\cos\theta}{2}\right) = \left(\dfrac{45 \times \cos 15° \times 14.30}{2}\right) = 311 \text{ kN}$

$$A_{st} = \left(\frac{311 \times 10^3}{150}\right) = 2074 \text{ mm}^2$$

Provide 8 bars of 20 mm diameter ($A_{st} = 2512 \text{ mm}^2$).
If A_c = cross-sectional area of ring beam

$$\left(\frac{311 \times 10^3}{A_c + 13 \times 2512}\right) = 1.2$$

$\therefore$ $A_c = 225610 \text{ mm}^2$

Provide a ring beam of size 600 mm × 400 mm.
Provide 8 mm diameter stirrups at 200 mm centres.

vi. *Design of conical shell*:

The geometry of the conical shell is shown in Fig. 8.13.

Thickness of shell at top = 200 mm

Thickness of shell at bottom = 300 mm

Average thickness = 250 mm, $\quad r_2 = 7.5$ m

Slope of wall = 45°, $\qquad\qquad r_3 = 2.5$ m

Height of cone = $h_3 = 5$ m

Self weight of slab = $(0.25 \times 24) = 6.0$ kN/m²

Length $L_c = \sqrt{(r_2 - r_3)^2 + h_3^2} = \sqrt{(7.5 - 2.5)^2 + 5^2} = 7.07$ m

Weight of conical wall $W_1 = 2 \times \pi \left(\dfrac{7.5 + 2.5}{2} \right) \times 0.25 \times 7.07 \times 24 = 1333$ kN

Weight of water over the conical wall $W_2 = \pi (5^2 - 2.5^2)\, 5 \times 10 = 2945$ kN

Self weight of top dome = $(\pi \times 7.5 \times 7.76 \times 3) = 549$ kN

Self weight of top ring beam = $(0.4 \times 0.6 \times \pi \times 15.4 \times 24) = 279$ kN

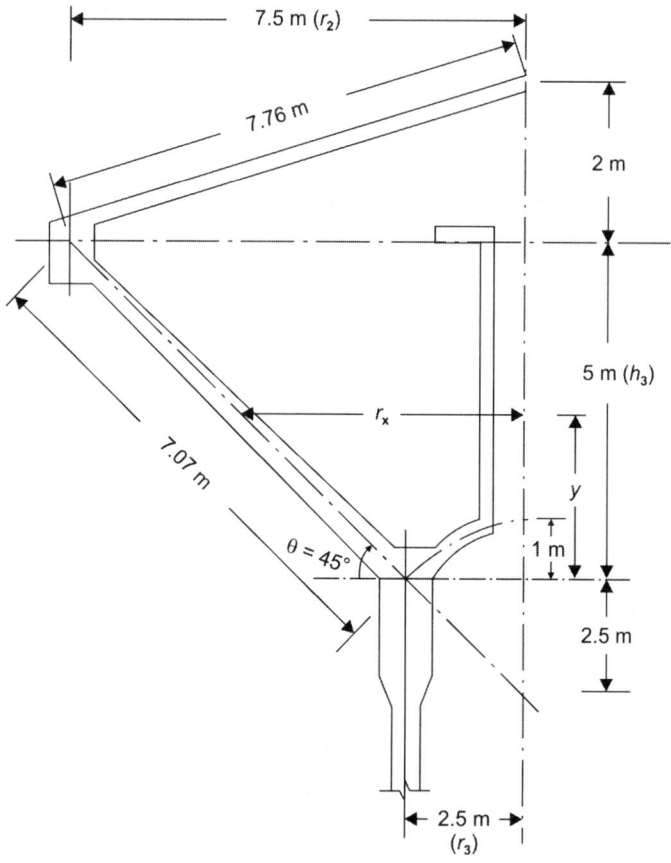

Fig. 8.13: Geometry of conical shell

∴   Total vertical load at base of conical shell

$$W = (549 + 279 + 1333 + 2945) \text{ kN} = 5106 \text{ kN}$$

∴   Meridional thrust $T = \left(\dfrac{W \cosec\theta}{2\pi r_3}\right) = \left(\dfrac{5106 \times \cosec 45°}{2\pi \times 2.5}\right) = 460 \text{ kN/m}$

Assuming thickness of conical shell at base, $t = 300$ mm

Meridional stress $= \left(\dfrac{460 \times 10^3}{1000 \times 300}\right) = 1.53 \text{ N/mm}^2 < 5 \text{ N/mm}^2$ (safe)

The hoop tension in the conical shell varies with the depth of water and is given by

$$F_t = (p\cosec\theta + q\cot\theta)D/2$$

where,

    D = diameter at any horizontal section

    p = intensity of water pressure normal to the inclined walls

      = $(10h)$ kN/m^2

    h = height of water above the section

    θ = angle made by the conical shell with the horizontal

    q = weight of conical slab per square metre of surface area

Hoop tension at bottom of conical shell is computed as follows:

    $h = 5$ m $D = 5$ m $\theta = 45°$

    $p = (10 \times 5) = 50$ kN/m^2

    $q = (0.3 \times 24) = 7.2$ kN/m^2

    $F_t = (50 \times \cosec 45° + 7.2 \times \cot 45°)\, 5/2 = 195$ kN

Hoop tension at 1 m above base is computed as follows:

    $h = 4$ m $D = 7$ m $\theta = 45°$

    $p = (10 \times 4) = 40$ kN/m^2

    $q = (0.28 \times 24) = 6.72$ kN/m^2

∴ $F_t = (40 \times \cosec 45° + 6.72 \times \cot 45°)\, 7/2 = 222$ kN

Hoop tension at 2 m above base is computed as follows:

    $h = 3$ m $D = 9$ m $\theta = 45°$

    $p = (10 \times 3) = 30$ kN/m^2

    $q = (0.26 \times 24) = 6.24$ kN/m^2

    $F_t = (30 \times \cosec 45° + 6.24 \times \cot 45°)\, 9/2 = 219$ kN

∴   Maximum hoop tension occurs at 1 m above the base section.

∴ $A_{st} = \left(\dfrac{222 \times 10^3}{150}\right) = 1480 \text{ mm}^2$

Provide 10 mm diameter bars at 100 mm centres on both faces ($A_{st} = 1570 \text{ mm}^2$)

Maximum tensile stress $= \left[\dfrac{222 \times 10^3}{(300 \times 1000) + (13 \times 1570)}\right]$

$$= 0.69 \text{ N/mm}^2 < 1.2 \text{ N/mm}^2 \text{ (safe)}$$

Distribution steel $= \left(\dfrac{0.3 \times 300 \times 1000}{100} \right) = 900 \text{ mm}^2$

Provide 10 mm diameter at 160 mm centres on both faces along the meridions.

vii. *Design of bottom dome and internal shaft*:

Diameter at base = 5 m

Rise of dome = 1 m

Thickness t = 150 mm

If R is the radius of the dome, then $(2R - 1) 1 = (5/2)^2 \quad \Rightarrow \quad R = 3.625 \text{ m}$

Self weight of dome slab = $(2 \times \pi \times 3.625 \times 1 \times 0.15 \times 24) = 82 \text{ kN}$

Internal diameter of vertical shaft = 800 mm

Thickness of walls = 100 mm

$\therefore$ External diameter = 1000 mm

Weight of water over bottom dome = $\pi (2.5^2 - 0.5^2) \, 4.5 \times 10 = 848 \text{ kN}$

Weight of vertical shaft = $(\pi \times 0.9 \times 0.1 \times 4 \times 24) = 28 \text{ kN}$

Total weight on dome = $(82 + 848 + 28) = 958 \text{ kN}$

$\therefore$ Load/unit area $w = \left(\dfrac{958}{\pi \times 2.5^2} \right) = 49 \text{ kN/m}^2$

$\therefore$ Meridional thrust $T_1 = \left(\dfrac{wR}{1 + \cos \theta} \right)$

$\cos \theta = \left(\dfrac{2.5}{3.625} \right) = 0.68 \quad \therefore \theta = 47°$

$\therefore \qquad T_1 = \left(\dfrac{49 \times 3.625}{1 + 0.68} \right) = 106 \text{ kN/m}$

$\therefore$ Meridional stress $= \left(\dfrac{106 \times 10^3}{150 \times 1000} \right) = 0.706 \text{ N/mm}^2 < 5 \text{ N/mm}^2$ (safe)

Circumferential force $= wR \left[\cos \theta - \dfrac{1}{(1 + \cos \theta)} \right]$

$= 49 \times 3.625 \left[0.68 - \dfrac{1}{1.68} \right] = 15 \text{ kN/m}$

$\therefore$ Hoop stress $= \left(\dfrac{15 \times 10^3}{150 \times 1000} \right) = 0.1 \text{ N/mm}^2$ (safe)

Provide nominal reinforcements of 0.3%.

$\therefore \qquad A_{st} = \left(\dfrac{0.3 \times 150 \times 1000}{100} \right) = 450 \text{ mm}^2$

Adopt 10 mm diameter at 160 mm centres both radially and circumferentially.

Maximum hoop compression in the internal shaft $= \left(\dfrac{10 \times 4 \times 0.9}{2} \right) = 18$ kN

Since hoop stresses are negligibly small, provide nominal reinforcements of 8 mm diameter at 160 mm centres in both directions.

viii. *Design of bottom ring beam*:

Horizontal component of thrust from conical shell $= H_1 = 460 \cos 45° = 326$ kN

Horizontal thrust from bottom dome $= H_2 = 106 \cos 47° = 72$ kN

Net compressive force in ring beam $= (H_1 - H_2) = (326 - 72) = 254$ kN

∴   Hoop compression $= \left(\dfrac{254 \times 5}{2} \right) = 635$ kN

Assume the ring beam size as 400 mm wide by 600 mm deep.

Compressive stress $= \left(\dfrac{635 \times 10^3}{400 \times 600} \right) = 2.64$ N/mm$^2 < 5$ N/mm^2 (safe)

Provide minimum reinforcement of 0.3%.

∴ $A_{st} = \left(\dfrac{0.3 \times 400 \times 600}{100} \right) = 720$ mm^2

Provide 8 bars of 12 mm diameter with 8 mm diameter two-legged stirrups at 200 mm centres. As the girder is supported by the shaft walls, there will be no shear in the ring beam.

ix. *Design of supporting cylindrical shaft*:

Diameter of shaft = 5 m

Height of shaft (above GL) = 20 m

Thickness of shaft walls above GL = 150 mm

Height of shaft (below GL) = 1.5 m

Thickness of shaft below GL = 300 mm

Self weight of shaft $= \pi \times 5 \times 24\,[(20 \times 0.15) + (1.5 \times 0.3)] = 1300$ kN

The vertical shaft has an external diameter of 5.30 m and an internal diameter of 4.70 m, thickness of walls being 150 mm.

Loads acting on shaft at ground level are as follows:

Load due to weight of	kN
a. the dome	549
b. top ring beam	279
c. conical shell	1333
d. bottom spherical dome	82
e. internal shaft	28
f. bottom ring beam	90
g. supporting shaft	1300
h. plastering, pipes, ladders, railings, etc. (lumpsum)	339
Total dead loads =	4000 kN
Weight of water =	3880 kN

Wind pressure:

Vertically projected area of overhead tank is

$$= \left(\frac{1}{2} \times 15.4 \times 2\right) + \frac{1}{2}(15.4 + 5.4)5 = 70 \text{ m}^2$$

∴   Wind force on tank = (0.7 × 1.5 × 70) = 73.5 kN

Acting at a distance of 24 m from base,

wind force on shaft = (0.7 × 1.5 × 5.3 × 20) = 112 kN

Acting at 10 m from the base,

total moment due to wind about base M = (73.5 × 24) + (112 × 10)

$$= 2884 \text{ kN·m}$$

Area of cross-section $A = \pi(2.65^2 - 2.35^2) = 4.71 \text{ m}^2$

Second moment of area $I = \dfrac{\pi}{4}(2.65^4 - 2.35^4) = 14.92 \text{ m}^4$

Stresses at base section

Tank empty condition:

$\quad P = 4000 \text{ kN} \qquad\qquad D = 5.3 \text{ m} \qquad\qquad M = 2884 \text{ kN·m}$

$$e = \left(\frac{M}{P}\right) = \left(\frac{2884}{4000}\right) = 0.72 \text{ m} < \left(\frac{D}{6}\right) = \left(\frac{5.3}{6}\right) = 0.88 \text{ m}$$

This section is under compression only.

$$\sigma_c = \left(\frac{P}{A} + \frac{My}{I}\right) = \left(\frac{4000 \times 10^3}{4.71 \times 10^6} + \frac{2884 \times 10^6 \times 2650}{14.92 \times 10^{12}}\right) = 1.361 \text{ N/mm}^2$$

∴   Compression < 5 N/mm² (hence safe)

Tank full condition:

$\quad P = (4000 + 3880) = 7880 \text{ kN}, \qquad M = 2884 \text{ kN·m}$

$$\sigma_c = \left(\frac{7880 \times 10^3}{4.71 \times 10^6} + \frac{2884 \times 10^6 \times 2650}{14.92 \times 10^{12}}\right) = 2.185 \text{ N/mm}^2 < 4 \text{ N/mm}^2 \text{ (safe)}$$

Provide minimum reinforcement of 0.8%.

$$\therefore \qquad A_{st} = \left(\frac{0.8 \times 150 \times 1000}{100}\right) = 1200 \text{ mm}^2$$

Provide 12 m diameter bars at 180 mm centres both vertically and circumferentially on both faces (A_{st} = 1256 mm²).

Check for seismic forces:

(Weight of tank portion + water) W_1 = (2361 + 3880) = 6241 kN

(Weight of shaft + finishes etc.) W_2 = (1300 + 339) = 1639 kN

Seismic coefficient $\alpha_n = 0.015$

Moment due to seismic force at ground level $M_q = \alpha_n (W_1 h_1 + W_2 h_2)$

$$= 0.015 \,(6241 \times 23 + 1639 \times 10)$$

$$= 2399 \text{ kN·m}$$

$\therefore$ Total load $P = 7880$ kN, $M = 2399$ kN·m

$$\sigma_c = \left(\frac{7880 \times 10^3}{4.71 \times 10^6} + \frac{2399 \times 10^6 \times 2650}{14.92 \times 10^{12}} \right)$$

$$= 2.09 \text{ N/mm}^2 < 5.00 \text{ N/mm}^2 \text{ (safe)}$$

x. *Design of raft foundations*:

Total load from tank and shaft = 7880 kN

Self weight of footing (10%) = 829 kN

Total load W = 8700 kN

SBC of the soil = 200 kN/m²

If A = area of footing

D = diameter of the footing

Direct load W = 8700 kN

Moment M = 2884 kN·m

then $\left(\dfrac{W}{A} + \dfrac{M}{Z} \right) = 200$

Hence $\left[\dfrac{W}{(\pi D^2 / 4)} + \dfrac{M}{(\pi D^3 / 32)} \right] = 200$

$$\left[\frac{8700}{(\pi D^2 / 4)} + \frac{2884}{(\pi D^3 / 32)} \right] = 200$$

Solving, $D = 8.54$ m

Adopt diameter of raft slab $D = 9$ m

Intensity of soil pressure $w = \left[\dfrac{8700}{(\pi \times 9^2)/4} \right] = 137$ kN/m²

The loading on the base is taken as annular loading on the mean diameter of the shaft. The design procedure is similar to that described in the design example in Section 5.9 (Refer to chimney foundations).

Diameter of raft slab = $2a$ = 9 m

Diameter of the shaft = $2b$ = 5 m

Maximum bending moment in the section is governed by the radial moment. Radial moment at centre of footing is given by

$$M_r = \frac{W}{8\pi} \left[2\log_e \left(\frac{a}{b} \right) + 1 - \left(\frac{b}{a} \right)^2 \right] - \frac{3}{16} wa^2$$

$$= \frac{8700}{8\pi} \left[2\log_e \left(\frac{4.5}{2.5} \right) + 1 - \left(\frac{2.5}{4.5} \right)^2 \right] - \left[\frac{3}{16} \times 137 \times 4.5^2 \right]$$

$$= 126 \text{ kN·m/m}$$

Moment at junction of footing and tank walls at a radius of 2.5 m is given by

$$M_r(\text{max.}) = \frac{W}{8\pi}\left[2\log_e\left(\frac{a}{b}\right)+1-\left(\frac{b}{a}\right)^2\right]-\frac{3}{16}w\,(a^2-b^2)$$

$$= \frac{8700}{8\pi}\left[2\log_e\left(\frac{4.5}{2.5}\right)+1-\left(\frac{2.5}{4.5}\right)^2\right]-\frac{3}{16}\times137\,(4.5^2-2.5^2)$$

$$= 287 \text{ kN·m/m}$$

Design ultimate moment $M_{ur} = (1.5 \times 287) = 430.5$ kN·m/m

Effective depth required $d = \sqrt{\dfrac{M_u}{0.138 f_{ck}b}} = \sqrt{\dfrac{430.5\times10^6}{(0.138\times20\times10^3)}} = 395$ mm

To resist shear stresses, depth required will be nearly 1.5 times that required based on moment considerations.

Hence adopt $d = 600$ mm and overall depth = 650 mm

Compute the parameter:

$$\left[\frac{M_u}{bd^2}\right] = \left[\frac{430.5\times10^6}{103\times600^2}\right] = 1.19$$

Refer to Table 2 of SP:16 and readout the percentage reinforcement as

$$p_t = \left[\frac{100\,A_{st}}{bd}\right] = 0.359$$

$$\therefore \qquad A_{st} = \left(\frac{0.359\times10^3\times600}{100}\right) = 2154 \text{ mm}^2/\text{m}$$

Adopt 25 mm diameter bars at 150 mm centres both ways ($A_{st} = 3272$ mm²)

Also provide 12 mm diameter bars at 200 mm centres both ways at the top of the footing. The details of reinforcements in the conical tank, supporting shaft and foundations are shown in Figs 8.14 and 8.15.

References

1. IS:456-2000, Indian Standard Code of Practice for Plain and Reinforced Concrete, Fourth Revision, Bureau of Indian Standards, 2000, p. 100.

2. ACI:318M-11, Building Code Requirements for Structural Concrete, American Concrete Institute, Farmington Hills, Michigan, 2005.

3. BS EN 1992-1-12004, Design of Concrete Structures, General Rules & Rules for Buildings, British Standards Institution, 2004.

4. Rowe RE, Cranston WB and Best BC, New concepts in the design of concrete, *Structural Engineer*, Vol. 43, 1965, pp. 339–403.

5. CEB Recommendations for International Code of Practice for Reinforced Concrete, American Concrete Institute and Cement & Concrete Association, London, 1964.

6. Bate SCC, Why limit state design, *Concrete*, March 1968, pp. 103–108.

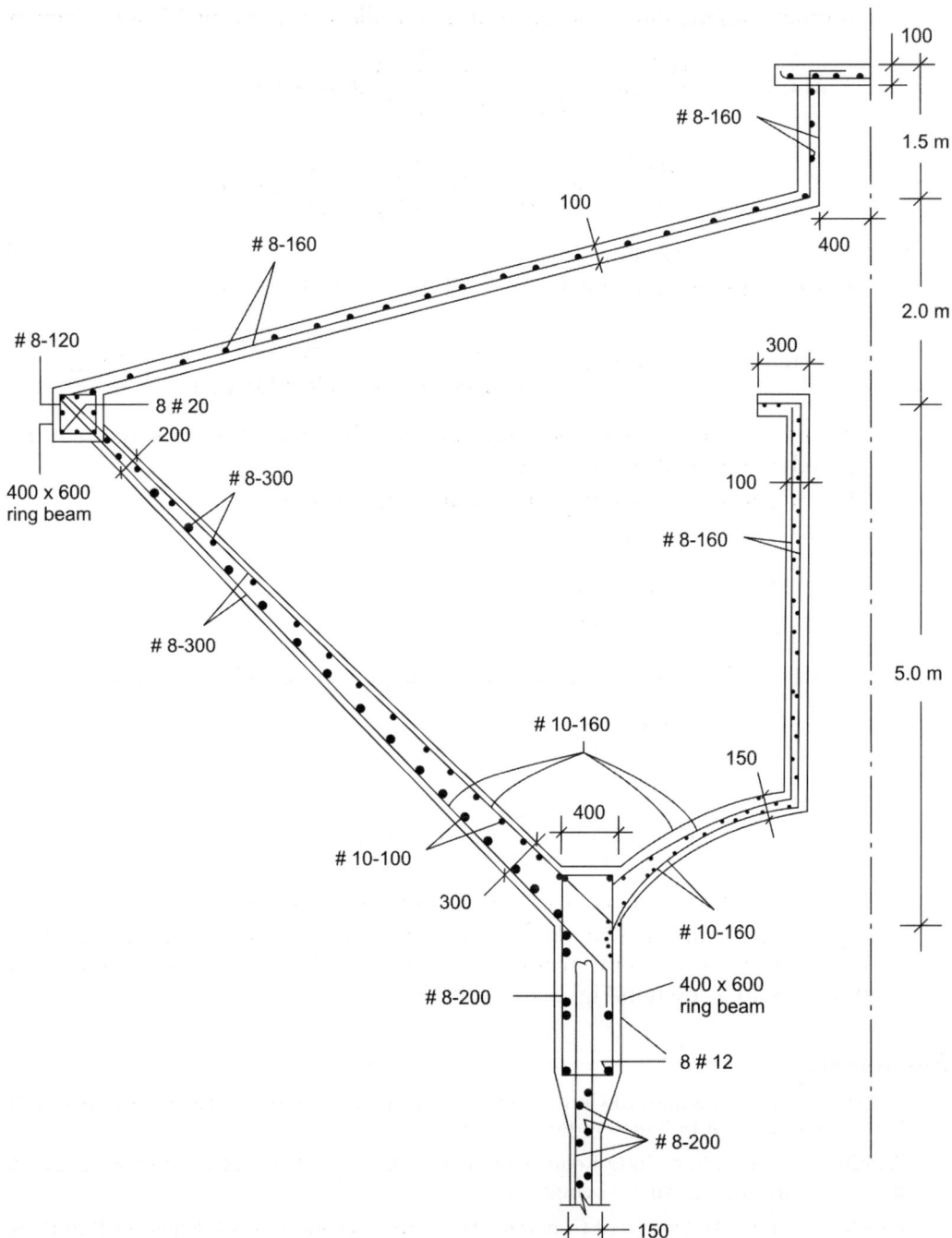

Fig. 8.14: Reinforcement details in funnel shaped overhead tank

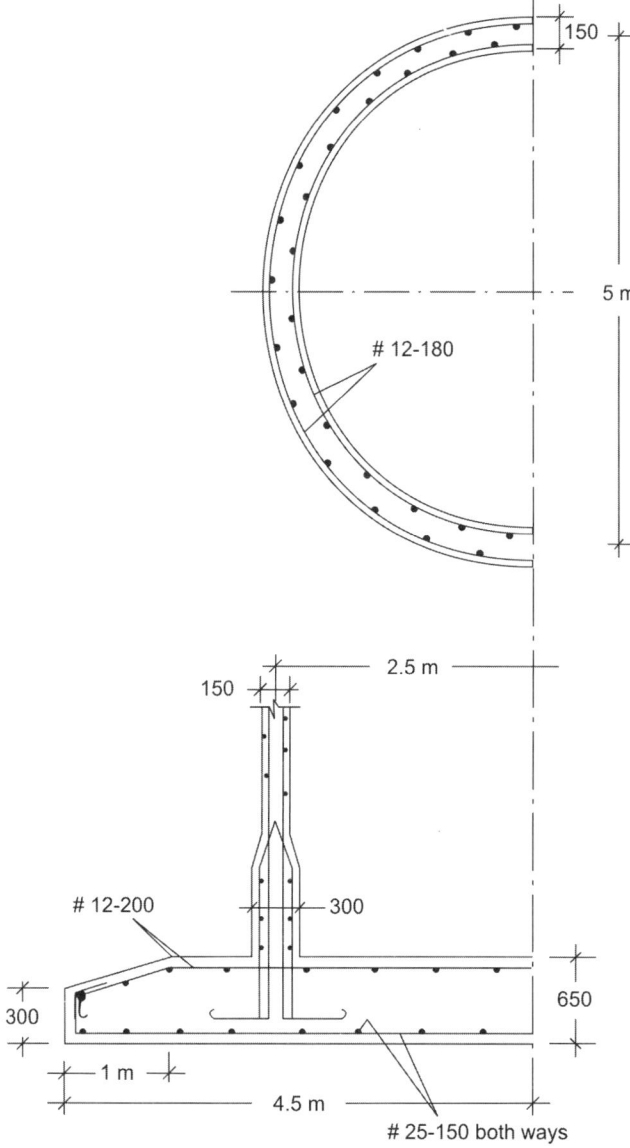

150

5 m

12-180

2.5 m

150

12-200

300

300

650

1 m

4.5 m

25-150 both ways

Fig. 8.15: Details of reinforcements in shaft and raft foundation

7. Krishna Raju N, *Structural Design & Drawing* (Reinforced Concrete & Steel), 3 edn, Universities Press, Hyderabad, 2009, pp. 87–107.

8. Tareq M Azabi, Behavior of Reinforced Concrete Conical Tanks Under Hydrostatic Loading, Post Doctoral Dissertation, University of Western Ontario, Canada, 2014, pp. 124.

9. Krishna Raju N and Pranesh RN, *Reinforced Concrete Design*, New Age International, New Delhi, 2003, pp. 494–509.

10. Peck RB, Hanson WH, Thornburn TH, *Foundation Engineering*, John Wiley & Sons, New York , 1974.

Assignment

1. Design an Intze type water tank to store 1.5 million litres of water. The height of the tank above ground level is 20 m. The site has soil of safe bearing capacity of 200 kN/m². Depth of foundation is 1.5 m below ground level. Basic wind pressure is 1.5 kN/m². The number of supporting columns is 8. Adopt M20 grade concrete and Fe 415 grade tor steel. The design of the tank should conform to the stresses specified in IS:3370 and IS:456 codes.

2. A reinforced concrete Intze type water tank is required to store 250000 litres of water. Height of staging is 12 m above ground level. The tank is supported on six columns. Safe bearing capacity of the soil is 150 kN/m². Basic wind pressure is 1.5 kN/m². Adopting M20 grade concrete and Fe 415 grade tor steel, design an Intze type tank and sketch the details of reinforcements in the various structural components of the tank.

3. A funnel shaped overhead water tank is required to store 500000 litres of water with the height of supporting tower being 25 m above ground level.

   Basic wind pressure = 1.5 kN/m²

   Depth of fundations = 2 m below GL

   Safe bearing capacity of the soil = 200 kN/m²

   Adopt M20 grade concrete and Fe 415 grade tor steel for all RCC work.

   Design the conical shaped tank, supporting cylindrical shaft and foundations for the tank. Sketch the details of reinforcements in the tank.

4. Design an Intze type overhead water tank to store 2 million litres of water. The height of tank above ground level is 15 m. The safe bearing capacity of soil at site is 180 kN/m². Depth of foundation is to be 1.6 m below ground level. Basic wind pressure is 1.5 kN/m². Number of supporting columns are 12. Adopt M25 grade concrete and Fe 415 HYSD bars. Design the Intze type tank and sketch the details of reinforcements in the various structural components.

5. A circular overhead water tank is to be designed to store 500000 litres of water at a height of 12 m above ground level. Depth of water storage is 4 m, free board is 500 mm, and the water tank has to be supported on 8 symmetrically arranged columns. Braces are to be provided at intervals of 3 m. Adopting M25 grade concrete and Fe 415 HYSD bars, assume wind intensity at site as 1.5 kN/m². Safe bearing capacity of soil at site is 250 kN/m².

   Design the following structural components of the elevated water tank.

   a. Spherical dome

   b. Ring beams

   c. Tower columns

   d. Braces

   e. Raft foundation

6. Design an Intze type tank for a town water supply system to store 750000 litres of water supported on 8 reinforced concrete columns at a height of 12 m above the ground. The wind intensity at site is estimated to be 1 kN/m². Depth of foundation is to be 1 m below ground level. The safe bearing capacity of soil at site is 300 kN/m². Adopt M25 grade concrete and Fe 500 grade HYSD reinforcements. Braces are spaced at intervals of 3 m. Design the tank, columns, braces and the raft foundation.

7. A funnel shaped water tank is to be designed to suit the following data:

   Capacity of the tank = 400000 litres

   Height of supporting tower = 16 m above ground level

   Intensity of wind at site = 2 kN/m²

   Depth of foundation = 2 m

   Safe bearing capacity of soil = 250 kN/m²

   Adopt M20 grade concrete and Fe 415 HYSD bars.

   Design the conical tank, supporting tower shaft and the raft foundations.

8. A funnel shaped water tank is required for a nuclear power station to store half a million litres of water. Design a suitable tank using the following data:

   Height of supporting tower = 20 m above ground level

   Intensity of wind at site = 1.5 kN/m²

   Depth of foundation = 2 m

   Safe bearing capacity of soil = 2 kN/m²

   Adopt M25 grade concrete and Fe 500 HYSD bars.

   Sketch the reinforcement details of the tank.

Review Questions

1. Explain the necessity of using elevated tanks in the water supply systems in towns.

2. What are the various types of overhead water tanks commonly used in the water supply systems in cities?

3. Distinguish between circular, Intze and conical or funnel shaped water tanks with sketches, mentioning the advantages of each of them.

4. Briefly outline the design principles of the various structural elements of overhead circular and Intze type water tanks.

5. What is the purpose of a ring beam in the case of circular water tanks and spherical domes?

6. What type of forces develop in conical domes? How do you design the conical dome of Intze type tanks?

7. List the various forces acting on an elevated water tank tower.

8. How do you design the supporting columns of a water tank?

9. What is the main purpose of using braces? Briefly outline the method of designing a typical brace in a overhead water tank.

10. What type of foundation is used for elevated water tanks? Briefly explain the method of designing the ring beam and raft slab of the foundation.

Objective Type Questions

1. In the design of overhead water tanks, the most economical design is possible by using
   a. conical shaped tanks
   b. rectangular tanks
   c. circular tanks

2. The various loads to be considered in the design of elevated water tanks are
   a. dead loads
   b. live loads
   c. dead, live and wind loads

3. In designing reinforced concrete overhead tanks in coastal zones, the most important load to be considered is
   a. dead load
   b. live load
   c. wind load

4. The ring beam at the base of circular domes covering the water tanks is subjected to
   a. compressive forces
   b. hoop tension
   c. bending and torsion

5. The minimum thickness of concrete dome covering the water tanks should be not less than
   a. 250 mm
   b. 50 mm
   c. 75 to 100 mm

6. The circular girder at the junction of conical and bottom dome of an Intze type water tank should be designed to resist
   a. hoop compression
   b. bending moment and shear force
   c. compression, bending moment, shear force and torsion

7. Circular girders supported on columns in an elevated water tank are subjected to torsional moments at
   a. centre of span
   b. support section
   c. sections nearer to the support

8. In the case of funnel shaped overhead water tanks, the conical shell is subjected to
   a. hoop compression
   b. bending moment and shear force
   c. hoop tension

9. Elevated water tanks located in seismic zones should be designed to resist
   a. dead and live loads
   b. dead, live and wind loads
   c. dead, live, wind and earthquake forces

10. The foundations of an overhead water tank generally comprises
    a. column footings
    b. pile foundations
    c. raft foundations

9

Box Culverts

9.1 GENERAL FEATURES

Reinforced concrete box culverts[1,2] are widely used in railway bridge crossings with high embankments crossing a stream with a limited flow. They are also used for pedestrian underground subway crossings of city roads. Box culvert consists of two horizontal and vertical slabs built monolithically to form a rigid box type frame with a square or rectangular cross-section. Box culverts are also used as underground service chambers to carry various types of service lines like electricity supply cables, sewage pipes, television, telephone service lines below major roads in urban areas.

Box culverts are economical due to their rigidity and monolithic action[3,4] and separate foundations are not required since the soffit slab rests directly on the stabilized soil bed. For small discharges, single cell box culvert is sufficient while for larger flows, multicelled box culverts can be employed. The barrel of the box culvert should be of sufficient length to accommodate the carriageway and the kerbs. Recent developments in the field of precast concrete technology has resulted in the use of modular precast box culvert units of desired cross-section and length. These units can be assembled at site to achieve the desired length of the box culvert.

9.2 ADVANTAGES OF PRECAST BOX CULVERTS

Box culverts are ideal for flows where hydraulic head is limited. For an equivalent waterway area to circular pipes, box culverts can be configured to have less impact on upstream water levels and downstream flow velocities than equivalent pipe structures. Precast construction means, traffic may use the installation immediately after placing and backfilling whereas *in situ* construction will require a period for curing prior to stripping forms ready for use. Due to their ability to tolerate heavy wheel loads even with no overfill in place, precast box culverts are superior to most alternative systems which require compacted overfill in place before loading is applied. Depending on the flow discharge required, single- or multicelled box culverts can be used without disruption of traffic for a longer period. Many construction companies[5,6] are manufacturing and supplying precast box culverts of desired requirements.

9.3 DESIGN LOADS

The structural design of a reinforced conceter box culvert comprises the detailed analysis of the rigid frame for moments, shear and thrusts due to various types of loading conditions outlined below:

1. *Concentrated loads*: In case, where the top slab forms the deck of the bridge, concentrated loads due to the wheel loads of the IRC: class AA or A type loading have to be considered.

   If W = concentrated load on the slab
      P = wheel load
      I = impact factor
      e = effective width of dispersion

   then $W = (PI/e)$

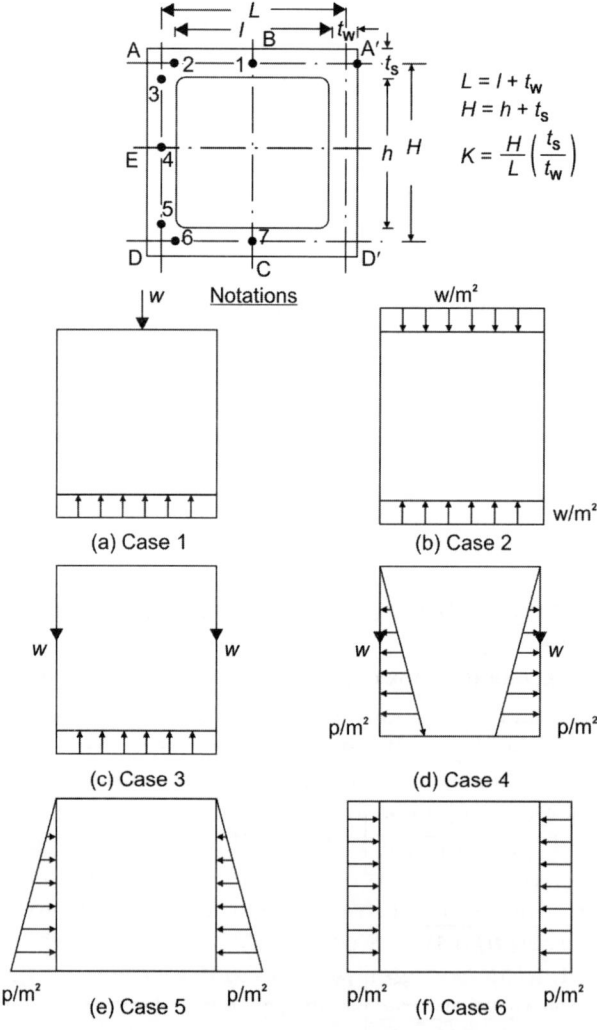

Fig. 9.1: Types of loading for box culverts

The soil reaction on the bottom slab is assumed to be uniform. The notations used for the box culvert and the type of loadings to be considered are shown in Fig. 9.1a–f.

2. *Uniformly distributed load*: The weight of embankment, wearing coat, deck slab, and the track load are considered to be uniformly distributed loads on the top slab with the uniform soil reaction on the bottom slab.

3. *Weight of side walls*: The self weight of two side walls acting as concentrated loads are assumed to produce uniform soil reaction on the bottom slab.

4. *Water pressure inside culvert*: When the culvert is full with water, the pressure distribution on side walls is assumed to be triangular with a maximum pressure intensity of $p = wh$ at the base, where w = density of water and h = depth of flow.

5. *Earth pressure on vertical side walls*: The earth pressure on the vertical side walls of the box culvert is computed according to the Coulomb's theory. The distribution of earth pressure on the side walls is shown in Fig 9.1e.

6. *Uniform lateral load on side walls*: Uniform lateral pressure on vertical side walls has to be considered due to the effect of live load surcharge. Also trapezoidal pressure distribution on side walls due to embankment loading can be obtained by combining the cases (5) and (6).

7. *Design moments, shears and thrusts*: The box culvert is analysed for moments, shear forces and axial thrusts developed due to the various loading conditions by any of the classical methods such as moment distribution, slope deflection or column analogy procedures. Alternatively coefficients for moments, shears and thrusts compiled by Victor[1] are very useful in the computation of various force components for the different loading conditions.

The fixed end moments developed for the six different loading cases are compiled in Table 9.1. The moment, shear and thrust coefficients for various loading cases are shown in Table 9.2, for two different ratios of $(L/H) = 1$ and 1.5

where,

L = span of the culvert

H = height of the culvert

Table 9.1: Fixed end moments in box culvert

Loading case	Fixed end moments	
	$M_A = M_{A'}$	$M_D = M_{D'}$
1	$-\dfrac{WL}{12}\left[\dfrac{2k+4.5}{(k+3)(k+1)}\right]$	$-\dfrac{WL}{24}\left[\dfrac{k+6}{(k+3)(k+1)}\right]$
2	$-\dfrac{wL^2}{12(k+1)}$	$-\dfrac{wL^2}{12(k+1)}$
3	$+\dfrac{WL}{6}\left[\dfrac{k}{(k+3)(k+1)}\right]$	$-\dfrac{WL}{6}\left[\dfrac{3+2k}{(k+3)(k+1)}\right]$

(Contd.)

Table 9.1: Fixed end moments in box culvert (*Contd.*)

Loading case	Fixed end moments	
	$M_A = M_{A'}$	$M_D = M_{D'}$
4	$+\dfrac{pH^2}{60}\left[\dfrac{k(2k+7)}{(k+3)(k+1)}\right]$	$+\dfrac{pH^2}{60}\left[\dfrac{k(3k+8)}{(k+3)(k+1)}\right]$
5	$-\dfrac{pH^2}{60}\left[\dfrac{k(2k+7)}{(k+3)(k+1)}\right]$	$-\dfrac{pH^2}{24}\left[\dfrac{k(3k+8)}{(k+3)(k+1)}\right]$
6	$-\dfrac{pkH^2}{12(k+1)}$	$-\dfrac{pkH^2}{12(k+1)}$

Table 9.2: Coefficients for moment, shear and thrust

$L:H$	Section	Forces	Loading case					
			1	*2*	*3*	*4*	*5*	*6*
		M	WL	wL^2	WL	pL^2	pL^2	pL^2
		N	W	wL	W	pL	pL	pL
		V	W	wL	W	pL	pL	pL
1 : 1	B 1	M	+0.182	+0.083	+0.021	+0.019	−0.019	−0.042
		N	0	0	0	−0.167	+0.167	+0.500
	A 2	M	−0.068	−0.042	0.021	+0.019	−0.019	−0.042
		N	0	0	0	−0.167	+0.167	−0.500
		V	+0.500	+0.500	0	0	0	0
	A 3	M	−0.068	−0.042	+0.021	+0.019	−0.019	−0.042
		N	+0.500	+0.500	0	0	0	0
		V	0	0	0	+0.167	−0.167	−0.500
	E 4	M	−0.052	−0.042	−0.042	−0.043	+0.043	+0.083
		N	+0.500	+0.500	+0.500	0	0	0
	D 5	M	−0.036	−0.042	−0.004	+0.023	−0.023	−0.042
		N	+0.500	+0.500	+1.000	−0.333	+0.333	0
		V	0	0	0	0	0	+0.500
	D 6	M	−0.036	−0.042	−0.104	+0.023	−0.023	−0.042
		N	0	0	0	0	0	+0.500
		V	−0.500	−0.500	−1.020	−0.333	+0.333	0
	C 7	M	+0.088	+0.083	+0.146	+0.023	−0.023	−0.042
		N	0	0	0	−0.333	+0.333	+0.500
1.5 : 1	B 1	M	+0.170	+0.075	+0.018	+0.015	−0.015	−0.033
		N	0	0	0	−0.167	+0.167	+0.500
	A 2	M	−0.079	−0.050	+0.018	+0.015	−0.015	−0.033
		N	0	0	0	−0.167	+0.167	+0.500
		V	+0.500	+0.500	0	0	0	0

(Contd.)

Table 9.2: Coefficients for moment, shear and thrust (*Contd.*)

L : H Section	Forces	Loading case 1	2	3	4	5	6
A 3	M	−0.079	−0.050	+0.018	+0.015	−0.015	−0.033
	N	+0.500	+0.500	0	0	0	0
	V	0	0	0	+0.167	−0.167	−0.500
E 4	M	−0.062	−0.050	−0.050	−0.047	+0.047	+0.092
	N	+0.500	+0.500	+0.500	0	0	0
D 5	M	−0.045	−0.050	−0.118	+0.018	−0.018	−0.033
	N	+0.500	+0.500	+1.000	0	0	0
	V	0	0	0	−0.333	+0.333	+0.500
D 6	M	−0.045	−0.050	−0.118	+0.018	−0.018	−0.033
	N	0	0	0	−0.333	+0.333	+0.500
	V	−0.500	−0.500	−1.000	0	0	0
C 7	M	+0.079	+0.075	+0.132	+0.018	−0.018	−0.033
	N	0	0	0	−0.333	+0.333	+0.500

Refer to Fig. 9.1 for details and notations.

Note: (1) Positive moment indicates tension on inside face. (2) Positive shear indicates the summation of force at the left of the section acts outward when viewed from within. (3) Positive thrust indicates compression on the section.

9.4 STRUCTURAL ANALYSIS OF BOX CULVERTS

Box culvert with rigid joints at the corners is treated as an indeterminate structure and analyzed for forces and moments developed under various loading conditions by using any of the well established classical methods[7] such as moment distribution, slope deflection, column analogy or matrix methods[8]. The various loading conditions identified in Section 9.3 (cases 1–6) are analysed using the moment, thrust and shear coefficients compiled in Table 9.2. The design forces resulting from the combination of various cases yielding maximum moments and forces at the support and mid span sections are determined to facilitate the structural design of the critical sections.

9.5 DESIGN OF CRITICAL SECTIONS

The critical sections to be designed in a typical box culvert are:
  i. Corner section of the soffit slab (D-6)
 ii. Corner section of the top slab (A-2)
iii. Centre span section of top slab (B-1)
 iv. Centre span section of soffit slab (C-7)
  v. Centre of the vertical side slab (E-4)

  The box culvert is analyzed for design maximum moments, thrust and shear force by considering the dead, live and earth pressure loads[9] by using the fixed end moments given in Table 9.1 and the coefficients given in Table 9.2 for different cases of loading.

  The maximum design moments resulting from the combination of various loading cases are determined. The moments at the centre of span of top, bottom and vertical wall slabs and the support sections are determined by suitably combining the different loading patterns. The maximum moments generally develop for the following critical load combinations:

1. When the slab supports the dead and live loads and the culvert is empty.
2. When the top slab supports the dead and live loads and the culvert is running full.
3. When the sides of the culvert do not carry the live load and the culvert is running full.

The various sections of the box culvert should be designed to conform to the provisions of the Indian standard code IS:456-2000[10]. The reinforcements designed to resist the various forces and moments in the culvert slabs should be distributed on both faces and fillets are provided at the inner corners of the culvert.

9.6 DESIGN EXAMPLE OF BOX CULVERT

Design a reinforced concrete box culvert having a clear vent way of (3 m × 3 m). The superimposed dead load on the culvert is 12.8 kN/m². The live load on the culvert is 50 kN/m². Density of soil at site is 18 kN/m² and angle of repose ϕ is 30°. Adopt M20 grade concrete mix and Fe 415 grade tor steel. Sketch the details of reinforcements in the box culvert.

i. *Data*:

Clear span $L = 3$ m
Height of vent $h = 3$ m
Dead load $= 12.8$ kN/m²
Live load $= 50$ kN/m²
Density of soil $= 18$ kN/m³
Angle of repose $\phi = 30°$
Concrete: M20 grade
Steel: Fe 415 tor steel

ii. *Permissible stresses*:

$\sigma_{cc} = 5$ N/mm²; $m = 13$
$\sigma_{cb} = 7$ N/mm²; $J = 0.86$
$\sigma_{st} = 150$ N/mm² (water face); $Q = 1.198$
$\sigma_{st} = 190$ N/mm² (away from water face)

iii. *Dimensions of box culvert*:

Adopting thickness of slab as 100 mm/metre span
Thickness $t_s = t_w = 300$ m
Effective span $= 3300$ mm

iv. *Loads*:

Self weight of top $= (0.3 × 24) = 7.2$ kN/m²
Superimposed dead load $= 12.8$
Live load $= 50.0$
Total load w $= 70.0$ kN/m²
Weight of vertical side walls $W = (0.3 × 3.3 × 24) = 24$ kN

Soil pressure $p = wh\left(\dfrac{1-\sin\phi}{1+\sin\phi}\right)$

At $\qquad h = 3.3$ m

$\qquad\qquad \phi = 30°$

$\qquad\qquad w = 18$ kN/m^3

$\therefore \qquad p = (18 \times 3.3 \times 1/3) = 20$ kN/m^2

Uniform lateral pressure due to the effect of superimposed dead load and live load surcharge is calculated as

$$p = (50+12.8)\left[\dfrac{1-\sin\phi}{1+\sin\phi}\right] = (62.8 \times 1/3) = 21 \text{ kN/m}^2$$

Uniform lateral pressure due to the effect of superimposed dead load surcharge only is given by

$$p = 12.8\left(\dfrac{1-\sin\phi}{1+\sin\phi}\right) = (12.8 \times 1/3) = 4.26 \text{ kN/m}^2$$

Intensity of water pressure is obtained as

$$p = wh = (10 \times 3.3) = 33 \text{ kN/m}^2$$

v. *Analysis of moments, shears and thrusts*:

The various loading patterns considered are shown in Fig. 9.2. The moments, shears and thrusts corresponding to different cases of loading (cases 1–6), evaluated using the coefficients given in Table 9.2 and are compiled in Table 9.3. The design forces resulting from the combination of the various cases yielding maximum moments and forces at the support and mid span sections are shown in Table 9.4.

The maximum positive moments develop at the centre of bottom and top slab for the condition that the sides of the culvert not carrying the live load and the culvert is running full with water.

The maximum negative moments develop at the support sections of the bottom slab for the condition that the culvert is empty and the top slab carries the dead and live load.

vi. *Design of reinforcements*:

Section C 7 (mid span of bottom slab)

$M = 76.10$ kN·m

$N = -7.43$ kN (tension)

$$A_{st} = \left(\dfrac{76.10\times10^6}{150\times0.86\times270}\right) = 2234 \text{ mm}^2/\text{m}$$

Provide 20 mm diameter at 140 mm centres.

Distribution steel $= \left(\dfrac{0.3\times300\times1000}{100}\right) = 900 \text{ mm}^2$

Section D 6 (support section)

$M = -54.43$ kN·m

$N = 34.65$ kN·m

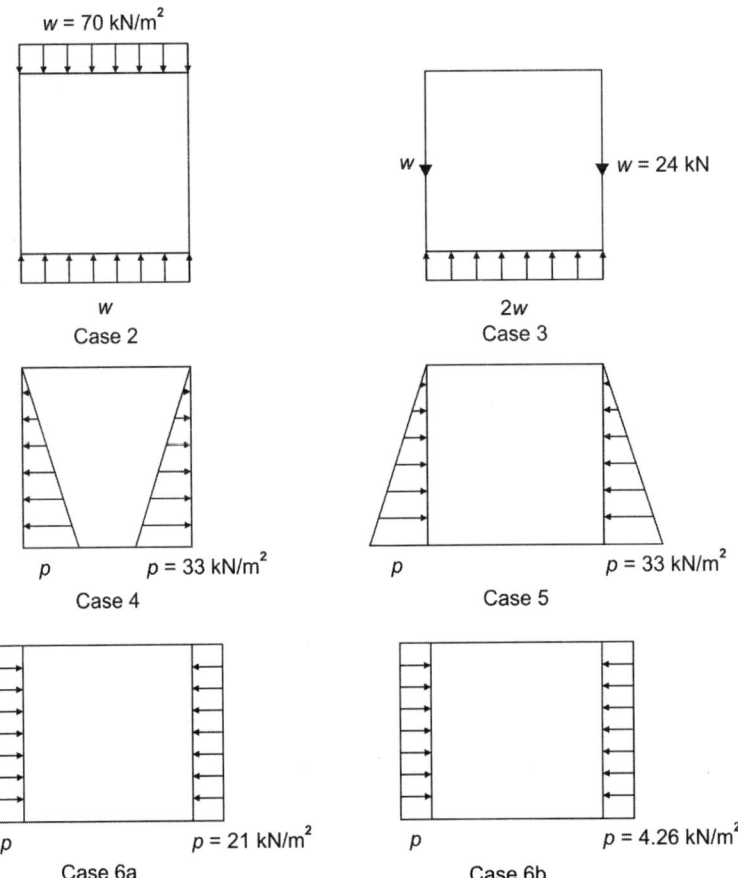

Fig. 9.2: Loading cases considered for box culvert

Table 9.3: Force components for different cases of loading case

Section	Forces	Case 2	Case 3	Case 4	Case 5	Case 6a	Case 6b
B 1	M	63.2	1.66	6.82	−4.13	−9.6	−1.92
	N	0	0	−18.18	+11.0	+34.65	+6.93
A 2	M	−31.6	1.66	6.82	−4.13	−9.6	−1.92
	N	0	0	−18.18	+11.0	−34.65	−6.93
	V	115.5	0	0	0	0	0
A 3	M	−31.6	1.66	6.82	−4.13	−9.6	−1.92
	N	115.5	0	0	0	0	0
	V	0	0	18.18	−11.0	−34.65	−6.93
E 4	M	−31.6	−0.317	8.26	−5.00	−9.6	−1.92
	N	115.5	+39.6	0	0	0	0
D 5	M	−31.6	−0.317	8.26	−5.00	−9.6	−1.92
	N	111.5	+79.2	−36.26	+21.9	0	0
	V	0	0	0	0	+34.65	+6.93

(Contd.)

Table 9.3: Force components for different cases of loading case *(Contd.)*

Section	Forces	Case 2	Case 3	Case 4	Case 5	Case 6a	Case 6b
D 6	M	−31.6	−8.23	8.26	−5.00	−9.6	−1.92
	N	0	0	0	0	+34.65	+6.93
	V	−115.5	−79.2	−36.26	+21.9	0	0
C 7	M	63.2	11.56	8.26	−5.00	−9.60	−1.92
	N	0	0	−36.26	+21.9	+34.65	+6.93

Moments are in kN/m; shear force and thrusts are in kN.

Table 9.4: Design moments and forces in box culvert

Section	Loading combination cases	Moment M (kN·m)	Thrust N (kN)	Shear force V (kN)
D 6	2 + 3 + 5 + 6(a)	−54.43	+34.65	−172.8
A 2	2 + 3 + 5 + 6(a)	−43.67	−23.65	+115.5
B 1	2 + 3 + 4 + 5 + 6(b)	65.63	−1.25	0
C 7	2 + 3 + 4 + 5 + 6(b)	76.10	−7.43	0
E 4	2 + 3 + 4 + 5 + 6(b)	−55.89	+155.1	0

$$A_{st} = \left(\frac{54.43 \times 10^6}{190 \times 0.86 \times 270} \right) = 1233 \text{ mm}^2/\text{m}$$

Provide 16 mm diameter at 150 mm centres and distribution bars of 10 mm diameter at 150 mm centres.

Section E 4 (vertical side wall)

$M = -55.89$ kN·m $\qquad \therefore \quad M_u = (1.5 \times 55.89) = 83.83$ kN·m

$N = 155.1$ kN $\qquad \therefore \quad N_u = (1.5 \times 155.1) = 232.5$ kN

$$\left(\frac{M_u}{f_{ck}bD^2} \right) = \left(\frac{83.83 \times 10^6}{20 \times 1000 \times (300)^2} \right) = 0.046 \quad \text{or} \quad f_y = 415 \text{ N/mm}^2$$

$$\left(\frac{N_u}{f_{ck}bD} \right) = \left(\frac{232.5 \times 10^3}{20 \times 1000 \times 300} \right) = 0.0387 \quad \text{or} \quad \left(\frac{d'}{D} \right) = \left(\frac{30}{300} \right) = 0.10$$

Referring to interaction curve (Chart 32) of SP:16, $\left(\dfrac{p}{f_{ck}} \right) = 0.02$

where,

$A_{sc} = A_{st} = 0.5 \, (pbD/100)$

$\qquad = 0.5 \, (0.02 \times 20 \times 1000 \times 300)/100$

$\qquad = 600 \text{ mm}^2 \quad \therefore \quad A_s = 1200 \text{ mm}^2$

But minimum reinforcement of 0.8% of cross-section has to be provided

$$\therefore \qquad A_s = \left(\frac{0.8 \times 300 \times 1000}{100} \right) = 2400 \text{ mm}^2$$

Provide 16 mm diameter at 150 mm centres on both faces in the vertical side walls. Distribution steel of 10 mm diameter at 150 mm centre is provided on both faces. The details of reinforcements in the box culvert are shown in Fig. 9.3.

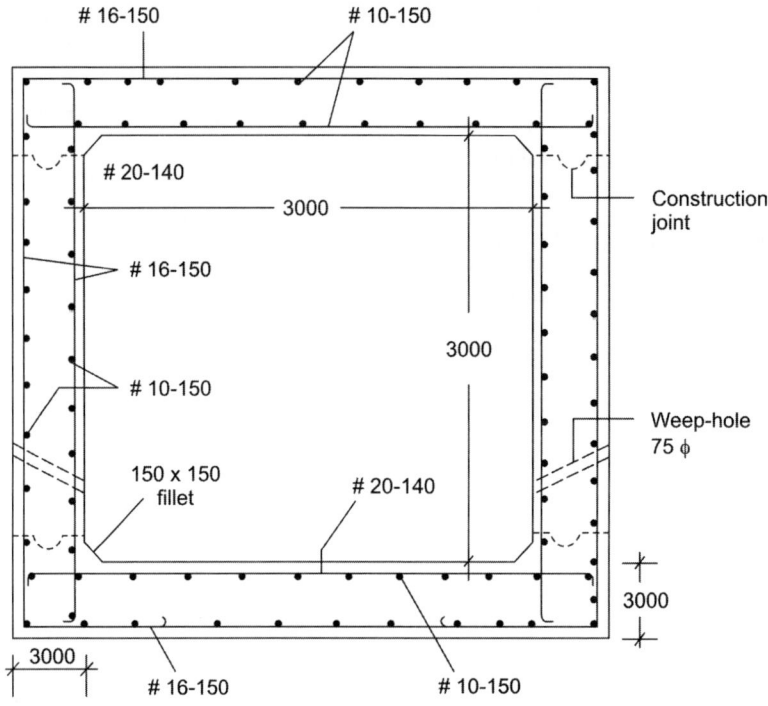

Fig. 9.3: Reinforcement details in box culvert

References

1. Johnson Victor D, *Essentials of Bridge Engineering*, 5 edn, Oxford & IBH, New Delhi, 2001, pp. 99–130.

2. Krishna Raju N, *Design of Bridges*, 4 edn, Oxford & IBH, New Delhi, 2009, pp. 100–109.

3. Reynolds CE, James C Steedman and Anthony J Threlfall, *Reinforced Concrete Designer's Handbook*, 11 edn, Taylor & Francis, London & New York, 2008.

4. Anil G, *Shear Behaviour of Reinforced Concrete Box Culverts*, (Behaviour subjected to highway vehicle loadings), Lambert Academic Publishing, August 2009, p. 280.

5. Humes A, Division of Holcim Company of Australia–Box Culvert Brochures–2010.

6. Hanson Pipe and Precast Products, United States Company, 2013.

7. Junarkar HB and Shah SJ, *Mechanics of Structures*. Vol. II, 13 edn, Charotar Publishing, Anand, 1989, pp. 629–724.

8. Martin HC, *Introduction to Matrix Methods of Structural Analysis*, McGraw-Hill, International Students Edition, Kogakusha Co, Japan, 1956, pp. 1–331.

9. Peck RB, Hanson WH, Thornburn TH, *Foundation Engineering*, John Wiley & Sons, New York, 1974.

10. IS:456-2000, *Indian Standard Code of Practice for Plain and Reinforced Concrete*, 4 edn, Bureau of Indian Standards, 2000, p. 100.

Assignment

1. Design a reinforced concrete box culvert with inside dimensions of 3 m height and 4.5 m width. The box culvert has to carry a superimposed dead load of 10 kN/m^2 and a live load of 50 kN/m^2. The density of the earth is 18 kN/m^3. Angle of repose of the soil is 30°. Adopt M20 grade concrete and Fe 415 HYSD bars. Sketch the details of reinforcements in the box culvert.

2. A reinforced concrete box culvert is required for a national highway crossing. The clear vent way of the box culvert is (4 m × 4 m). Design the box culvert assuming a superimposed dead load of 12 kN/m^2 and a live load of 50 kN/m^2. The density of soil is 16 kN/m^3. Angle of repose of the soil is 30°. Adopt M20 grade concrete and Fe 415 HYSD bars. Sketch the details of reinforcements in the box culvert.

3. A reinforced concrete box culvert of prismatic form with a clear vent way of (3.5 m × 3.5 m) is required for a road crossing. The box culvert has to support a superimposed dead load of 8 kN/m^2, and a live load of 50 kN/m^2. Density of the soil is 18 kN/m^3, and the angle of repose of the soil is 30°. Adopting M20 grade concrete and Fe 415 HYSD bars, design the box culvert and sketch the details of reinforcements in the box culvert.

4. A reinforced concrete box culvert with a clear vent way of 4.5 m wide by 3 m deep is required for a national highway crossing. Design the box culvert assuming a superimposed dead load of 40 kN/m^2. The density of soil at site is 18 kN/m^3. The angle of repose of soil is 30°. Adopt M25 grade concrete and Fe 500 HYSD reinforcements. Sketch the details of reinforcements in the box culvert.

5. A single-celled reinforced concrete box culvert is required for a national highway crossing. Design the box culvert to suit the following data and sketch the details of reinforcements in the box culvert.

   Dimensions of the vent way: 4 m wide by 3.5 m deep

   Superimposed load = 15 kN/m^2

   Distributed live load = 60 kN/m^2

   Density of the soil at site = 16 kN/m^3

   Angle of repose = 35°

   Grade of concrete = M25

   Type of reinforcements = Fe 500 HYSD bars

6. A two-celled reinforced concrete box culvert is to be designed for a state highway crossing. Design the box culvert to suit the following data:

   Twin box vents of 4 m width and 3 m height

   Load due to earth fill = 16 kN/m^2

   Live load due to traffic = 50 kN/m^2

   Density of the soil at site = 18 kN/m^3

   Angle of repose = 30°

   Grade of concrete = M25

   Type of reinforcements = Fe 500 HYSD bars

   Design the twin-cell box culvert and sketch the details of reinforcements in the section.

Review Questions

1. Under what situations would you recommend the use of reinforced concrete box culverts for highway crossings?
2. What are the advantages of using box culverts in preference to the adoption of slab and beam type bridges for highway crossings?
3. Briefly outline with sketches the salient structural elements of a box culvert.
4. List the various loads to be considered in the structural design of box culverts.
5. Explain with sketches the different loading conditions to be considered in the structural design of box culverts.
6. What methods would you adopt to analyse the box culvert subjected to the various types of loads?
7. How do you determine the design maximum moments and force components in the various structural components of a box culvert?
8. Identify and list the various critical cross-sections of a box culvert likely to be subjected to maximum design moments and forces.
9. Explain the method of designing reinforcements in the critical cross-sections of a reinforced concrete box culvert.
10. Sketch the arrangement of reinforcements in the top, soffit and vertical wall slabs of a typical box culvert.

Objective Type Questions

1. Reinforced concrete box culverts are ideally suited for highway crossings of
   a. major rivers
   b. streams with high embankments
   c. via ducts

2. In general, the vent dimensions of a typical reinforced concrete box culvert is
   a. 2 m × 3 m
   b. 6 m × 8 m
   c. 4 m × 3 m

3. Box culverts are generally subjected to various loads like
   a. dead and live loads
   b. hydraulic and earth pressure
   c. dead, live, water and earth loads

4. Structural analysis of box culverts under various loading conditions is done by considering the box culvert as
   a. simply supported structure
   b. determinate structure
   c. indeterminate structure

5. The hydrostatic pressure in the vertical walls of a box culvert due to water flow will be
   a. uniform
   b. trapezoidal
   c. triangular

6. The critical moments, shear forces and thrusts in a typical box culvert depends upon the
   a. length of the horizontal slabs
   b. depth of the vertical slabs
   c. ratio of length to height

7. The maximum positive design moment due to various load combinations in a box culvert usually develops at the
   a. mid span of top slab
   b. centre of vertical slabs
   c. mid span of soffit slab

8. The maximum shear force due to various load combinations in a box culvert usually develops at the
   a. middle of vertical slabs
   b. mid span of horizontal slabs
   c. bottom of vertical slabs

9. Quality control and economy can be achieved in box culverts by adopting
   a. cast *in situ* construction
   b. assembling the structural components at site
   c. precast technology

10. Box culverts should be reinforced with the designed reinforcements distributed on
    a. the outer faces of the structural elements
    b. the inner faces of the structural elements
    c. both faces of the structural elements

10

Portal Frames

10.1 INTRODUCTION

Reinforced concrete portal frames[1,2] are widely used in highway bridge crossings and in the construction of large span industrial structures. They can be built as single or multibay units. Reinforced concrete portal frames have more or less replaced the steel portal frames due to their inherent advantages like fire resistance, durability and negligible maintenance costs. In addition, they are economical in comparison with steel frames. Portal frames are rigid structures and are grouped under indeterminate structures requiring detailed analysis to determine the design forces and moments when subjected to superimposed loads.

A typical reinforced concrete portal frame comprises a horizontal beam referred to as transom, with built in columns at the ends. The columns are supported on ground with suitable foundations like isolated footing. The columns may be either hinged or fixed at the supporting ends. The different types of portal frames[3,4] commonly employed are illustrated in Fig. 10.1. For workshops and storage sheds, portal frames with sloping transoms are preferred for drainage of rain water on the roof. For highways and buildings, portal frames with horizontal transom are generally used to facilitate further construction.

10.2 ADVANTAGES OF PORTAL FRAMES

Reinforced concrete portal frames have the following advantages in comparison with other types of structural members:

i. Portal frames provide uninterrupted space for housing materials and machinery.

ii. Rigid frame bridges comprising portal frames[5,6], being monolithic structures, eliminate the use of separate abutments. The vertical sides of the rigid continuous frame serve as retaining wall to retain earth in highway crossings of embankments.

iii. Slab type rigid frame bridges can be easily cast *in situ* since plain moving formwork can be used for rapid construction work.

iv. Rigid portal frame bridges can be adavantageously adopted in flyover crossings where roads criss-cross at different levels. Typical rigid frame bridges are shown in Fig. 10.2.

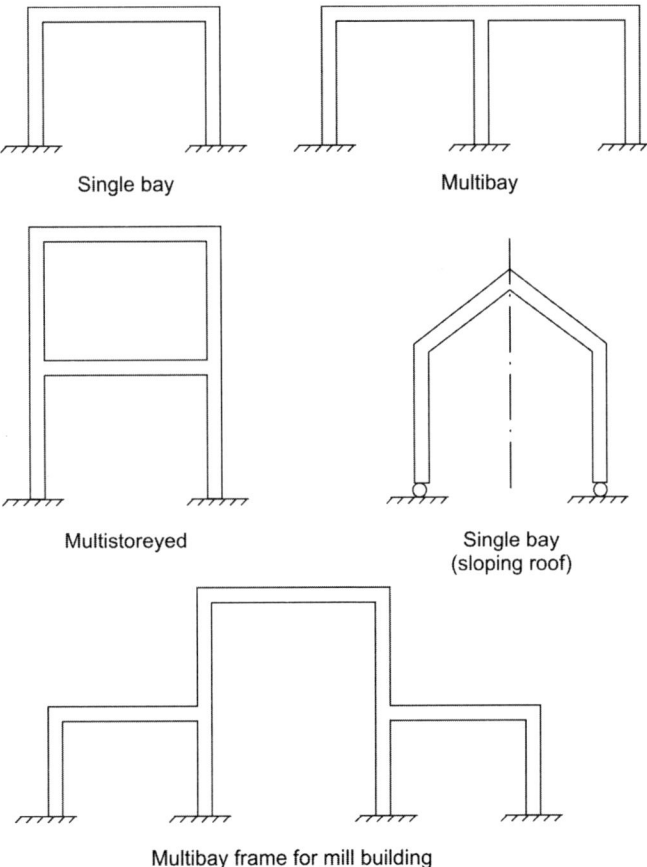

Single bay

Multibay

Multistoreyed

Single bay
(sloping roof)

Multibay frame for mill building

Fig. 10.1: Types of portal frames

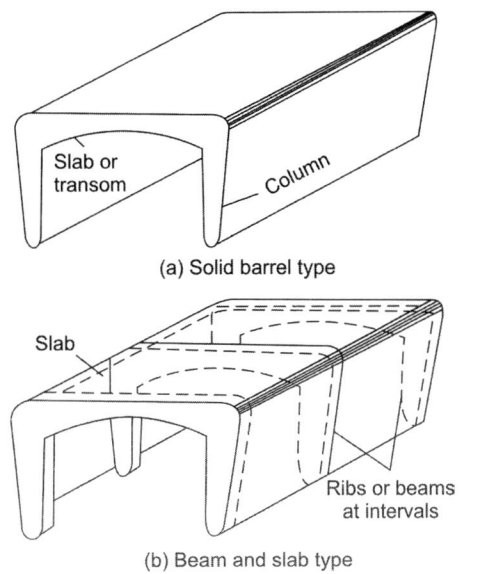

Slab or
transom

Column

(a) Solid barrel type

Slab

Ribs or beams
at intervals

(b) Beam and slab type

Fig. 10.2: Types of rigid frame bridges

v. Rigid portal frame structures have high stability against forces like wind, earthquake and soil pressure.

vi. Portal frame structures do not require separate bearings since hinged or fixed supports provide structural advantages.

vii. Portal frame structures are aesthetically superior and consume less quantity of materials in comparison with other traditional types of construction.

10.3 ANALYSIS AND DESIGN OF PORTAL FRAMES

Portal frames are indeterminate structures and hence they have to be analyzed using the various standard methods such as moment distribution[7], slope deflection, column analogy or matrix methods[8]. In case of buildings, the portal frames are generally spaced at intervals of 3 to 4 m with a reinforced concrete slab cast monolithically between the frames. Frames used for warehouse sheds and workshop structures are provided with sloping roof comprising of purlins and roofing sheets between the portal frames.

The base of the columns of the portal frames are either fixed or hinged depending upon the end conditions. if individual column footings are provided, the base is assumed to be hinged while raft or pile foundations are provided, the base can treated as fixed for purposes of structural analysis of the portal frame. The roof slab cast between the frames is analyzed as a continuous slab. The mid span section of the transom behaves as a tee section while the support section is designed as a rectangular section. Design aids like interaction diagrams of SP:16-1980[9] are very useful in the design of beams and columns subjected to axial thrust and bending moment. The structural design should conform to the provisions of the Indian standard code IS:456-2000[10].

Portal frames precast in the horizontal position on ground are economical in situations where a large number of similar frames are used in a building project and the precast units considerably reduce the construction time besides resulting in overall economy due to the better quality and efficient utilization of the materials. The analysis and design of the different types of portal frames are illustrated by the following design examples.

10.4 DESIGN EXAMPLE OF A HINGED PORTAL FRAME

Design a portal frame hinged at base to suit the following data:

Spacings of portal frames = 4 m

Height of columns = 4 m

Distance between column centres = 10 m

Live load on roof = 1.5 kN/m²

RCC slab continuous over portal frames. Safe bearing capacity of soil at site = 200 kN/m². Adopt M20 grade concrete and Fe 415 HYSD bars. Design the slab, portal frame and foundations, and sketch the details of reinforcements.

    i. *Data*:

        Spacing of portal frames = 4 m

        Span of portal frame = 10 m

Height of columns = 4 m

Live load on roof = 1.5 kN/m²

Materials: M20 grade concrete and Fe 415 HYSD bars

ii. *Characteristic strength*:

$f_{ck} = 20$ N/mm²

$f_y = 415$ N/mm²

iii. *Design of continuous slab*:

Since the slab serves as a roof, loading is light, hence assume an overall depth of 120 mm and effective depth $d = 100$ mm.

Dead load of slab = $(0.12 \times 24) = 2.88$ kN/m²

Roof finishes = 0.50 kN/m²

Ceiling finish = 0.25 kN/m²

Dead load $(g) = 3.63$ kN/m²

Live load $(q) = 1.50$ kN/m²

Negative service load moment at interior support

$$= \left[\frac{gL^2}{10} + \frac{qL^2}{9}\right] = \left[\frac{3.63 \times 4^2}{10} + \frac{1.5 \times 4^2}{9}\right] = 8.5 \text{ kN·m/m}$$

Factored moment M_u (−ve) = $(1.5 \times 8.5) = 12.75$ kN·m/m

Limiting moment of resistance is computed as

$$M_{u\,lim} = (0.138 f_{ck} bd^2) = (0.138 \times 20 \times 1000 \times 100^2)\, 10^{-6} = 27.6 \text{ kN·m}$$

Since $M_u < M_{u\,lim}$, the section is under-reinforced.

Compute the parameter:

$$\left(\frac{M_u}{bd^2}\right) = \left(\frac{12.75 \times 10^6}{1000 \times 100^2}\right) = 1.275$$

Refer to Table 2 of SP:16 and readout the percentage of reinforcement as

$$\therefore \qquad p_t = \left(\frac{100 A_{st}}{bd}\right) = 0.384$$

$$\because \qquad A_{st} = \left(\frac{0.384 \times 1000 \times 100}{100}\right) = 384 \text{ mm}^2$$

Provide 10 mm diameter bars at 200 mm centres (A_{st} = 393 mm²). The same reinforcement is provided at the centre of span section. Distribution reinforcement:

$$= (0.0012 \times 120 \times 10000) = 144 \text{ mm}^2$$

Provide 6 mm diameter bars at 130 mm centres.

iv. *Design of portal frame*:

Effective span of beam = 10 m

If d = effective depth

$$\left(\frac{\text{span}}{\text{effective depth}}\right)=\left(\frac{L}{d}\right) = 12 \text{ to } 15 \text{ (heavy loading on beam)}$$

$$\therefore \qquad d =\left(\frac{10\times10^3}{12}\right) = 833 \text{ mm to}\left(\frac{10\times10^3}{15}\right) = 666 \text{ mm}$$

Adopt effective depth $d = 700$ mm

and overall depth $D = 750$ mm

Width of beam $b = 450$ mm

Column section is assumed as (450 mm × 600 mm)

a. *Loads on frame*:

Load from slab = (5.13 × 4) = 20.52 kN/m

Self weight of beam = (0.45 × 0.63 × 24) = 6.80 kN/m

Finishes on beam = 0.68 kN/m

Total load $w = 28.00$ kN/m

$$h = \text{height of centre line of beam above hinge} =\left(4.0+0.10-\frac{0.75}{2}\right) = 3.72 \text{ m}$$

The moments in the portal frame are analysed by moment distribution:

$$\text{AB} = 3.72 \text{ m and}$$
$$\text{BC} = 10.00 \text{ m}$$

$$I_{AB} =\left(\frac{450\times600^3}{12}\right), \qquad I_{BC} =\left(\frac{450\times750^3}{12}\right)$$

$$\therefore \qquad I_{AB} : I_{BC} =1 :\left(\frac{750}{600}\right)^3 = 1 : 1.95$$

b. *Relative stiffness*:

$$k_{BA} =\left(\frac{3}{4}\times\frac{I}{3.72}\right) = 0.201I$$

$$k_{BC} =\left(\frac{1.95I}{10}\right) = 0.195I \text{ or } 0.20I$$

c. *Distribution factors*:

$$d_{BC} = d_{BA} =\left(\frac{0.20I}{0.20I +0.20I}\right) = 0.5$$

d. *Fixed end moments*:

$$F_{BC} =-\left(\frac{wL^2}{12}\right)=-\left(\frac{28\times10^2}{12}\right) = -234 \text{ kN·m}$$

$$F_{BC} =+\left(\frac{wL^2}{12}\right)=+\left(\frac{28\times10^2}{12}\right) = 234 \text{ kN·m}$$

The moment distribution is carried out to get the resultant moments for the frame loaded as shown in Fig. 10.3.

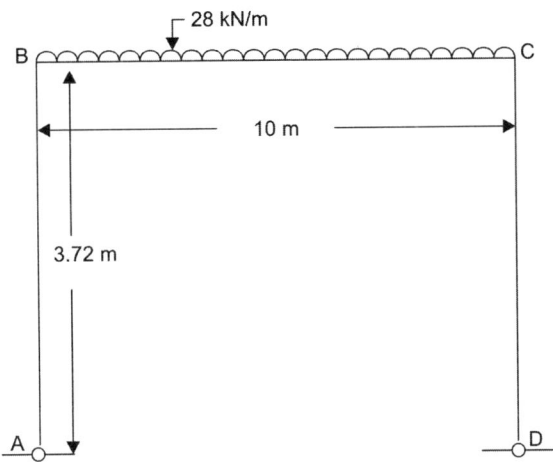

Fig. 10.3: Portal frame

e. *Moment distribution:*

A	B		C		D
	0.5	0.5	0.5	0.5	
		−234	+234		
	+117	+117	−117	−117	
		−58.5	+58.5		
	+29.25	+29.25	−29.25	−29.25	
		−14.6	+14.6		
	+7.3	+7.3	−7.3	−7.3	
		−3.65	+1.83	−1.83	
	+1.83	+1.83	1.83	−1.83	
		−0.92	+0.92		
	+0.46	+0.46	−0.46	−0.46	
0	+156	−156	+156	−156	kN·m

f. *Design moments and shear forces:*

Maximum negative working moment at support B = 156 kN·m

Factored negative moment = (1.5 × 156) = 234 kN·m

Maximum positive working moment at centre of span BC

$$= (0.125 \ wL^2 - 156) = [(0.125 \times 28 \times 10^2) - 156] = 194 \ \text{kN·m}$$

Factored positive bending moment = (1.5 × 194) = 291 kN·m

Maximum working shear force at B = V = (0.5 × 28 × 10) = 140 kN

Factored shear force at B = V_u= (1.5 × 140) = 210 kN

Working shear force at A (194/3.72) = 52.15 kN

Factored shear force at A = (1.5 × 52.15) = 78.23 kN

g. *Design of beam sections:*

Mid span section:

The mid span section is designed as a T-section.

Factored design moment $M_u = 291$ kN·m

Effective width of flange is computed as

$$b_f = [(10/6) + 0.45 + (6 \times 0.12)] = 2.83 \text{ m} < 4 \text{ m}$$

∴ $b_f = 2830$ mm

Limiting neutral axis depth for $f_y = 415$ N/mm² is computed as

$$x_u = 0.48d = (0.48 \times 700) = 336 \text{ mm}$$

$$\left(\frac{D_f}{d}\right) = \left(\frac{120}{700}\right) = 0.17 < 0.2$$

$$\left(\frac{D_f}{x_u}\right) = \left(\frac{120}{336}\right) = 0.35 < 0.43$$

Since the moment value is small, determine the value of neutral axis depth, x_u falling within the flange using the equation

$$M_u = 0.36 f_{ck} x_u b_f (d - 0.42 x_u)$$
$$(291 \times 10^6) = (0.36 \times 20 \times x_u \times 2830)(700 - 0.42 x_u)$$

Solving, $x_u = 22$ mm $< D_f = 120$ mm

Hence considering the section as rectangular, the area of reinforcement required is given by

∴ $$M_u = 0.87 f_y A_{st} d \left[1 - \frac{A_{st} \cdot f_y}{bdf_{ck}} \right]$$

$$(201 \times 10^6) = (0.87 \times 415 \times A_{st} \times 700)\left[1 - \frac{415 A_{st}}{(2830 \times 700 \times 20)} \right]$$

Solving, $A_{st} = 1168$ mm²

Provide 4 bars of 22 mm diameter ($A_{st} = 1520$ mm²)

Support section:

Factored design moment $M_u = 234$ kN·m

The support section is designed as rectangular section, i.e.

$$(234 \times 10^6) = (0.87 \times 415 \times A_{st} \times 700)\left[1 - \frac{415 A_{st}}{(450 \times 700 \times 20)} \right]$$

Solving, $A_{st} = 1000$ mm²

Provide 4 bars of 22 mm diameter ($A_{st} = 1520$ mm²) since the support section has to resist shear force also.

$$\tau_v = \left(\frac{V_u}{bd}\right) = \left(\frac{210 \times 10^3}{450 \times 700}\right) = 0.66 \text{ N/mm}^2$$

$$p_t = \left(\frac{100 A_{st}}{bd}\right) = \left(\frac{100 \times 1520}{450 \times 700}\right) = 0.48$$

Refer to Table 19 (IS:456-2000) and readout the permissible shear stress

$$\tau_c = 0.47 \text{ N/mm}^2 < \tau_v$$

Hence shear reinforcements are required.

Balance shear force $= [V_u - \tau_c bd] = [210 - (0.47 \times 450 \times 700) 10^{-3}] = 62$ kN

Using 6 mm diameter two-legged stirrups, spacing is given by

$$S_v = \left[\frac{A_{sv}\, 0.87\, f_y\, d}{V_{us}}\right] = \left[\frac{0.87\times415\times2\times28\times700}{62\times1000}\right] = 228 \text{ mm}$$

Adopt 6 mm diameter two-legged stirrups at 140 mm centres at support and increase the spacing to 400 mm towards the centre of span.

h. *Design of column section*:

The column section is subjected to a moment

$$M = 156 \text{ kN·m and a thrust } P = 140 \text{ kN}$$

Using a load factor of 1.5

$$M_u = (156 \times 1.5) = 234 \text{ kN·m}$$
$$P_u = (140 \times 1.5) = 210 \text{ kN}$$

Section is of size (450 mm × 600 mm)

$$\therefore \quad b = 450, \qquad D = 600 \text{ mm}$$
$$d = 550 \text{ mm}, \qquad d' = 50 \text{ mm}$$

or $\quad \left(\dfrac{d'}{D}\right) = \left(\dfrac{50}{600}\right) = 0.10$

$$\left(\frac{M_u}{f_{ck}bD^2}\right) = \left(\frac{234\times10^6}{20\times450\times(600)^2}\right) = 0.07$$

$$\left(\frac{P_u}{f_{ck}bD}\right) = \left(\frac{210\times10^3}{20\times450\times600}\right) = 0.04$$

Using interaction curves of SP:16, we get

$$\left(\frac{p}{f_{ck}}\right) = 0.04 \quad \therefore \quad p = (20 \times 0.04) = 0.8$$

$$\therefore \qquad A_s = (pbD/100) = (0.8 \times 450 \times 600)/100 = 2160 \text{ mm}^2$$

But minimum area of steel $= \left(\dfrac{0.8\times450\times600}{100}\right) = 2160 \text{ mm}^2$

Provide 4 bars of 20 mm diameter on each face, ($A_{st} = 2512 \text{ mm}^2$), and 8 mm diameter ties at 300 mm centres throughout the column.

i. *Design of hinge*:

At the hinge portion, concrete is under triaxial stress and can withstand higher permissible stresses.

Permissible compressive stress in concrete at hinge is:

$$= (2 \times 0.4\, f_{ck}) = (2 \times 0.4 \times 20) = 16 \text{ N/mm}^2$$

Factored thrust $P_u = 210 \text{ kN}$

Cross-sectional area of hinge required $= \left[\dfrac{210\times10^3}{16}\right] = 13125 \text{ mm}^2$

Area provided = (200 mm × 100 mm) [$A_c = (200 \times 100) = 20000 \text{ mm}^2$]

Ultimate shear force at hinge = 78.23 kN

If A_{st} = area of reinforcements required to resist the shear force at hinge section

θ = inclination of bars to vertical as shown in Fig. 10.4.

then $(A_{st} \sin \theta \times 0.87 \times 415) = (78.23 \times 10^3)$

$$\therefore \qquad A_{st} = \left[\frac{78.23 \times 10^3}{0.87 \times 415 \times \sin 31°} \right] = 421 \text{ mm}^2$$

Provide 4 bars of 16 mm diameter (A_{st} = 840 mm²)

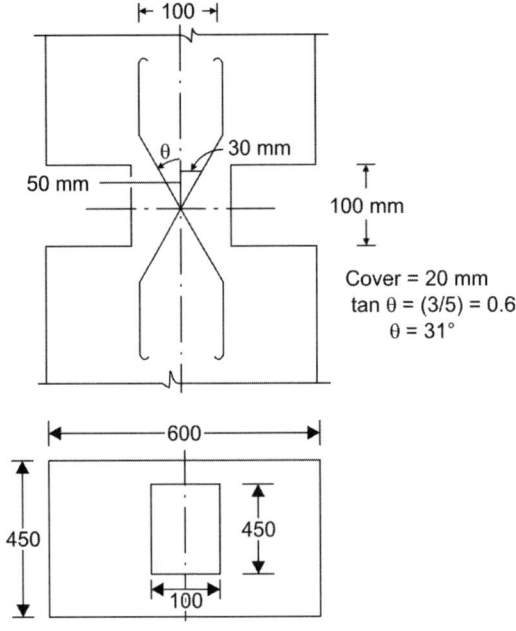

(a) RC hinge at base of column

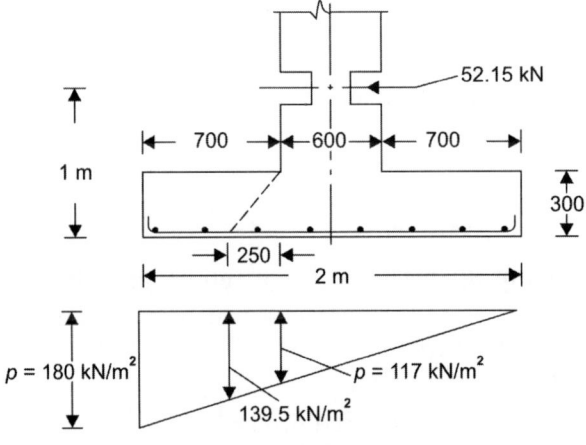

(b) RC footing for column

Fig. 10.4: Details of hinge and footing

j. *Design of foundations*:

Working axial load on column = 140 kN

Self weight of column = (0.4 × 0.6 × 3.72 × 24) = 24 kN

Self weight of foundations @ 10% = 16 kN

Total working load P = 180 kN

Factored axial load P_u = (1.5 × 180) = 270 kN

Working moment about base [Refer to Fig. 10.4b]
$$M = (52.15 \times 1) = 52.15 \text{ kN·m}$$

∴   Eccentricity $e = \left[\dfrac{M}{P}\right] = \left[\dfrac{52.15 \times 10^6}{180 \times 10^3}\right] = 290$ mm

To avoid tension in the foundations:

Breadth $b = 6e = (6 \times 290) = 1740$ mm

Provide a foundation area of (1 m × 2 m)

Intensity of pressure $p = \left[\dfrac{2 \times 180}{1 \times 2}\right] = 180$ kN/m² < 200 kN/m² (safe)

$$p' = \left[\dfrac{180 \times 1.3}{2}\right] = 117 \text{ kN/m}^2$$

Hence total soil pressure on the cantilever portion is given as

$$P = \left[\dfrac{180 + 117}{2}\right] 0.7 = 104 \text{ kN acting at 0.4 m from column edge}$$

Walking moment at edge of column = (104 × 0.4) = 42 kN·m

Ultimate moment M_u = (1.5 × 42) = 62.4 kN·m

Effective depth required for footing slab is computed as

$$d = \sqrt{\dfrac{M_u}{0.138 f_{ck} b}} = \sqrt{\dfrac{62.4 \times 10^6}{(0.138 \times 20 \times 1000)}} = 150 \text{ mm}$$

Depth required to resist shear force is greater than that obtained from moment considerations.

Hence provide overall depth = (2 × 150) = 300 mm

and effective depth d = 250 mm

Area of reinforcements are obtained by computing the parameter as

$$\left(\dfrac{M_u}{bd^2}\right) = \left(\dfrac{62.4 \times 10^6}{1000 \times 250^2}\right) = 0.998$$

Refer to Table 2 of SP:16 (Design Tables) and readout the reinforcement percentage as

$$p_t = \left(\dfrac{100 A_{st}}{bd}\right) = 0.295$$

∴ $A_{st} = \left(\dfrac{0.295 \times 1000 \times 250}{100}\right) = 737.5 \text{ mm}^2$

Provide 16 mm diameter bars at 150 mm centres (A_{st} = 1341 mm²)

Distribution reinforcement = (0.0012 × 300 × 1000) = 360 mm²

Provide 10 mm diameter bars at 180 mm centres (A_{st} = 436 mm²)

Check for shear:

Working shear force acting at a distance of 250 mm from the face of column is given by

$$V = \left[\frac{180 + 139.5}{2}\right] 0.45 = 72 \text{ kN}$$

Ultimate shear force V_u = (1.5 × 72) = 108 kN

$$\tau_v = \left(\frac{V_u}{bd}\right) = \left(\frac{108 \times 10^3}{1000 \times 250}\right) = 0.432 \text{ N/mm}^2$$

$$\left(\frac{100 A_{st}}{bd}\right) = \left(\frac{100 \times 1341}{1000 \times 250}\right) = 0.54$$

Refer to Table 19 of IS:456-2000 and readout the design shear strength of concrete as

$$\tau_c = 0.48 \text{ N/mm}^2$$

Since $\tau_c > \tau_v$, shear reinforcements are not required.

The details of reinforcements in the portal frame are shown in Fig. 10.5.

10.5 DESIGN EXAMPLE OF A FIXED PORTAL FRAME

The roof of a 8 m wide hall is supprted on a portal frame spaced at 4 m intervals.

The height of the portal frame is 4 m. The continuous slab is 120 mm thick. Live load on roof = 1.5 kN/m². Bearing capacity of the soil is 150 kN/m². The columns are connected with a plinth beam and the base of the column may be assumed as fixed. Design the column and beam members and a suitable foundation footing for the columns of the portal frams.

Adopt M20 grade concrete and Fe 415 HYSD bars.

i. *Data*:

Spacing of portal frame = 4 m

Span of portal frame = 8 m

Height of columns = 4 m

Live load on roof = 1.5 kN/m²

Materials: M20 grade concrete and Fe 415 HYSD bars

ii. *Characteristic strength*:

$$f_{ck} = 20 \text{ N/mm}^2 \text{ and } f_y = 415 \text{ N/mm}^2$$

iii. *Design of slab*:

The slab design is similar to that presented in Design Example 8.4, provide 120 mm slab with 10 mm bars at 200 mm centres at supports and mid span sections. Distribution steel is 6 mm diameter at 130 mm centres.

iv. *Design of portal frame*:

Effective span of beam = 8 m

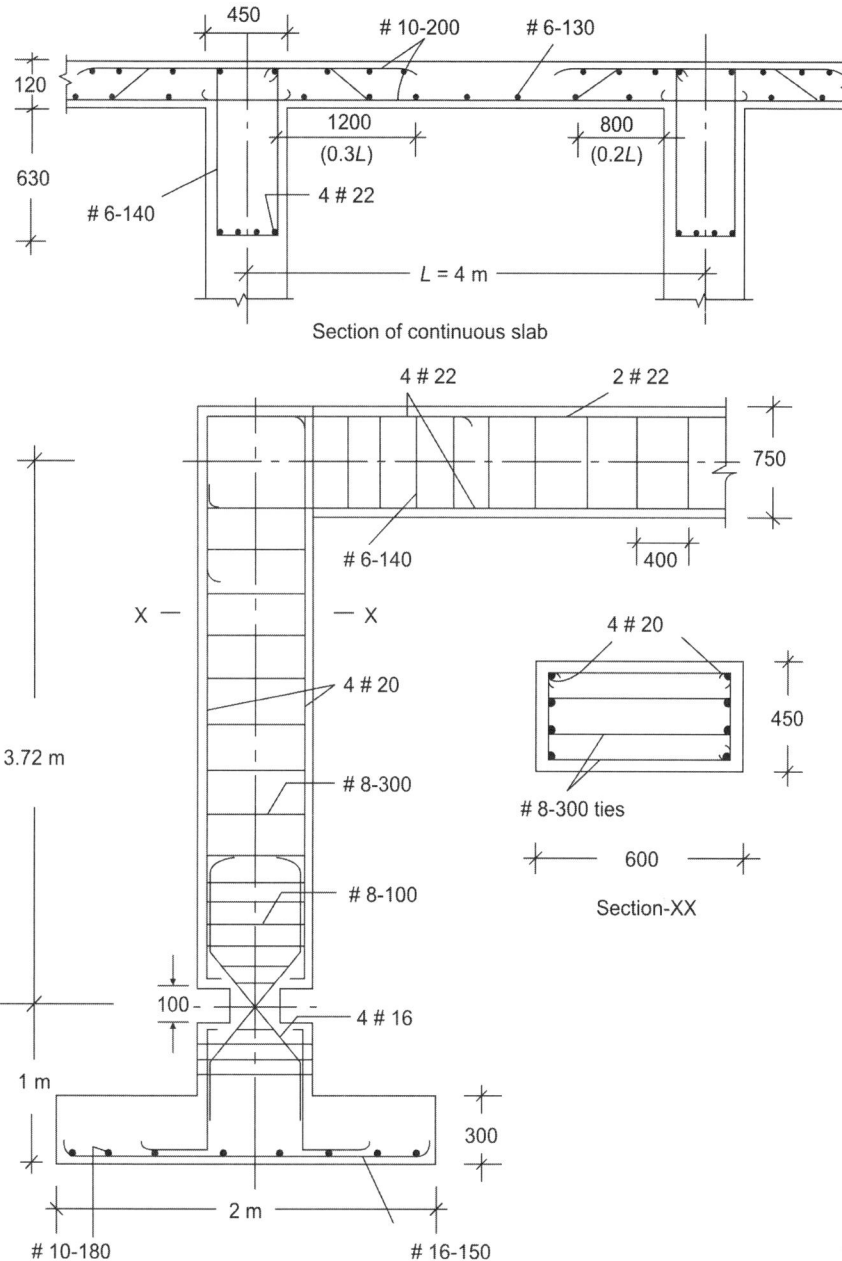

Fig. 10.5: Reinforcement details in portal frame

Effective depth $d = \left(\dfrac{8000}{12}\right) = 666$ mm

Adopt $d = 650$ mm and overall depth = 700 mm, breadth $b = 400$ mm. Column section is assumed as (400 mm × 600 mm).

a. *Load on frame*:

Self weight of slab	$= (0.12 \times 24)$	$= 2.88 \text{ kN/m}^2$
Roof finish		$= 0.50$
Ceiling finish		$= 0.25$
Live load		$= 1.50$
Total load		$= 5.13 \text{ kN/m}^2$
Load from slab	$= (5.13 \times 4)$	$= 20.52 \text{ kN/m}$
Self weight of beam	$= (0.4 \times 0.58 \times 24)$	$= 5.56$
Finishes of beam		$= 0.92$
Total load on beam	$= w$	$= 27.00 \text{ kN/m}$

The moments in the portal frame fixed at base and loaded as shown in Fig. 10.6a are analysed by moment distribution.

$$AB = 4 \text{ m}, \quad BC = 8 \text{ m}$$

$$I_{AB} = \left(\frac{400 \times 600^3}{12} \right), \qquad I_{BC} = \left(\frac{400 \times 700^3}{12} \right)$$

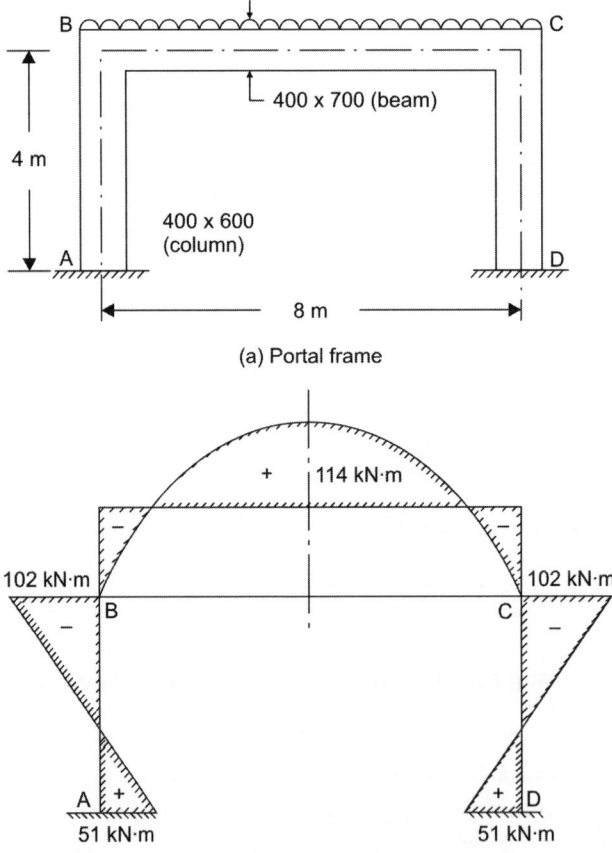

(a) Portal frame

(b) Bending moment diagram

Fig. 10.6: Portal frame (fixed at base)

$\therefore$ $\qquad I_{AB} : I_{BC} = \left(\dfrac{700}{600}\right)^3 = 1 : 1.57$

b. *Relative stiffness*:

$$k_{BA} = \left(\dfrac{I}{4}\right) = 0.25I$$

$$k_{BC} = \left(\dfrac{1.57I}{8}\right) = 0.20I$$

c. *Distribution factors*:

$$d_{BA} = \left(\dfrac{0.25I}{0.25I + 0.20I}\right) = 0.55$$

$$d_{BC} = \left(\dfrac{0.20I}{0.25I + 0.20I}\right) = 0.45$$

d. *Fixed end moments*:

$$F_{BC} = -\left(\dfrac{wL^2}{12}\right) = -\left(\dfrac{27 \times 8^2}{12}\right) = -144 \text{ kN·m}$$

$$F_{BC} = +\left(\dfrac{wL^2}{12}\right) = +\left(\dfrac{27 \times 8^2}{12}\right) = +144 \text{ kN·m}$$

$$F_{BA} = F_{AB} = 0$$

e. *Moment distribution*:

A	B		C		D
	0.55	0.45	0.45	0.55	
0	0	−144	+144		0
39.5	+79	+65	−65	−79	−39.5
		−32.5	+32.5		
8.9	+17.8	+14.7	−14.7	−17.8	−8.9
		−7.4	+7.4		
2.0	+4.0	+3.4	−3.4	−4.0	−2.0
		−1.7	+1.7		
0.47	+0.93	0.77	−0.77	−0.93	−0.47
51	+102	−102	+102	−102	−51 kN·m

f. *Design moments and shear forces*:
   Service load moments: $M_B = 102$ kN·m and $M_A = 51$ kN·m
   Design ultimate moments are computed as
   $$M_{uB} = (1.5 \times 102) = 153 \text{ kN·m}$$
   $$M_{uA} = (1.5 \times 51) = 76.5 \text{ kN·m}$$

Maximum positive service load moment at centre of span BC
$$= [0.125 \, wL^2 - 102]$$
$$= [(0.125 \times 27 \times 8^2) - 102] = 114 \text{ kN·m}$$

Ultimate moment at centre of span BC = M_u = (1.5 × 114) = 171 kN·m

The service load bending moment diagram for the portal frame is shown in Fig. 10.6b.

Maximum working shear force at B = (0.5 × 27 × 8) = 108 kN

Maximum ultimate shear force at B = V_u = (1.5 × 108) = 162 kN

Working load shear force at A $= \left(\dfrac{M_A + M_B}{4} \right) = \left(\dfrac{51 + 102}{4} \right) = 38.25 \text{ kN}$

Ultimate shear force at A = (1.5 × 38.25) = 57.37 kN

g. *Design of beam section*:

Mid span section is designed as a T-section.

Factored design moment M_u = 171 kN·m

Effective width of flange (b_f) is given by the relation
$$b_f = [(L_o/6) + b_w + 6D_f]$$
$$= [(8/6) + 0.4 + (6 \times 0.12)] = 2.45 \text{ m} < 4 \text{ m}$$

Hence, b_f = 2450 mm and effective depth d = 650 mm
$$x_u = 0.48 \, d = (0.48 \times 650) = 312 \text{ mm}$$

$$\left[\frac{D_f}{d} \right] = \left[\frac{120}{650} \right] = 0.18 < 0.2$$

$$\left[\frac{D_f}{x_u} \right] = \left[\frac{120}{312} \right] = 0.38 < 0.43$$

Since the moment value is small, determine the value of neutral axis depth, x_u falling within the flange using the equation
$$M_u = 0.36 \, f_{ck} x_u b_f \, (d - 0.42 \, x_u)$$
$$(171 \times 10^6) = [(0.36 \times 20 \times x_u \times 2450)(650 - 0.42 x_u)]$$

Solving, x_u = 15 mm < D_f = 120 mm

Hence, considering the section as rectangular, the area of reinforcement required is computed using the equation

$$M_u = 0.87 f_y A_{st} d \left[1 - \frac{A_{st} f_y}{b d f_{ck}} \right]$$

$$(171 \times 10^6) = (0.87 \times 415 \, A_{st} \times 650) \left[1 - \frac{415 \, A_{st}}{(2450 \times 650 \times 20)} \right]$$

Solving, A_{st} = 973 mm²

Provide 4 bars of 20 mm diameter (A_{st} = 1256 mm²)

Support section:

Factored design moment M_u = 153 kN·m

The support section is designed as rectangular section and the reinforcement area is computed using the equation

$$(153 \times 10^6) = (0.87 \times 415\, A_{st} \times 650)\left[1 - \frac{415\, A_{st}}{400 \times 650 \times 20}\right]$$

Solving, $A_{st} = 687$ mm^2 > $A_{st\ min} = 512$ mm^2

Since the section has to resist shear force also, provide 4 bars of 20 mm diameter providing an area of 1256 mm^2.

Nominal shear stress $\tau_v = \left(\dfrac{V_u}{bd}\right) = \left(\dfrac{162 \times 10^3}{400 \times 650}\right) = 0.62$ N/mm^2

$$p_t = \left(\frac{100\, A_{st}}{bd}\right) = \left(\frac{100 \times 1256}{400 \times 650}\right) = 0.48$$

Refer to Table 19 (IS:456-2000) and readout the permissible shear stress as

$$\tau_c = 0.47 \text{ N/mm}^2 < \tau_v$$

Hence shear reinforcements are required.

Balance shear force $V_{us} = [162 - (0.47 \times 400 \times 650)\, 10^{-3}] = 40$ kN

Using 6 mm diameter two-legged stirrups, spacing is given by the relation

$$S_v = \left[\frac{0.87\, f_y\, A_{sv}\, d}{V_{us}}\right] = \left[\frac{0.87 \times 415 \times 2 \times 28 \times 650}{40 \times 10^3}\right] = 328 \text{ mm}$$

Adopt 300 mm spacing near the supports, gradually increasing to 400 mm towards the centre of span.

h. *Design of column section*:

Section at top (B):

Moment $\quad M = 102$ kN·m

Thrust $\quad\ \ P = 108$ kN

Using a load factor of 1.5

$$M_u = (102 \times 1.5) = 153 \text{ kN·m}$$
$$P_u = (108 \times 21.5) = 162 \text{ kN}$$

Column section is of size (400 mm × 600 mm)

$\therefore \quad b = 400$ mm, $\quad d = 550$ mm

$$D = 600 \text{ mm}, \quad d' = 50 \text{ mm}, \quad \left(\frac{d'}{D}\right) = 0.10$$

$f_{ck} = 20$ N/mm^2

$$\left(\frac{M_u}{f_{ck}bD^2}\right) = \left(\frac{153 \times 10^6}{20 \times 400 \times 600^2}\right) = 0.053$$

$$\left(\frac{P_u}{f_{ck}bD}\right) = \left(\frac{162 \times 10^3}{20 \times 400 \times 600}\right) = 0.033$$

Using interaction curves of SP:16, we get

$$\left(\frac{p}{f_{ck}}\right) = 0.03 \quad \therefore \quad p = (20 \times 0.03) = 0.6$$

$\therefore \qquad A_s = (pbD/100) = (0.6 \times 400 \times 600)/100 = 1440 \text{ mm}^2$

But minimum area of steel $= \left(\dfrac{0.8 \times 400 \times 600}{100} \right) = 1920 \text{ mm}^2$

Provide 4 bars of 20 mm diameter on each face ($A_{st} = 2512 \text{ mm}^2$) and 8 mm diameter ties at 300 mm centres throughout the column.

Same reinforcements are provided at the base section of the column.

i. *Design of foundations*:

Factored design moment $= (1.5 \times 25.6) = 38.4 \text{ kN·m}$

Effective depth required for footing slab is computed as

$$d = \sqrt{\frac{M_u}{0.138 \, f_{ck} \, b}} = \sqrt{\frac{38.4 \times 10^6}{0.138 \times 20 \times 1000}} = 118 \text{ mm}$$

Depth required to resist shear force is greater than that obtained from moment considerations. Hence adopt overall depth of 300 mm.

Effective depth $d = 250$ mm

Area of reinforcements are obtained by computing the parameter

$$\left(\frac{M_u}{bd^2} \right) = \left(\frac{38.4 \times 10^6}{1000 \times 250^2} \right) = 0.614$$

Refer to Table 2 of SP:16 (Design Tables) and readout the percentage steel as

$$p_t = \left(\frac{100 \, A_{st}}{bd} \right) = 0.178$$

$\therefore \qquad A_{st} = \left(\dfrac{0.178 \times 1000 \times 250}{100} \right) = 445 \text{ mm}^2 > A_{st\,min}$

Provide 12 mm diameter bars at 150 mm centres bothways ($A_{st} = 754 \text{ mm}^2$).

Check for shear:

Shear force acting at a distance of 250 mm from the face of the column is given by:

Working shear force $V = \left(\dfrac{103 + 79}{2} \right) 0.5 = 45.5 \text{ kN}$

Ultimate shear force $V_u = (1.5 \times 45.5) = 68.25 \text{ kN}$

$$\tau_v = \left(\frac{V_u}{bd} \right) = \left(\frac{68.25 \times 10^3}{1000 \times 250} \right) = 0.273$$

$\therefore \qquad \left(\dfrac{100 \, A_{st}}{bd} \right) = \left(\dfrac{100 \times 754}{1000 \times 250} \right) = 0.30$

Refer to Table 19 of IS:456-2000 and readout the design shear strength of concrete τ_c as

$$\tau_c = 0.38 \text{ N/mm}^2$$

Since $\tau_c > \tau_v$, the footing slab safely resist the shear stresses.

The details of reinforcements in the portal frame is shown in Fig. 10.7.

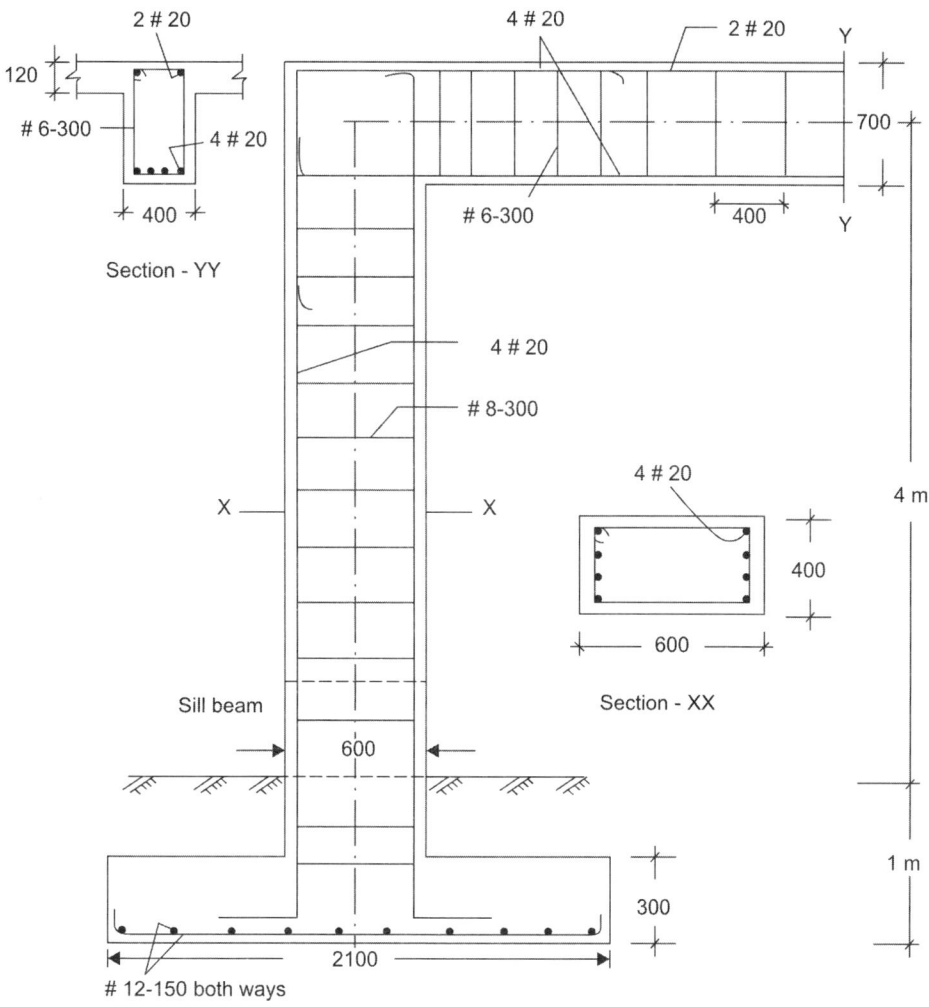

Fig. 10.7: Reinforcement details in portal frame

10.6 DESIGN EXAMPLE OF A RIGID FRAME HIGHWAY BRIDGE

A rigid frame bridge for a highway crossing has a span 15 m with a road width of 7.5 m. The selected portal frame dimensions based on empirical relations is shown in Fig. 10.5. The bridge was analyzed for IRC Class AA tracked vehicle loading, earth pressure from sides acting on the columns of the portal frame together with temperature effects up to 30 °C. The bridge had to be designed using M20 grade concrete and Fe 415 HYSD bars. The abstract of design moments and thrusts developed at the salient sections like crown, point 5 (near junction of transom and column) and point 2 (middle of column) determined by rigorous analysis is compiled in Table 10.1. Design suitable reinforcements at the salient sections of the portal frame conforming to IS:456-2000 code specifications of limit state design. Sketch the details of reinforcements in half of the portal frame.

Table 10.1: Abstract of moments and thrusts in portal frame

Particulars of loading	Crown		Point 5		Point 2	
	BM (kN·m)	Thrust (kN)	BM (kN·m)	Thrust (kN)	BM (kN·m)	Thrust (kN)
Dead loads	227.3	36.8	−88.88	36.8	−112.9	36.8
Live loads	168.6	19.74	−76.44	18.33	−59.0	19.43
Earth pressure	EP from		EP from			
−ve or +ve or both	left only		right only			
Whichever produces:						
Worst effect	6.55	−15	−78.20	−15	−41.83	−15
Temperature action causing						
worst moment	19.71	−3.18	−1914	3.18	−8.58	3.18
Total values	422.16	38.36	−261.66	43.31	−222.31	44.41

The salient sections of the portal frame are designed for the maximum moments and thrusts compiled in Table 10.1.

a. *Crown section*:

Overall depth of the section D = 450 mm

Width b = 1000 mm

Cover d' = 50 mm

Design moment M = 422.16 kN·m

Design thrust P = 38.36 kN

Ultimate moment M_u = (1.5 × 422.16) = 633.24 kN·m

Ultimate thrust P_u = (1.5 × 38.36) = 57.54 kN

Using M20 grade concrete f_{ck} = 20 N/mm²

The parameters are computed as

$$\left[\frac{P_u}{f_{ck}bD}\right] = \left[\frac{57.54 \times 10^3}{20 \times 1000 \times 450}\right] = 0.0063$$

$$\left[\frac{M_u}{f_{ck}bD^2}\right] = \left[\frac{633.24 \times 10^6}{20 \times 1000 \times 450^2}\right] = 0.156$$

Using Chart 32 of SP:16 (Design Aids to IS:456)

For (d'/D) = 0.10, the value of (p/f_{ck}) = 0.10 or p = 2.00

A_s = $(pbD/100)$ = [(2.00 × 1000 × 450)/100] = 9000 mm²

Use 25 mm diameter bars at 100 mm centres both at top and bottom.

b. *Section 5*:

Overall depth of section D = 1000 mm

Cover d' = 50 mm

Design moment $M = -261.66$ kN·m

Design thrust $P = 43.31$ kN

Ultimate moment $M_u = (1.5 \times 261.66) = 392.5$ kN·m

Ultimate thrust $P_u = (1.5 \times 43.31) = 65$ kN

Using M20 grade concrete, $f_{ck} = 20$ N/mm²

$$\left[\frac{P_u}{f_{ck}bD}\right] = \left[\frac{65 \times 10^3}{20 \times 1000 \times 1000}\right] = 0.0032$$

$$\left[\frac{M_u}{f_{ck}bD^2}\right] = \left[\frac{392.5 \times 10^6}{20 \times 1000 \times (1000)^2}\right] = 0.0196$$

Using Chart 31 of SP:16 (Design Aids to IS:456)

For $(d'/D) = 0.050$, the value of $(p/f_{ck}) = 0.01$ ∴ $p = (0.01 \times 20) = 0.2$

$A_s = (pbD/100) = [(0.20 \times 1000 \times 1000)/100] = 2000$ mm²

Minimum reinforcement $= [(0.8/100) \times 1000 \times 1000] = 8000$ mm²

Use 25 mm diameter bars at 100 mm centres both at top and bottom.

c. *Section 2*:

Overall depth of section $D = 840$ mm

Cover $d' = 50$ mm

Design moment $M = -222.316$ kN·m

Design thrust $P = 44.41$ kN

Ultimate moment $M_u = (1.5 \times 222.316) = 334$ kN·m

Ultimate thrust $P_u = (1.5 \times 44.41) = 67$ kN

Using M20 grade concrete, $f_{ck} = 20$ N/mm²

$$\left[\frac{P_u}{f_{ck}bD}\right] = \left[\frac{67 \times 10^3}{20 \times 1000 \times 840}\right] = 0.004$$

$$\left[\frac{M_u}{f_{ck}bD^2}\right] = \left[\frac{334 \times 10^6}{20 \times 1000 \times (840)^2}\right] = 0.023$$

Using Chart 31 of SP:16 (Design Aids to IS:456)

For $(d'/D) = 0.050$, the value of $(p/f_{ck}) = 0.01$

$A_s = (pbD/100) = [(0.01 \times 1000 \times 840)/100] = 84$ mm²

Minimum reinforcement $= [(0.8/100) \times 1000 \times 840] = 6720$ mm²

Use 25 mm diameter bars at 100 mm centres on both faces.

Distribution bars $= [(0.12/100) \times 1000 \times 840] = 1080$ mm²

Provide 16 mm diameter bars at 300 mm centres.

The reinforcement details in the portal frame is shown in Fig. 10.8.

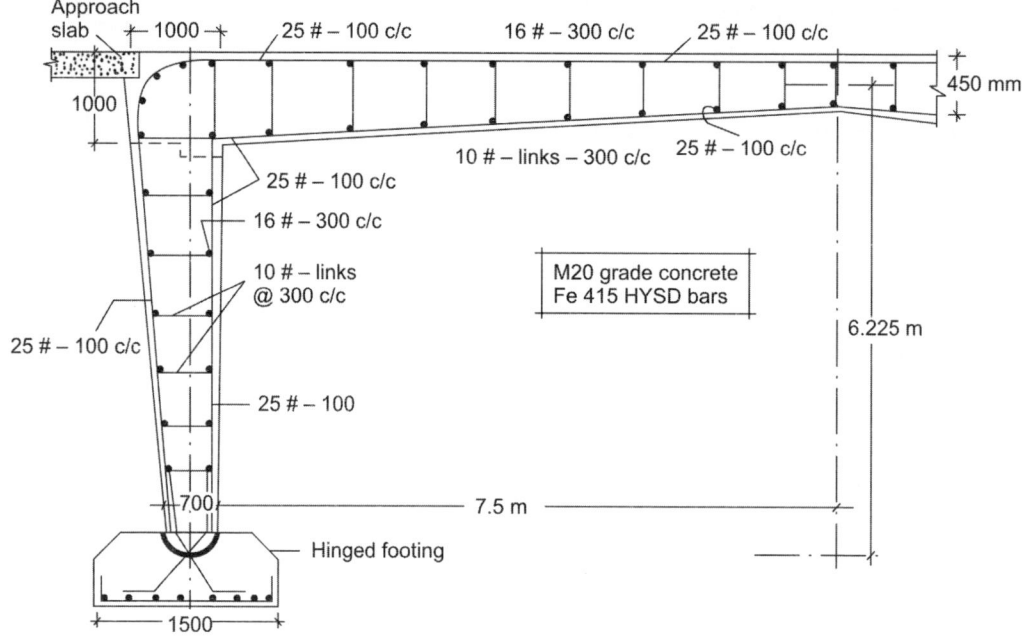

Fig. 10.8: Longitudinal section of rigid frame

References

1. Taylor FW, Thompson SE and Smulski E, *Reinforced Concrete Bridges*, John Wiley & Sons, New York, 1955.

2. Chettoe CS and Adams HC, *Reinforced Concrete Bridge Design*, Chapman & Hall, London, 1952, pp. 1–416.

3. Krishna Raju N, *Design of Bridges*, 4 edn, Oxford & IBH, New Delhi, 2009, pp. 265–294.

4. Raina VK, *Concrete Bridge Practice, Analysis, Design & Economics*, Tata McGraw-Hill, New Delhi, 1991, pp. 196–200.

5. Krishna Raju N, *Prestressed Concrete Bridges*, CBS Publishers, New Delhi, 2009, pp. 165–186.

6. Rowe RE, *Concrete Bridge Design*, CR Books Ltd, London, 1962, p. 336.

7. Marshall WT and Nelson HM, *Structures*, Pitman, London, 1970, pp. 1–442.

8. Reddy CS, *Basic Structural Analysis*, Tata McGraw-Hill, New Delhi, 1981, pp. 382–435.

9. SP:16-1980, *Design Aids for Reinforced Concrete to IS:456*, Bureau of Indian Standards, New Delhi, 1980.

10. IS:456-2000, *Indian Standard Code of Practice for Plain and Reinforced Concrete*, Fourth Revision, Bureau of Indian Standards, 2000, p. 100.

Assignment

1. A portal frame ABCD has fixed supports at A and D. The columns AB and CD are 5 m in height while the transom BC is 10 m in length. The frames are spaced at 3.5 m intervals. The live load on the roof slab which is 10 cm thick may be taken as

1.5 kN/m^2. Design the transom BC and sketch the details of reinforcements. Adopt M20 concrete and Fe 415 HYSD bars.

2. A hall 50 m long and 10 m wide has to be covered by a continuous RCC slab over portal frames spaced at 3 m intervals. The height of the hall is 7 m. Design the slab and one intermediate portal frame. Adopt M20 grade concrete and Fe 415 grade tor steel. Sketch the details of reinforcements.

3. The roof of an assembly hall 30 m long and 12 m wide between centres of columns, consists of a continuous reinforced concrete slab over rectangular portal frames spaced 3 m apart. The columns are provided with independent footings and are hinged at the bottom. The ceiling height is 3.5 m above the hinge level. Adopting M20 grade concrete and Fe 415 grade tor steel, design the continuous roof slab, the portal frame and foundation footing for the columns. Assume safe bearing capacity of the soil as 150 kN/m^2. Sketch the details of reinforcements in the portal frame.

4. A reinforced concrete two bay portal frame is used for covering a hall 20 m wide by 40 m length. Distance between centre line of columns is 10 m. The portals are spaced at 4 m intervals. The height of the columns is 5 m. Design the continuous slab, and one of the intermediate portal frames. Assume hinged condition at the base of the columns. Assume safe bearing capacity of the soil as 200 kN/m^2. Sketch the details of reinforcements in the portal frame. Adopt M25 grade concrete and Fe 500 HYSD bars.

5. A reinforced concrete portal frame is to be designed for an effective span of 18 m with height of columns being 4 m. The portal frame is fixed at the supports. If the imposed dead and live loads on the transom is 30 kN/m, design the reinforcements in the portal frame members. Assume overall dimensions of both the column and beam as 400 mm by 800 mm. Adopt M25 grade concrete and Fe 415 HYSD bars. Design also a suitable foundation footing assuming the safe bearing capacity of the soil at site as 200 kN/m^2 and the depth of the foundation as 1.5 m below average ground level. Sketch the details of reinforcements in the portal frame and foundations.

6. A rigid frame highway bridge crossing is required for a span of 10 m. The portal frame comprises a transom of 500 mm deep by 1000 mm width. The column section is 800 mm by 1000 mm. The analysis of highway loadings and other forces indicate the maximum design moments are 500 and 250 kN·m at crown and centre of column sections respectively. The corresponding thrusts at these sections are 50 and 40 kN respectively. Adopting M25 grade concrete and Fe 500 HYSD reinforcements, design suitable reinforcements at these sections using the limit state provisions of IS:456 code.

Review Questions

1. What are the advantages of using portal frames for structural construction? In what way they are superior to other types of structural members?

2. Briefly explain the various structural elements of a portal frame mentioning their main function in resisting loads.

3. Under what type of structures do you classify the portal frames?

4. What are the various standard methods adopted for the analysis of portal frames?

5. What are the structural advantages of using reinforced concrete portal frames with hinged supports at the base?

6. What are the specific advantages of using reinforced concrete portal frames in bridge structures?

7. How do you fix the dimensions of the transom and column members of a reinforced concrete portal frame proposed for a national highway crossing?

8. In the case of portal frames spaced at certain intervals, what type of roof you would propose to cover the space in between the frames?

9. Explain with sketches the different types of rigid frame bridges generally adopted when using portal frames.

10. Explain with sketches the detailing of reinforcements in the transom, vertical columns and at the supports of a typical portal frame.

Objective Type Questions

1. Reinforced concrete portal frames of single and multibay with different types of support conditions are classified as
   a. statically determinate structures
   b. complex structures
   c. statically indeterminate structures

2. Portal frame structures are generally preferred for
   a. short spans
   b. medium spans
   c. very long spans

3. Maximum positive moments in a portal frame under dead and live loads develop at
   a. the support section
   b. the base of the columns
   c. the crown section

4. In the case of portal frames subjected to dead, live and earth pressures, the salient sections of the frame are subjected to
   a. thrust only
   b. moment only
   c. moment and thrust

5. Portal frames hinged at the supports will have
   a. larger sections
   b. smaller sections
   c. same section as that of crown

6. The crown section of a reinforced concrete portal frame is designed by considering the section as
   a. rectangular
   b. tee section
   c. inverted tee section

7. The hinge portion at the base of the supports of a portal frame is designed by considering the stress distribution at the hinge as
   a. uniaxial
   b. biaxial
   c. triaxial

8. In comparison with other traditional types of structures, portal frame structures consume
   a. more materials
   b. same amount of materials
   c. lesser quantity of materials

9. Maximum shear force due to loads in a portal frame structure develops at
   a. the centre of crown section
   b. at the bottom of column section
   c. near the junction of transom and column

10. Precast portal frames can be conveniently cast by arranging the formwork in the
    a. vertical position
    b. horizontal position
    c. inclined position

11
Multistorey Building Frames

11.1 ■ GENERAL ASPECTS

A typical multistorey building frame comprises of columns, beams and slabs forming the floors and roof in the structural system. The first reinforced concrete multistorey building frame of 15 storeys shown in Fig. 11.1 was built by EL Ransome in 1903[1]. Multistorey building framed structures are generally composed of columns and beams with slabs in between them forming the floors and roof. In most of the populated cities of the world, the demand for space in the heart of the city necessitated vertical growth of buildings. Most of the metropolis in the world resort to multistorey framed buildings catering to the residential and commercial needs of the community. Advanced research on the behaviour of multistorey framed structures to wind loads has helped in evolving novel types of lateral load resisting systems in multistorey structures.

New concept lateral load resisting systems include the shear walls which usually extend over the full height of the building and commonly located at the lift and staircase zones. Shear walls can be located along the transverse direction of a building either as interior or exterior walls having considerable depth in the direction of lateral loads. They resist the lateral wind loads by bending like vertical cantilevers with fixed base. Shear wall–frame systems can be used up to 40 storeys and for taller buildings framed tube system comprising closely spaced columns with deep spandrel beams connecting the columns along the periphery of the building are generally preferred. For taller buildings, it is economical to adopt concepts of tube-in-tube in which an inner tube like the central shear core is connected with the outer framed tube. For sky scrappers exceeding 80 storeys, bundled tube or multi-cell framed tube system is found to be more effective in resisting wind loads. An excellent summary of different systems used for multistorey structures presented by Unnikrishna Pillai and Devdas Menon[2] is illustrated in Fig. 11.2.

11.2 ■ ANALYSIS OF MULTISTOREY FRAMES

Reinforced concrete multistorey building frames with number of rigid joints between columns and beams are classified as indeterminate structures. Rigorous analysis of multistorey frames subjected to dead and imposed loads involves complex computations. There are several well established methods of analysis such as moment

Fig. 11.1: First reinforced concrete sky scrapper (1903) built by EL Ransome

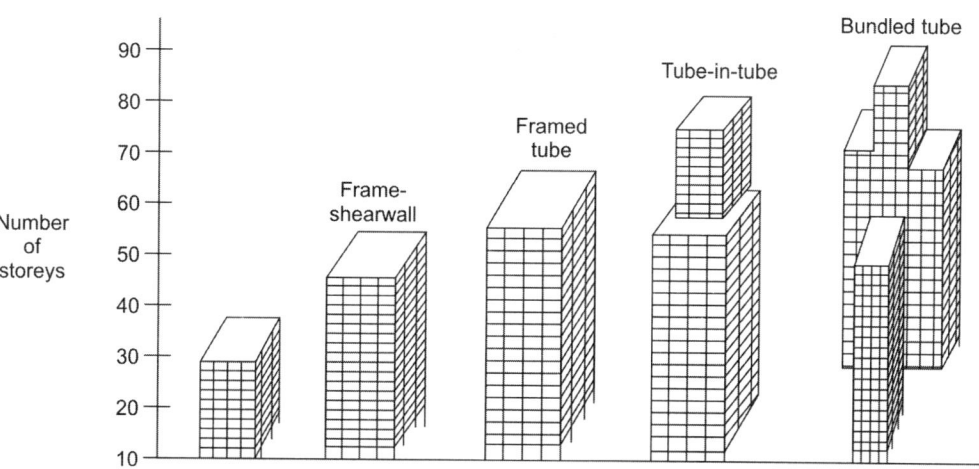

Fig. 11.2: Systems used in multistorey frames

distribution[3], slope deflection, Kani's rotation contribution method[4], Takabeya's method and matrix methods. However, the classical methods like moment distribution and Kani's rotation contribution are suitable for small frames involving lengthy computations when used for multistorey and multibay frames.

Matrix methods[5,6,7] with the use of computers are ideally suited for the analysis of multistorey frames with large number of redundants. However, certain reasonable assumptions will result in approximate methods which are simpler in computational effort, although resulting in conservative estimates of the redundant forces. The most commonly used method for the analysis of vertical loads comprising dead and live loads is the substitute frame method. For the analysis of wind loads, the portal and cantilever methods are generally used. The following section illustrates the use of substitute frame method for multistorey and multibay frames.

At present advanced computer software is available for analysis and design of multi-storey building frames subjected to dead, wind and seismic loads. The reader may refer to excellent monographs by Taranath[8] and Smith *et al*[9] for analysis and design of tall buildings. Reynold's Reinforced Concrete Designer's Handbook[10] is very useful for simpler methods of analysis and design of multistorey buildings.

11.3 ░ METHOD OF SUBSTITUTE FRAMES

A substitute frame consists of a small portion of the multistorey, multibay frame generally comprising the floor beams, with the columns above and below the floor assumed to be fixed at the far ends as shown in Fig. 11.3.

It is sufficient to consider the loads on the two nearest spans on each side of the joint under consideration. The continuous beam is analysed for vertical loads by moment distribution to compute the maximum span and support moments using the following criterion:

i. The maximum positive bending moment at mid point of any particular span develops when the load is placed on the span under consideration and on the alternate span as shown in Fig. 11.4a.

ii. The maximum negative bending moment at any particular support develops when the loads are placed on two spans adjacent to the support under consideration as shown in Fig. 11.4b.

iii. The maximum negative bending moment at mid point of any particular span develops when the loads

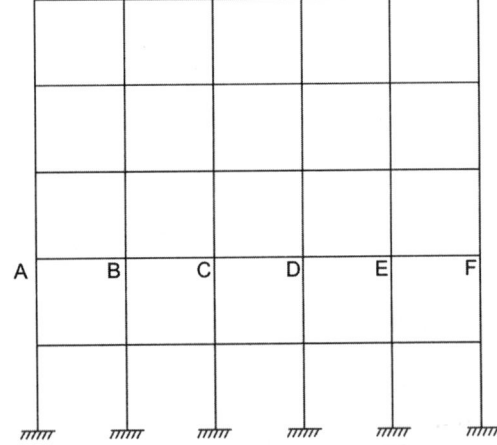

(a) Multibay building frame (on the alternate span)

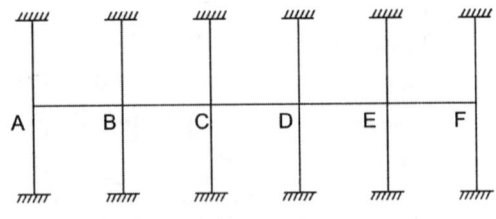

(b) Substitute frame

Fig. 11.3: Multistorey–multibay building frame

are placed on the spans adjacent to the span under consideration as shown in Fig. 11.4c.

The computation of moments in beams and columns by using a substitute frame is illustrated by the following design example.

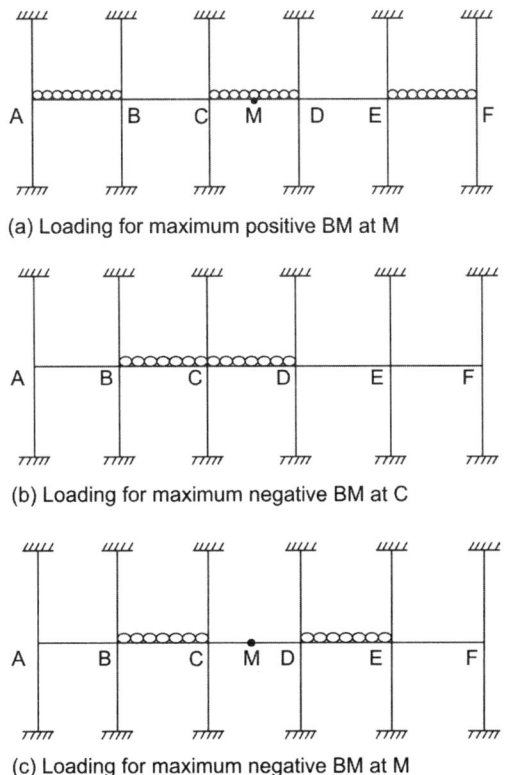

(a) Loading for maximum positive BM at M

(b) Loading for maximum negative BM at C

(c) Loading for maximum negative BM at M

Fig. 11.4: Loading pattern for maximum moments

11.4 DESIGN EXAMPLE

The substitute frame shown in Fig. 11.5 has to be analysed for maximum positive and negative moments in the beams AB, BC and CD. Use the following data to estimate the maximum moments in beams and columns. The beams are spaced at 3 m intervals.

Thickness of floor = 100 mm

Live load (residential flats) = 2 kN/m²

Floor finish = 0.6 kN/m²

Size of beams = 200 mm × 400 mm

Size of column = 200 mm × 400 mm

   i. *Loads on beams*:

     Self weight of slab = (0.1 × 24) = 2.4

     Floor finish   = 0.6

     Dead load   = 3.0 kN/m²

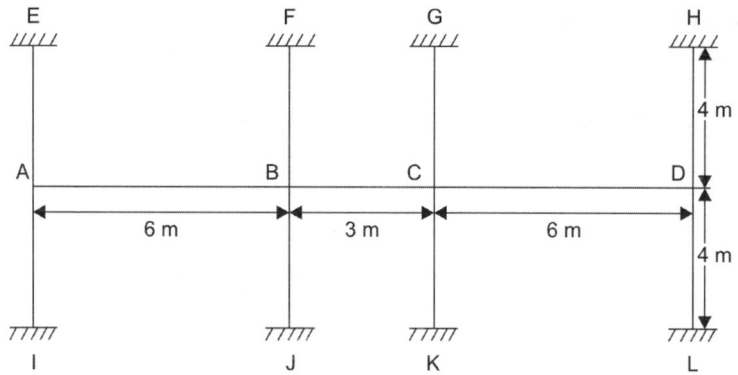

Fig. 11.5: Substitute frame for analysis of moments

Live load $= q = 2 \text{ kN/m}^2$
Self weight of beam $= (0.2 \times 0.4 \times 24) = 2 \text{ kN/m}$
Dead load from slab $= (3 \times 3) = 9 \text{ kN/m}$
Live load $q = (2 \times 3) \quad = 6 \text{ kN/m}$
$\therefore$ Total DL on beam $= (9 + 2) = 11 \text{ kN/m}$
     Total LL on beam $= 6 \text{ kN/m}$

ii. *Stiffness and distribution factors:*
   Since the beam and column section is same (200 mm × 400 mm)
   $\therefore \qquad I_{\text{beam}} = I_{\text{column}}$

$$k_{AB} = k_{CD} = \left(\frac{I}{6}\right), \quad K_{AE} = \left(\frac{I}{4}\right) - k_{AI} \text{ and } k_{BC} = \left(\frac{I}{3}\right)$$

$$\therefore \qquad d_{AB} = \left[\frac{I/6}{\dfrac{I}{6}+\dfrac{I}{4}+\dfrac{I}{4}}\right] = 0.25$$

$$d_{AE} = d_{AI} = \left[\frac{I/4}{\dfrac{I}{6}+\dfrac{I}{4}+\dfrac{I}{4}}\right] = 0.375$$

$$d_{BA} = \left[\frac{I/4}{\dfrac{I}{6}+\dfrac{I}{3}+\dfrac{I}{4}+\dfrac{I}{4}}\right] = 0.166$$

$$d_{BC} = \left[\frac{I/3}{\dfrac{I}{6}+\dfrac{I}{3}+\dfrac{I}{4}+\dfrac{I}{4}}\right] = 0.333$$

$$d_{EF} = d_{BJ} = \left(\frac{I/4}{1.0}\right) = 0.25$$

iii. *Fixed end moments*:

Dead load FEM

$$F_{AB} = -\left(\frac{11 \times 6^2}{12}\right) = -33 \text{ kN·m}$$

$$F_{BC} = -\left(\frac{11 \times 3^2}{12}\right) = -8.25 \text{ kN·m}$$

Live load FEM

$$F_{AB} = -\left(\frac{6 \times 6^2}{12}\right) = -18 \text{ kN·m}$$

$$F_{BC} = -\left(\frac{6 \times 3^2}{12}\right) = -4.5 \text{ kN·m}$$

For span AB, positive bending moment at mid span

$$= \left(\frac{wL^2}{8}\right) = \left(\frac{17 \times 6^2}{8}\right) = 76.5 \text{ kN·m}$$

For span BC, positive bending moment at mid span

$$= \left(\frac{wL^2}{8}\right) = \left(\frac{17 \times 3^2}{8}\right) = 19.12 \text{ kN·m}$$

iv. *Moments in beams*:

Case 1: Maximum positive BM at mid span of AB. The continuous beam ABCD is loaded as shown in Fig. 11.6 and moment distribution is carried out to determine the support moments.

Negative moments at supports A and B are

$$M_{AB} = 41 \text{ kN·m}$$
$$M_{BA} = 48 \text{ kN·m}$$

Positive BM ordinate at mid span of AB

$$= \left(\frac{17 \times 6^2}{8}\right) = 76.5 \text{ kN·m}$$

Maximum postive BM at mid span of AB

$$= \left[76.5 - \left(\frac{41 + 48}{2}\right)\right] = 32 \text{ kN·m}$$

Case 2: Maximum positive BM at mid span of BC

Referring to Fig. 11.7, the support moments obtained from moment distribution are given by

$$M_{BC} = M_{CB} = 18.6 \text{ kN·m}$$

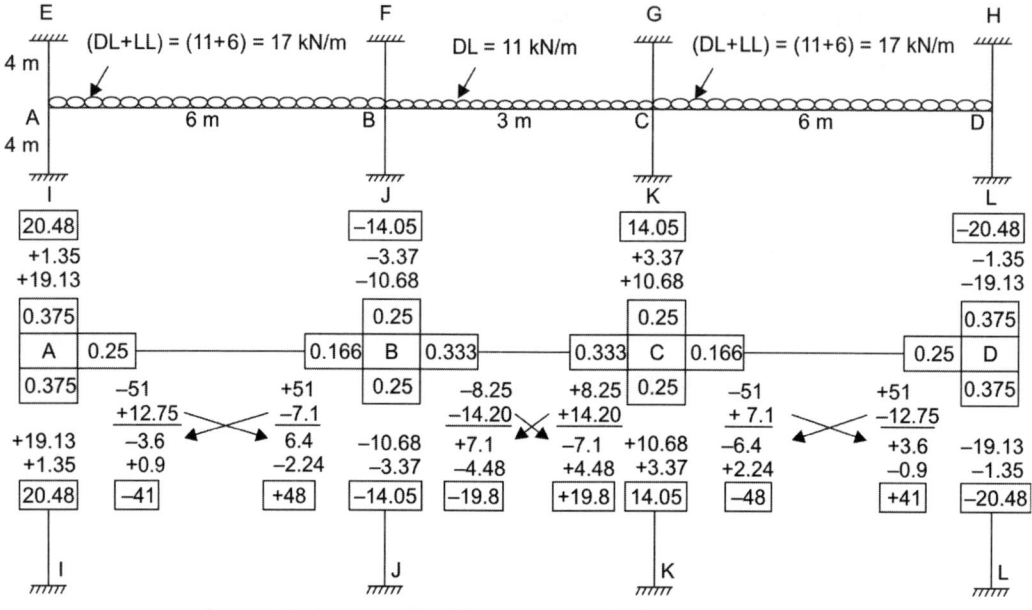

Case 1: Maximum positive BM at mid span of AB (moments in kN.m)
(a) DL on ABCD   (b) LL on AB and CD

Fig. 11.6: Moment distribution of substitute frame

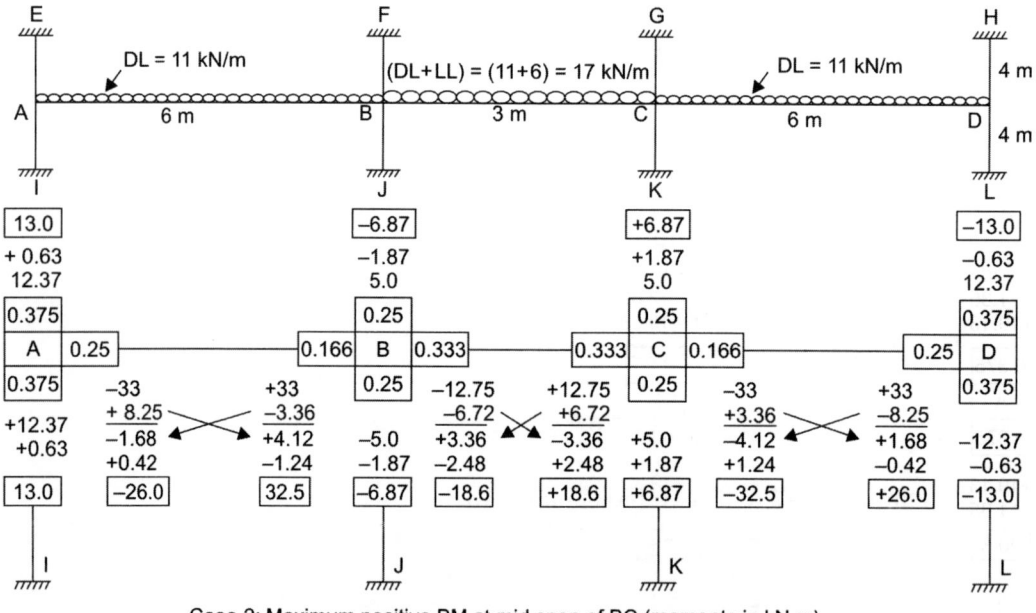

Case 2: Maximum positive BM at mid span of BC (moments in kN·m)
(a) DL on ABCD   (b) LL on BC

Fig. 11.7: Moment distribution of substitute frame

Positive BM ordinate at mid span of BC

$$= \left(\frac{17 \times 3^2}{8}\right) = 19.1 \text{ kN·m}$$

∴ Maximum positive BM at mid span of BC

$$= (19.1 - 18.6) = 0.5 \text{ kN·m}$$

Case 3: Maximum negative BM and SF at A

Referring to Fig. 11.8, the maximum negative BM at support section A of beam AB is obtained as

$$M_{AB} = -41 \text{ kN·m}$$

By analysing the free body diagram (FBD), the shear force at support A is computed as

$$V_A = \left(\frac{17 \times 6}{2}\right) + \left(\frac{41 - 48.5}{6}\right) = 49.75 \text{ kN}$$

Case 4: Maximum negative BM and SF at B

Referring to Fig. 11.9, the maximum negative BM at support section B of beam BA is obtained as

$$M_{BA} = 49.4 \text{ kN·m}$$

By analysing the FBD, the maximum shear force at support B of beam BA is computed as

$$V_B = \left(\frac{17 \times 6}{2}\right) + \left(\frac{49.4 - 40.6}{6}\right) = 55.46 \text{ kN}$$

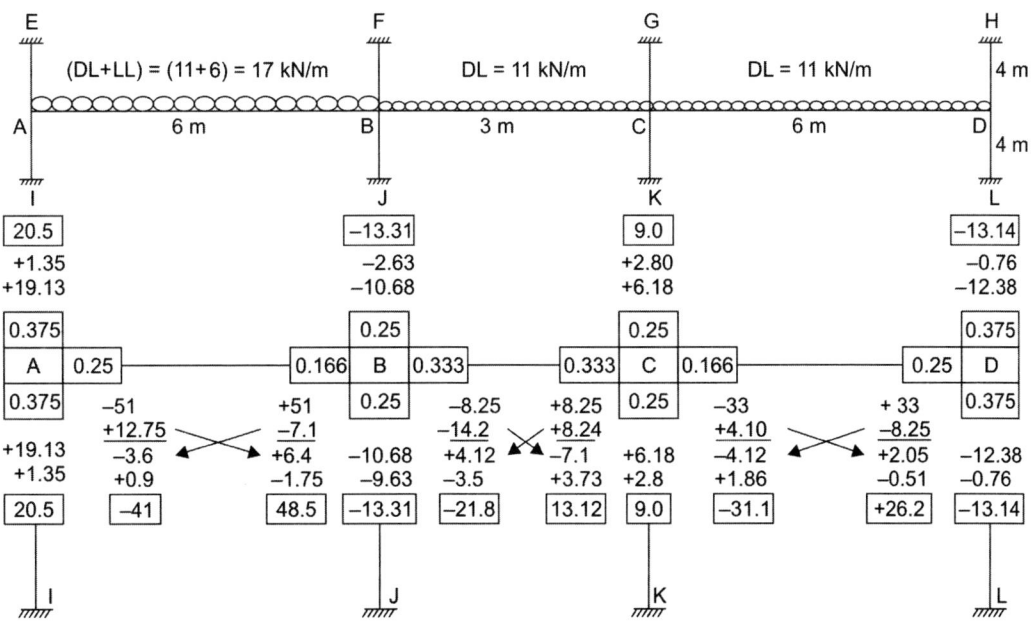

Case 3: Maximum negative BM at A (moments in kN·m)
(a) DL on ABCD   (b) LL on AB

Fig. 11.8: Moment distribution of substitute frame

v. *Moments and axial thrust in columns*:

a. Exterior column AI

From Fig. 11.8,

Moment at top of column AI = 20.5 kN·m

Axial thrust = V_A = 46.75 kN

b. Interior column BJ

From Fig. 11.9,

Moment at top of column BJ = 12.01 kN·m

$$\text{Axial thrust} = \left[\left(\frac{49.4 - 40.6}{6}\right) + \left(\frac{17 \times 6}{2}\right)\right] + \left[\left(\frac{25.2 - 16.6}{3}\right) + \left(\frac{17 \times 3}{2}\right)\right] = 84 \text{ kN}$$

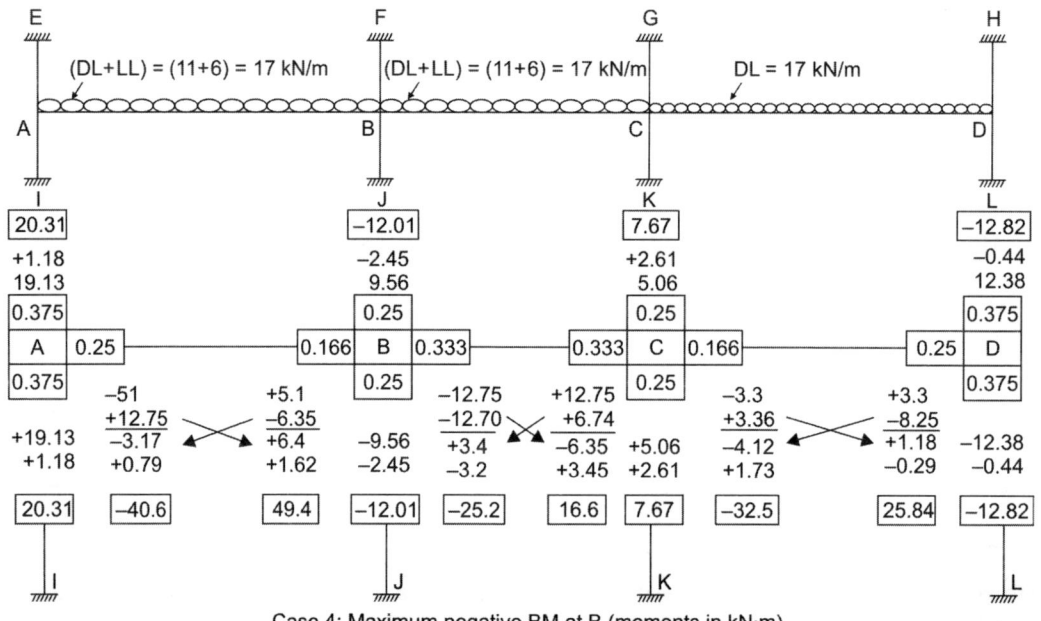

Case 4: Maximum negative BM at B (moments in kN·m)

Fig. 11.9: Moment distribution of substitute frame

11.5 BENDING MOMENTS IN COLUMNS

Bending moments developed in the columns of a multistorey, single and multibay frame can be evaluated by an approximate method which takes into account the stiffness of the columns and beams in the frames. This method gives reasonably good result and is sufficiently accurate for general designs.

The moments in columns of single and multibay are computed by the expressions given in Table 11.1.

Table 11.1: Bending moments in columns

	Moments for frames of one bay	Moments for frames of two or more bays
1. *External columns*		
a. Moment at foot of upper column	$M_e\left(\dfrac{k_u}{k_u + k_L + 0.5k_b}\right)$	$M_e\left(\dfrac{k_u}{k_u + k_L + k_b}\right)$
b. Moment at head of lower column	$M_e\left(\dfrac{k_L}{k_u + k_L + 0.5k_b}\right)$	$M_e\left(\dfrac{k_L}{k_u + k_L + k_b}\right)$
2. *Internal columns*		
a. Moment at foot of upper column	—	$M_{es}\left(\dfrac{k_u}{k_L + k_u + k_{b1} + k_{b2}}\right)$
b. Moment at head of lower column	—	$M_{es}\left(\dfrac{k_L}{k_L + k_u + k_{b1} + k_{b2}}\right)$

Note:

M_e = bending moment at the end of the beam framing into the column assuming fixity at the connection.

M_{es} = maximum difference between the moments at the ends of the two beams framing into opposite sides of the column, each calculated on the assumption that the ends of the beams are fixed and assuming one of the beams unloaded.

k_u = stiffness factor of the upper column

k_L = stiffness factor of the lower column

k_{b1} = stiffness of the beam on one of the side of the column

k_{b2} = stiffness of the beam on other side of the column

The application of this approximate method to compute moments in the exterior columns of Example 11.4, indicates that this methods results in moments about 6% less than those obtained by the substitute frame method. However, this approximate method is simpler and yields reasonably accurate values of moments for design purposes.

11.6 ANALYSIS OF MULTISTOREY FRAMES SUBJECTED TO HORIZONTAL FORCES

Multistorey building frames with the ratio of height to the least lateral dimension greater than 2, have to be analysed for moments developed in the members due to the effect of horizontal wind forces acting on the building. The horizontal forces due to wind are assumed to act at each of the floor levels and they induce axial forces in the columns and bending moments in all the members of the frame. These moments and forces can be analysed by the following approximate methods: (a) portal method (b) cantilever method.

a. *Portal Method*

The portal method is based on the following assumptions:

   i. Points of contraflexure are assumed to develop at the mid point of beams and columns.

  ii. The interior columns are assumed to resist double the shear force taken by each of the external columns.

   This method is illustrated with reference to the multistorey, multibay frame shown in Fig. 11.10.

   The horizontal shear forces resisted by the columns in each of the floors are shown above and below the contraflexure points in Fig. 11.10.

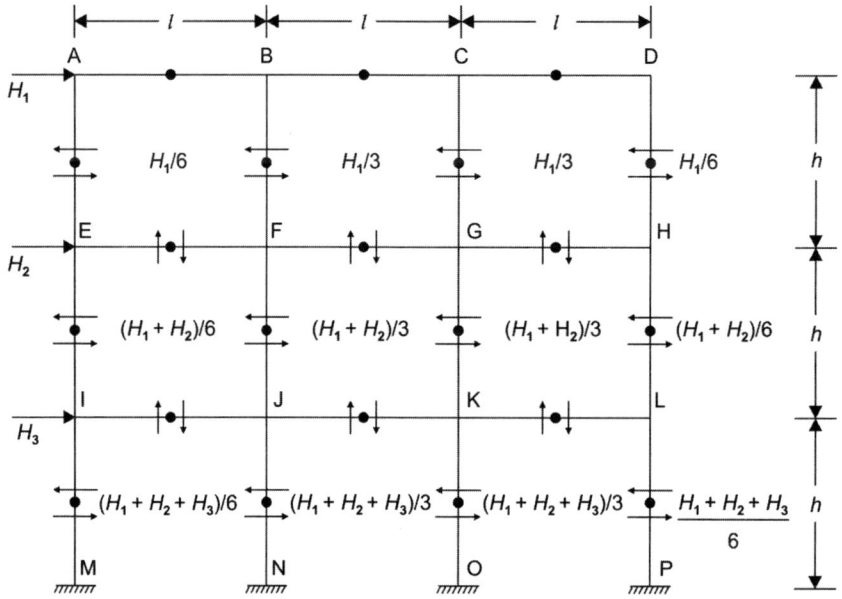

Fig. 11.10: Portal method of analysis for wind loads

The moments in the columns are computed as shown below:

$$M_{EA} = \left(\frac{H_1}{6} \times \frac{h}{2}\right) = \left(\frac{H_1 h}{12}\right)$$

$$M_{EI} = \left(\frac{H_1 + H_2}{6}\right) \times \left(\frac{h}{2}\right) = \frac{(H_1 + H_2)h}{12}$$

$$M_{MI} = \frac{(H_1 + H_2 + H_3)}{6} \times \left(\frac{h}{2}\right) = \frac{(H_1 + H_2 + H_3)h}{12}$$

$$M_{FB} = \left(\frac{H_1}{3} \times \frac{h}{2}\right) = \left(\frac{H_1 h}{6}\right)$$

$$M_{FJ} = \left(\frac{H_1 + H_2}{3}\right) \times \left(\frac{h}{2}\right) = \frac{(H_1 + H_2)h}{6}$$

$$M_{NJ} = \left(\frac{H_1 + H_2 + H_3}{3}\right) \times \left(\frac{h}{2}\right) = \frac{(H_1 + H_2 + H_3)h}{6}$$

The moments in the beams are computed by using the FBDs shown in Fig. 11.11.

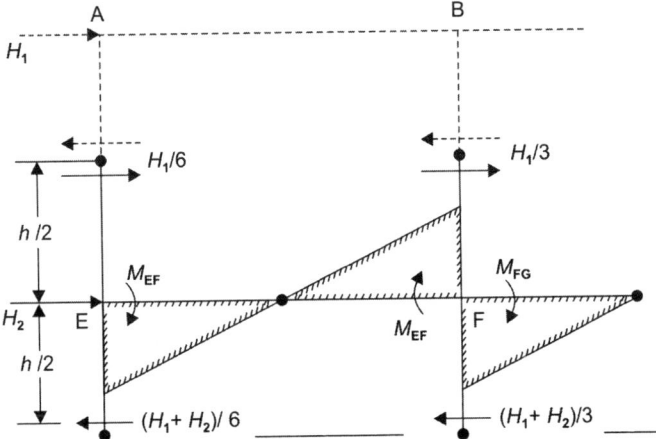

Fig. 11.11: Moments in a beam

Moment in column just above $E = \left(\frac{H_1}{3} \times \frac{h}{2}\right)$

Moment in column just below $E = \left(\frac{H_1 + H_2}{6}\right) \times \left(\frac{h}{2}\right)$

Therefore, for equilibrium of moments at joint E, the beam is subjected to a moment

$$M_{EF} = \left[\frac{H_1 h}{12} + \frac{(H_1 + H_2)h}{12}\right]$$

Similarly at point F, the clockwise moment is given by

$$M_F = \left[\left(\frac{H_1}{3}\right)\left(\frac{h}{2}\right) + \frac{(H_1 + H_2)}{3} \times \frac{h}{2}\right]$$

This moment is sheared equally by the beams FE and FG.

$$\therefore \qquad M_{FE} = M_{FG} = \frac{1}{2}\left[\frac{H_1 h}{6} + \frac{(H_1 + H_2)h}{6}\right] = \left[\frac{H_1 h}{12} + \frac{(H_1 + H_2)h}{12}\right]$$

Vertical reactions in columns develop only for the exterior columns. If M is moment at the ends of beams and L = length of beam, the windward column is subjected to a pull of $(2M/L)$. The reactions neutralise for interior columns.

b. *Cantilever Method*

The following assumptions are made in the cantilever method:

 i. Points of contraflexure occur at the mid point of various members of the frame.
 ii. Direct stresses in the columns are proportional to their distances from the centroidal vertical axis of the frame.

Refer to the multistorey frame shown in Fig. 11.12. Let V_1, V_2, V_3 and V_4 be the vertical reactions developed in the top storey of the frame due to the horizontal force H_1 acting at the joint A_1.

If the cross-sectional area of the columns are the same, then the reactive forces in the columns are proportional to their distances from the centroidal axis of the frame.

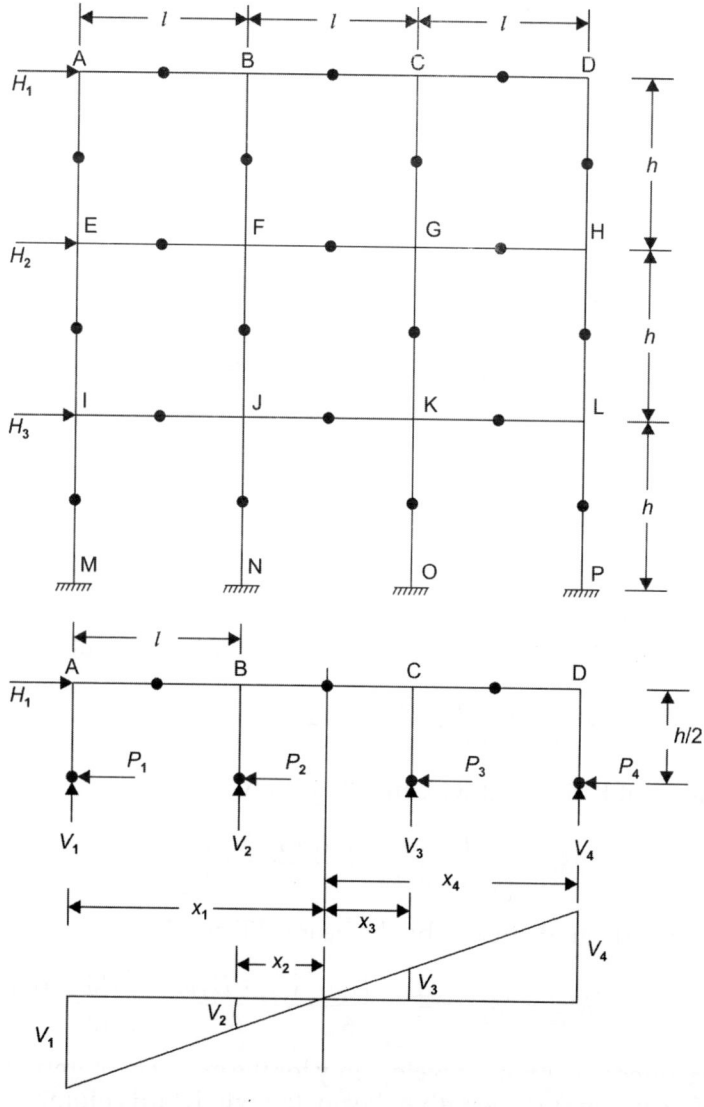

Fig. 11.12: Cantilever method of analysis for wind loads

Let P_1, P_2, P_3 and P_4 be the horizontal shears in the top storey. Then using the assumption specified, the following equations can be obtained.

$$\left(\frac{V_1}{x_1}\right) = \left(\frac{V_2}{x_2}\right) = \left(\frac{V_3}{x_3}\right) = \left(\frac{V_4}{x_4}\right) \qquad ...(11.1)$$

$$V_1 x_1 + V_2 x_2 + V_3 x_3 + V_4 x_4 = \left(\frac{H_1 h}{2}\right) \qquad ...(11.2)$$

$$(P_1 + P_2 + P_3 + P_4) = \frac{h}{2}\left(\frac{H_1 h}{2}\right) \qquad ...(11.3)$$

From Eqs (11.1) and (11.2), the vertical reaction components V_1, V_2, V_3 and V_4 can be computed. By taking moments about the point of contraflexure in beam AB (Fig. 11.13), we have

$$\left(\frac{P_1 h}{2}\right) = \left(\frac{V_1 L}{2}\right) \quad \therefore P_1 = \left(\frac{L}{h} \cdot V_1\right) \qquad ...(11.4)$$

By taking moments about the contraflexure point at the mid point R of BC, we have

$$(P_1 + P_2)\frac{h}{2} = V_1(L + L/2) + V_2\left(\frac{L}{2}\right) = V_1 L + (V_1 + V_2)\frac{L}{2}$$

$$(P_1 + P_2) = \left[\frac{2V_1 L + (V_1 + V_2)L/2}{h}\right] \qquad ...(11.5)$$

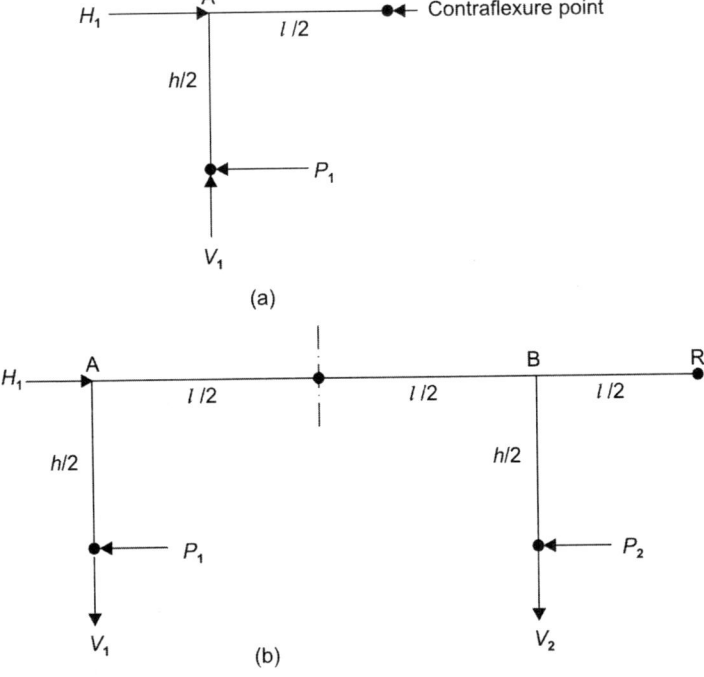

Fig. 11.13: Free body diagrams

Since P_1 is evaluated from Eq. (11.4), P_2 can be obtained from Eq. (11.5). Similarly, by taking moments about the contraflexure points, the horizontal shears P_3 and P_4 can be determined. The analysis is repeated for different storeys. Knowing the horizontal shears, the moments in the columns and beams can be computed in every floor of the frame.

11.7 DESIGN EXAMPLE

Analyse the multistorey frame shown in Fig. 11.10 for moments in the ground floor columns IM, JN, KO and LP and the beams IJ, JK and KL. Adopt the following data: $L = 6$ m, $h = 3$ m, wind loads are $H_1 = 6$ kN, $H_2 = 12$ kN, $H_3 = 12$ kN. The columns have the same cross-section. Compare the results of the portal and cantilever methods of analysis.

Portal Method

Referring to Fig. 11.10, the horizontal shears in the first storey are computed as below:

$$\left(\frac{H_1 + H_2 + H_3}{6}\right) = \left(\frac{6 + 12 + 12}{6}\right) = 5 \text{ kN}$$

$$\left(\frac{H_1 + H_2 + H_3}{3}\right) = \left(\frac{6 + 12 + 12}{3}\right) = 10 \text{ kN}$$

The moments in columns are obtained as:

$$M_{IM} = M_{MI} = (5 \times 1.5) = 7.5 \text{ kN·m} = M_{LP}$$
$$M_{JN} = M_{NJ} = (10 \times 1.5) = 15.0 \text{ kN·m} = M_{KO}$$

The moments in the beams are computed as follows:

$$M_u = \left[\left(\frac{H_1 + H_2}{6} \times \frac{h}{2}\right) + \left(\frac{H_1 + H_2 + H_3}{6} \times \frac{h}{2}\right)\right]$$

$$= \left[\left(\frac{6 + 12}{6} \times \frac{3}{2}\right) + \left(\frac{6 + 12 + 12}{6} \times \frac{3}{2}\right)\right] = 12 \text{ kN·m}$$

$$M_{JI} = M_{JK} = \frac{1}{2}\left[\left(\frac{H_1 + H_2}{3} \times \frac{h}{2}\right) + \left(\frac{H_1 + H_2 + H_3}{3} \times \frac{h}{2}\right)\right]$$

$$= \frac{1}{2}\left[\left(\frac{6 + 12}{3} \times \frac{3}{2}\right) + \left(\frac{6 + 12 + 12}{3} \times \frac{3}{2}\right)\right] = 12 \text{ kN·m}$$

Cantilever Method

Referring to Fig. 11.10, if V_1, V_2, V_3 and V_4 are the reactions developed in the columns, we have by symmetry $V_1 = V_4$ and $V_2 = V_3$.

Also $\qquad \left(\frac{V_1}{x_1}\right) = \left(\frac{V_2}{x_2}\right) \quad \therefore V_2 = V_1\left(\frac{x_2}{x_1}\right) = V_1\left(\frac{3}{9}\right) = \frac{V_1}{3}$

Let P_1, P_2, P_3 and P_4 are the horizontal shears, in the second storey.
Then $(P_1 + P_2 + P_3 + P_4) = (H_1 + H_2) = (6 + 12) = 18$ kN

Also $(P_1 + P_2 + P_3 + P_4)\dfrac{h}{2} = V_1 x_1 + V_2 x_2 + V_3 x_3 + V_4 x_4$

$\therefore \qquad \left(\dfrac{18 \times 3}{2}\right) = [V_1 9 + (V_1/3)3 + (V_1/3)3 + V_1 \times 9]$

$$27 = 20 V_1$$

$\therefore \qquad V_1 = 1.35 \text{ kN} = V_4, \ V_2 = \left(\dfrac{1.35}{3}\right) = 0.45 \text{ kN} = V_3$

$$P_1 = \left(\dfrac{L}{h}\right) V_1 = (6/3)\ 1.35 = 2.70 \text{ kN}$$

$$P_1 + P_2 = 2\left[\dfrac{V_1 L + (V_1 + V_2)L/2}{h}\right]$$

$$= 2\left[\dfrac{(1.35 \times 6) + (1.35 + 0.45) \times 3}{3}\right] = 9$$

$\therefore \qquad P_2 = (9 - 2.70) = 6.30 \text{ kN}$

$\therefore \qquad P_1 = P_4 = 2.70 \text{ kN}, P_2 = P_3 = 6.30 \text{ kN}$

Similarly, considering the first storey (ground floor), if P_1, P_2, P_3 and P_4 are the horizontal shears at the contraflexure point, we have

$$(P_1 + P_2 + P_3 + P_4) = (H_1 + H_2 + H_3) = (6 + 12 + 12) = 30 \text{ kN}$$

Also $\quad [(P_1 + P_2 + P_3 + P_4)\ h/2 = (V_1 x_1 + V_2 x_2 + V_3 x_3 + V_4 x_4)]$

$\therefore \qquad \left(\dfrac{30 \times 3}{2}\right) = V_1 \times 9 + \left(\dfrac{V_1}{3}\right) \times 3 + \left(\dfrac{V_1}{3}\right) \times 3 + V_1 \times 9$

$$45 = 20 V_1$$

$\therefore \qquad V_1 = 2.25 \text{ kN} = V_4 \qquad \therefore \ V_2 = \left(\dfrac{2.25}{3}\right) = 0.75 \text{ kN} = V_3$

$$P_1 = (L/h)V_1 = (6/3)\ 2.25 = 4.5 \text{ kN}$$

$$(P_1 + P_2) = 2\left[\dfrac{V_1 L + (V_1 + V_2)L/2}{h}\right] = 2\left[\dfrac{2.25 \times 6 + (2.25 + 0.75)3}{3}\right]$$

$$= 15.00$$

$\therefore \qquad P_2 = (15 - 4.5) = 10.5 \text{ kN}$

Column moments:

$M_{IE} = M_{LH} = (2.70 \times 1.5) = 4.05 \text{ kN·m}$

$M_{JF} = M_{KG} = (6.3 \times 1.5) = 9.45 \text{ kN·m}$

$M_{IM} = M_{LP} = (4.5 \times 1.5) = 6.75 \text{ kN·m}$

$M_{JN} = M_{KO} = (10.5 \times 1.5) = 15.75 \text{ kN·m}$

Beam moments:

$$M_{IJ} = (M_{IE} + M_{IM}) = (4.05 + 6.75) = 10.80 \text{ kN·m}$$

$$M_{JI} = M_{JK} = \left[\frac{(M_{JF} + M_{JN})}{2}\right] = \left[\frac{(9.45 + 15.75)}{2}\right] = 12.6 \text{ kN·m}$$

The portal method results in slightly higher values of moments in the exterior columns where as the cantilever method gives higher values of moments for the interior columns. Also, the shears resisted by the interior columns are higher according to the cantilever method of analysis.

References

1. Frederick WT and Sanford ET, *A Treatise on Concrete, Plain & Reinforced*, 3 edn, 1916.
2. Pillai U and Menon D, *Reinforced Concrete Design*, 3 edn. Tata McGraw-Hill Education Pvt Ltd, New Delhi, 2009, pp. 9–21.
3. Coates RC, Coutie MG, and Kong F, *Structural Analysis*, ELBS, London, 1975, pp. 183–207.
4. Reddy CS, *Basic Structural Analysis*, Tata McGraw-Hill, New Delhi, 1981, pp. 382–435.
5. Martin HC, *Introduction to Matrix Methods of Structural Analysis*, McGraw-Hill Book, International Student's Edition, Kogakusha Co, Japan, 1966, pp. 1–131.
6. Bhatt P, *Problems in Structural Analysis by Matrix Methods*, Wheeler & Co India, 1989, pp. 1–465.
7. Ley S, Computation of influence coefficients for aircraft structures with discontinuous and sweepback, *Journal of Aeronautical Science*, Vol. 14, No.10, Oct. 1947, pp. 547–560.
8. Taranath BS, *Structural Analysis and Design of Tall Buildings*, McGraw-Hill, New York, 1984.
9. Smith BS and Coull A, *Tall Building Structure—Analysis & Design*, John Wiley & Sons Inc, New York, 1991.
10. Reynolds CE, James C Steedman and Anthony J Threlfall, *Reinforced Concrete Designer's Handbook*, 11 edn, Taylor & Francis, London & New York, 2008.

Assignment

1. The substitute frame of a multistorey building having three bays has a continuous beam ABCD with AB = 4.0 m, BC = 2.5 m, and CD = 4.0 m. The beams are spaced at 3 m intervals. Thickness for floor slab is 120 mm. Live load (office floor) is 4 kN/m². Floor finish is 0.6 kN/m². Size of beams is (250 mm × 400 mm). Size of columns is (250 mm × 400 mm). Height between floors is 4 m. Analyse the substitute frame and estimate the maximum design moments in the beams and columns.

2. A four-bay multistoreyed frame has the following details:

   Continuous beam ABCDE with AB = BC = CD = DE = 4 m. Height between floors is 4 m. Size of beams is (300 mm × 500 mm). Size of columns is (300 mm × 400 mm). Thickness of floor slabs is 150 mm. Floor finish is 1 kN/m². Live load is 2 kN/m². Estimate the maximum design moments in the beams and columns. Assume four storeys in the building.

3. A four storeyed building frame has four equal bays of 4 m each and the height between floors is 4 m. The wind loads acting at roof level and various floor levels are:

   $H_1 = 5$ kN, $H_2 = 10$ kN, $H_3 = 10$ kN and $H_4 = 10$ kN.

   The columns have the same cross-section. Estimate the moments in the columns and beams using (a) portal method, and (b) cantilever method.

4. A three-bay multistoreyed frame has to be analysed for design bending moments under dead and imposed live loads. Determine the design bending moments using the following data:

Exterior span length = 6 m

Interior span length = 4 m

Height between floors = 4 m

Size of beams = 300 mm × 450 mm

Size of columns = 250 mm × 350 mm

Thickness of floor slabs = 150 mm

Floor finishes = 1 kN/m²

Live load = 2.5 kN/m²

5. A six storeyed building frame has 6 bays of 4 m each and the height between the floors is 3.5 m The wind loads acting at roof level and various floor levels are

$H_1 = 8$ kN and $H_2 = H_3 = H_4 = H_5 = H_6 = 16$ kN

The columns have the same cross-section

Estimate the moments in the columns and beams using (a) portal method, and (b) cantilever method.

6. The substitute frame of a four bay multistoreyed frame has a continuous beam ABCDE with AB = 5 m, BC = 4 m, CD = 4 m and DE = 5 m. The beams are spaced at 4 m intervals. Thickness of the floor slab is 150 mm, live load (office floor) is 4 kN/m², load from floor finishes is 0.6 kN/m². The beams are of uniform size with a width of 300 mm and depth of 450 mm. Height between floors is 4 m. Analyse the substitute frame for dead and imposed loads and determine the design moments in the beams and columns.

Review Questions

1. Explain the necessity of using multistorey building frames. Under what situations and purpose you would resort to the use of multistorey frames?

2. What are the various methods of analysis of multistorey building frames? Briefly discuss the merits and demerits of the various methods.

3. What are the advantages of using matrix methods for the analysis of multistorey building frames? Mention the advantages of using this method.

4. What are substitute frames? Why should we use them for the analysis of multistorey building frames.

5. What are the various assumptions made in the substitute frame method of analyzing the multistorey buildings?

6. What are the various methods you would use for analyzing multistorey building frames subjected to wind loads?

7. What assumptions you would make in analyzing the multistorey building frames subjected to wind loads when using the portal and cantilever methods?

8. Briefly discuss the variations in the values of moments and shear forces resulting from the use of portal and cantilever methods.

9. Explain the method of evaluating the bending moments in the internal and external columns of a multistorey building frames using the stiffness factors.

10. How do you determine the axial thrusts developed in the columns of a multistorey building frame subjected to dead and live loads?

Objective Type Questions

1. Multistorey building frames are classified as
   a. simply supported structures
   b. determinate structures
   c. indeterminate structures

2. In multibay building frames with continuous beams, the maximum positive bending moment develops in the mid span, when
   a. adjacent spans are loaded
   b. alternate spans are loaded
   c. all the spans are loaded

3. In multibay building frames with continuous beams, the maximum negative bending moment at support develops, when
   a. adjacent spans are loaded
   b. alternate spans are loaded
   c. all the spans are loaded

4. The most commonly used method for the analysis of moments in a multistorey building frame subjected to vertical loads is the
   a. portal method
   b. matrix method
   c. substitute frame method

5. In the substitute frame method of analysis of moments using moment distribution
   a. all the storeys in the frame are considered
   b. only the two adjacent spans in a bay are considered
   c. storey's above and below the main continuous beam are considered

6. In a three bay multistorey frame, the maximum positive bending moment due to dead and live loads at the mid span of the end bay occurs, when
   a. all the spans carry the dead and live loads
   b. alternate spans carry the dead and live loads
   c. the end spans support the total loads and the central span carries only the dead load

7. In a three bay multistorey frame, the maximum negative bending moment due to dead and live loads at the end support occurs when
   a. all the spans carry the dead and live loads
   b. alternate spans carry the dead and live loads
   c. the end spans support the total load and the other spans support only the dead load

8. Analysis for moments in a multistorey frame for wind loads, the method to be used is
   a. moment distribution
   b. column analogy
   c. portal or cantilever method

9. When a three bay, three storey frame subjected to wind loads only, the maximum moment develops in the
   a. exterior columns of the middle storey
   b. interior columns of the ground storey
   c. exterior columns of the ground storey

10. In the analysis of moments in a multistorey building frame due to wind loads using portal
    a. the support junctions
    b. base support points
    c. mid point of beams and columns

12

Cylindrical Shell Structures

12.1 INTRODUCTION

The most economical structural solution for covering large spaces is possible by adoption of shell structures[1,2]. In addition, they can be grouped under lightweight structural options in comparison to the traditional types of structures like beam and slab, grids, and flat slabs. Most of these flat type structures are comparatively heavier and they cannot be used to cover large column free spaces. Concrete shell roofs have been widely used to cover large floor spaces of industrial structures. Reinforced concrete shells[3,4] are ideally suited to cover floor spaces over medium to long range spans of up to 30 m. The 43 m diameter world famous concrete dome of Pantheon in Rome built in AD 125 is a remarkable feat of engineering and it is still standing to prove the durability of shell structures.

Many pioneers like Pier Luig Nervi (1891–1979), Anton Tedesko (1903–1994), and Edward Torroja (1891–1961) have contributed immensely to the research, development and use of thin concrete shells in Europe and the United States to construct outstanding structures. Nervi's famous exhibition hall Pallazo Delle Esposizioni in Turin, Italy, built in 1948 is a classic example of early thin shell concrete structure. During the middle of 20th century, research workers like Lundgren[5], Jenkins[6], Tottenham[7] and McNamee[8] have contributed to the development of theoretical procedures for the design of various types of shell structures. The ASCE method[9] associated with the design tables is immensely useful in the analysis and design of cylindrical shells with any ratio of width to span length.

The earliest examples of shells built in India include those constructed at Meerut in 1941[10], covering spans of 36 m with a chord width of 10.5 m and the thickness of the shell being 63 mm. The prestressed concrete shell roof of the aircraft hangar at Karachi was built in 1942 over spans of 40 m. At present, shell structures are commonly used for the construction of indoor sports stadiums, exhibition halls and various types of industrial structures requiring large column free spaces. Shell structures are usually thin with the thickness ranging from 75 to 150 mm and the entire section effectively resists the imposed loads by developing compressive stresses.

The criteria for the design and construction of shell structures should conform to the Indian standard code IS:2210-1988[11]. The code has set detailed guidelines for the analysis, design and construction of shells. In the case of circular cylindrical shell roofs having spans exceeding 30 m, the depth of reinforced concrete edge beams can be reduced by resorting to prestressing the edge beams. The reader may refer to detailed publications[12,13] for the analysis, design and construction of prestressed concrete shell structures.

12.2 SHELL TERMINOLOGY

The following terms are generally used in the study of shell structures according to the Indian standard code IS:2210.

Shell: A curved surface having small thickness compared to the radius and other dimensions.

Shell of revolution: These are obtained when a plane curve is rotated about the axis of symmetry. The common examples of shells of revolution are the circular dome, the cone and the paraboloid of the revolution shown in Fig. 12.1.

Shell of translation: Shells of translation are formed when one curve moves parallel to itself along another curve, the planes of the two curves being at right angles to each other. Elliptic paraboloid and hyperbolic paraboloid are common examples of shells of translation. These are shown in Fig. 12.2.

Cylindrical shells: These are shells in which the generatrix (moving curve) or the directrix (stationary curve) is a straight line. For cylindrical shells, the common curves used are the arc of a circle, semiellipse, parabola or catenary. The various structural components of a cylindrical shell are the thin shell, edge beam and end frame or traverse as shown in Fig. 12.3.

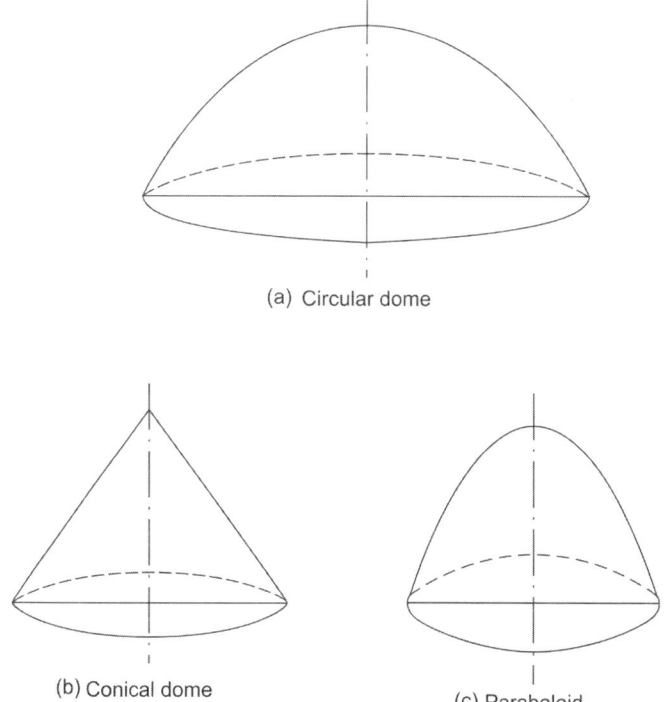

(a) Circular dome

(b) Conical dome

(c) Paraboloid

Fig. 12.1: Shells of revolution

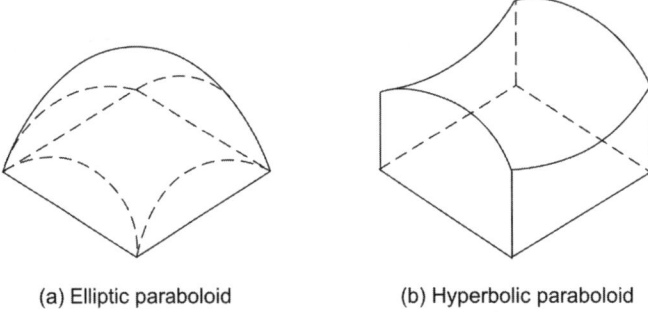

(a) Elliptic paraboloid   (b) Hyperbolic paraboloid

Fig. 12.2: Shells of translation

Multiple cylindrical shells: A series of parallel cylindrical shells which are transversely continuous are termed multiple cylindrical shells. Generally they are used for hangers, warehouses and factory buildings. A typical multiple cylindrical shell is shown in Fig. 12.4.

Continuous cylindrical shells: These are cylindrical shells which are longitudinally continuous over the traverses. Multiple and continuous cylindrical shells are widely used for market hall and warehouse roofs. Cylindrical shells which are symmetrical about the crown are termed *barrel shells*.

North light shells: North light shells comprise cylindrical shells with two springings at different levels and built in single or multiple bays and have provisions for north light glazing.

North light shells are commonly used for factory shed roofs. A typical north light shell of multibay is shown in Fig. 12.5.

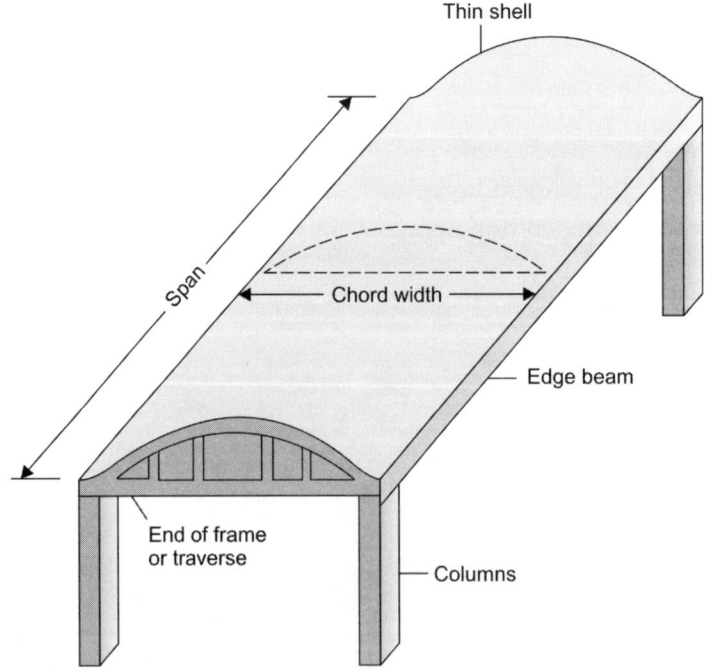

Fig. 12.3: A cylindrical shell

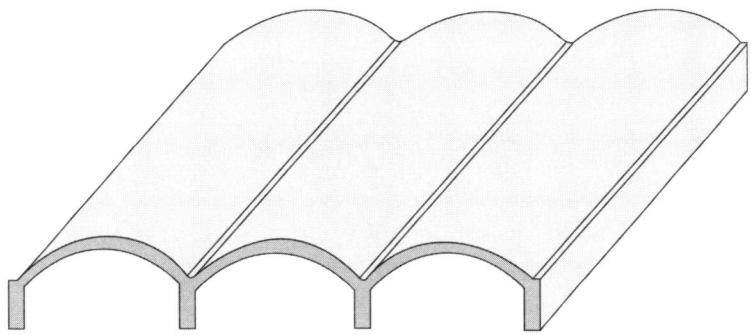

Fig. 12.4: Multiple cylindrical shells

Butterfly shell: A butterfly shell is formed when two parts of a cylindrical shell are joined together at their lower edges as shown in Fig. 12.6. This type of shell is commonly used for railway platforms and bus shelters.

Ruled surface: These are surfaces which can be generated entirely by straight lines. If at every point, a single straight line can be ruled, the surface generated is termed singly ruled surface. Common examples of singly ruled surfaces are cylindrical shells and conical shells.

If at every point, two straight lines can be ruled, the surface generated is termed doubly ruled surface. Common examples of doubly ruled surfaces are hyperbolic paraboloid and hyperboloid of revolution of one sheet.

Radius of the shell: The general expression for the radius at any point of a shell surface is given by

$$R = R_0 \cos^n \phi$$

where,

R = radius at any point

R_0 = radius at crown

ϕ = slope of the tangent to the curve at that point

n = a constant depending upon the type of curve

= 0 for a circle

= 1 for cycloid

= –2 for catenary

= –3 for parabola

For an ellipse, we have

$$R = \left[\frac{a^2 b^2}{(a^2 \sin^2 \phi + b^2 \cos^2 \phi)^{3/2}} \right]$$

where,

a = semimajor axis

b = semiminor axis

ϕ = slope of tangent at that point

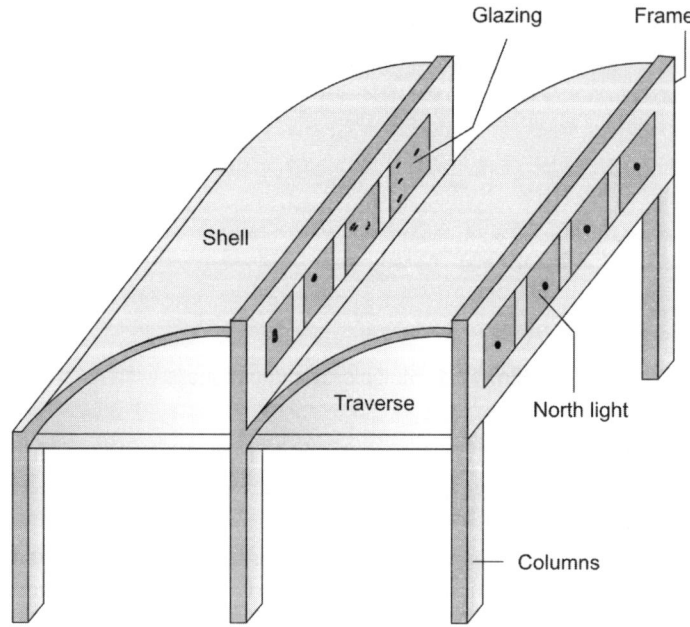

Fig. 12.5: North light shells

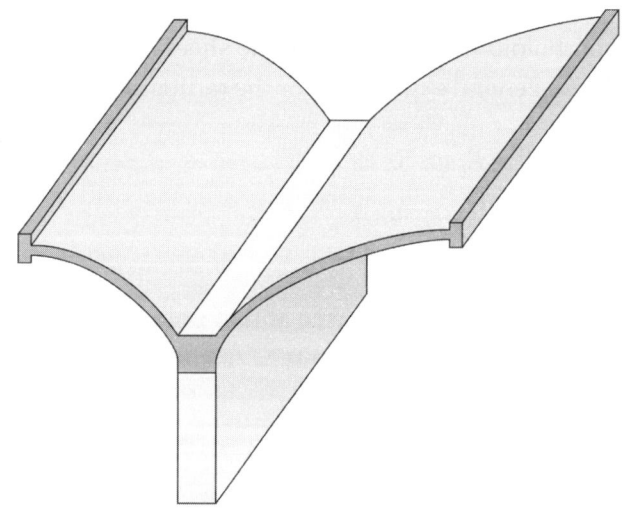

Fig. 12.6: Butterfly shell

Gauss Curvature

Gauss curvature is the product of the two principal curvatures $1/R_1$ and $1/R_2$ at any point on the surface of the shell. For singly curved developable shells, Gauss curvature is zero. For doubly curved non-developable shells, Gauss curvature is positive for synclastic shells and negative for anticlastic shells.

Span of shell: The distance between the two adjacent end frames or traverses is termed the span of the shell.

Chord width: Horizontal projection of the arc of the shell.

Rise: Vertical distance between the apex of the curve and the springing.

End frames or traverses: Frames provided at the ends to support and preserve the geometry of the shell.

Edge member or beam: The horizontal beam supported on columns and supporting the longitudinal edges of the shell.

12.3 █ CLASSIFICATION OF SHELLS

Shells are broadly classified into two major groups as follows:

    a. Singly curved shells which are developable

    b. Doubly curved shells which are non-developable

    Under the singly curved shells we have the conical and cylindrical shells. The common examples of doubly curved shells are the circular domes, paraboloid, ellipsoid, hyperbolic paraboloid and elliptic paraboloid. These shells are generally grouped under the three categories designated as below:

    a. Shells of revolution

    b. Shells of translation

    c. Ruled surfaces

    Figure 12.7 shows the general classification of shells with examples.

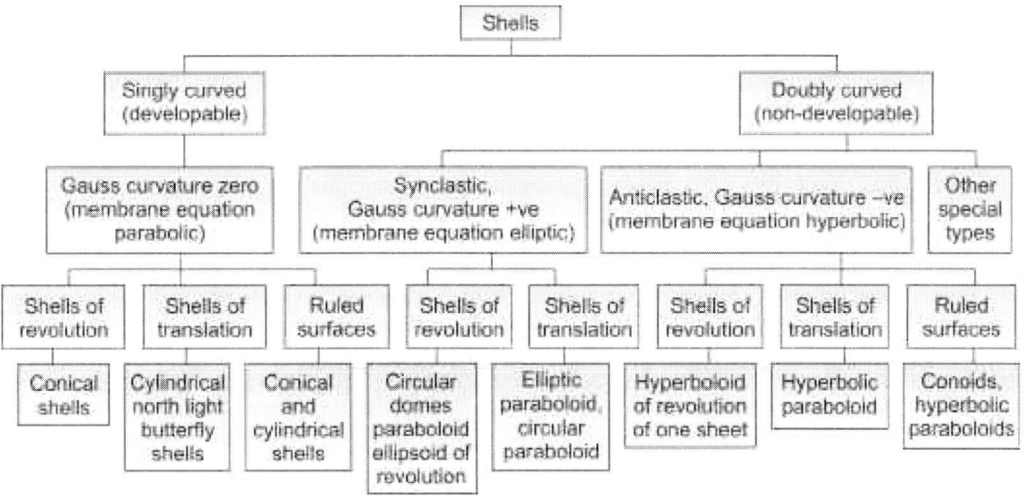

Fig. 12.7: General classification of shells

    Shells are also classified as thin and thick shells. A shell can be considered as thin if the ratio of the radius to the thickness of the shell is greater than 20. In general, most of the shells used in practice come under the category of thin shells.

    Thin cylindrical shells are generally classified in two groups:

    a. Long shells

    b. Short shells

According to the ASCE manual on design of cylindrical concrete shell roofs, shells having ratios of radius to span (R/L) less than 0.6 are classified as long shells. If the ratio exceeds this value, the shells are considered as short barrels.

12.4 GENERAL SPECIFICATIONS OF SHELLS

The salient dimensions of the shells have to be selected before a rigorous mathematical analysis is carried out for the computation of membrane forces in the shell. The general guidelines followed for selecting the dimensions of the various structural components of the shell are detailed below:

i. *Thickness:* The overall thickness of a reinforced concrete shell should not be less than 50 mm for singly curved shells, 40 mm for doubly curved shells and 25 mm for precast shells. Generally, the thickness is in the range of 80 to 120 mm for most of the shells based on practical considerations.

At the junction of the edge beams, the thickness of the shell is increased by 30% over a length of $0.38\sqrt{Rd}$ to $0.76\sqrt{Rd}$ for singly curved shells, where R = radius of curvature of the shells and d = overall thickness of the shell. For doubly curved shells, this distance depends upon the geometry of the shell and boundary conditions.

ii. *Span and chord width:* The span of reinforced concrete shells should not be greater than 30 m to limit the size and reinforcement within practicable limits in the edge beams. For longer spans, prestressed edge beams can be used. The width of the edge member is limited to 2 to 3 times the thickness of the shell.

In shells with chord width much larger than the span, the chord width shall be preferably 3 to 5 times the span.

iii. *Depth of shells:* For large span shells, depth = 1/6 to 1/12 span, larger figures are applicable to small spans. For shells without edge members, depth $< \left(\dfrac{\text{span}}{10} \right)$. For shells with chord width much larger than the span, depth < 1/10 chord width.

iv. *Semicentral angle:* The semicentral angle should be in the range of 30° to 45°. If the angle is less than 45°, the effect of wind load may be ignored. For larger angles with steep slopes, back forms may be necessary for casting.

v. *Reinforcements in shell:* The diameter of reinforcements should not exceed 10 mm for 50 mm thick shells and 12 mm for 65 mm thick shells and 16 mm for shells having thickness greater than 65 mm. In the junction zones where the shell is thickened, larger diameter bars are permissible. The spacing of the bars should not be more than five times the thickness of the shell. Minimum clear cover must be 12 mm or the nominal size of the reinforcing bar. Generally, a minimum reinforcement of 0.15% of the gross cross-section in the principal direction is recommended for thin shell structures.

vi. *Shell joints:* In case of shells with lengths exceeding 40 m, expansion joints have to be provided. The construction joints are provided along the curve lengths of the shell where the shear forces are minimum. Shells have to be well cured for a minimum period of two weeks before decentering. The end beams and diaphragms have to be cured for 3 to 4 weeks before decentering.

12.5 ANALYSIS OF SHELLS

Membrane Theory

The membrane theory was formulated by Dischinger with the assumption that the shell is regarded as a perfectly flexible membrane of infinite extent, carrying direct forces in its plane only. Over a limited zone at sufficient distance away from the boundaries, the stresses in the shell slab approach a distribution which is statically determinate and may be found by the membrane theory. This procedure is applicable to shells whose span to radius ratio is less than 0.5.

The equilibrium of the shell element shown in Fig. 12.8 is examined under the given set of direct forces with the following notations:

x: Direction of generatrix

y: Direction of tangent to directrix at A

z: Direction of the outward normal at A

T_x, T_y and S: Forces per unit length

R: Radius of the singly curved x, y and z are the components of external loads per unit area on the element.

$$dy = R \cdot d\phi$$

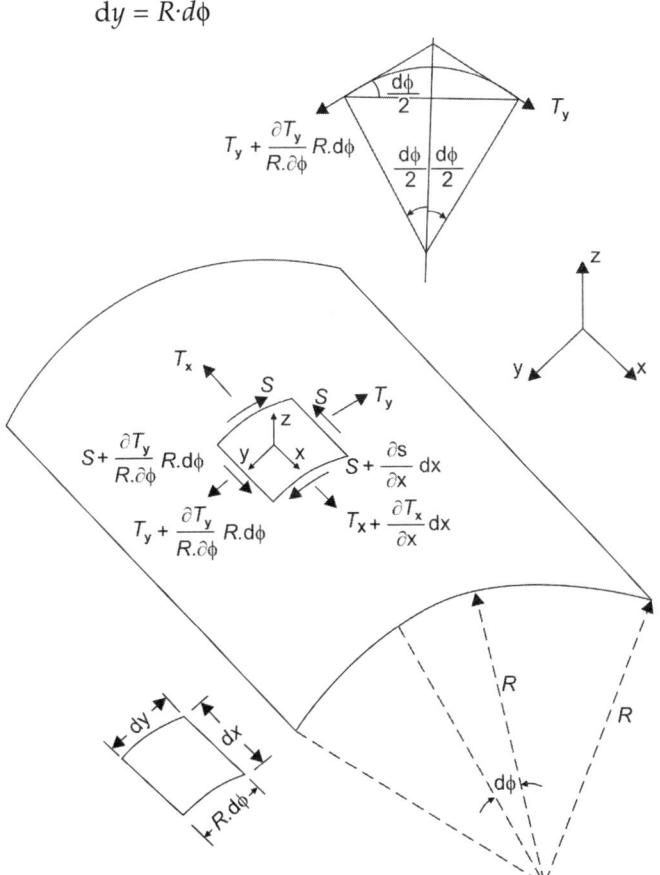

Fig. 12.8: Equilibrium of shell element

Equations of equilibrium:

$$\Sigma x = 0, \quad \Sigma y = 0, \quad \Sigma z = 0$$

a. *Forces in x-direction*:

$$\frac{\partial T_y}{\partial x} Rd\phi dx + \frac{\partial S}{R\partial\phi} Rd\phi dx + x dx Rd\phi = 0$$

$$\frac{\partial T_x}{\partial x} Rd\phi dx + \frac{\partial S}{R\partial\phi} + x = 0 \qquad \qquad \ldots(12.1)$$

b. *Forces in y-direction*:

$$\frac{\partial T_y}{R\partial x} Rd\phi dx + \frac{\partial S}{R\partial\phi} dx Rd\phi + y dx Rd\phi = 0$$

$$\frac{\partial T_y}{R\partial\phi} + \frac{\partial S}{\partial x} + Y = 0 \qquad \qquad \ldots(12.2)$$

c. *Forces in z-direction*:

$$2T_y dx \sin\frac{d\phi}{2} + \frac{\partial T_y}{R\partial\phi} Rd\phi dx \sin\frac{d\phi}{2} - z dx Rd\phi = 0$$

Since $d\phi$ is small, $\sin\dfrac{d\phi}{2} = \dfrac{d\phi}{2}$

and $\dfrac{\partial T_y}{R\partial\phi} Rd\phi dx \sin\dfrac{d\phi}{2} = 0$

$$\therefore \quad T_y - Rz = 0 \qquad \qquad \ldots(12.3)$$

The values of T_x and S are to be evaluated by integrating Eqs (12.1) and (12.2).

$$\therefore \qquad S = -\int \frac{\partial T_y}{R\partial\phi} dx - \int y dx + F_1(\phi)$$

$$T_x = -\int \frac{\partial S}{R\partial\phi} dx - \int x dx + F_2(\phi)$$

where, $F_1(\phi)$ and $F_2(\phi)$ are functions of ϕ only which have to be determined from boundary conditions.

In most cases in practice x, y, and z are functions of ϕ only and do not vary along x. Then T_y is a function of ϕ only and does not depend on x.

Hence,

$$S = -x\left(\frac{T_y}{R\partial\phi} + Y\right) + F_1(\phi)$$

$$= xK + F_1(\phi)$$

$$T_x = \frac{x^2}{2}\left(\frac{\partial K}{R\partial \phi}\right) - \frac{x\partial[F_1(\phi)]}{R\partial \phi} - xX + F_2(\phi)$$

For a shell, simply supported at the end frames and symmetrically loaded, shear $S = 0$ at $x = 0$

$\therefore \qquad F_1(\phi) = 0$

The traverses at ends cannot withstand forces normal to their planes. Hence, $T_x = 0$ at $x \pm L$ and also in most cases $x = 0$.

$\therefore \qquad F_2(\phi) = -\left(\frac{L^2}{2}\right)\left(\frac{\partial K}{R\partial \phi}\right)$

$\therefore \qquad T_y = ZR$

$$S = -x\left(\frac{\partial T_y}{R\partial \phi} + Y\right) = xK$$

$$T_x = -\left(\frac{L^2 - x^2}{2}\right)\left(\frac{\partial K}{R\partial \phi}\right)$$

These three equations can be used to calculate membrane stresses for any type of directrix.

Equations for membrane stresses under various types of loads:

a. *Self weight* (Fig. 12.9a)

$y = g\sin\phi$

$z = -g\cos\phi$

$x = 0$

$\therefore \qquad T_y = -gR\cos\phi$

Hence $\qquad K = \left(Y + \frac{\partial T_y}{R\partial \phi}\right)$

$$= g\sin\phi + \frac{1}{R}\left(gR\sin\phi - g\cos\phi\frac{dR}{d\phi}\right)$$

$$= 2g\sin\phi - \left(\frac{g\cos\phi}{R}\right)\left(\frac{dR}{d\phi}\right)$$

b. *Snow load* (Fig. 12.9b)

Consider the snow load to be uniform per unit length of projection.

$y = p_0\cos\phi\sin\phi$

$z = -p_0\cos^2\phi$

$x = 0$

$\therefore \qquad T_y = ZR = -p_0R\cos^2\phi$

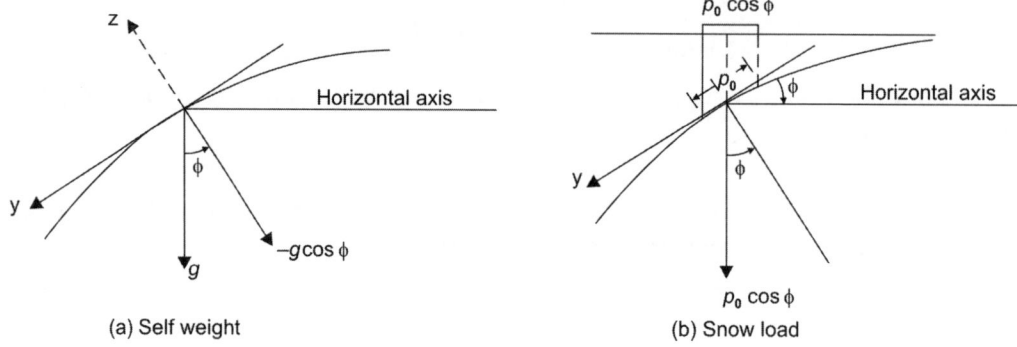

(a) Self weight   (b) Snow load

Fig. 12.9: Types of loads on membrane

$$K = \left(y + \frac{\partial T_y}{R\partial\phi} \right)$$

$$= 3p_0 \sin\phi\cos\phi - p_0 \frac{\cos^2\phi}{R}\left(\frac{dR}{d\phi} \right)$$

Membrane stresses for various directrices:

General equation for the family of curves is

$$R = R_0 \cos^n\phi$$

where,

R_0 = radius of curvature at crown

R = radius of curvature at any point

ϕ = angle made by the tangent to the curve at any point with the horizontal

n = constant having values

= 0 for circle

= 1 for cycloid

= –2 for caternary

= –3 for parabola

Substituting the value of R in the membrane equation, we have

$$T_y = -gR\cos\phi = -gR_0\cos^n\phi\cos\phi = -gR_0\cos^{n+1}\phi$$

As $\quad R = R_0\cos^n\phi$

∴ $\quad \dfrac{dR}{d\phi} = -nR_0\cos^{n-1}\phi\sin\phi$

$$K = \left(y + \frac{\partial T_y}{R\partial\phi} \right) = g\sin\phi + (n+1)g\sin\phi = (n+2)g\sin\phi$$

$$\frac{\partial K}{R\partial\phi} = \frac{1}{R}(n+2)g\cos\phi = \left[\frac{(n+2)g}{R_0\cos^{n-1}\phi} \right]$$

Similarly, the membrane forces can be calculated. The membrane forces due to self weight (g) and snow load ($p_0 \cos\phi$) are compiled in Table 12.1.

Table 12.1: Membrane forces due to self weight and snow loads

Membrane force	Due to self weight (g)	Due to snow load ($p_0\cos\phi$)
T_x	$-\dfrac{(n+2)g(L^2-x^2)}{2R_0\cos^{n-1}\phi}$	$-\dfrac{(n+3)p_0(L^2-x^2)(\cos^2\phi-\sin^2\phi)}{2R_0\cos^n\phi}$
T_y	$-gR_0\cos^{n+1}\phi$	$-p_0R_0\cos^{n+2}\phi$
S	$-(n+2)gx\sin\phi$	$-(n+3)p_0x\sin\phi\cos\phi$

The membrane forces for circular cylindrical shells ($n = 0$) are compiled in Table 12.2.

Table 12.2: Membrane forces in circular cylindrical shells

Membrane force	Due to self weight (g)	Due to snow load ($p_0\cos\phi$)
T_x	$-\dfrac{g(L^2-x^2)\cos\phi}{R}$	$-\dfrac{3p_0(L^2-x^2)(\cos^2\phi-\sin^2\phi)}{2R}$
T_y	$-gR\cos\phi$	$-p_0R\cos^2\phi$
S	$-2gx\sin\phi$	$-3p_0x\sin\phi\cos\phi$

Design Examples

1. A reinforced concrete shell having semicircular directrix is freely supported at the ends. The data given is as follows:

Radius of the shell $R = 8$ m.

Length of shell $2L = 36$ m

Thickness of shell $t = 60$ mm

Calculate the membrane forces at $x = 0$, 9 and 18 m and $\phi = 0°$, 30°, 60° and 90° under its own self weight.

Density of concrete = 25 kN/m³

The self weight per unit area of shell $g = (0.06 \times 25) = 1.5$ kN/m²

For a circle $R = R_0 = 8$ since $n = 0$

The membrane forces are given by the following equations:

$$T_x = -\left[\frac{g(L^2+x^2)\cos\phi}{R}\right] = -\left[\frac{1.5(18^2-x^2)\cos\phi}{8}\right]$$

$$= -(60.75 - 0.1875x^2)\cos\phi$$

$$T_y = -gR\cos\phi = -1.5\times8\times\cos\phi = -12\cos\phi$$

$$S = -2gx\sin\phi = -2\times1.5\times x\times\sin\phi = -3x\sin\phi$$

Using these equations, the membrane forces are computed for different values of ϕ as:

$$\phi = 0°, 30°, 60° \text{ and } 90° \text{ and } x = 0, 9 \text{ and } 18 \text{ m.}$$

The values are compiled in Table 12.3.

Maximum unit stress occurs at $x = 0$ and $\phi = 0$, i.e. at centre of span and crown.

$$T_x = 60.75 \text{ kN/m}$$

$$\text{Maximum compressive stress} = \left(\frac{60.75 \times 10^3}{1000 \times 60}\right) = 1.01 \text{ N/mm}^2$$

Table 12.3: Membrane forces in semicircular shells

ϕ	Membrane forces (kN/m)									(−ve) compression		
	0°			30°			60°			90°		
x	T_x	T_y	S	T_x	T_y	S	T_x	T_y	S	T_x	T_y	S
0	60.75	12.00	0	52.61	10.39	0	30.37	0	6.00	0	0	0
9 m	45.56	12.00	0	39.46	10.39	13.50	22.78	6.00	23.38	0	0	27.00
18 m	0	12.00	0	0	10.39	27.00	0	6.00	46.77	0	0	54.00

Forces T_x and T_y are compressive.

2. A reinforced concrete shell with circular directrix has the following dimensions:

$$R = 6 \text{ m}$$
$$2L = 24 \text{ m}$$
$$t = 50 \text{ mm}$$
$$\phi = 60°$$
$$g = (0.05 \times 25) = 1.25 \text{ kN/m}^2$$

Calculate:

i. the maximum stress in the shell

ii. the maximum bending moment and tension developed in the edge beams.

Maximum stress is developed at the crown and centre of span for value of $x = 0$ and $\phi = 0$.

$$\therefore \qquad T_x = -\left(\frac{gL^2}{R}\right) = -\left(\frac{1.25 \times 12^2}{6}\right) = -30 \text{ kN/m}$$

$\therefore$ Maximum compressive stress is given by

$$\sigma_{max} = \frac{30 \times 10^3}{1000 \times 50} = 0.6 \text{ N/mm}^2 \text{ (compression)}$$

Forces acting on edge beam at $\phi = 60°$ is given by

$$T_y = -gR\cos\phi = -1.25 \times 6 \times \cos 60° = -3.75 \text{ kN/m}$$

$\therefore$ Load acting on the edge beam is 3.75 kN/m over a span of 24 m.

$\therefore$ Maximum bending moment at centre of span of edge beam of length $L = 24$ m,

$$M_{max} = \frac{3.75 \times 24^2}{8} = 270 \text{ kN·m}$$

Tension in the edge beam (axial force) at a distance x from the centre is:

$$= -\int_x^L S dx = -\int_x^L -2gx\sin\phi = g(L^2 - x^2)\sin\phi$$

Maximum tension occurs at $x = 0$ (centre of span section)

$\therefore$ $\quad N_{max} = gL^2 \sin\phi = (1.25 \times 12^2 \times \sin 60°) = 156$ kN

3. A reinforced concrete shell of circular directrix has the following dimensions:

ϕ = semicentral angle = $60°$

span $2L = 24$ m

radius $R = 6$ m

thickness $t = 50$ mm

The shell is subjected to the action of snow load only of intensity 1 kN/m² per unit length of curved surface of shell.

　i. Determine the maximum stresses in the shell.

　ii. Also calculate the maximum bending moment and tension in edge beams.

Membrane forces are given by

$$T_x = -\left[\frac{3p_0(L^2 - x^2)(\cos^2\phi - \sin^2\phi)}{2R}\right] = -\left[\frac{3\times 1(12^2 - x^2)\cos 2\phi}{2\times 6}\right]$$

$$= -(36 - 0.25x^2)$$

$$T_y = -p_0 R\cos^2\phi = -1\times 6\times\cos^2\phi = -6\cos^2\phi$$

$$S = -3p_0 x\sin\phi\cos\phi = -3\times 1\times x\times\frac{\sin 2\phi}{2}$$

$$= -1.5x\sin 2\phi$$

Maximum compressive stress occurs at $x = 0$ and $\phi = 0$.

$\therefore$ $\quad T_x = -36$ kN/m

$\quad\quad T_y = -6$ kN/m

$\therefore$ $\quad \sigma_{max} = \left(\frac{36\times 10^3}{1000\times 50}\right) = 0.72$ N/mm²

At the edges, $\phi = 60°$

$\therefore$ $\quad T_y = -6\cos^2 60° = -1.5$ kN/m

$\therefore$ BM in edge beam $= \left(\frac{1.5\times 24^2}{8}\right) = 108$ kN·m

Tension in the edge beam at a distance x from the centre is:

$$= -\int_x^L S dx = -\int_x^L -3p_0 x\sin\phi\cos\phi$$

$$= 3p_0 \sin\phi\cos\phi\left(\frac{L^2 - x^2}{2}\right)$$

Maximum tension occurs at $x = 0$ and $\phi = 60°$.

$$\therefore \qquad N_{max} = \left(3 \times 1 \times \sin 60° \times \cos 60° \times \frac{12^2}{2}\right) = 94 \text{ kN}$$

Beam Theory

In the beam theory developed by Lundgren, the shell is analysed as a beam of curved cross-section spanning between the end frames or traverses. In the case of long shells, the longitudinal force components are predominant and hence the beam theory is ideally suited for the analysis. The beam theory is generally applicable for cylindrical shells of L/R ratio exceeding the value of π. In the beam theory, the cross-section of the shell is assumed with or without edge beams and the sectional properties are determined and the stresses are computed using the beam theory. An arch analysis is also conducted to determine the transverse moments and thrusts so that suitable reinforcements are designed.

Referring to Fig. 12.10,

$$\bar{x} = \frac{\sum Ax}{\sum A} = \frac{2\int_0^\phi R \cdot d\theta \cdot tR \cdot \cos\theta}{2\int_0^\phi R \cdot d\theta \cdot t}$$

where,

R = radius of the shell

t = thickness of the shell

$$\therefore \qquad \bar{x} = \frac{\int_0^\phi R^2 d\theta \cdot t \cdot \cos\theta}{\int_0^\phi R^2 d\theta \cdot t}$$

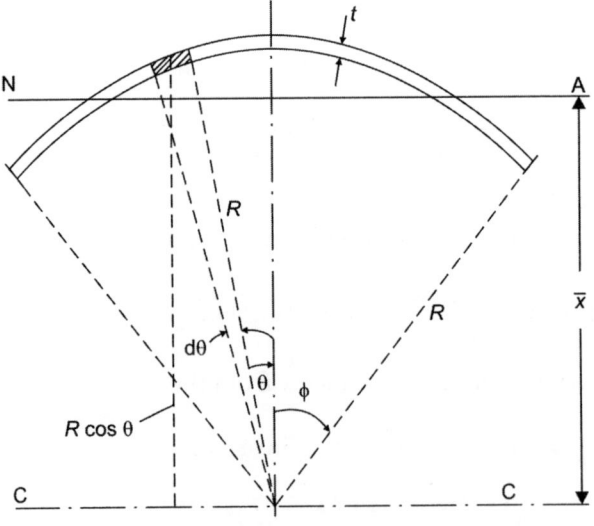

Fig. 12.10: Shell properties

Second moment of area

$$I_{CC} = 2\int_0^\phi R\cdot d\theta \cdot t(R\cos\theta)^2 = 2R^3t\int_0^\phi \frac{(1+\cos2\theta)}{2}\,d\theta$$

$$= 2R^3t\left[\frac{\theta}{2} + \frac{\sin2\theta}{4}\right]_0^\phi$$

Area of segment of shell $\Sigma A = 2\int_0^\phi R\cdot d\theta \cdot t$

$\therefore$

$$I_{NA} = [I_{cc} - A\bar{x}^2]$$

$$I_{NA} = 2R^3t\left[\frac{\theta}{2} + \frac{\sin2\theta}{4}\right]_0^\phi - \left(2\int_0^\phi R\cdot d\theta \cdot t\right)\bar{x}^2$$

The use of these equations is illustrated with the help of the following example.

4. A reinforced concrete circular shell has the following particulars:

Radius $R = 6$ m

Span $2L = 24$ m

Semicentral angle $\phi = 60°$

Thickness $t = 50$ mm

Calculate the maximum stress due to self weight only in the shell by beam theory and compare the values with the results of membrane theory.

Rise of shell $= (R - R\cos60°) = 6(1 - \cos60°) = 3$ m

Area of segment $= 2\int_0^{\pi/3}(6d\theta)(0.05)$

$\therefore$ $\qquad \Sigma A = (2 \times \pi/3 \times 6 \times 0.05) = 0.628$ m^2

$$\bar{x} = \frac{2\int_0^{\pi/3}R^2d\theta t\cos\theta}{2\int_0^{\pi/3}Rd\theta t} = \left[\frac{2\times6^2\times0.05\times\sin60°}{0.628}\right] = 4.97 \text{ m}$$

$$I_{CC} = 2R^3t\left[\frac{\theta}{2} + \frac{\sin2\theta}{4}\right]_0^\phi = 2\times6^3\times0.05\left[\frac{\pi}{6} + \frac{\sin120}{4}\right] = 15.99 \text{ m}^4$$

$\therefore$ $\qquad I_{NA} = [I_{CC} - A\bar{x}^2] = [15.99 - (0.628)(4.97)^2] = 0.50 \text{ m}^4$

Weight/metre run of shell $= (0.628 \times 25) = 15.71$ kN/m

$\therefore$ Maximum BM $= \left(\dfrac{15.71\times24^2}{8}\right) = 1131$ kN-m

Extreme fibre stress at crown

$$\sigma_c = \left[\frac{1131\times10^6\times1.03\times10^3}{0.50\times10^{12}}\right] = 2.33 \text{ N/mm}^2$$

The value of compressive stress σ_c obtained by membrane theory in Design Example 2 is 0.6 N/mm². This indicates that when membrane theory is applied for long shells, the stresses are underestimated.

5. A circular cylindrical shell with edge beams has the following particulars:

Radius of the shell R = 10.10 m
Central rise   = 2.35 m
Chord width   = 13.00 m
Span   = 30.00 m
Thickness of shell   = 80 mm
Semi-central angle   = 40°
Edge beam size   = 200 mm × 1880 mm
Reinforcements in edge beam   = 16 bars of 32 diameter
Width of edge beams   = 200 mm
Modular ratio   = 13
Effective cover   = 300 mm

Analyse the shell for stress in concrete and steel if the live load on the shell is 1 kN/m². The shell with edge beams is shown in Fig. 12.11.

Let the neutral axis cut the shell at an angle α. Taking moments of effective areas about neutral axis, we have:

$$2\int_0^\alpha R \cdot d\theta \cdot t(R\cos\theta - R\cos\alpha) = m2A_t(1.58 + R\cos\alpha - 7.75)$$

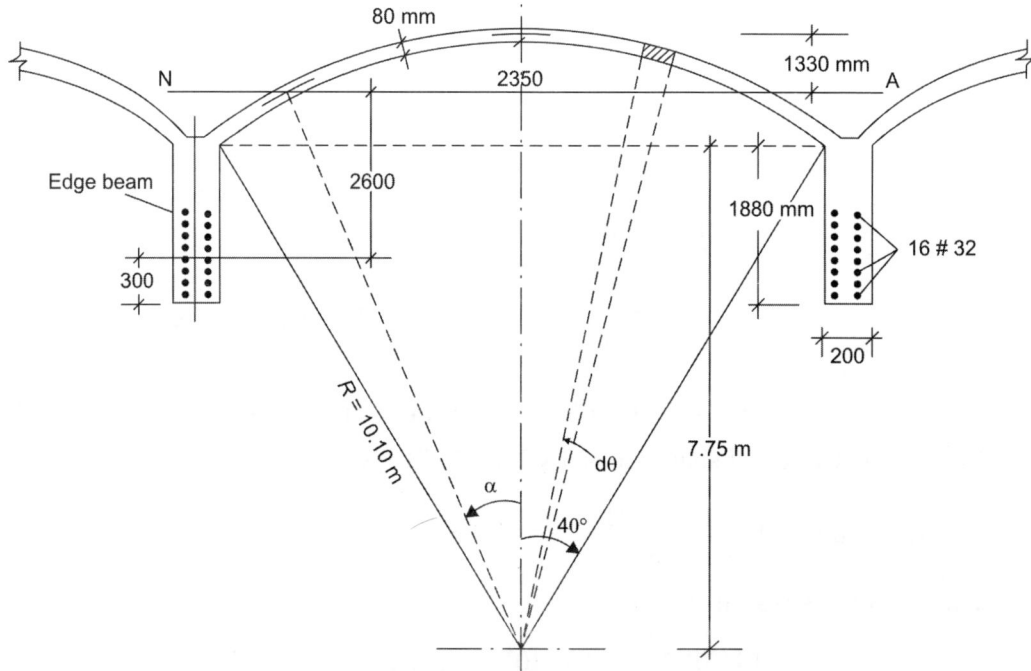

Fig. 12.11: Cylindrical shell with edge beams

Putting $\quad\quad m = 13$

$\quad\quad\quad\quad R = 10.10$

$\quad\quad\quad\quad A_t = (8 \times 8.04) = 64.32 \text{ cm}^2 = 0.0064 \text{ m}^2$

$\quad\quad\quad\quad\quad R^2 t \,(\sin\alpha - \alpha\cos\alpha) = mA_t\,(R\cos\alpha - 6.17)$

$\quad\quad\quad\quad\quad 10.10^2 \times 0.08\,(\sin\alpha - \alpha\cos\alpha) = 13 \times 0.0064\,(10.10\cos\alpha - 6.17)$

$\quad\quad\quad\quad\quad (\sin\alpha + 0.063) = \cos\alpha\,(\alpha + 103)$

∴ $\quad\quad\quad\quad \alpha = 26°15'$

$$I_{NA} = 2\int_0^\alpha Rd\theta\, tR^2\,(\cos\theta - \cos\alpha)^2 + m2A_t\,(1.58 + R\cos\alpha - 7.75)^2$$

$$= 1.835 \text{ m}^4$$

Self weight of shell $= (0.08 \times 25) = 1.92 \text{ kN/m}^2$

Water proofing and live load $\quad = 1.00 \text{ kN/m}^2$

Total load $\quad\quad\quad\quad\quad\quad\quad = 2.92 \text{ kN/m}^2$

∴ Total weight per metre run $= 2\int_0^{40} 2.92(Rd\theta) = 40 \text{ kN·m}$

Weight of edge beam $= 2(0.1 \times 1.88 \times 24) = 9.0 \text{ kN/m}$

Weight of filling in the valley portion $\quad = 1.5 \text{ kN/m}$

∴ Total load $= (40 + 10.50) \quad\quad\quad = 50.5 \text{ kN/m}$

Maximum bending moment $= \left(\dfrac{50.5 \times 30^2}{8}\right) = 5700 \text{ kN-m}$

Maximum shear force $= (0.5 \times 50.5 \times 30) = 757.5 \text{ kN}$

Maximum compressive stress at crown:

$$\sigma_c = \left(\frac{5700 \times 10^6 \times 1330}{1.835 \times 10^{12}}\right) = 3.52 \text{ N/mm}^2$$

Maximum tensile stress at centre of gravity of steel is given by:

$$\sigma_t = \left(\frac{3.62 \times 2600 \times 23}{1330}\right) = 89.5 \text{ N/mm}^2$$

Shear stress $\quad \tau = \dfrac{V}{Ib}(Ay)$

$\quad\quad\quad A.y = (16 \times 804 \times 13 \times 2600) = 435 \times 10^6 \text{ mm}^3$

∴ Maximum horizontal shear stress at neutral axis:

$$\tau_y = \left(\frac{757.7 \times 10^3 \times 435 \times 10^6}{1.835 \times 10^{12} \times 2 \times 80}\right) = 1.12 \text{ N/mm}^2$$

Using 8 mm diameter bars inclined at 45° to the longitudinal axis of the shell, spacing is given by:

$$S_v = \left(\frac{\sqrt{2} \times 50 \times 230}{80 \times 1.12}\right) = 180 \text{ mm near supports}$$

Towards the centre of span where span stress is less, adopt 6 mm diameter bars at 250 mm centres.

Arch Analysis

Transverse moments M_ϕ and thrust N_ϕ are obtained by treating the shell as an arch fixed between the longitudinal beams. The given shell arch is divided into a suitable number of parts at 5° to 6° intervals as shown in Fig. 12.12a. In the present example, the shell arch is divided into 16 parts at 5° intervals.

$$\text{Area of each element} = (t \cdot R \cdot d\theta) = \left(\frac{0.08 \times 10.1 \times 5 \times 2\pi}{360} \right) = 0.0705 \text{ m}^2$$

Area of edge beam at each end = $(1.58 \times 0.10) = 0.158$ m²

Specific shears acting on different elements determined by shear analysis. Shear stress at each elements is given by:

$$q = \left(\frac{\partial F}{2tI} \right) A\bar{Z}$$

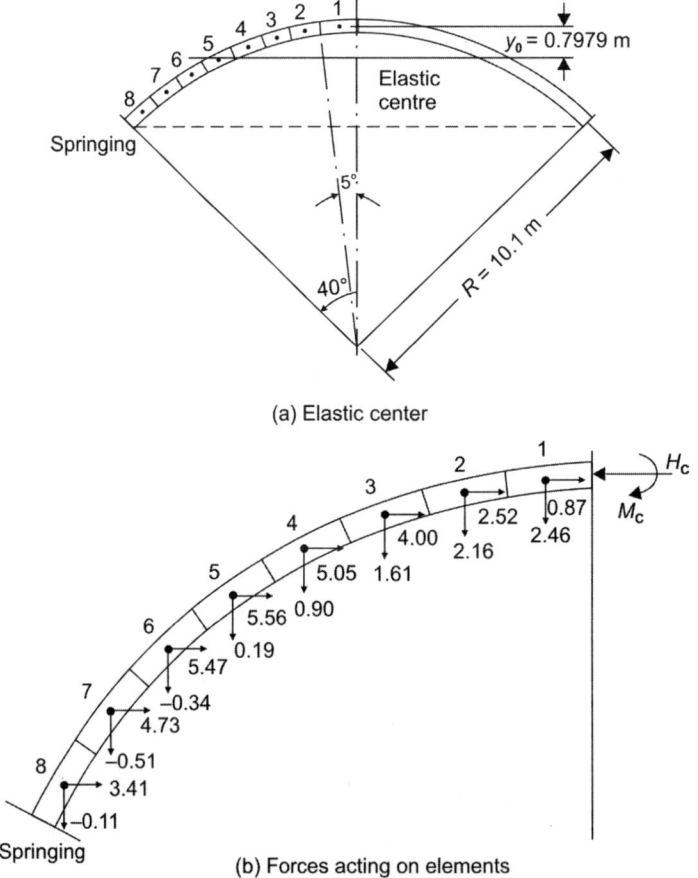

(a) Elastic center

(b) Forces acting on elements

Fig. 12.12: Arch analysis of shell

In this example:

∂F = load on one metre length of shell

$\qquad$ = 50.5 kN

$\qquad t$ = 80 mm

$\qquad I$ = 183.5 × 10^{12} mm^4 = 1.835 m^4

$$A\bar{Z} = 2\int_0^\theta r \cdot d\theta \cdot tz$$

where, $\qquad\qquad Z = R(\cos\theta - \cos 26°15')$

$\therefore \qquad\qquad\qquad A\bar{Z} = 2\int_0^\theta R \cdot d\theta \cdot t \cdot R(\cos\theta - \cos 26°15')$

$$= 2R^2 t \,(\cos\theta - 0.8970\,\theta)$$

$\therefore \qquad\qquad\qquad q = \left(\dfrac{50.5}{2 \times 0.08 \times 1.835}\right)(2 \times 10.1^2 \times 0.08)\,(\sin\theta - 0.8970\,\theta)$

$$q = 2800\,(\sin\theta - 0.8970\,\theta)$$

Using this equation, shear stress at each section is calculated.

The computations are shown in Table 12.4.

The specific shear force is given by:

$$T = (q \times A)$$

where, $\qquad\qquad A$ = area of each element

Net vertical force V $\qquad$ = $(W - T\sin\theta)$

Net horizontal force H = $T\cos\theta$

The thrusts at various sections are resolved into vertical and horizontal components. The vertical component with the dead load at each section gives the net vertical force at each section.

$$W = \text{self weight of each element}$$

Total weight of shell per metre run = 40 kN

$\therefore \qquad\qquad\qquad\qquad W = \left(\dfrac{40}{16}\right) = 2.5 \text{ kN}$

The force acting on the elements is shown in Fig. 12.12b. The elastic centre of the arch shell is determined as shown in Table 12.5.

Elastic centre $y_0 = \left(\dfrac{\Sigma y}{N}\right) = \left(\dfrac{6.3839}{8}\right) = 0.7979$ m

If $\quad M_c$ = moment at crown

$\qquad H_c$ = thrust at crown

$\qquad m$ = cantilever moment about the point due to applied loads V and H between the points and the crowns

$\qquad y_0$ = position of the elastic centre from crown

$\qquad y_1$ = vertical distance of the point from the elastic centre

Table 12.4: Specific shear at different sections

1 Section	2 θ Degrees	Radians	3 $\sin\theta$	4 0.8970θ	5 $q = 2800$ $[3-4]$ (kN/m^2)	6 $T = (q \times A)$ $(q \times 0.0705)$ (kN)	7 W (kN)	8 $T \cdot \sin\theta$ (kN)	9 $\cos\theta$	10 $H = T$ $\cos\theta$ (kN)	11 $V = (W - T$ $\sin\theta)$ (kN)
1	2.5	0.0436	0.0436	0.0391	12.60	0.88	2.5	0.3383	0.9990	0.8791	2.4617
2	7.5	0.1308	0.1300	0.1173	35.56	2.60	2.5	0.3380	0.9914	2.5776	2.1620
3	12.5	0.2181	0.2164	0.1956	58.24	4.10	2.5	0.8872	0.9763	4.0028	1.6128
4	17.5	0.3054	0.3007	0.2739	75.04	5.30	2.5	1.5930	0.9537	5.0546	0.9070
5	22.5	0.3926	0.3826	0.3521	85.40	6.02	2.5	2.3032	0.9239	5.5618	0.1968
6	27.5	0.4800	0.4617	0.4305	87.64	6.17	2.5	2.8486	0.8870	5.4727	−0.3486
7	32.5	0.5672	0.5372	0.5087	79.80	5.62	2.5	3.0190	0.8433	4.7393	−0.5190
8	37.5	0.6544	0.6087	0.5869	61.14	4.30	2.5	2.6174	0.7933	3.4111	−0.1174

Table 12.5: Determination of elastic centre

Section	θ	$\sin\theta$	$\cos\theta$	$(1-\cos\theta)$	$x = R\sin\theta$ (m)	$y = R(1-\cos\theta)$ (m)
1	2.5	0.0436	0.9990	0.0010	0.4403	0.0101
2	7.5	0.1300	0.9914	0.0086	1.3130	0.0868
3	12.5	0.2164	0.9736	0.0237	2.1856	0.2393
4	17.5	0.3007	0.9537	0.0463	3.0370	0.4676
5	22.5	0.3826	0.9239	0.0761	3.8642	0.8786
6	27.5	0.4617	0.8870	0.1130	4.6631	1.1413
7	32.5	0.5372	0.8433	0.1567	5.4257	1.5826
8	37.5	0.6087	0.7933	0.2067	6.1478	2.0876
						$\Sigma y = 6.3839$
Springing	40.0	0.6427	0.7660	0.2340	6.4912	2.36

Then, the horizontal thrust at crown is given by:

$$H_c = \left(\frac{\Sigma my_1}{\Sigma y^2}\right)$$

and the moment at crown M_c is given by:

$$M_c = \left(\frac{\Sigma m}{N} - H_c y_c\right)$$

where, N = number of elements

The net moment M at any section is $M = M_c + H_c y - m$

The net thrust at any section is $P = \Sigma V \sin\theta + (H_c - \Sigma H)\cos\theta$

The computations of moment at various sections is shown in Table 12.6.

Maximum transverse moment of –3.102 kN·m, occurs at springing. In general the magnitude of transverse moments are small in long shells. Using M20 grade concrete and Fe 415 HYSD bars. The effective depth of shell is computed.

Working moment M_w = 3.102 kN·m

Design ultimate moment $M_u = (1.5 \times 3.102) = 4.653$ kN·m

Effective depth $d = \sqrt{\dfrac{M_u}{0.138\,f_{ck}b}} = \sqrt{\dfrac{4.653\times10^6}{0.138\times20\times10^3}} = 41$ mm

Thickness of shell adopted = 80 mm

Adopt effective depth d = 60 mm

Areas of reinforcement are obtained by computing the parameter:

$$\left(\frac{M_u}{bd^2}\right) = \left(\frac{4.653\times10^6}{10^3\times60^2}\right) = 1.30$$

Table 12.6: Computation of transverse moments and thrusts

Section	x (m)	Δx	V (kN)	ΣV	$m_v = \Delta x\,\Sigma v$	y	Δy	H (kN)	ΣH	$m_H = \Delta y\,\Sigma H$	$m_v + m_H$	$m = \Sigma(m_v + m_H)$	$y_1 = y - y_0 = y - 0.798$	$m y_1$	y_1^2	Net moment $M = M_c + H_c y - m$ (kN·m)	Net thrust $= \Sigma V \sin\theta + (H_c - \Sigma H)\cos\theta$ (kN)
1	2	3	4	5	6	7	8	9	10	11	12	13	14	15	16	17	18
Crown	0	0	0	–	–	–	–	–	–	–	–	–	–	–	–	–1.430	33.80
1	0.44	0.44	2.46	2.46	–	0.010	–	0.87	–	–	–	–	–0.788	–	0.620	–1.092	–
2	1.31	0.87	2.16	4.62	2.14	0.086	0.076	2.57	0.87	0.066	2.206	2.206	–0.712	–1.570	0.506	–0.729	–
3	2.18	0.87	2.61	6.23	4.01	0.239	0.153	4.00	3.44	0.526	4.536	6.742	–0.559	–3.768	0.312	–0.093	–
4	3.03	0.85	0.90	7.13	5.29	0.467	0.228	5.05	7.44	1.696	6.986	13.728	–0.331	–4.543	0.109	+0.626	–
5	3.86	0.83	0.19	7.32	5.91	0.768	0.301	5.56	12.49	3.759	9.669	23.397	–0.030	–0.701	0.0009	+1.131	–
6	4.66	0.80	–0.34	6.98	5.85	1.141	0.373	5.47	18.05	6.732	12.582	35.979	0.343	12.340	0.177	+1.157	–
7	5.42	0.76	–0.51	6.47	5.30	1.582	0.441	4.73	23.52	10.372	15.672	51.651	0.784	40.494	0.614	+0.390	–
8	6.14	0.72	–0.11	6.36	4.65	2.087	0.505	3.41	28.25	14.266	18.916	70.567	1.289	90.960	1.661	–1.456	–
Springing	6.49	0.35	–	6.36	2.23	2.360	0.237	–	31.66	8.643	10.873	81.440	–	–	–	–3.102	5.726
												$\Sigma m = 204.27$		$\Sigma m y_1 = 133.21$	$\Sigma y_1^2 = 3.94$		

$$H_c = \frac{\Sigma m y_1}{\Sigma y_1^2} = \frac{133.21}{3.94} = 33.80 \text{ kN}, \qquad M_c \left[\frac{\Sigma m}{N} - H_c y_0\right] = \left[\frac{204.27}{8} - 33.8 \times 0.798\right] = -1.43 \text{ kN·m}$$

Refer to Table 2 of SP:16 and readout the percentage of reinforcement as:

$$p_t = \left(\frac{100 A_{st}}{bd}\right) = 0.392$$

$$\therefore \qquad A_{st} = \left[\frac{0.392 \times 10^3 \times 60}{100}\right] = 235 \text{ mm}^2 > A_{st(min)}$$

Provide 8 mm diameter bars at 200 mm centres in the transverse direction and distribution reinforcement of 6 mm diameter bars at 200 mm centres.

The details of reinforcement are shown in Fig. 12.13.

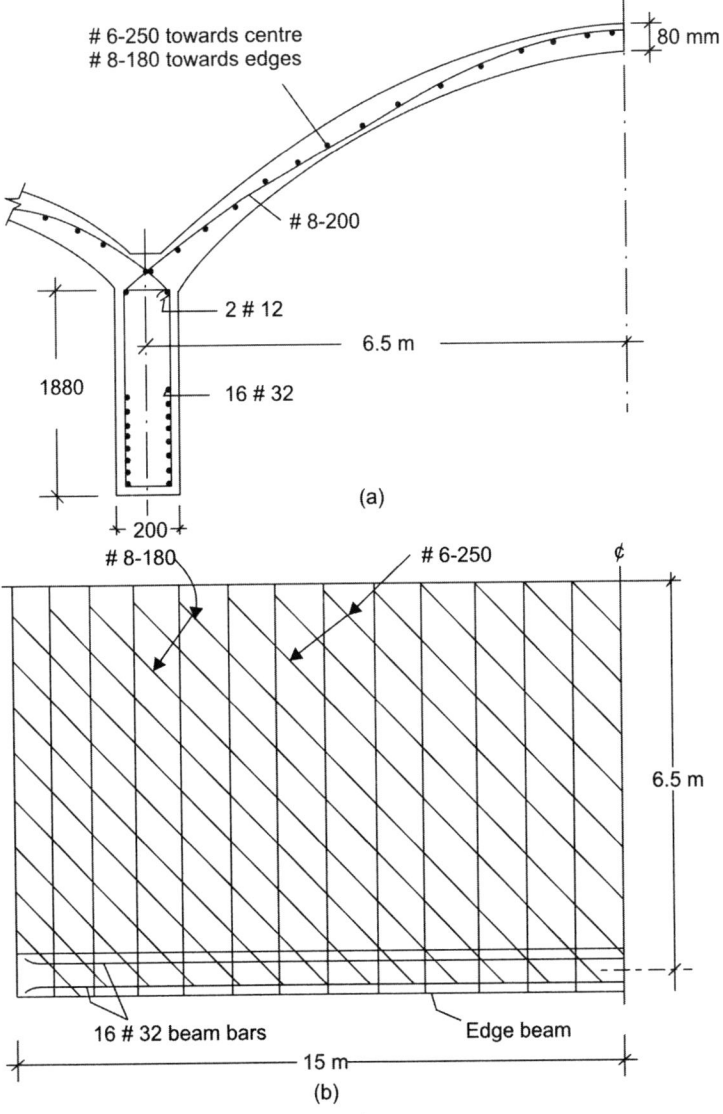

Fig. 12.13: Reinforcement details in a shell

Rigorous Theories of Analysis

The rigorous mathematical analysis of thin curved shells is governed by a differential equation of 8th order involving the main shell parameters and deformations. Solutions based on simplifying approximations have been developed by several investigators and a comparative analysis of various analytical approximation is reported by McNamee. It is pertinent to note that an exact mathematical solution of the 8th order differential equation, would be itself be no more than an approximate solution due to the inelastic properties of concrete.

The Donnel–Karman–Jenkins theory is considered to be the simplest among the rigorous theories which is applicable for both long and short shells. The ASCE method associated with the design tables is immensely useful in the analysis and design of cylindrical shells with any ratio of width to length. The Schorer–Tottenham formulation is by far the simplest among the various theories and sufficiently accurate for purposes of design in the range of medium to long shells. The rigorous methods generally involve lengthy computations to estimate the redundant reactions between the shell and edge beam.

ASCE manual 31, presents design coefficients of forces and deformations of the shell for different types of load conditions. These coefficients are useful in the design of cylindrical shells.

References

1. Haas AM, *Design of Thin Concrete Shells*, John Wiley & Sons Inc , New York, 1962.
2. Ramaswamy GS, *Design & Construction of Concrete Shell Roofs*, McGraw-Hill Book Co. New York, 1968.
3. Chatterjee BK, *Theory and Design of Concrete Shells*, Chapman & Hall, London, 1988.
4. Billington D, *Thin Shell Concrete Structures*, 2 edn, McGraw-Hill Book Co, New York, 1982.
5. Lundgren H, *Cylindrical Shells* (Vol. 1), *Cylindrical Roofs*, The Institution of Danish Civil Engineers, The Danish Technical Press, Copenhagen, 1949, Third Printing, January 1960.
6. Jenkins RS, *Theory and Design of Cylindrical Shell Structures*, The ON Arup Group of Consulting Engineers, Bulletin 1, London, 1947.
7. Tottenham H, Simplified method of design for cylindrical shell roofs, *The Structural Engineer*, London, Vol. 32, June 1954, pp. 161–180.
8. McNamee, Existing Methods for the Analysis of Concrete Shell Roofs, Proceeding of the Symposium on Concrete Shell Roof Construction, Cement and Concrete Association, London 1954, pp. 45–50.
9. Design of Cylindrical Shell Roofs, Manual of Engineering Practice, No. 31, American Society of Civil Engineers, 1952.
10. Krishna Raju N, *Prestressed Concrete*, 5 edn, McGraw-Hill Education (India) Pvt, Ltd, New Delhi, 2012, pp. 549–573.
11. Indian Standard, IS:2210-1988 (First Revision), Criteria for the Design and Construction of Reinforced Concrete Shell Structures and Folded Plates, Bureau of Indian Standards, New Delhi, 1992.
12. Dabrowiski R, Analysis of prestressed cylindrical shell roofs, *Journal of the Structural Division, Proceedings of ASCE*, ST-5, October 1963.
13. Ramaswamy GS, Important shell structures in India, Symposium on Shell Structures, Central Building Research Institute, Roorkee, Indian Concrete Journal, Vol. 33, Dec. 1959, pp. 469–471.

Assignment

1. A reinforced concrete shell with a circular directrix has the following dimensions:

   Distance between the traverses = 30 m

   Radius of the shell = 8 m

   Thickness of shell = 60 mm

   Semicentral angle = 60°

   If the water proofing course and occasional live load is 1 kN/m² of roof surface calculate:

   a. The maximum compressive stress in the shell.

   b. Maximum bending moment and tension in the longitudinal edge of the shell.

2. A reinforced concrete shell having a semicircular directrix is freely supported between the traverses separated by a distance of 35 m. If the radius and thickness of shell are 10 m and 60 mm respectively, calculate the membrane forces at the crown and edge due to its own self weight and a snow load of 1 kN/m². Also calculate the maximum compressive stress in concrete and the maximum tension in the edge members.

3. A circular cylindrical reinforced concrete shell spanning 25 m, has a radius of 8 m, with the central angle being 90°. The thickness of the shell is 70 mm and the density of concrete may be assumed as 24 kN/m³. If the weight of water proofing and occasional live load on the shell may be taken 1.2 kN/m² of the surface of shell, calculate the maximum compressive stress in concrete at the crown of the shell using beam theory.

4. A cylindrical shell roof consists of 10 parallel shells, each shell having a chord width of 10 m and a span of 25 m. The central angle is 80° and the thickness of the shell is 70 mm. The intermediate edge beams are 180 mm thick and 1.5 m in depth. The beams are reinforced with 20 bars of 28 mm diameter with their effective depth 30 cm from the soffit of the beams. The superimposed load due to water proofing cover and occasional live load may be taken as equivalent to 1.2 kN/m² of the surface of the shell. If M20 grade concrete is used with modular ratio = 13, calculate the maximum stresses in concrete and steel and also the maximum shear stress. Design suitable shear reinforcements. Also determine the transverse moments by arch analysis and design suitable reinforcements in the shell.

5. A circular cylindrical shell with edge beams has the following details:

   Span of edge beams = 25 m

   Radius of the shell = 8 m

   Chord width = 10 m

   Thickness of shell = 75 mm

   Size of edge beam = 250 mm × 1600 mm

   Reinforcements in edge beam = 12 bars of 25 mm diameter

   Grade of concrete = M20

   Type of reinforcement = Fe 415 HYSD bars

   Effective cover of edge beam reinforcements = 300 mm

   Analyse the shell for stresses in concrete and steel, if the service live load on the shell is 1 kN/m². Also design suitable reinforcements in the shell and sketch the details of reinforcements.

6. A circular cylindrical reinforced concrete shell spanning 20 m has a chord width of 10 m and central rise of 2 m. The thickness of the shell is 75 mm. The edge beams are 200 mm wide × 1600 mm deep and reinforced with 16 bars of 20 mm diameter. Using M25 grade concrete and Fe 415 HYSD bars, calculate the maximum stresses developed in the shell and design suitable reinforcements in the shell portion and sketch the details of reinforcements.

Review Questions

1. Briefly outline the specific advantages of reinforced shell structures in comparison with the traditional types of reinforced concrete structures.
2. In what type of structures, do you prefer to use shells as an alternative to the traditional beam and slab type of structures?
3. Briefly explain different types of shells used in the construction industry with neat sketches mentioning their specific applications.
4. Distinguish between shells of rotation and shells of translation and mention their suitability for adoption in different situations.
5. Explain briefly the various types of classification of shells with examples.
6. What specific guidelines do you follow in selecting the dimensions of various structural components of a shell.
7. Briefly explain the analysis of shells with specific reference to the membrane theory. What are the types of stresses developed in the shell when it is supporting the dead and inposed loads?
8. What is beam theory? Under what situations do you adopt the beam theory?
9. What is the necessity of using arch analysis in shell structures? Explain the necessity of using this procedure in the analysis of shells.
10. Briefly outline the rigorous theories of analysis of shells mentioning the specific advantage of using the ASCE method.

Objective Type Questions

1. Reinforced concrete shell structures are ideally suited for
   a. domestic buildings
   b. office buildings
   c. aircraft hangers
2. In cylindrical shells, the common curves used are the arc of a
   a. hyperbola
   b. circle
   c. hyperbolic paraboloid
3. The common example of a shell of revolution is
   a. elliptic paraboloid
   b. hyperbolic paraboloid
   c. circular dome
4. An excellent example of a shell of translation is
   a. conical dome
   b. circular dome
   c. hyperbolic paraboloid

5. A circular cylindrical shell is classified as long shell, if the ratio of radius to span is
   a. equal to 0.8
   b. greater than 0.6
   c. less than 0.6

6. The overall thickness of a reinforced concrete shell from practical considerations should be in the range of
   a. 75 to 120 mm
   b. 30 to 50 mm
   c. 150 to 200 mm

7. The beam theory is generally applicable for long shells having the ratio of span to radius exceeding the value of
   a. 2
   b. 3
   c. π

8. The span of a long reinforced concrete cylindrical shell should preferably be not greater than
   a. 40 m
   b. 30 m
   c. 50 m

9. The cross-section of a thin concrete shell, when supporting the dead and live loads is subjected to
   a. tension
   b. bending and compression
   c. compression

10. In case of multiple cylindrical shell roof used for an aircraft hanger, the edge beams are subjected to
    a. pure compression
    b. pure tension
    c. flexural stresses

13

Hyperbolic Paraboloid Shells

13.1 GENERAL ASPECTS

Hyperbolic paraboloid shells[1,2] grouped under the category of doubly curved anticlastic shells, were first successfully used as roofing unit by Silberkhul[3] in Germany by the middle of 20th century. Hypar shells are widely used for a variety of structural applications like exhibition halls, swimming pools, cycle stands and roofs of industrial structures. Hyperboloidal structures have a negative Gaussian curvature meaning they curve inward rather than outward direction. As doubly ruled surfaces, they can be generated by straight lines. Consequently, they are easier to build compared to the other types of traditional shells of translation and rotation. According to Cowan[4], hyperboloid structures are superior in stability for resisting external forces and are widely used as aesthetic structures.

Analysis of hyperboloidal shells have been reported by Candela[5] and Ramaswamy Rao[6], applied the membrane theory to hyperbolic paraboloidal shells to examine the nature of stresses developed in the shells when subjected to dead and imposed loads. The main advantage of the hypar shell is that its surface can be generated by two systems of straight lines so that the form work for casting the shell can be provided by straight boards, warped only slightly over their length. Tedesko[7] has reported the design and construction of hyperbolic paraboloidal shell of wide span at Denver, Colorado. Structural engineers have used this form of shell structure for a variety of situations like roofs of industrial structures, large areas of exhibition halls, storage structures and even as foundations when the safe bearing capacity of the soil is very low.

Ramaswamy *et al*[8] have reported a simplified method of analysis of hyperboloid roofing units. The design of a pretensioned hyperboloid shell roof element for a warehouse having span of 15 m has been reported by the author elsewhere[9]. The units designed to support loads as prescribed in IS:874-1964 using M50 grade concrete and 5 mm diameter high tensile steel wires, have a shell thickness of 60 mm and overall width of 2100 mm. The design and construction of hyperbolic paraboloid shells for foundations, should conform to the specifications prescribed in the Indian standard code IS:9456-1980[10].

13.2 GEOMETRY OF HYPAR SHELL

The generation of hypar shell surface is shown with the aid of Fig. 13.1. Consider a plane rectangle OXBY. If B is now elevated by an amount BB' = h, the warped surface OXB' Y is an hyperbolic paraboloid. Each pair of opposite sides are divided into equal parts and the corresponding points joined by straight lines to obtain the hypar surface.

The surface between X and Y is parabolic (convex parabola) and is well suited to resist compressive forces similar to that of arch elements. The surface between O and B' is an inverted parobola (concave parabola or catenary) and is well suited to take tensile forces much similar to that of a catenary.

The hypar surface combining these shapes has a great stiffness and resistance to buckling and except for secondary bending effects, normal applied loads are carried by membrane or direct forces within the shell gather along the four edges, transferring the applied loads to the positions where support is provided.

13.3 ANALYSIS OF MEMBRANE FORCES

The equation of the hypar shell in cartesian coordinates (Fig. 13.1) is given by

$$Z = c\left(\frac{y}{b}\right) = \left(h\frac{x}{a}\right)\left(\frac{y}{b}\right) = \left(\frac{h}{ab}\right)xy = kxy$$

In the above equation, the constant $k = \left(\dfrac{h}{ab}\right)$ determines the intensity of warping of the surface and is referred to as the coefficient of specific distortion.

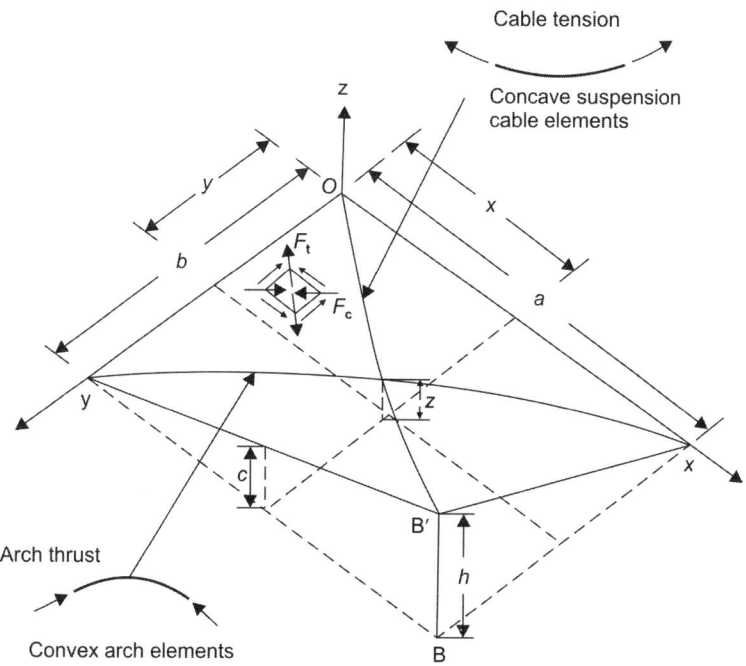

Fig. 13.1: Geometry of hypar shell

Under vertical loads in the direction z, any element with edges parallel to the generatrix (boundaries of shell) is subjected to only shear forces along the edges. The state of stress is defined as:

$$N_x = 0, N_y = 0, Q = \left(\frac{wab}{2h}\right)$$

where, Q = shear force/metre.

The principal stresses are also of same magnitude as shear stress and occur along lines at 45° to the shell boundaries. There will be compression F_c along XY, and tension F_t along OB'. The shears which collect along OX and OY are resisted by these edges acting in compression and shears along XB' and YB' are resisted by these edges acting in tension.

$$Q = \left(\frac{wab}{2h}\right) \text{ kN/m}$$

∴   Shear stress $q = \left(\frac{Q}{1000t}\right) \text{ N/mm}^2$

where, t = thickness of the shell (mm)

13.4 INVERTED UMBRELLA ROOF

The inverted umbrella type of roof consisting of hypar units is widely used for cycle and bus stands, race tracks and market halls. The umbrella roof consists of four hypar units joined together at the centre where the main supporting column is provided. The structural components of a typical inverted type umbrella roof shown in Fig. 13.2 are as follows:

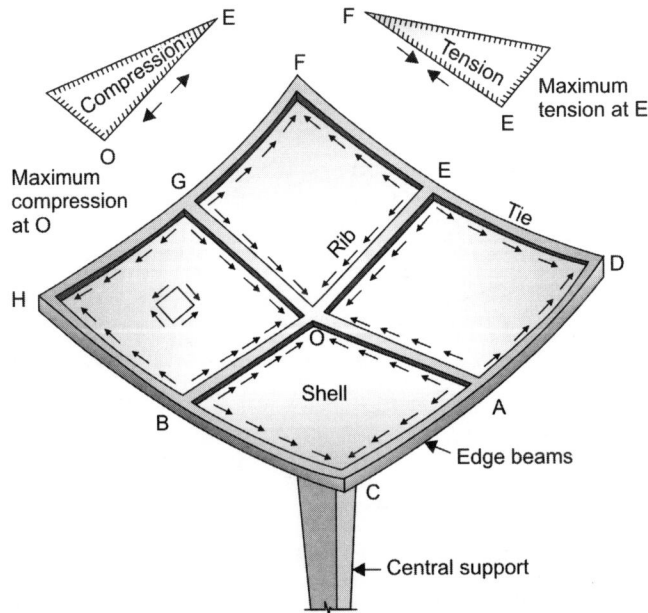

Fig. 13.2: Forces in inverted umbrella roof

a. Central supporting column
b. Edge beams
c. Sloping ribs connecting the column
d. Shell surface

The edge beams FE, ED, DA, AC, CB, BH, HG and GF are subjected to tensile force. The tensile forces are zero at the corners D, F, H and C and maximum at A, E, G and B.

The sloping ribs GO, EO, AO and BO are subjected to compression, varying from zero at the edges A, E, G and B to a maximum at the junction O.

13.5 DESIGN EXAMPLES

1. Design a hypar shell roof of the inverted umbrella type to suit the following data:
   Area to be covered in plan = 12 m × 12 m
   M20 grade concrete and Fe 415 grade tor steel
   Sketch the details of reinforcements in the shell and edge beams (Fig. 13.3).

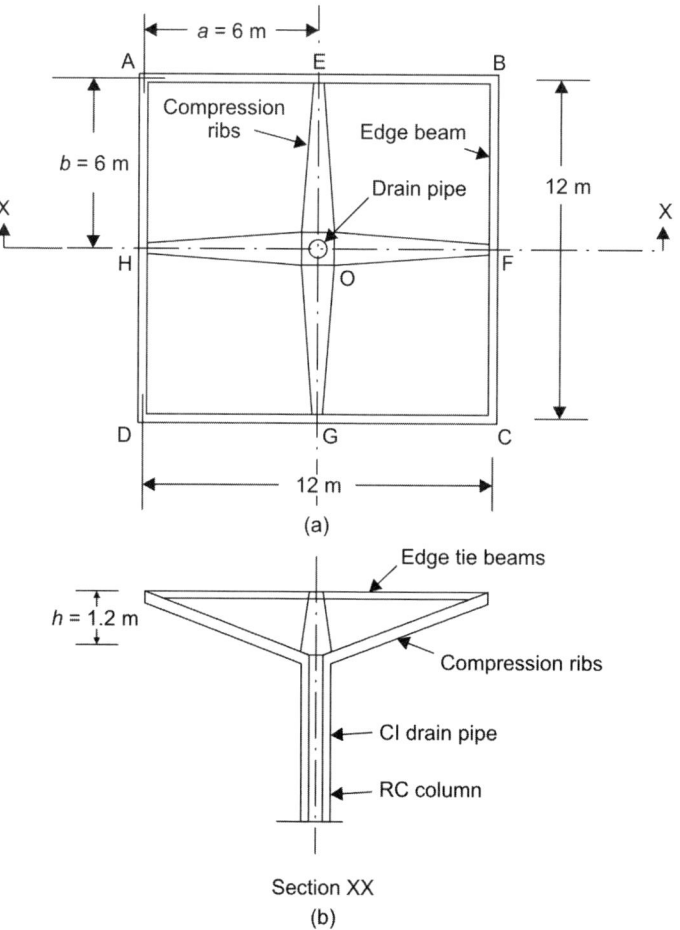

Fig. 13.3: Inverted umbrella roof

$$\text{Central dip} = \left(\frac{a}{5}\right) = \left(\frac{6}{5}\right) = 1.2 \text{ m}$$

Let the thickness of shell $t = 70$ mm

i. *Loads*:

Dead load $= (0.7 \times 24)$ $\qquad\qquad\qquad\qquad = 1.68$

Live load $\qquad\qquad\qquad\qquad\qquad\qquad\qquad = 0.50$

Load due to edge beams (water proofing etc.) $= \underline{0.22}$

Total load w $\qquad\qquad\qquad\qquad\qquad\qquad = 2.40 \text{ kN/m}^2$

$\therefore\qquad w = 2.40 \text{ kN/m}^2$

$\qquad\qquad a = b = 6 \text{ m}$

$\qquad\qquad h = 1.2 \text{ m}$

$$\text{Membrane shear force} = \left(\frac{wab}{2h}\right) = \left(\frac{2.4 \times 6 \times 6}{2 \times 1.2}\right) = 36 \text{ kN/m}$$

$$\text{Shear stress} = \left(\frac{36 \times 10^3}{1000 \times 70}\right) = 0.51 \text{ N/mm}^2$$

$\therefore\qquad$ Principal stress $= 0.51 \text{ N/mm}^2$

Using nominal reinforcement for temperature and shrinkage at 0.2% of gross cross-section,

$$A_{st} = \left(\frac{0.2}{100} \times 1000 \times 70\right) = 140 \text{ mm}^2/\text{m}$$

But $\qquad A_{st} = \left(\frac{36 \times 10^3}{230}\right) = 157 \text{ mm}^2/\text{m}$

Using 8 mm diameter at 300 mm centres both ways (A_{st} provided $= 167 \text{ mm}^2$).

$$\therefore \text{ Tensile stress} = \left[\frac{36 \times 10^3}{(1000 \times 70) + (13-1)167}\right] = 0.50 \text{ N/mm}^2 < 2.8 \text{ N/mm}^2$$

ii. *Sloping compression ribs:*

Length of member OF = OE = OH = OG = $\sqrt{6^2 + 1.2^2} = 6.12$ m

Maximum compression in sloping rib at O $= (2 \times 36 \times 6.12) = 441$ kN

Length of compression member $= 6.12$ m

Assuming least lateral dimension of $(70 + 130) = 200$ mm, $b = 0.2$ m

Since $\qquad \left(\dfrac{L_{ef}}{b}\right) = \left(\dfrac{6.12}{0.2}\right) = 30.6 > 12$

Hence, the compression member should be designed as long column.

$$\text{Reduction coefficient } C_r = \left[1.25 - \frac{L_{ef}}{48b}\right] = \left[1.25 - \frac{30.6}{48}\right] = 0.612$$

$\therefore$ Using 1% steel in the compression member

$$P = (\sigma_{cc}A_c + \sigma_{sc}A_{sc})$$
$$441 \times 10^3 = 0.612(5 \times 0.99\, A_c + 190 \times 0.01A_c)$$

$\therefore$ $\qquad A_c = 105195$ mm^2

$\therefore$ Width of beam $= \left(\dfrac{105195}{200}\right) = 525$ mm

Provide 600 mm wide by 200 mm deep compression ribs

$$A_{st} = (0.01 \times 600 \times 200) = 1200 \text{ mm}^2$$

Provide 4 bars of 20 mm diameter (A_{st} = 1256 mm^2), as main steel and nominal binders of 6 mm diameter at 200 mm centres.

iii. *Edge beams (tension members)*:

The tension members are AE, EB, BF, FC, CG, GD, DH and HA

Maximum tension develops at the centre of the beam (at points E, F, G and H)

Maximum tension = (36 × 6) = 216 kN

$\therefore$ $\qquad A_{st} = \left(\dfrac{216 \times 10^3}{230}\right) = 939$ mm^2

Using 4 bars of 20 mm, A_{st} = 1256 mm^2

For M20 grade concrete, permissible tensile stress = 2.8 N/mm^2

If area of concrete A_c

$$\left[\frac{216 \times 10^3}{A_c + (13-1)1256}\right] = 2.8$$

$\therefore$ $\qquad A_c = 62071$ mm^2

Adopting a depth of 200 mm.

Width of beam $= \left(\dfrac{62071}{200}\right) = 310$ mm

Adopt edge beams of size 320 mm × 200 mm. The details of reinforcements in the hypar shell is shown in Fig. 13.4.

2. Design a hypar shell roof of tilted inverted umbrella type to cover an area of 24.4 m × 24.4 m. The edges AHD and BFC are 3.66 m and 2.44 m respectively above the central valley point O as shown in Fig. 13.5. Adopting M20 grade concrete and Fe 415 tor steel, design the various structural components of the hypar shell roof.

   i. *Data*:

   $\qquad a = 12.2$ m
   $\qquad b = 12.2$ m $\qquad$ M20 grade concrete and
   $\qquad h_{min} = 2.44$ mm $\qquad$ Fe 415 HYSD bars
   $\qquad h_{max} = 3.66$ mm

   ii. *Shell surface*:

   Assume thickness of shell t = 80 mm

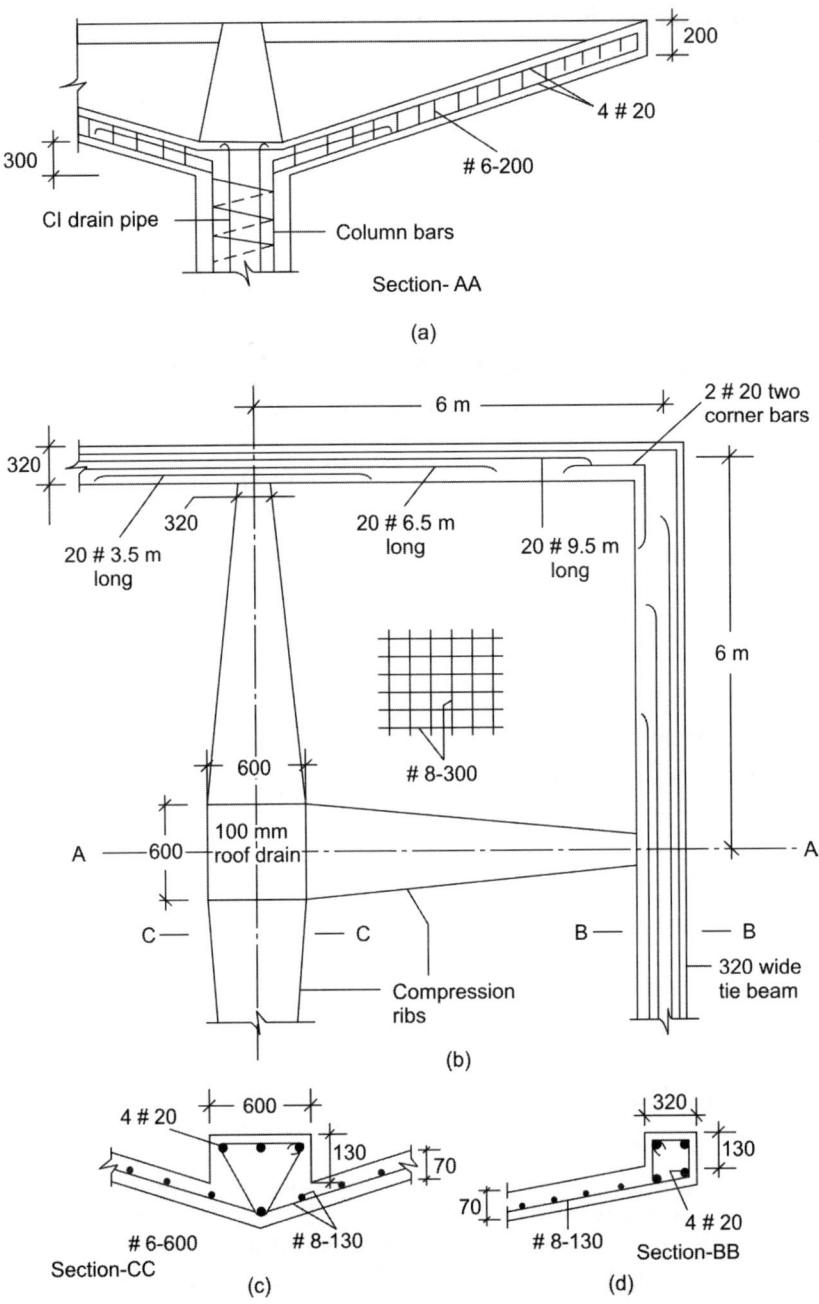

Fig. 13.4: Reinforcement details in hypar shells

Self weight of shell = (0.08×24) = 1.92
Live load   = 0.30
Water proofing, weight of ribs, etc. = 0.20
Total load w = 2.42 kN/m^2

$$\text{Maximum shear force} = \left(\frac{wab}{2h_{min}}\right) = \left(\frac{2.42 \times 12.2 \times 12.2}{2 \times 2.44}\right) = 72 \text{ kN/m}$$

$$\therefore \quad \text{Maximum shear stress} \quad = \left(\frac{72 \times 10^3}{1000 \times 80}\right) = 0.9 \text{ N/mm}^2$$

$\therefore$ Principal stress = 0.9 N/mm^2

Using nominal reinforcement for temperature and shrinkage of 0.2% of gross area.

$$A_{st} = \left(\frac{0.2}{100} \times 1000 \times 80\right) = 160 \text{ mm}^2/\text{m}$$

But $\quad A_{st} = \left(\frac{72 \times 10^3}{230}\right) = 313 \text{ mm}^2/\text{m}$

Provide 10 mm diameter at 250 mm centres both ways.

iii. *Sloping compression ribs*:

The lengths of the various sloping ribs are computed as shown in Fig. 13.5.

$$OH = \sqrt{3.66^2 + 12.2^2} = 12.7 \text{ m}$$

$$OF = \sqrt{2.44^2 + 12.2^2} = 12.45 \text{ m}$$

$$OG = \sqrt{3.05^2 + 12.2^2} = 12.6 \text{ m}$$

$$GD = \sqrt{12.2^2 + 0.61^2} = 12.25 \text{ m}$$

Shear force for left half of structure

$$= \left(\frac{wab}{2h_{max}}\right) = \left(\frac{2.42 \times 12.2 \times 12.2}{2 \times 3.66}\right) = 49 \text{ kN/m}$$

Shear force in the right half panel = 72 kN/m

$$\therefore \quad \text{Average shear force at junction} = \left(\frac{2.42 \times 12.2 \times 12.2}{2 \times 3.05}\right) = 59.2 \text{ kN/m}$$

iv. *Forces in compression and tension members*:

OH = 2(12.7 × 49) = 1244 kN ⎫
OF = 2(12.45 × 72) = 1790 kN ⎬ Compression
OG = 2(12.60 × 59.2) = 1498 kN ⎭

GD = (12.25 × 72) = 880 kN ⎫
GC = (12.25 × 49) = 600 kN ⎬ Tension
AH = (12.20 × 72) = 878 kN ⎪
BF = (12.20 × 49) = 597 kN ⎭

v. The tension and compression members are designed for the forces according to the method explained in design Example 13.5.

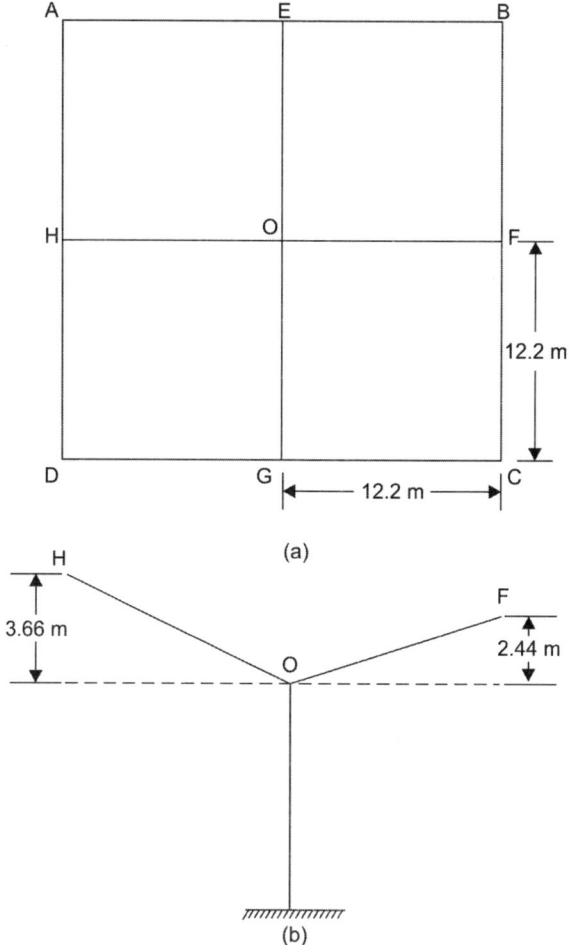

Fig. 13.5: Tilted inverted umbrella type roof

13.6 TYPES OF HYPERBOLIC PARABOLOID ROOFS

a. The hipped type roof consisting of 4 units of hypar shells as shown in Fig. 13.6, are ideally suited for covering large rectangular grids for market halls or storage sheds. In this type of roof the edge members BE and BF are in compression. The supports are located at A, B, C and D. The members OE, OF, OG and OH are subjected to tension.

b. The turkey shed shown in Fig. 13.7 is a combination of four basic hypar units with the supports located at the mid points of the sides. In this type of roof the edge members are in tension and the interior members OA, OE, OC and OG are in compression. Ties are required along AE and GC to take up the thrust from the arch elements AOE and GOC.

c. Hypar shell roofs formed by depressing the two opposite corners of a square grid and elevating the remaining two opposite corners is widely popular for swimming pools. The side are covered by glazing. A typical hypar shell of this type is shown in Fig. 13.8. A tie is required connecting the two corners A and C.

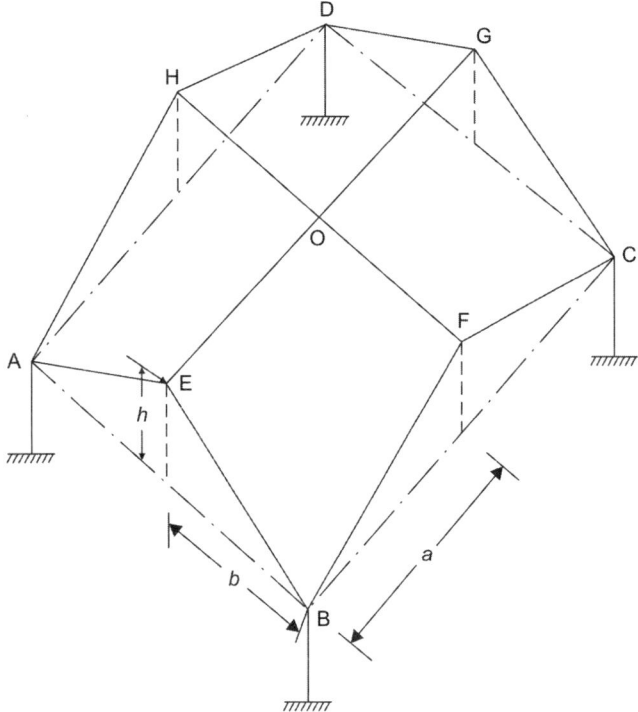

Fig. 13.6: Hipped type roof

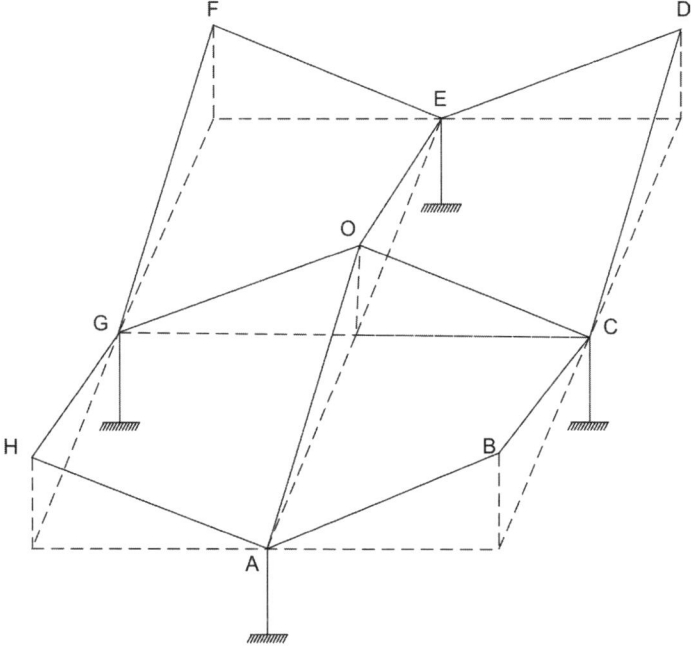

Fig. 13.7: Turkey shed using hypar units

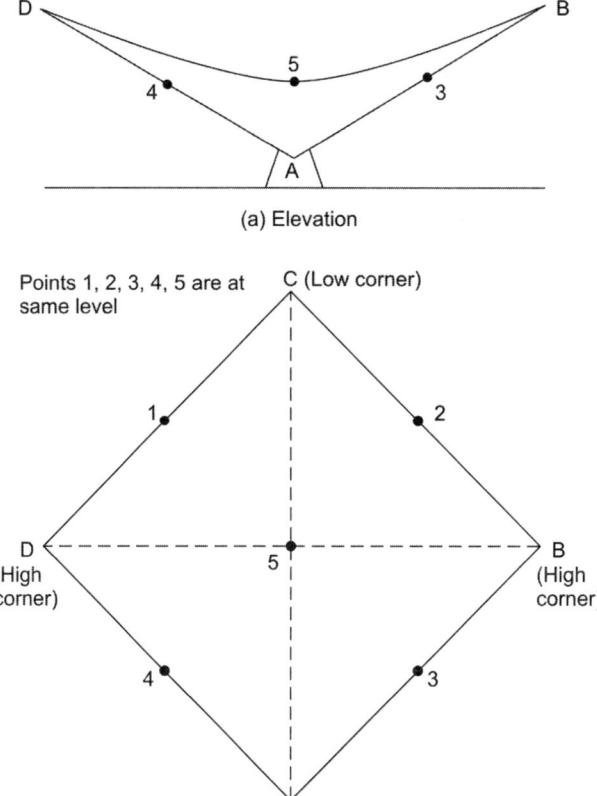

(a) Elevation

Points 1, 2, 3, 4, 5 are at same level

(b) Plan

Fig. 13.8: Hypar shell roof

References

1. Shukhov V (1853–1939), *The World's First Hyperboloid Structure*, Wikipedia.

2. Janburg N, Hyperboloid Water Tower, International Database and Gallery of Structures, ICS, Moscow, 2007.

3. Haeusler E, Precast prestressed hyperboloids of revolution, *International Association for Shell and Spatial Structures (IASS) Bulletin*, No.6, A-7, Madrid, 1959.

4. Cowan HJ, *Handbook of Architectural Technology*, Von Nostrand, Reinhold, Germany, 1991, p. 175.

5. Candela F, General formulae for membrane stresses in hyperbolic paraboloid shells, *Journal of the American Concrete Institute*, 1960.

6. Rao R, The membrane theory applied to hyperbolic paraboloid shells, *Indian Concrete Journal*, Bombay, India, 1961.

7. Tedesko A, Shell at Denver–hyperbolic paraboloidal structure of wide span, *Journal of the American Concrete Institute*, Vol. 57, No. 4, Oct. 1960, pp. 403–412.

8. Ramaswamy GS and Sayeed NR, Simplified analysis of pretensioned hyperboloid roffing units, *Bulletin of the International Association for Shell and Spatial Structures (IASS)*, Vol. 55, Madrid.

9. Krishna Raju N, *Prestressed Concrete*, 5 edn, McGraw-Hill Education (India), New Delhi, 2012, pp. 549–573.

10. IS:9456-1980, Code of Practice for Design and Construction of Conical and Hyperbolic Paraboloid Types of Shells for Foundations, Bureau of Indian Standards, New Delhi, 1980.

Assignment

1. Design a hypar shell roof of inverted umbrella type to suit the following data; area to be covered in plan is 20 m × 20 m, M20 grade concrete and Fe 415 grade tor steel are available for use. Sketch the details of reinforcements in the shell, edge tie beams and compression ribs.

2. A hypar shell roof of tilted inverted umbrella type cover with an area of 20 m × 20 m is required. The opposite edges of the roof are 3 and 2 m respectively above the central valley point. Adopting M20 grade concrete and Fe 415 grade tor steel, design the various structural elements of the shell.

3. A market hall measuring 20 m × 30 m is to be covered by a turkey shed using hypar shell units. Design the hypar shell roof using the supports at the mid points of the sides of the gird. Adopt M20 grade concrete and Fe 415 HYSD bars.

4. An exhibition pavilion is planned using hyperbolic paraboloid units of inverted umbrella type to cover an area of 15 m × 15 m with a single supporting column at the centre. The edges of the roof element should be 2 m above the central valley point. Adopting M25 grade concrete and Fe 500 HYSD steel bars, design the various structural components of the shell.

5. A hypar shell roof is proposed to cover an area of 25 m × 25 m. The columns have to be located at the centre of the sides. The corners and the central point should be 2 m above the top of the columns which are 4 m in height above the ground. Design the shell roof elements using M25 grade concrete and Fe 500 HYSD bars.

6. A hipped type hyperbolic paraboloid shell roof is to be designed for an exhibition pavilion for covering an area of 30 m × 30 m. The 4 m long columns are to be located at the four corners. The centre of the sides should be 2 m above that of the top of the columns. Design the hypar shell with the edge beams using M25 grade concrete and Fe 415 HYSD bars.

Review Questions

1. List the various structural advantages of using hyperbolic paraboloid elements in roofs of industrial structures and exhibition pavilions.

2. What are the various structural components of a hypar shell? Explain the various elements with sketches.

3. Explain with sketches the structural action of a hypar shell showing the forces developed in the shell when it supports its self weight and imposed loads.

4. What are the various types of hyperbolic paraboloid roofs used in practice? Explain the structural elements associated with them by sketches.

5. What is coefficient of specific distortion? How do you compute its value in a hypar shell? What is the significance of this term in a hypar shell?

6. Briefly explain the load carrying mechanism in a hypar shell subjected to vertical loads.

7. Sketch the various structural components in an inverted umbrella type hypar shell showing the types of forces developed in these elements.

8. How do you compute the shear forces and stresses developed in a hypar shell? Evaluate the maximum shear stress developed in the edge beams of a hypar shell of overall dimensions 20 m × 30 m and thickness 80 mm and the difference in height between adjacent corners is 2 m.

9. Identify and list the various structural elements in a hypar shell, which are subjected to maximum compression and tension. Also specify the critical sections for design.

10. What are the sections subjected to maximum shear force in a hypar shell. How do you design this section?

Objective Type Questions

1. Hyperbolic paraboloid shells are ideally suited for roofing units of
   a. office floors
   b. domestic buildings
   c. factory sheds

2. The quantity of steel and concrete used in the cross-sections of members is least in
   a. flat slabs
   b. beam and slab floors
   c. hypar shells

3. The thickness of slabs in a hyperbolic paraboloid shell structure is of the order of
   a. 150 to 180 mm
   b. 30 to 50 mm
   c. 60 to 100 mm

4. The casting of a hypar shell unit requires the form work to be arranged using
   a. curved boards to suit the curvature
   b. straight boards arranged to get the curved surface
   c. inclined boards to achieve the desired curvature

5. The total cost per unit area covered is the least by adopting roofs built using
   a. beam and slab units
   b. grid floors
   c. hypar shells

6. The slab of a hypar shell structure under vertical loads develops
   a. flexural and compressive stresses
   b. torsional and tensile stresses
   c. compressive and tensile stresses

7. In a hypar shell with edge beams under vertical loads, the maximum tensile forces develop in
   a. outer edge beams
   b. the main shell slab
   c. inner edge beams meeting at the top of supporting column

8. Among the various types of structures used in exhibition halls, aesthetically superior structure is obtained by using
   a. beam and slabs
   b. trussed units
   c. hypar shells

9. Among the various types of reinforced concrete structural units, the best type to be used both for roofs and foundations is
   a. slab with beams in perpendicular directions
   b. flat slabs
   c. hyperbolic paraboloid shells

10. Hypar shells are classified under the group
    a. singly curved shells
    b. synclastic shells
    c. anticlastic shells

14

Hyperbolic Cooling Towers

14.1 INTRODUCTION

The development of thermal and atomic power stations, refineries and other industrial plants during the early 19th century required large quantities of water for cooling the power plants. The hot water from the power stations is cooled before reuse by circulating in a natural draught cooling tower. In the beginning, steel circular cooling towers similar to chimney stacks were used followed by reinforced concrete circular cooling towers according to Wikipedia reports. Van Iterson[1] developed the hyperbolic cooling tower for power stations of Dutch State Mines during 1915–17. The first successful hyperbolic natural draught cooling tower was built in the province of Limbug in the Netherlands during 1917–18 with Van Iterson as the principal engineer and Kuypers as the contractor. Continuous research and design by engineers[2,3,4] throughout the world has contributed to the evolution of shape of the doubly curved shell structure and its behavior under static, dynamic wind seismic loads.

Gupta *et al*[5] and Noh[6] have reported their investigations on the limit state of collapse of hyperbolic cooling towers subjected to ultimate loads. Sudden failure of 375 ft tall Ferry Bridge[7] cooling towers in Yorkshire, UK during 1965 under severe wind loads indicated that more studies were necessary to understand the behavior of cooling towers under dynamic wind loads. The structural behavior of cooling towers under dynamic wind loads has been reported by Armitt[8] and strengthening of natural draught cooling tower shells with stiffening rings have been examined by Boseman[9]. The problem of repairing damaged cooling towers has been examined in detail by Gould and Guedethoefer[10].

Analysis and design aspects of hyperbolic cooling towers are investigated by Gurfinkel[11] and the Indian standard code IS:11504-1985[12] has specified useful criteria for structural design of reinforced concrete natural draught cooling towers. Hyperbolic cooling towers are widely used in various thermal power stations in India. The prominent among them are the tall cooling towers at Raichur, Karnataka (Refer to Fig. 14.1). India's major superthermal power stations spread over different states listed in Table 14.1 invariably use hyperbolic cooling towers in their power plants. A typical section of a hyperbolic cooling tower of 100 m height is shown in Fig. 14.2.

Fig. 14.1: Hyperbolic cooling towers at Raichur thermal power station in Karnataka

Table 14.1: Prominent super thermal power stations of India

S. No.	Name of thermal power station	State	Capacity (MW)
1.	Vindyachal thermal power station	Madhya Pradesh	4260
2.	Mundhra thermal power station	Gujarat	4260
3.	Tiroda thermal power station	Maharashtra	3300
4.	Talchar thermal power station	Odisha	3000
5.	Sipat thermal power station	Chattisgarh	2980
6.	National thermal power corporation	Uttar Pradesh	2637
7.	Ramagundam thermal power station	Andhra Pradesh	2600
8.	Simhadri super thermal power station	Andhra Pradesh	2000
9.	Rihand thermal power station	Uttar Pradesh	2000
10.	Raichur thermal power station	Karnataka	1470

14.2 HYPERBOLOID OF REVOLUTION

The hyperboloid of revolution of one sheet is generally used for cooling towers of thermal stations. The great advantage of this type of shell is that it is generated by two families of intersecting straight lines and the form work can be achieved by straight boards warped only slightly over the lengths.

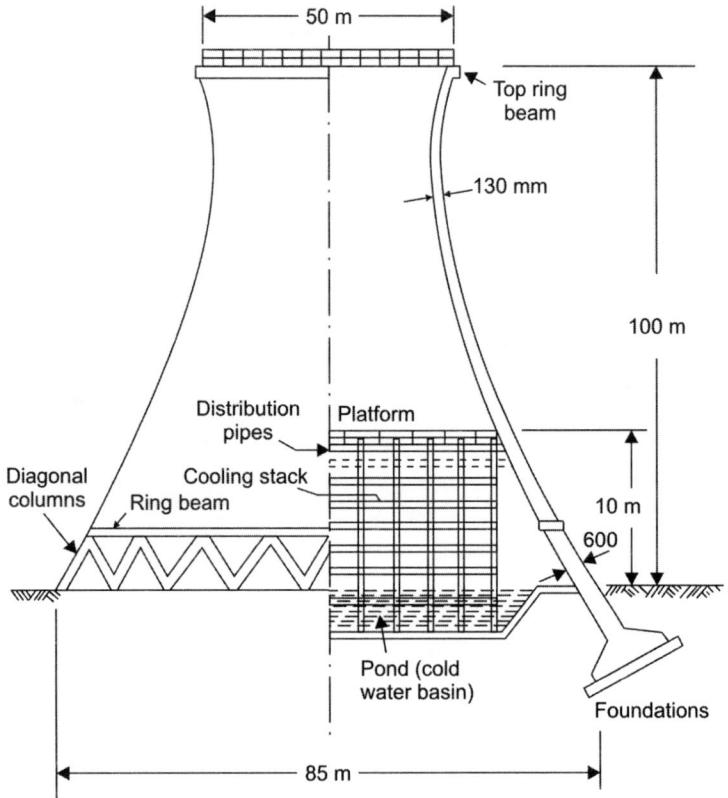

Fig. 14.2: Hyperbolic cooling tower

The intersecting grid of straight lines from rhombuses of intersection. The shell surface can also be built of precast rhombic elements which are repeated along the complete circumferences at fixed heights. The generation of the hyperboloid of revolution by intersecting straight lines is shown in Fig. 14.3.

14.3 ANALYSIS OF MEMBRANE FORCES

The equation of the hyperboloid is given by

$$\left[\frac{r_0^2}{a^2} - \frac{z^2}{b^2}\right] = 1 \qquad \dots(14.1)$$

Principal radii of curvature r_1 and r_2 are given by the equations:

$$r_1 = a^2 b^2 \left[\frac{r_0^2}{a^4} + \frac{z^2}{b^4}\right] \qquad \dots(14.2)$$

$$r_2 = a \left[1 + \frac{z^2}{a^2}\left\{\frac{a^2}{b^2} + \left(\frac{a^2}{b^2}\right)^2\right\}\right]^{\frac{1}{2}} \qquad \dots(14.3)$$

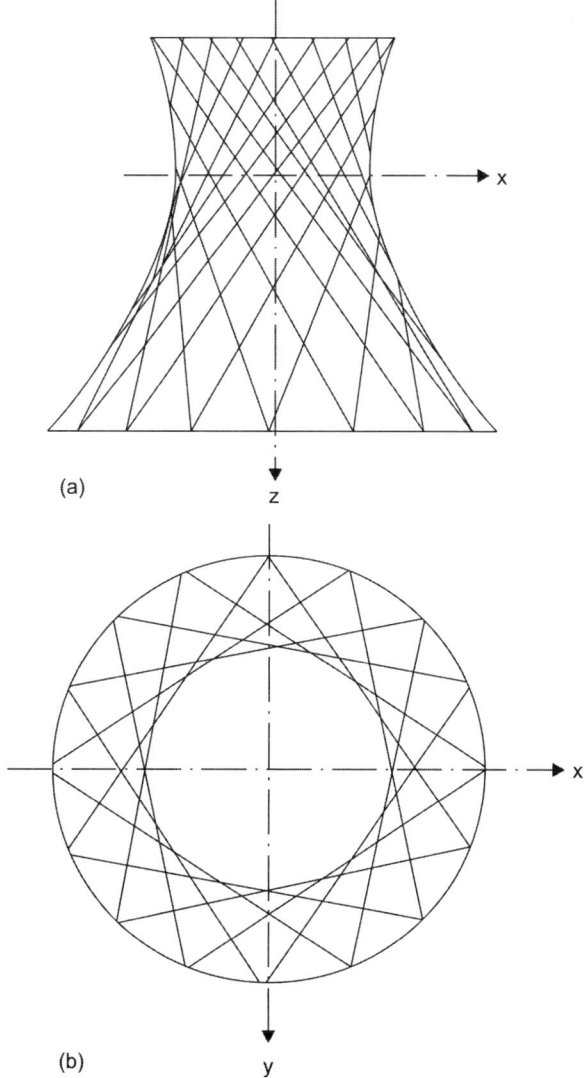

(a)

(b)

Fig. 14.3: Hyperboloid of revolution

Also, $$\alpha = \left(\frac{a^2}{b^2}\right)$$

The geometry of the hyperboloid is shown in Fig. 14.4

Relation between r_1 and r_2 is given by

$$r_1 = \left(\frac{r_2^3}{\alpha a^2}\right) \qquad \qquad ...(14.4)$$

Also from Eq. (14.1)

$$z = \pm b/a\sqrt{r_0^2 - a^2}$$

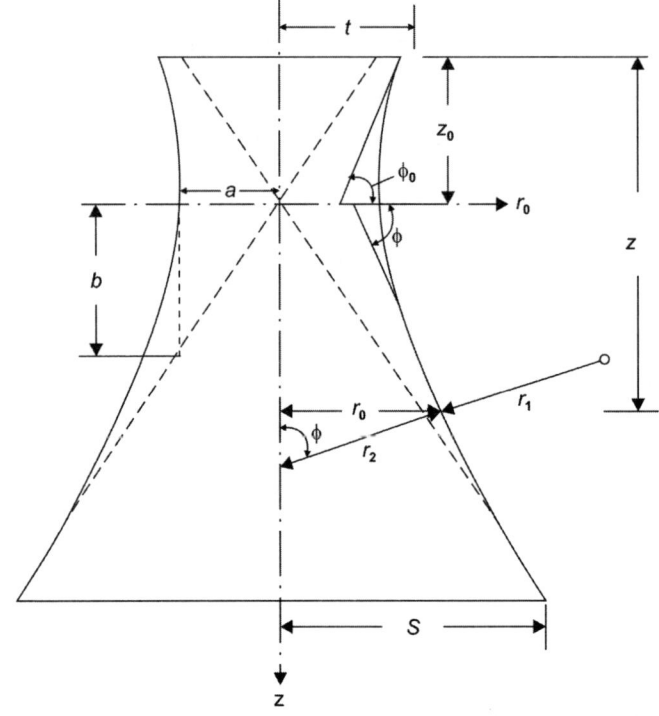

Fig. 14.4: Geometry of hyperboloid

$\therefore$
$$\frac{dz}{dr_0} = \tan\phi = +b/a\sqrt{\frac{r_0^2}{r_0^2 - a^2}}$$

$\therefore$
$$\tan^2\phi = \frac{b^2}{a^2}\left(\frac{r_0^2}{r_0^2 - a^2}\right)$$

$$\cot^2\phi = \frac{a^2}{b^2}\left(1 - \frac{a^2}{r_0^2}\right)$$

Simplifying,
$$r_0 = \frac{a^2\sin^2\phi}{(a^2\sin^2\phi - b^2\cos^2\phi)^{1/2}} \qquad \dots(14.5)$$

Similarly,
$$z = \frac{b^2\cos^2\phi}{(a^2\sin^2\phi - b^2\cos^2\phi)^{1/2}} \qquad \dots(14.6)$$

$$r_2 = \left(\frac{r_0}{\sin\phi}\right) = \frac{a^2}{(a^2\sin^2\phi - b^2\cos^2\phi)^{1/2}} \qquad \dots(14.7)$$

From Eq. (14.4)
$$r_1 = \frac{-a^2b^2}{(a^2\sin^2\phi - b^2\cos^2\phi)^{3/2}} \qquad \dots(14.8)$$

If W = total vertical load above the level ϕ and

        g = density of material

Then,

$$W = g \int 2\pi r_0 r_1 \cdot d\phi$$

$$= 2\pi g a^4 b^2 \int_\phi^{\phi_0} \frac{\sin\phi \cdot d\phi}{(a^2 \sin^2\phi - b^2 \cos^2\phi)^2} \qquad \qquad \ldots(14.9)$$

where,

        ϕ corresponds to height z

        ϕ_0 corresponds to height z_0

For the purpose of integration,

substitute, $\cos\phi = \left[\dfrac{a}{\sqrt{a^2 + b^2}} \xi \right]$ $\ldots(14.10)$

Then

$$W = \frac{2\pi g a b^2}{\sqrt{a^2 + b^2}} \int_\varepsilon^{\varepsilon_0} \frac{d\xi}{(1 - \xi^2)^2}$$

$$= \frac{\pi g}{2} \cdot \frac{a b^2}{\sqrt{a^2 + b^2}} \left[\frac{2\xi}{1 - \xi^2} + \log\left(\frac{1 - \xi}{1 + \xi} \right) \right]_\varepsilon^{\varepsilon_0} \qquad \ldots(14.11)$$

For different values of ξ, the function is defined as

$$f(\xi) = \left[\frac{2\xi}{1 - \xi^2} + \log\left(\frac{1 - \xi}{1 + \xi} \right) \right]$$

The values of $f(\xi)$ are compiled in Table 14.2.

If N_ϕ = meridional thrust at level ϕ

        W = total load above level ϕ

Then, $N_\phi \sin\phi = \left(\dfrac{W}{2\pi r_2 \sin\phi} \right)$ $\therefore$ $N_\phi = -\left(\dfrac{W}{2\pi r_2 \sin^2\phi} \right)$

The negative value indicates compression. Substituting the value of W from Eq. (14.11):

$$N_\phi = \frac{g}{4} b^2 \sqrt{a^2 + b^2} \left\{ \frac{\sqrt{1 - \xi^2}}{a^2 + b^2 - a^2\xi^2} \right\} [f(\xi) - f(\xi_0)] \qquad \ldots(14.12)$$

The circumferential thrust N_ϕ is given by

$$\left[\frac{N_\phi}{r_1} + \frac{N_\phi}{r_2} \right] = -g \cos\phi$$

Table 14.2: Values of function $f(\xi)$

ξ	$f(\xi)^*$	ξ	$f(\xi)^*$
0.000	0.275	0.275	1.159
0.010	0.040	0.300	1.278
0.020	0.080	0.325	1.401
0.030	0.120	0.350	1.529
0.040	0.160	0.375	1.661
0.050	0.200	0.400	1.800
0.060	0.241	0.425	1.945
0.070	0.281	0.450	2.111
0.080	0.321	0.475	2.261
0.090	0.362	0.500	2.432
0.100	0.403	0.550	2.814
0.110	0.444	0.600	3.261
0.120	0.485	0.650	3.802
0.130	0.526	0.700	4.480
0.140	0.567	0.750	5.374
0.150	0.609	0.800	6.642
0.175	0.725	0.850	8.638
0.200	0.822	0.900	12.418
0.225	0.932	0.950	23.151
0.250	1.044	1.000	∞

$$ {}^*f(\xi) = \left[\frac{2\xi}{1-\xi^2} + \log\left(\frac{1-\xi}{1+\xi}\right) \right] $$

Substituting the values of N_ϕ from Eq. (14.12)

$$ N_\phi = \frac{ga^2}{\sqrt{a^2+b^2}}\left(\frac{\xi}{1-\xi^2}\right) + N_\phi \frac{a^2}{b^2}(1-\xi^2) \qquad \text{...(14.13)} $$

14.4 DESIGN EXAMPLE

A hyperbolic cooling tower of height 84 m has the following data:

Top diameter = 45 m

Throat diameter = 42 m

Density of concrete = 24 kN/m³

Analyse the membrane forces at the base section of the tower and design suitable thickness and reinforcements for the bottom section. Referring to Fig. 14.5,

$t = 22.5$ m	$z_t = 20$ m
$a = 21.0$ m	$z_b = 64$ m

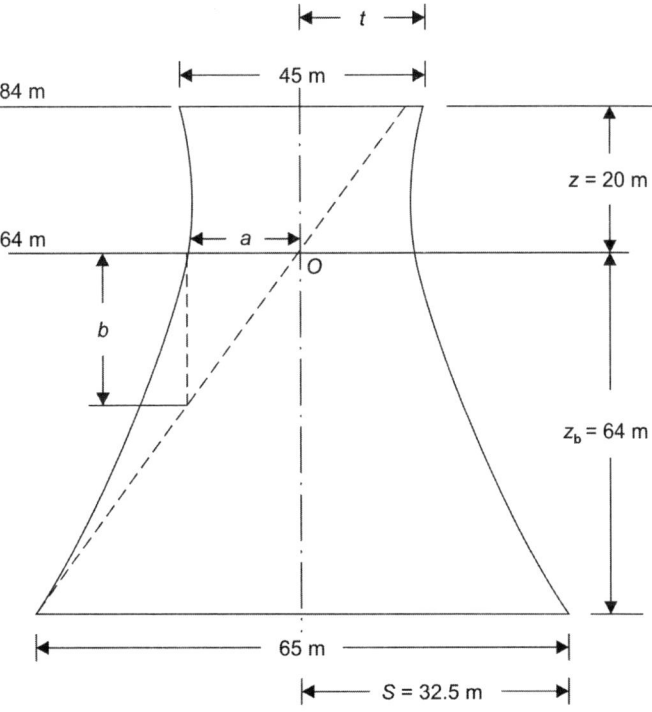

Fig. 14.5: Geometry of cooling tower

From the relation $\left(\dfrac{r_0^2}{a^2} - \dfrac{z^2}{b^2} \right)$,

$$\left(\frac{t^2}{a^2} - \frac{z^2}{b^2} \right) = 1$$

where, r_0 = radius at any section

$\therefore \qquad b = \left[\dfrac{az_t}{\sqrt{t^2 - a^2}} \right]$ $\qquad\qquad \therefore \quad b = \left(\dfrac{21 \times 20}{\sqrt{22.5^2 - 21^2}} \right) = 53$ m

If S = radius of base section

$$S = a\sqrt{1 + \frac{z_b^2}{b^2}} \text{ or } 21\sqrt{1 + \frac{64^2}{53^2}} = 32.5 \text{ m}$$

$\therefore$ Diameter at base = 65 m

For base section, we have:

$$\tan\phi = \frac{b}{a}\sqrt{\frac{r_0^2}{r_0^2 - a^2}} \text{ or } \frac{53}{21}\sqrt{\frac{32.5^2}{32.5^2 - 21^2}} = 3.32$$

$$\cos\phi = \left(\frac{1}{\sqrt{1 + \tan^2\phi}} \right) = 0.28$$

From Eq. (14.10)

$$\cos\phi = \frac{a}{\sqrt{a^2 + b^2}}\xi$$

$$\therefore \qquad \xi = \left[\frac{\cos\phi\sqrt{a^2 + b^2}}{a}\right] \text{ or } \left[\frac{0.28\sqrt{21^2 + 53^2}}{21}\right] = 0.75$$

From Table 14.1, for $\xi = 0.75$, $f(\xi) = 5.374$.

For the top section:

$$t = \sqrt{1 + \frac{z^2}{b^2}} \text{ or } 21\sqrt{1 + \frac{20^2}{53^2}} = 22.5 \text{ m}$$

$$\tan\phi = \frac{b}{a}\sqrt{\frac{r_0^2}{r_0^2 - a^2}} \text{ or } \frac{53}{21}\sqrt{\frac{22.5^2}{22.5^2 - 21^2}} = 7.1$$

$$\therefore \qquad \cos\phi = 0.14$$

$$\therefore \qquad \xi_0 = \left[\frac{0.14\sqrt{21^2 + 53^2}}{21}\right] = 0.38$$

$\therefore$ From Table 14.2 for $\xi_0 = 0.38$,

$$f(\xi_0) = 1.661$$

Membrane forces:

At top of section:

$$N_\phi = 0, \text{ because } f(\xi) - f(\xi_0) = 0 \text{ and}$$

$$N_\theta = -\left(\frac{ga^2}{\sqrt{a^2 + b^2}}\right)\left(\frac{\xi_0}{\sqrt{1 - \xi_0^2}}\right) = -\left(\frac{24 \times 21^2}{\sqrt{21^2 + 53^2}}\right)\left(\frac{0.375}{\sqrt{1 - 0.375^2}}\right)$$

$$= -74 \text{ kN/m}$$

At base section

$$N_\phi = -\frac{g}{4}b^2\sqrt{a^2 + b^2}\left\{\frac{1 - \xi^2}{a^2 + b^2 - a^2\xi^2}\right\}[f(\xi) - f(\xi_0)]$$

where,

$g = 24 \text{ kN/m}^3$

$a = 21 \text{ m}$

$b = 53 \text{ m}$

$\xi = 0.75$

$f(\xi) = 5.374$

$f(\xi_0) = 1.661$

$$\therefore \qquad N_\phi = -\left(\frac{24}{4}\right)53^2\sqrt{21^2 + 53^2}\left\{\frac{\sqrt{1 - 0.75^2}}{21^2 + 53^2 - (21^2 \times 0.75^2)}\right\}[5.374 - 1.661]$$

$$= -810 \text{ kN/m (compression)}$$

$$N_\theta = \left(\frac{ga^2}{\sqrt{a^2+b^2}}\right)\left(\frac{\xi}{1-\xi^2}\right) + N_\phi\left(\frac{a^2}{b^2}\right)(1-\xi^2)$$

$$= \left(\frac{24\times21^2}{\sqrt{21^2+53^2}}\right)\left(\frac{0.75}{1-0.75^2}\right) + \left[-810\times\frac{21^2}{53^2}(1-0.75^2)\right]$$

$$= -264 \text{ kN/m (compression)}$$

Design of shell section and reinforcement:

At base section $N_\phi = 810$ kN/m

Using M20 grade concrete and Fe 415 grade tor steel, permissible stresses $\sigma_{cc} = (0.5 \times 5) = 2.5 \text{ N/mm}^2$

If t = thickness of the concrete shell at base.

Then, $\qquad \left(\frac{N_\phi}{1000t}\right) = \sigma_{cc}$

$\therefore \qquad\qquad t = \left(\frac{810\times10^3}{1000\times2.5}\right) = 324 \text{ mm}$

Adopt a shell thickness of 350 mm at base gradually reducing to 150 mm at the top.

Minimum reinforcement = 1% of cross-sectional area

$$= \left(\frac{1}{100}\times1000\times350\right) = 3500 \text{ mm}^2$$

Provide 20 mm diameter bars at 175 mm centres on both faces in meridional direction $A_{st} = 3600 \text{ mm}^2$

Minimum reinforcement in the circumferential direction

$$= 0.15\% \text{ of gross area} = \left(\frac{0.15}{100}\times350\times1000\right) = 525 \text{ mm}^2$$

Provide 10 mm diameter bars at 300 mm centre on both faces in the circumferential direction at hoops. The reinforcements can be decreased towards the top of the tower.

14.5 ANALYSIS OF HYPERBOLIC COOLING TOWERS FOR WIND LOADS

Hyperbolic cooling towers being tall structures are subjected to wind loads, which result in meridional and circumferential forces in towers. Based on wind tunnel experiments, Rish and Steel (Refer to Design and selecton of hyperbolic cooling towers, *Journal of Power Division, ASCE proceedings*, Vol. 85. October 1959, pp. 89–117),[13] have recommended the following type of wind pressure distribution in hyperbolic cooling towers, as shown in Fig. 14.6.

The windward side of the tower is subjected to positive wind pressure over an area bounded by an angle 95.12°. The leeward side is subjected to suction pressure of intensities as shown in Fig. 14.6. Due to the action of wind, membrane forces N_ϕ, N_θ and $N_{\theta\phi}$ develop with maximum values at the base section.

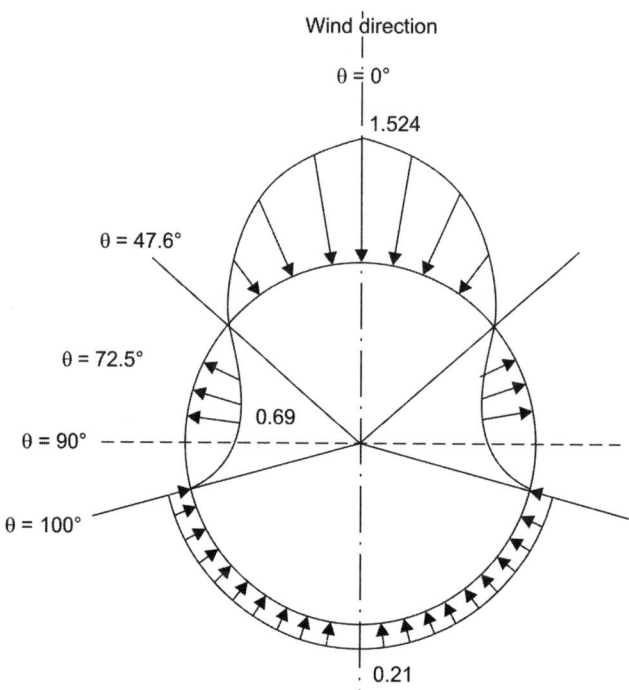

Fig. 14.6: Wind pressure distribution

Typical distribution of the various force components at the base of a hyperbolic cooling tower are shown in Figs 14.7–14.9, for the following tower parameters:

$$\left(\frac{a}{t}\right) = 0.90, \qquad \left(\frac{a}{s}\right) = 0.55$$

$$K^2 = \left(1 + \frac{a^2}{b^2}\right) = 1.18, \qquad \phi = 1.0$$

$$P = \text{wind intensity}$$

The force distribution analysis indicates that the maximum tensile meridional force develops at $\theta = 0$ and compressive force at $\theta = 70°$. The maximum hoop tension develops at $\theta = 70°$, while the maximum hoop compression develops at an angle $\theta = 0°$. The maximum shear force develops at angles $\theta = 90°$ and $\theta = 50°$.

Based on membrane analysis, Gould and Lee (Refer to Hyperbolic cooling towers under wind load, *Journal of Structural Division, ASCE Proceedings*, Vol. 93, October 1967, pp. 487–514)[14] have developed design tables which give the meridional force (N_ϕ) and circumferential force (N_θ) in the cooling towers subjected to the action of wind loads. The design tables have been presented in the form of $\left(\dfrac{N_\phi}{Pa}\right), \left(\dfrac{N_\theta}{Pa}\right)$ and $\left(\dfrac{N_{\phi\theta}}{Pa}\right)$ for different values of the variables, (a/s), (a/t), K^2, θ and ϕ, as shown in Fig. 14.10. The range of values covered in the tables are as follows:

(a/s) from 0.45 to 0.65
(a/t) from 0.85 to 0.95
K^2 from 1.05 to 1.50
θ at 0° to 70°
ϕ from 0 to 1.0

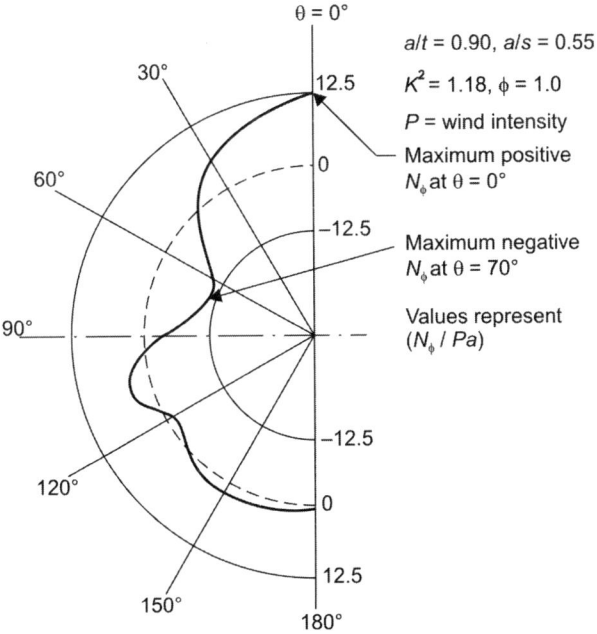

Fig. 14.7: Meridional thrust

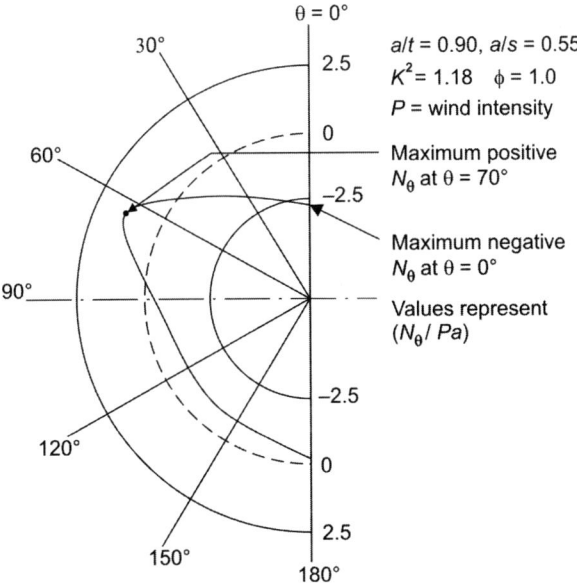

Fig. 14.8: Circumferential thrust

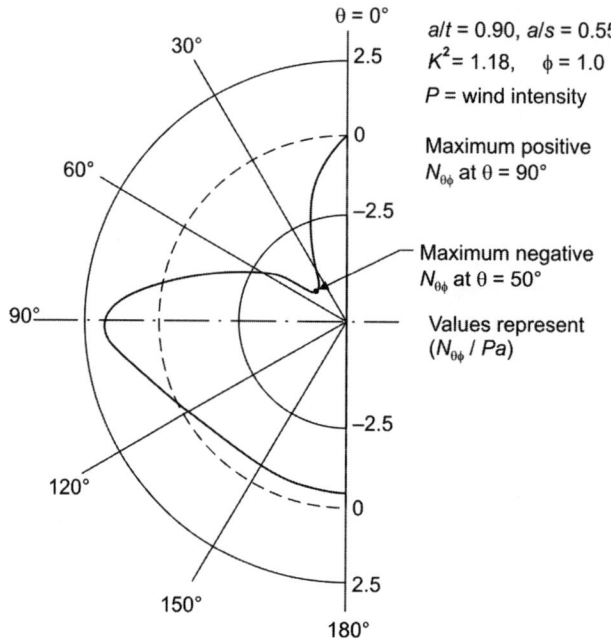

Fig. 14.9: Shear forces

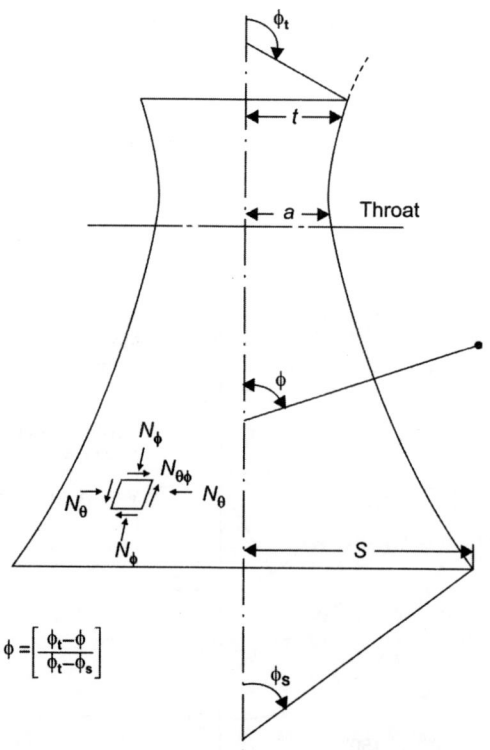

Fig. 14.10: Cooling tower parameters

14.6 ▮ DESIGN EXAMPLE

A hyperbolic cooling tower has the following parameters:

Throat diameter = 42 m

Base diameter = 65 m

$b = 53$ m

Top diameter = 42 m

Wind intensity $P = 1$ kN/m².

Compute the maximum design membrane forces at the base section due to the action of wind and self weight of shell.

The design parameters are:

$a = 21$ m $P = 1$ kN/m²

$b = 53$ m $a/s = 0.65$

$s = 32.5$ m $a/t = 0.95$

$t = 22.5$ m $K^2 = \left(1 + \dfrac{a^2}{b^2}\right) = 1.155$

Referring to design tables for maximum values of N_ϕ, N_θ, we have for $\theta = 0$ and $\phi = 1.00$ (base section):

$$\left(\frac{N_\phi}{Pa}\right) = 12.35$$

$\therefore$ $N_\phi = (12.35 \times 1 \times 21) = 260$ kN/m

$$\left(\frac{N_\theta}{Pa}\right) = -1.914$$

$\therefore$ $N_\theta = (-1.914 \times 1 \times 21) = -40$ kN/m

Maximum negative value of N_ϕ occurs at $\theta = 70°$ and $\phi = 1.00$ (base section)

$$\left(\frac{N_\phi}{Pa}\right) = -11.202$$

$\therefore$ $N_\phi = (-11.202 \times 1 \times 21) = -235$ kN/m

Maximum positive value of N_θ occurs at $\theta = 70°$ and $\phi = 1.00$ (base section)

$$\left(\frac{N_\theta}{Pa}\right) = 0.60$$

$\therefore$ $N_\theta = (0.6 \times 1 \times 21) = 12.6$ kN/m

The forces N_ϕ and N_θ due to self weight for the shell computed as shown in design Example 14.4.

Hence, the design forces are shown in table given below:

Force	Due to DL	Due to WL	Maximum design force
N_ϕ	−810	−235	−1045 kN/m
N_θ	−264	−40	−304 kN/m

The base section is designed for the maximum meridional and circumferential forces.

References

1. Mungan I and Wittek W (Eds), Natural Draught Cooling Towers, Proceedings of the Symposium on Natural Draught Cooling Towers, organized by International Association for Shell and Spatial Structures (IASS), London, 2004, p. 378.

2. EL Ansary AM, EL Damatty AA and Naseef AO, Optimum shape and design of cooling towers, *World Academy of Science, Engineering and Technology*, Vol. 5, No. 12–21, 2011.

3. Zienkiwicz OC and Campbell JC, Shape optimization and sequential linear programming, *Optimum Structural Design*, John Wiley & Sons Inc, New York, 1973, pp. 109–126.

4. Gould PL, Lee SL, Hyperbolic cooling towers under seismic design load, *Journal of Structural Engineering*, ASCE, Vol. 96, No. 9, 1970, pp. 1889–1902.

5. Gupta AK and Maestrini S, Investigations on hyperbolic cooling towers–ultimate load behaviour, *Engineering Structures*, Vol. 8, No. 2, 1986, pp. 87–92.

6. Noh HC, Ultimate strength of large scale reinforced concrete thin shell structures, *Thin Walled Structures*, Vol. 43, No. 9, 2005, pp. 1418–1443.

7. Proceedings of the Conference on Ferry Bridge Natural Draught Cooling Towers Failure, Institution of Civil Engineers, London, June 1967.

8. Armitt J, Wind Loadings on Cooling Towers, *Journal of Structural Engineering*, ASCE, Vol. 106, No. 3, 1980, pp. 623–641.

9. Boseman PM, Strengthening of natural draught cooling tower shells with stiffening rings, *Engineering Structures*, Vol. 20, No. 10, 1998, pp. 909–914.

10. Gould PL and Guedethoefer OC, Repair and completion of damaged cooling towers, *Journal of Structural Engineering*, ASCE, Vol. 115, No. 3, 1989, pp. 576–593.

11. Gurfinkel G, Analysis and design of hyperbolic cooling towers, *Journal of the Power Division*, ASCE, Vol. 98, No. 1, June 1972, pp, 133–152.

12. IS:11504-1985, Indian Code of Practice–Criteria for Structural Design of Reinforced Concrete Natural Draught Cooling Towers, Bureau of Indian Standards, New Delhi, 1985.

13. Rish RF and Steel RF, Design and selection of hyperbolic cooling towers, *Journal of the Power Division*, ASCE Proc. Paper 2227, October 1959, pp. 89–117.

14. Gould PL and Lee SL, Hyperbolic cooling towers under wind load, *Journal of the Structural Division*, ASCE, Vol. 93, No. ST-5, Oct. 1967, pp. 487–514.

Assignment

1. A hyperbolic cooling tower of overall height 100 m has a throat radius of 23 m. The throat section is located 25 m from the top of the shell. The diameter of the top section is 50 m. Analyse for membrane forces at the base section of the tower and design suitable concrete section and reinforcements. Adopt M20 grade concrete and Fe 415 grade HYSD bars.

2. A hyperbolic reinforced concrete cooling tower has the following parameters:

   Top diameter = 50 m   Throat diameter = 46 m

   Base diameter = 68 m   Height of throat section z_t = 25 m

   Total height of tower = 100 m.

   The tower is subjected to a wind pressure of 1.5 kN/m^2. Estimate the maximum membrane forces in the shell due to self weight and wind loads and design the section. Adopt M30 grade concrete and Fe 415 HYSD bars.

3. A hyperbolic cooling tower has the following salient dimensional details:

   Top diameter = 40 m

Bottom diameter = 60 m

Total height of tower = 120 m

Throat diameter = 35 m

Distance of throat section from top = 35 m

The tower is subjected to a wind pressure intensity of 1.6 kN/m^2. Estimate the maximum membrane forces developed in the shell due to self weight and wind loads and design the sections. Adopt M25 grade concrete and Fe 415 HYSD bars.

4. The Niederaussen hyperbolic cooling tower in Germany is 200 m tall and has the following dimensions at various sections:

Diameter at top = 88 m

Throat diameter = 85 m

Base diameter = 135 m

Density of reinforced concrete = 25 kN/m^3

Analyse the membrane forces at the base section of the tower and design suitable section and reinforcements using M25 grade concrete and Fe 500 HYSD bars.

5. In a superthermal power station, the hyperbolic cooling tower used has the following dimensions:

Diameter at top = 50 m

Diameter at bottom = 75 m

Diameter at throat = 45 m

Distance of throat section from bottom = 100 m

Total height of tower = 150 m

The tower is subjected to a wind intensity of 1.5 kN/m^2.

Compute the membrane forces developed in the shell due to self weight and wind loads and design the sections. Adopt M25 grade concrete and Fe 500 HYSD bars.

6. Evaluate the maximum design membrane forces developed at the base section of a hyperbolic cooling tower due to the action of wind using the following parameters of the tower:

Diameter at base = 70 m

Throat diameter = 40 m

Diameter at top = 42 m

Intensity of wind = 1.5 kN/m^2

Review Questions

1. What is the necessity of using hyperbolic cooling towers in thermal power plants?

2. What are the structural advantages in using hyperbolic cooling towers in preference to circular towers?

3. List the various forces to be considered in the design of hyperbolic cooling towers.

4. What are the critical sections to be checked in cooling towers?

5. Explain with sketches the various structural components of a hyperbolic cooling tower and mention their functions.

6. Briefly explain the method of analysing membrane forces in hyperbolic cooling towers when subjected to different type of loads.

7. Write a note on the wind pressure distribution in hyperbolic cooling towers.

8. Explain with sketches the distribution of meridional and circumferential thrust and the shear forces developed under wind loads in cooling towers.

9. Mention briefly the various types of reinforcements generally used in hyperbolic cooling towers.

10. Sketch the reinforcement details at the base of a typical hyperbolic cooling tower. How do you arrange the reinforcements in the meridional and circumferential directions at critical cross-sections of the tower?

Objective Type Questions

1. The hyperboloid shell of revolution can be cast at site using
   a. curved form work
   b. straight sheet form work
   c. cylindrical type of form work

2. In thermal power stations, cooling towers are used to
   a. provide cool air to the power plant
   b. cool the hot water from the power plant
   c. store cool water for the power plant

3. The main type of force component developed in a hyperbolic cooling tower under loads is
   a. flexural
   b. torsional
   c. membranous

4. Reinforced concrete hyperbolic cooling towers are generally supported at base using
   a. ring beams
   b. inclined columns
   c. straignt columns

5. The critical section to be checked in a hyperbolic cooling tower under loads is the
   a. top section
   b. throat section
   c. base section

6. In a reinforced concrete hyperbolic cooling tower, the minimum reinforcement as a percentage of gross area of cross-section is
   a. 0.05%
   b. 0.55%
   c. 0.15%

7. In a reinforced concrete hyperbolic cooling tower, the least diameter of the tower is at the
   a. bottom
   b. top
   c. throat

8. Due to the action of wind on a hyperbolic cooling tower, the magnitude of forces developed will be maximum along the
   a. circumferential direction
   b. horizontal direction
   c. meridional direction

9. In the design of reinforced concrete cooling towers, the critical limit state to be examined is
   a. deflection
   b. cracking
   c. collapse

10. In general, the meridional and circumferential reinforcements in a tall cooling tower is distributed at the
    a. outer face
    b. inner face
    c. both faces

Folded Plates

15.1 GENERAL FEATURES

Reinforced concrete folded plates are often referred to as faltwerke or hipped plates and their origin could be traced to Germany where they were first used for the construction of large coal bunkers by G Ehlers in 1924–25. The structural behaviour of folded plates is akin to that of thin shells and they can be considered as examples of stressed skin structural elements. In contrast to shells, folded plates are prismatic plates without any curvature meeting at rigid joints. The thickness of folded plate structures is very small in comparison to their length and width. The main advantage of folded plate construction is that they can be cast using plane form work in contrast to the curved surface of the shells. Also, they are amenable for precast work since individual plates can be precast in a factory and assembled by forming the rigid joints at site resulting in an integrated structure. When used as roof elements for long spans, they can be prestressed by housing the high tensile cables housed in the valley junctions.

15.2 STRUCTURAL TYPES OF FOLDED PLATES

Folded plates can be built in various structural configurations depending upon the type of structure and its functional requirements. Initially they were used for large coal bunkers by the pioneering German engineers. Folded plates have specific advantages in comparison to other types of shell structures. The various types of folded plate structural roof elements are shown in Fig. 15.1 and listed below:

 i. Prismatic, V- and trough type, generally used for warehouses and factory sheds
 ii. Pyramidal and prismoidal shapes for covering exhibition halls and pavilions
 iii. North light shell roof elements for aircraft, automobile and other industrial structures
 iv. Staircase elements
 v. Coal bunkers

15.3 ANALYSIS OF FOLDED PLATES

Analysis of folded plates was first attempted by G Ehlers of Germany with the first publication in 1930, in which the longitudinal joints were assumed to be hinged,

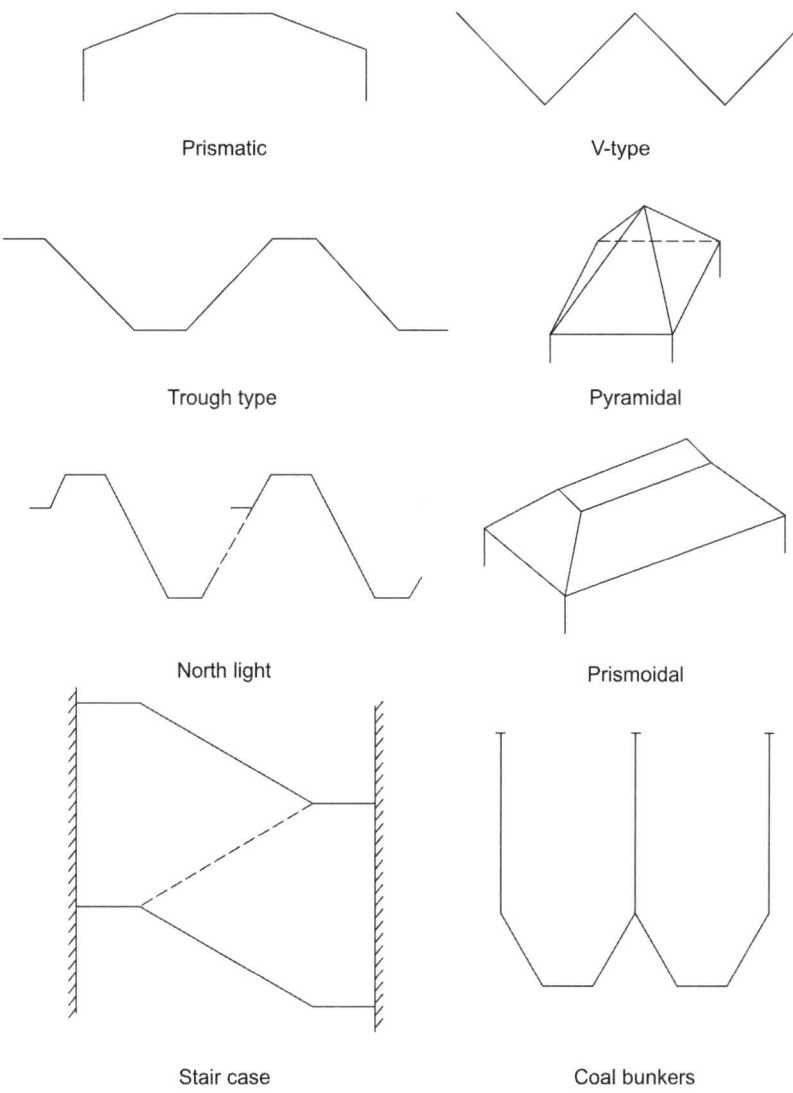

Fig. 15.1: Types of folded plates

neglecting the transverse moments at the joints. This assumption is similar to that made in the analysis of steel trusses under joint loads. The displacements of the joints were also ignored. Later, more exact methods by considering transverse continuity and joint displacements were developed by Gruber, Vlassov and Yitzhaki according to the task committee report[1] of the ASCE. They developed a solution in the form of simultaneous differential equations of the 4th order which were solved by the use of rapidly converging series. This approach involves $(7n + 2)$ unknowns for $(n + 1)$ plates. The analysis indicated that assumption of hinged or rigid joints would significantly influence the resulting moments and stresses in the plates.

In the year 1953, Creamer published a paper setting out the tough limits for classification of folded plates as long and short, based on the length to width ratio of the individual plates. Winter and Pei[2] in 1947 modified the algebraic solution into a stress distribution procedure which is very much similar to the traditional moment distribution method. But this method neglects joint displacements and is useful only for the analysis of short folded plates. For long folded plates, the joint displacements cannot be ignored. In 1954, Gaafar[3] published a modification to the Winter and Pei's method including the effect of joint displacements. In 1958, Simpson[4] published a new method involving a number of moment and stress distributions. Whitney[5] in 1959 presented a simplified version of the modified Girkmann's method.

A comparative analysis of various methods indicate that Winter and Pei's method is by far the simplest but it is applicable only to short folded plates for which joint displacements can be ignored without appreciable error. In general Whitney and Simpson's methods are applicable to folded plates of all proportions and are ideally suited for use in design office. For long folded plates of span to width ratio 10, Whitney has shown that a simplified analysis can be used based on beam theory[6] assuming straight line distribution of longitudinal stress. Generally, the analysis of folded plates is based on the following assumptions:

i. The structure is monolithic with rigid joints

ii. The material is homogeneous, elastic and isotropic

iii. The length of each plate is more than twice its width

iv. Plane sections remain plane after deformation in all the plates

15.4 STRUCTURAL BEHAVIOUR OF FOLDED PLATES

Reinforced concrete folded plates resist the system of transverse loads by two significant actions identified as (a) slab action and (b) plate action. The loads acting normal to each plate causes transverse bending between the junctions of the plates, which can be considered as imaginary supports of a continuous slab. This transverse bending is termed 'slab action'. The transverse moments developed in the plate can be determined by the continuous beam analysis assuming imaginary supports at the junctions of the plates. The slab action mechanism developed in a folded plate is shown in Fig. 15.2a.

The plates being supported at the ends of span on deep beams termed traverses, bend under the action of loads in their own plane as shown in Fig. 15.2b. The longitudinal bending of the plates in their own plane is termed 'plate action'. The flexural stresses resulting from the plate action may be considered to have a linear distribution across each plate, with maximum intensity at the centre of span section. The salient features of Whitney and Simpson's method of analysis of folded plates are outlined in the following sections.

15.5 VARIOUS METHODS OF ANALYSIS OF FOLDED PLATES

A number of methods have been developed by various research investigators and engineers during the last several decades for the analysis and design of reinforced concrete folded plates. A critical review of the various methods of analysis of folded plates has been presented by Evans and Rockey[7]. Continuous folded plates have been investigated by Rao[8] in his paper published in the IASS bulletin. Aroni[9] summarises the architectural aspects of folded plates suitable for different types of buildings. Analysis of folded plates by using finite strip method has been attempted by Cheung[10].

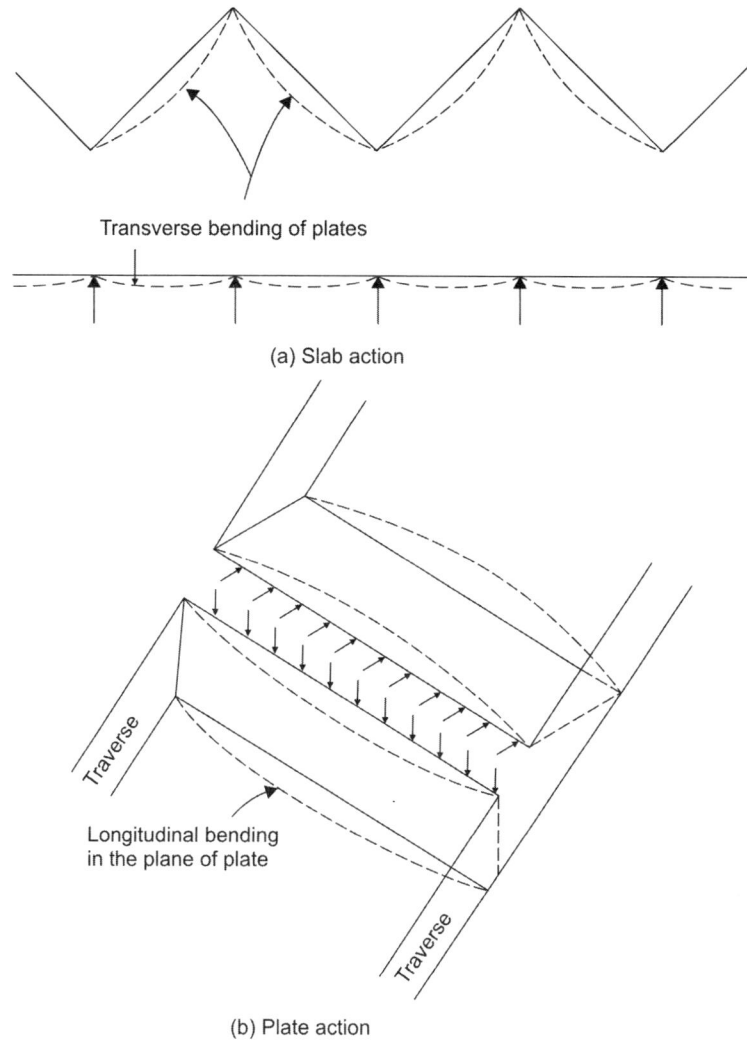

(a) Slab action

Transverse bending of plates

(b) Plate action

Fig. 15.2: Structural behaviour of folded plates

Guralnick *et al*[11,12] have reported their investigations on the buckling and stability aspects of folded plates. The most commonly used methods are presented in the following sections.

15.6 WHITNEY'S METHOD

The Whitney method is applicable for folded plates with width and thickness of the plates and the intensity of loading uniform along the length of the plate. Mathematical computations are greatly simplified by replacing the uniform load by Fourier loading and considering only the first term of the series. The plate moments, stresses and deformations, therefore, vary as sine functions along the length, having maximum values at mid span. The salient steps of this method are as follows:

i. The joint loads are computed by assuming each plate to be simply supported spanning between the adjacent plates. The joint loads are replaced by their components in the plane of the plates. These form the initial plate loads.

ii. The continuity of the plates due to the rigidity of the joints induces transverse moments causing additional joint loads in terms of unknown transverse moments and hence additional plate loads are added to the initial plate loads.

iii. Under the action of plates loads, each plate bends in the longitudinal direction in its own plane between the traverses. The bending moments and the longitudinal stresses at the common edges of the adjacent plates will not be equal. Longitudinal edge shear forces are now introduced to establish the compatibility of stresses at the junctions.

iv. The individual plate deflections caused by the transverse loads and the edge shear forces are calculated in terms of the applied loads and the unknown transverse moments.

v. For each of the rigid joints, the condition and plate deflections resulting in n equations for n plates.

vi. The equations are solved for the unknown transverse moments and other stress resultants are computed.

15.7 ■ SIMPSON'S METHOD

The prominent steps to be followed in Simpson's method of analysis of folded plates are outlined below:

i. A transverse section of the folded plate of unit width is treated as a continuous beam on rigid supports and the joint loads obtained are resolved into components in the planes of the plates.

ii. The plates are assumed to bend independently between the diaphragms due to plate loads and the corresponding moments and stresses are calculated.

iii. Compatibility conditions require that the longitudinal stresses at common edges of adjacent plates should be equal. The stresses are corrected by distribution procedure. (At this stage, the results obtained will be indentical to those obtained by Winter and Pei solution).

iv. The stresses obtained are now further corrected to allow for the joint displacements. An arbitrary rotation is given to each joint in turn at mid span, the rotation along the span varying as a sine curve. The joint loads and the corresponding longitudinal stresses are calculated in each case and the compatibility of the stresses established by a distribution procedure. This procedure is repeated for each joint and the joint deflections are calculated.

v. The final deflections are then expressed as the sum of the deflections due to the no rotation solutions, each multiplied by an unknown factor α_n. The final plate rotations are equal to the values for the no rotation solution plus α_n times the arbitrarily assumed rotations.

vi. The total angle changes at the junctions are equated to zero resulting in a set of simultaneous equations which are solved for α_n.

vii. The final moments and stresses in the plates are obtained as the sum of the no rotation solution and the rotation solutions each multiplied by its corresponding α_n.

15.8 ITERATION METHOD

The iteration method developed by Brielamaier[13] is applicable for symmetrical V-type folded plates with simpler computational effort in comparison with the Whitney and Simpson's methods. However, this method is not universally applicable to folded plates of all types. The following steps are involved in the analysis of folded plates by iteration method.

i. The folded plate is analysed for slab action as a continuous beam and the ridge loads are computed from the reactions. The longitudinal stresses developed in the plates are computed assuming the plates to bend in their own plane due to the action of plate loads. The stresses at the junctions are computed and the no rotation solution obtained by the stress distribution procedure as in the case of Simpson's method.

ii. The plate deflections corresponding to the no rotation solution is computed and from these the relative displacements of the joints can be obtained.

iii. The transverse moments developed due to the relative joint displacements are computed and the moments are distributed to estimate the ridge loads and plate loads. The longitudinal stresses due to the plate loads are computed and corrected by stress distribution procedure. The resulting stresses are compared with those obtained from the no rotation. If the corrections are relatively small, the iteration process is stopped at this stage and the final moments and stresses are obtained by adding the results of the no rotation solution and those due to the first cycle of iteration.

iv. If the corrections are not small, the deflections caused by correction stresses are computed and the cycle of iteration repeated until the desired results are obtained.

The convergence of the iteration method depends upon the relative rigidities of longitudinal plate action and transverse slab action and on the geometry of the structure. The iteration method has been found to converge rapidly for symmetrical V-type folded plates while it diverges when applied to north light folded plates.

15.9 BEAM METHOD

The beam method is based on the linear stress variation across the section and the folded plate is assumed to bend longitudinally as a beam between the traverses. This method is applicable for long folded plates with span to depth ratios exceeding 10. The beam method is by far the simplest due to the minimum computational effort involved when compared to other methods.

15.10 WINTER AND PEI'S METHOD

For short folded plates having span to depth ratio less than 4, Winter and Pei's method can be adopted. In the case of short folded plates, the displacements of junctions of hipped plates can be ignored without any appreciable error. In this method the folded plate is treated as a continuous beam on rigid supports and the joint loads obtained are resolved into plate loads. Under the action of the plate loads, the individual plates are assumed to bend in their own plane in the longitudinal direction and the corresponding moments and stresses are computed in the plates. Compatibility conditions require that the longitudinal stresses at common edges of adjacent plates should be equal. The stresses are corrected by a stress distribution procedure.

15.11 EQUATION OF THREE SHEARS

Consider a folded plate of two-folds only having depths of plates as h_1 and h_2 and the cross-sectional area of the plates being A_1 and A_2 as shown in Fig. 15.3.

Z_1 and Z_2 are the section moduli of the two plates.

M_1 and M_2 are the moments acting on the plates computed from the plate loads.

σ_A, σ_B and σ_C are the edge stresses, T_A, T_B and T_C are the shear forces at the junctions A, B, and C, considered as positive when acting in opposition to beam moments M_1 and M_2.

Moments are considered as positive when they develop tension at bottom and compression at top.

The stresses developed in plates 1 and 2 can be written as,

stress at edge B in plate 1 is

$$\sigma_B = \left[\frac{M_1}{Z_1} + \frac{T_B}{A_1} + \frac{T_B h_1}{2Z_1} - \frac{T_A}{A_1} + \frac{T_A h_1}{2Z_1} \right] \qquad \text{...(15.1)}$$

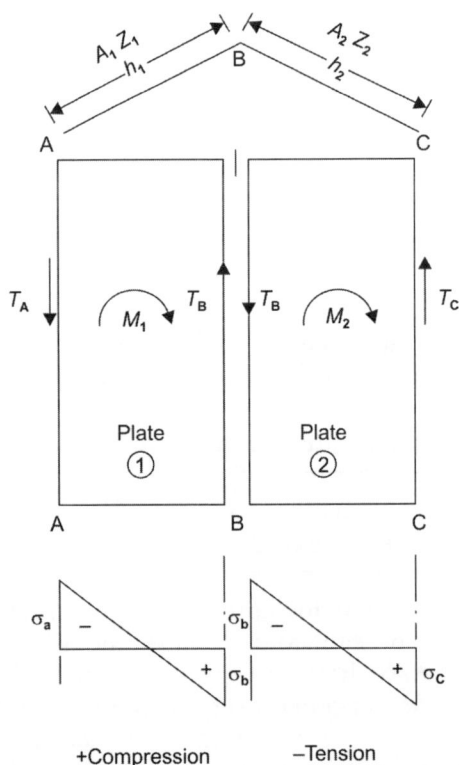

+Compression   −Tension

Fig. 15.3: Force components in folded plates

stress at edge B in plate 2 is:

$$\sigma_B = \left[-\frac{M_1}{Z_2} - \frac{T_B h_2}{2Z_2} - \frac{T_B}{A_2} + \frac{T_C}{A_2} - \frac{T_C h_2}{2Z_2} \right] \qquad \ldots(15.2)$$

If $\quad t_1 =$ thickness of plate 1

$\quad\quad t_2 =$ thickness of plate 2

$\quad\quad A_1 = t_1, h_1 \quad\quad\quad\quad\quad\quad\quad Z_1 = (t_1 h_1^2)/6$

$\quad\quad A_2 = t_2, h_2 \quad\quad\quad\quad\quad\quad\quad Z_2 = (t_2 h_2^2)/6$

$\therefore \quad \left(\dfrac{T_A h_1}{2Z_1} \right) = \dfrac{T_A h_1}{2\left(\dfrac{t_1 h_1^2}{6} \right)} = \dfrac{3T_A}{t_1 h_1} = \dfrac{3T_A}{A_1}$

Similarly

$$\frac{T_B h_1}{2Z_1} = \frac{3T_B}{A_1}, \quad \frac{T_B h_2}{2Z_2} = \frac{3T_B}{A_2}, \quad \frac{T_C h_2}{2Z_2} = \frac{3T_C}{A_2}$$

Substituting these values in Eqs (15.1) and (15.2) and equating them, we have the condition:

$$\left[\frac{M_1}{Z_1} + \frac{4T_B}{A_1} + \frac{2T_A}{A_1} \right] = \left[-\frac{M_2}{Z_2} - \frac{4T_B}{A_2} - \frac{2T_C}{A_2} \right]$$

$$\therefore \quad \left[\frac{2T_A}{A_1} + \frac{4T_B}{A_1} + \frac{4T_B}{A_2} + \frac{2T_C}{A_2} \right] = -\left[\frac{M_1}{Z_1} + \frac{M_2}{Z_2} \right]$$

The equation of three shears is written as

$$\frac{T_A}{A_1} + 2T_B \left[\frac{1}{A_1} + \frac{1}{A_2} \right] + \frac{T_C}{A_2} = -\frac{1}{2} \left[\frac{M_1}{Z_1} + \frac{M_2}{Z_2} \right]$$

In a given problem, the equation of three shears is applied to two adjacent folded plates at a time and the algebraic equations thus obtained are solved for the unknown shears. Once the shears are computed, the stresses in the plates can be computed.

15.12 ░ STRESS CONDITIONS IN HIPPED PLATES

The nature of stresses developed in hipped plate depend upon the sign of moments in the individual plates, the different combination of moments such as sagging or hogging results in different types of stress distribution in individual and hipped plate structure as shown in Fig. 15.4. The nature of moments in the plates depend upon whether the junction of the two plates form a ridge or a valley. If the two plates meet at a ridge, the moments developed are such that compressive stresses develop at the ridge. Alternatively, if the two plates meet in a valley, the moments develop tensile stresses in the valley.

The design of folded plate structures should conform to the specifications prescribed in the Indian standard code IS:2210-1988 (reaffirmed 2003)[14]. Folded plate structures

Stresses

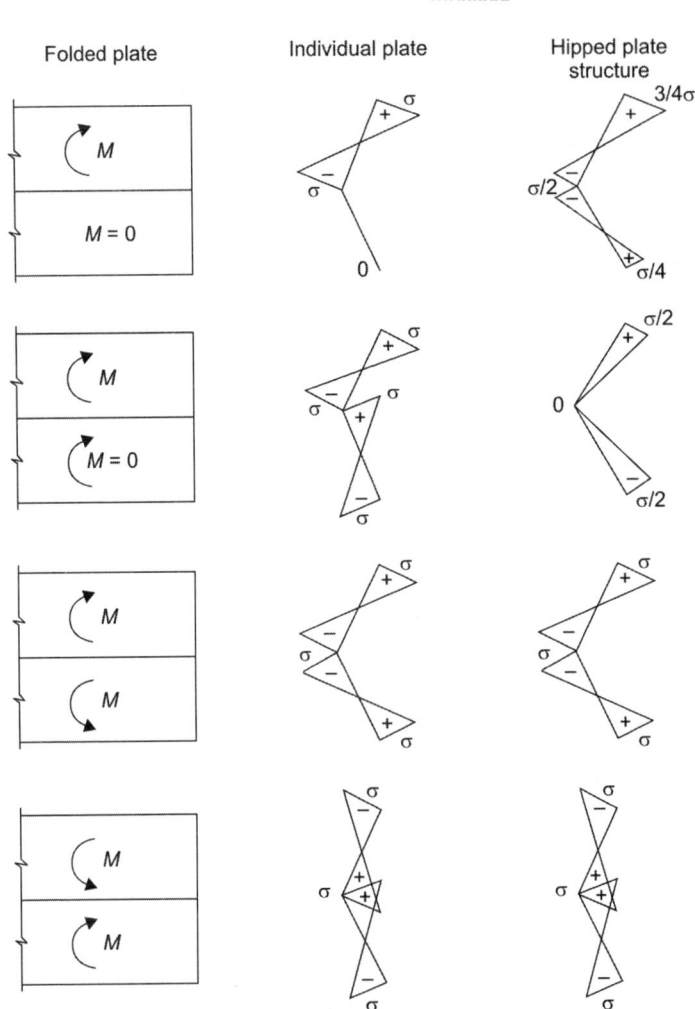

Fig. 15.4: Stress conditions in hipped plate structures

with spans exceeding 15 m to 20 m are generally prestressed to reduce the depth of the plates. The reader may refer to a monograph by the author[15] for design details of a prestressed concrete folded plate structure used for the roof of a biscuit factory measuring 30 m × 90 m.

15.13 ▌ DESIGN EXAMPLES

1. A folded plate with two folds AB and BC is subjected to moments in the plane of the plates. Using the following data, calculate the stress in the folded plate.

    i. *Data*:

      Thickness of plates = 100 mm

      Depth of plates $h_1 = h_2 = 2$ m

Moment in plates $M_1 = M_2 = 333$ kN·m

$$Z_1 = Z_2 = \left(\frac{100 \times 2000^2}{6} \right) = 66.66 \times 10^6 \text{ mm}^2$$

$$A_1 = A_2 = (100 \times 2000) = 2 \times 10^5 \text{ mm}^2$$

$$\left(\frac{M_1}{Z_1} \right) = \left(\frac{M_2}{Z_2} \right) = \frac{333 \times 10^6}{66.66 \times 10^6} = 5 \text{ N/mm}^2$$

ii. *Edge shear forces*:

Let T_A, T_B and T_C be the edge shear forces, since edges A and C are free edges

$$T_A = T_C = 0$$

Using the equation of three shears:

$$2T_B \left[\frac{1}{A_1} + \frac{1}{A_2} \right] = -\frac{1}{2} \left[\frac{M_1}{Z_1} + \frac{M_2}{Z_2} \right]$$

$$2T_B \left[\frac{1}{2 \times 10^5} + \frac{1}{2 \times 10^5} \right] = -\frac{1}{2}(5 + 5)$$

∴ $$T_B = -250000 \text{ N}$$

iii. *Resultant stresses*:

The resultant stresses at plate edges are obtained from Eqs (15.1) and (15.2).

$$\sigma_B = \left[\frac{M_1}{Z_1} + \frac{T_B}{A_1} + \frac{T_B h_1}{2Z_1} \right] = 5 + \left(\frac{-250000}{2 \times 10^5} \right) + \left(\frac{-250000 \times 2000}{2 \times 66.66 \times 10^6} \right) = 0$$

$$\sigma_A = \left[-\frac{M_1}{Z_1} + \frac{T_B}{A_1} - \frac{T_B h_1}{2Z_1} \right] = -5 - 1.25 + 3.75 = -2.5 \text{ N/mm}^2 \text{ (tension)}$$

The resultant stress distribution with the force components are shown in Fig. 15.5.

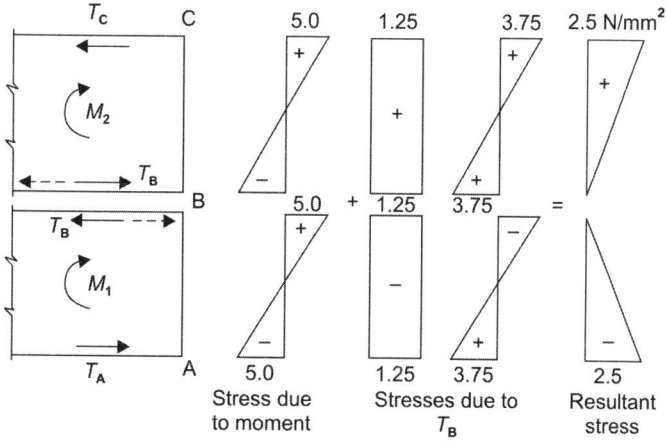

Fig. 15.5: Stress in folded plate

$$\sigma_C = \left[\frac{M_2}{Z_2} - \frac{T_B}{A_2} + \frac{T_B h_2}{2Z_2} \right] = 5 + 1.25 - 3.75 = +2.5 \, \text{N/mm}^2 \, \text{(compression)}$$

2. A prismatic folded plate ABCDE shown in Fig. 15.6, supports a load of 0.4 kN/m²
in addition to the self weight. Estimate the stresses developed in the plate at mid
span section if the plates BC, CD and DE are 120 mm thick and plates AB and EF
are 250 mm thick. Span of the folded plate = 8 m.

   i. *Loads*:

Self weight of plates BC, CD and DE = (0.12 × 24) = 2.88 kN/m²
Self weight of plates AB and EF = (0.25 × 24) = 6.00 kN/m²
Live load = 0.40 kN/m²
∴ Total load/m on CD = (2.88 + 0.4) = 3.28 kN/m

Total load/m on BC and DE = $\left(\dfrac{2.88 \times 2.25}{2} \right)$ = 3.65 kN/m

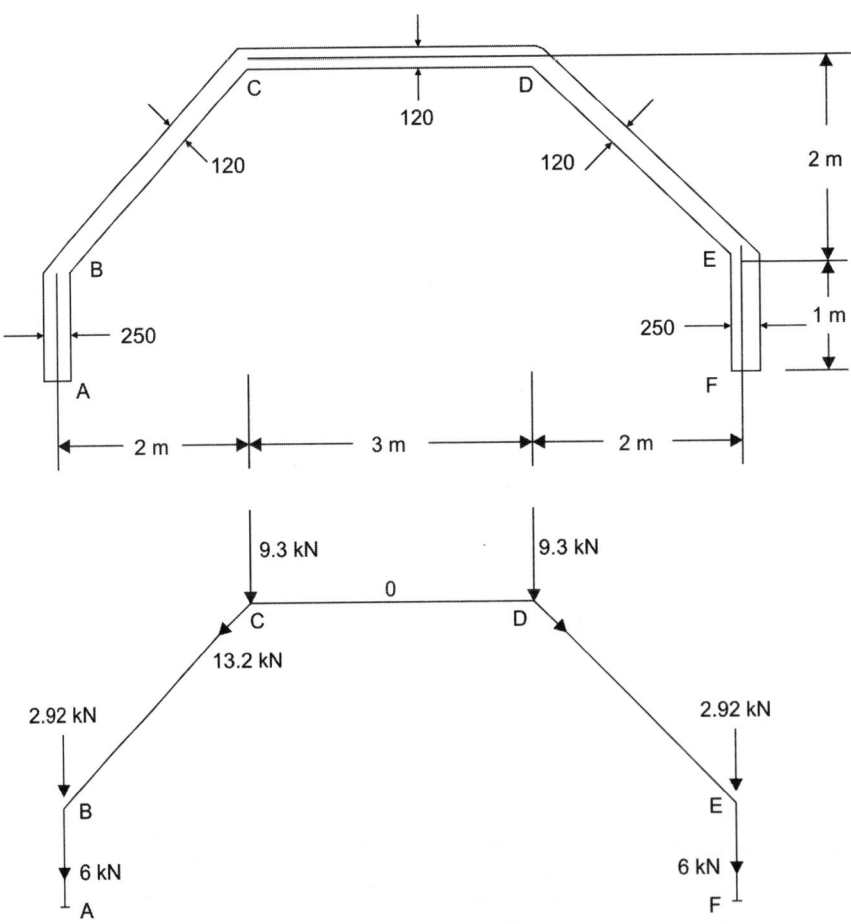

Fig. 15.6: Forces acting on folded plates

ii. *Reactions*:

The approximate value of reactions at B, C, D and E, treating the folded plate as a continuous beam BCDE, are:

$$R_B = R_E = (0.4 \times 3.65 \times 2) = 2.92 \text{ kN}$$
$$R_C = R_D = (0.6 \times 3.65 \times 2) + (0.5 \times 3.28 \times 3) = 9.30 \text{ kN}$$

The plate loads are shown in Fig. 15.6

iii. *Section properties*:

Plate 1 (AB) area $A_1 = (0.25 \times 1) = 0.25 \text{ m}^2$

Plate 2 (BC), area $A_2 = (2.84 \times 0.12) = 0.34 \text{ m}^2$

Plate 3 (CD), area $A_3 = (3 \times 0.12) = 0.36 \text{ m}^2$

$$Z_1 = \left(\frac{0.25 \times 1^2}{6} \right) = 0.041 \text{ m}^3$$

$$Z_2 = \left(\frac{0.12 \times 2.84^2}{6} \right) = 0.162 \text{ m}^3$$

$$Z_3 = \left(\frac{0.12 \times 3^3}{6} \right) = 0.18 \text{ m}^3$$

iv. *Moments in plates*:

$$M_1 = \left(\frac{8.92 \times 8^2}{8} \right) = 71.36 \text{ kN·m}$$

$$M_2 = \left(\frac{13.20 \times 8^2}{8} \right) = 105.6 \text{ kN·m}$$

$$M_3 = 0$$

v. *Stresses in plates*:

$$\left(\frac{M_1}{Z_1} \right) = \left(\frac{71.36 \times 10^6}{0.041 \times 10^9} \right) = 1.74 \text{ N/mm}^2$$

$$\left(\frac{M_2}{Z_2} \right) = \left(\frac{105.6 \times 10^6}{0.162 \times 10^9} \right) = 0.65 \text{ N/mm}^2$$

vi. *Equation of three shears*:

Let T_B and T_C be the unknown shear at junctions B and C

$$T_A = T_F = 0 \qquad\qquad T_C = -T_D$$

Applying the equation of three shears to plates AB and BD, we have

$$\left(\frac{T_A}{0.25} \right) + 2T_B \left[\frac{1}{0.25} + \frac{1}{0.34} \right] + \frac{T_C}{0.34} = -\left(\frac{1.74 + 0.65}{2} \right) 10^3$$

For plates BC and CD

$$\left(\frac{T_B}{0.34}\right) + 2T_C\left[\frac{1}{0.34} + \frac{1}{0.36}\right] + \frac{T_D}{0.36} = -\left(\frac{0.650+0}{2}\right)10^3$$

Also $\quad T_C = T_D$

Solving the two equations, the unknown shears are:

$$T_C = -8.7 \text{ kN}$$
$$T_B = -84.0 \text{ kN}$$

vii. *Stresses in plates due to shear forces*:

Plate 1

$$\sigma_A = \left[\frac{84\times10^6 \times 0.5}{0.041\times10^9} - \frac{84\times10^3}{0.25\times10^6}\right] = 0.684 \text{ N/mm}^2$$

$$\sigma_B = \left[-\frac{84\times10^6 \times 0.5}{0.041\times10^9} - \frac{84\times10^3}{0.25\times10^6}\right] = -1.356 \text{ N/mm}^2$$

Plate 2

$$\sigma_B = \left[\frac{(84+8.7)10^6 \times 2.84}{2\times0.162\times10^9} + \frac{(84-8.7)10^3}{0.34\times10^6}\right] = 1.037 \text{ N/mm}^2$$

$$\sigma_C = \left[-\frac{(84+8.7)10^6 \times 2.84}{2\times0.162\times10^9} + \frac{(84-8.7)10^3}{0.34\times10^6}\right] = -0.594 \text{ N/mm}^2$$

Plate 3

$$\sigma_C = \left[\frac{8.7\times2\times10^3}{0.36\times10^6}\right] = 0.048 \text{ N/mm}^2$$

The stresses due to shears in various plates are shown in Fig. 15.7.

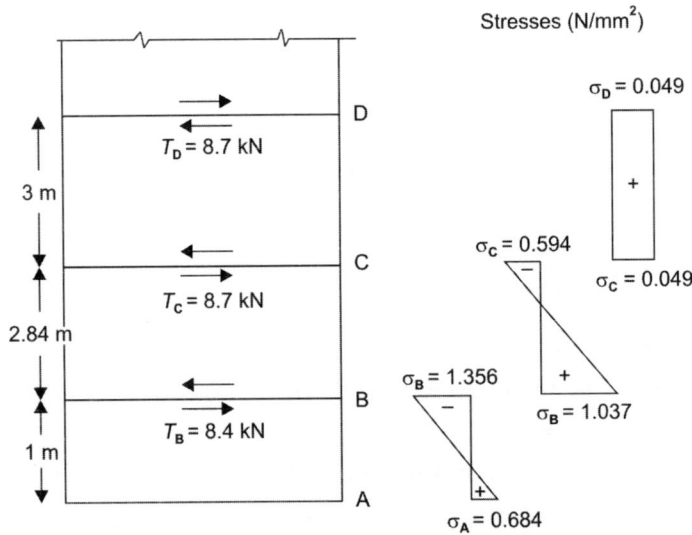

Fig. 15.7: Stresses in folded plates due to joint shear forces

viii. *Resultant stresses*:

The resultant stresses in plates are obtained by superposing the stresses due to plate moments and joint shears.

Plate 1

$$A = (-1.74 + 0.684) = -1.056 \text{ N/mm}^2$$
$$B = (1.74 - 1.356) = +0.384 \text{ N/mm}^2$$

Plate 2

$$B = (-0.65 + 1.037) = +0.387 \text{ N/mm}^2$$
$$C = (+0.65 - 0.594) = +0.056 \text{ N/mm}^2$$

Plate 3

$$C = (0 + 0.049) = +0.049 \text{ N/mm}^2$$

The resultant stresses are shown in Fig. 15.8.

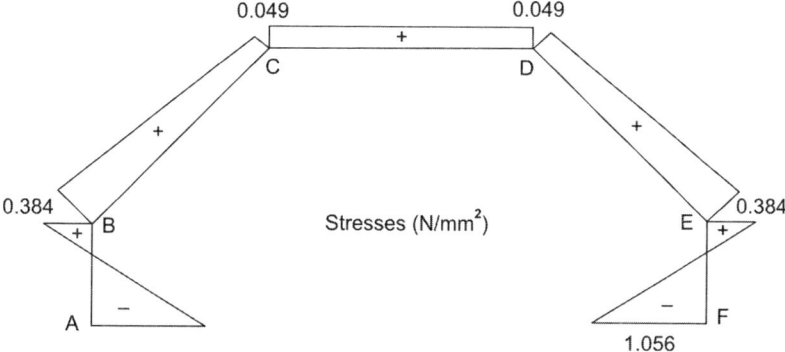

Fig. 15.8: Resultant final stresses in hipped plates

3. Analyse the symmetrical V-shaped folded plate shown in Fig. 15.9 and design reinforcements using the following data:

Span of the folded plate = 20 m

Thickness of plates = 100 mm

Live load = 0.6 kN/m²

Concrete = M20 grade

Steel = Fe 415 HYSD bars

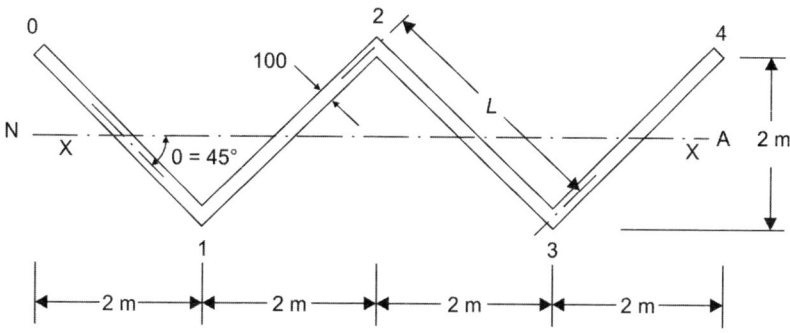

Fig. 15.9: V-shaped folded plate

The symmetrical V-shaped plate is analysed by using iteration method.

i. *Loads*:

Self weight = (0.1×24) = 2.4 kN/m²
Live load   = 0.6 kN/m²
Total load w = 3.0 kN/m²

ii. *Fixed end moments*:

Considering the transverse section of the folded plate as a continuous beam on rigid supports as shown in Fig. 15.10, the fixed end moments are computed and tabulated as shown below:

Plate	Load (kN/m)	Fixed end moment (kN/m/m)	Remarks
0–1	3.0	$\dfrac{wL^2}{2} = 6.00$	Cantilever
1–2	4.0	$\dfrac{wL^2}{12} = 1.00$	Fixed beam
2–3	3.0	$\dfrac{wL^2}{12} = 1.00$	Fixed beam
3–4	3.0	$\dfrac{wL^2}{2} = 6.00$	Cantilever

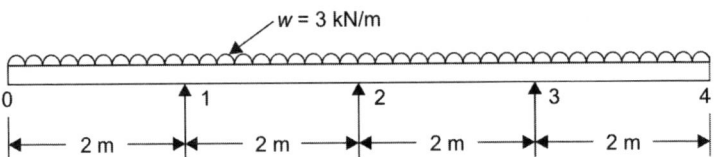

Fig. 15.10: Analysis for slab action

iii. *Moment distribution (no rotation solution)*:

		1		2		3	
	0	1	0.5	0.5	1	0	
Initial moments (kN·m)	+6.00	−1.00	+1.00	−1.00	+1.00	−6.00	
		−5.00 →	−2.50	+2.50	← +5.00		
Final moments (kN·m)	+6.00	−6.00	−1.50	+1.50	+6.00	−6.00	
Reaction due to moments (kN)		↑ 3.75	3.75 ↓	↓ 3.75	3.75 ↑		
Reaction due to loads	6.00 ↑	↑ 3.00	3.00 ↑	↑ 3.00	3.00 ↑	↑ 6.00	
Total reactions		↑ 12.75 ↓		↓ 1.50 ↑		↑ 12.75 ↓	
Joint loads (kN)		12.75		1.50		12.75	

iv. *Plate loads*:

The joint loads are resolved into components along the directions of the plates resulting in plate loads as shown in Fig. 15.11

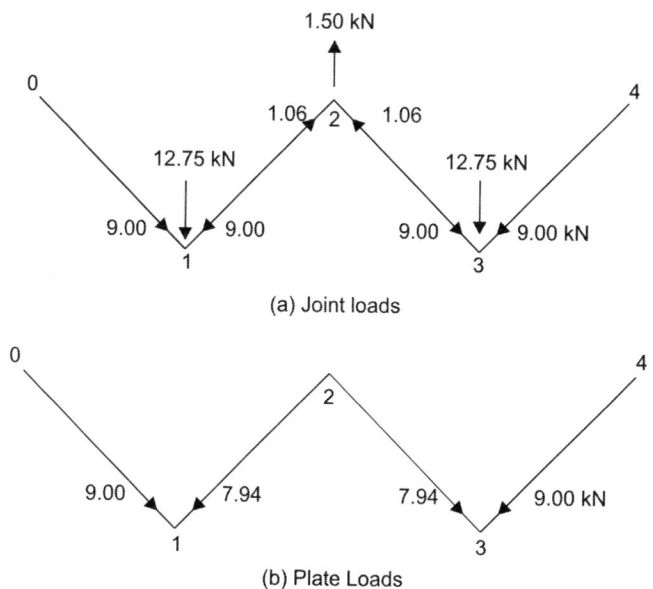

(a) Joint loads

(b) Plate Loads

Fig. 15.11: Joint loads and plate loads

v. *Moments and stresses in plates*:

The longitudinal moments developed in the plates due to the plate loads are computed assuming the plates to bend longitudinally in their own plane between the reverses. The stresses developed at the edges of the plates are compiled in Table 15.1.

Table 15.1: Longitudinal stresses

Plate	Load (R) (kN/m)	Moment (kN·m) ($RL^2/8$)	Section modulus (m^3)	Stresses (N/mm^2)	Free edge stresses (N/mm^2)
0–1, 20 m	9.00	450	0.133	3.40	+3.4 / 3.4 −
1–2	7.94	397	0.133	3.00	3.0 / +3.0
2–3	7.94	397	0.133	3.00	+3.0 / 3.0
3–4	9.00	450	0.133	3.40	3.0 / +3.4

+ Compression   − Tension

vi. *Stress distribution for no rotation*:
Solution to stress distribution for no rotation is shown in Table 15.2.

Table 15.2: Stress distribution (no rotation)

	0		1		2		3		4
Distribution factors		0.5	0.5	0.5	0.5	0.5	0.5		
Free edge	+3.4	−3.4	−3.0	+3.0	−3.0	−3.4	−3.4	+3.4
Stresses (N/mm²)	−0.1	+0.2	−0.2	+0.1	+0.1	−0.2	+0.2	−0.1
	+3.3	−3.2	−3.2	+3.1	+3.1	−3.2	−3.2	+3.3
$(\sigma_b - \sigma_t)$ (N/mm²)		6.5		6.3		6.3		6.5	

vii. *Plate deflections due to no rotation solution*:
The maximum deflection of a plate of length 'L' and depth 'h' can be computed as detailed below:

$$\text{Deflection } \delta = \frac{5}{384}\left(\frac{wL^4}{EI}\right) = \frac{5}{48}\left(\frac{wL^2}{8}\right)\frac{L^2}{EI} = \frac{5}{48}\left(\frac{\sigma_b - \sigma_t}{2}\right)Z\frac{L^2}{EI}$$

$$= \frac{5}{48}\left(\frac{\sigma_b - \sigma_t}{2}\right)\left(\frac{I}{h/2}\right)\left(\frac{L^2}{EI}\right) = \frac{5}{48}\frac{(\sigma_b - \sigma_t)L^2}{Eh}$$

The plate deflections and relative displacements of joints are shown in Fig. 15.12.

$$V_1 = V_4 = \left(\frac{5}{48}\right)\frac{(6.5\times10^6)20^2}{2\sqrt{2}E} = \left(\frac{1\times10^8}{E}\right) \text{ m}$$

$$V_2 = V_3 = \left(\frac{5}{48}\right)\frac{(6.5\times10^5)20^2}{2\sqrt{2}E} = \left(\frac{0.9\times10^8}{E}\right) \text{ m}$$

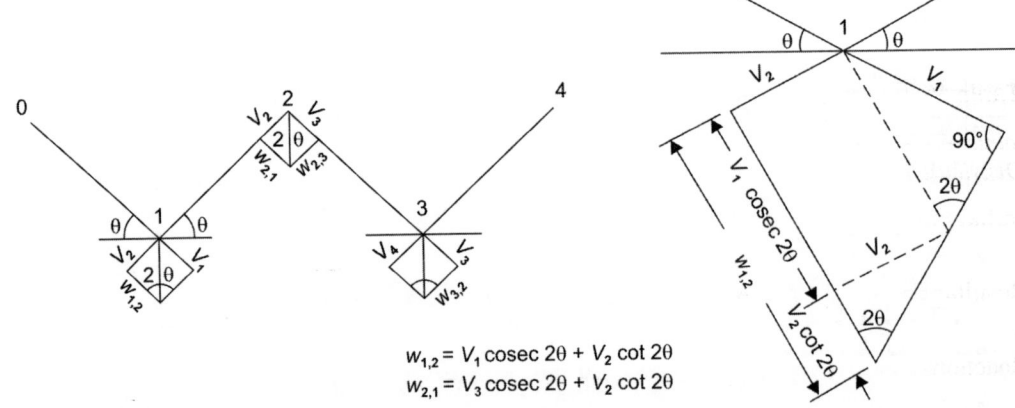

$$w_{1,2} = V_1 \cosec 2\theta + V_2 \cot 2\theta$$
$$w_{2,1} = V_3 \cosec 2\theta + V_2 \cot 2\theta$$

Fig. 15.12: Relative displacement of joints

viii. *Joint displacements*:

$$w_{1,2} = V_1 \operatorname{cosec} 2\theta + V_2 \cot 2\theta$$

since, $\theta = 45°$ $\quad w_{1,2} = \left(\dfrac{1\times10^8}{E}\right)$ m

$$w_{2,1} = V_3 \operatorname{cosec} 2\theta + V_2 \cot 2\theta$$

since, $V_2 = V_3$ and $\theta = 45°$, $\quad w_{2,1} = \left(\dfrac{0.9\times10^8}{E}\right)$ m

$\therefore$ Transverse moment $= \left(\dfrac{6EI_2\Delta_2}{h_2^2}\right)$

$$\Delta_2 = (w_{1,2} - w_{2,1}) = \left(\dfrac{0.1\times10^8}{E}\right) \text{ m}$$

For plate 2, transverse moment is given by

$$\left(\dfrac{6EI_2\Delta_2}{h_2^2}\right) = \left[\dfrac{6\times E\times1\dfrac{(0.1)^3}{12}\times\dfrac{0.1\times10^8}{E}}{(2\sqrt{2})^2}\right] = 625 \text{ N.m} = 0.625 \text{ kN·m}$$

Similarly for plate 3, by symmetry

the transverse moment $= \left(\dfrac{6EI_3\Delta_3}{h_3^2}\right)$

where, $\quad \Delta_3 = (w_{2,3} - w_{3,2}) = -\left(\dfrac{0.1\times10^8}{E}\right)$ m

$\therefore$ Transverse moment $= -0.625$ kN·m

ix. *Determination of joint loads*:

Using these moments, we can determine the joint loads developed by the distribution of moments as shown in Table 15.3.

Table 15.3: Distribution of moments caused by joint displacement (first cycle of iteration)

Joints		1		2	3
Distribution factor	0	1			
Initial moments (kN·m)		+0.625	+0.625	−0.625	−0.625
		−0.625 →	−0.312	+0.312 ←	+0.625
Resultant moments		0	+0.312	−0.312	0
		↓	↑	↑	↓
Reactions		0.156	0.156	0.156	0.156
		↑			↑
Loads (kN)		0.156	0.312		0.156

The joint loads and plate loads are shown in Fig. 15.13.

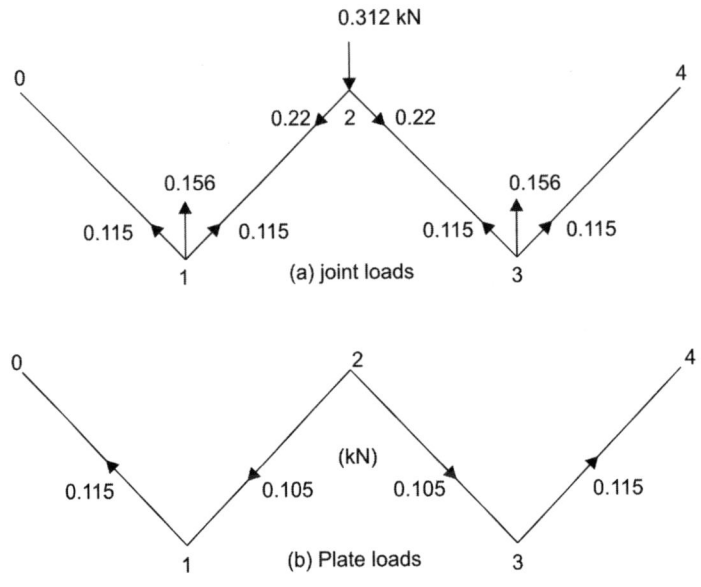

Fig. 15.13: Plate loads due to joint displacements

x. *Longitudinal stresses:*

The longitudinal stresses developed due to these plate loads are shown in Table 15.4.

Table 15.4: Longitudinal stresses

Plate	Load (R) (kN/m)	Moment (kN·m) (RL²/β)	Section modulus (m³)	Stresses (N/mm²)	Free edge stresses (N/mm²)
0–1	0.115	5.75	0.133	0.043	−0.043 / +0.043
1–2	0.105	5.25	0.133	0.039	−0.039 / +0.039
2–3	0.105	5.25	0.133	0.039	+0.039 / −0.039
3–4	0.115	5.75	0.133	0.043	+0.043 / −0.043

+ Compression   − Tension

xi. *Stress distribution:*

The stresses obtained in Table 15.4 are distributed as shown in Table 15.5.

Table 15.5: Stresses distribution

Joints	0		1		2		3		4
Free edge stresses (N/mm²)	−0.043	+0.043	−0.039	+0.039	+0.039	−0.039	+0.043	−0.043	
	+0.020 ← −0.041		+0.020 ← −0.020		+0.02 ← +0.041		−0.041 → +0.02		
Resultant stresses	−0.023	+0.002	+0.002	+0.019	+0.019	+0.002	+0.002	−0.23	

xii. *Find moments and stresses:*

The final moments and stresses are compiled in Tables 15.6 and 15.7 respectively.

Table 15.6: Final transverse moments (values in kN·m)

Joints	No rotation	First cycle of iteration	Final moments
0	0	0	0
1	+6.00	0	+6.000
2	−1.50	+0.312	−1.188
3	+6.00	0	+6.00
4	0	0	0

Table 15.7: Final longitudinal stresses (values in N/mm²)

Joints	No rotation	First cycle of iteration	Final stresses
0	+3.3	−0.023	+3.277
1	−3.2	+0.002	−3.198
2	+3.1	+0.019	+3.119
3	−3.2	+0.002	−3.198
4	+3.3	−0.023	+3.277
+ Compression		−Tension	

xiii. *Design of reinforcements:*

The design of reinforcement in the folded plate should conform to the recommendations made in IS:2210-1962. The transverse reinforcement is designed to resist the moments in the transverse direction computed at mid span to the support section based on the magnitude of the transverse bending moment.

The longitudinal reinforcements in the direction of the span are designed to resist the total tensile force developed in the tension zone. However, minimum percentage of reinforcements as recommended in the codes should be provided at all cross-sections.

The shear reinforcements are designed to resist the principal tensile stresses. When principal tensile stress exceeds $1.7\sqrt{f_{ck}}$, reinforcement will be necessary to take up the entire tension; elsewhere, nominal reinforcement at spacing not exceeding five times the thickness of the plate is to be provided. The maximum shear stresses are likely to develop at quarter span sections.

a. Transverse reinforcement:

At joints 1 and 3, the transverse moments cause tension at top of slabs.

Thickness of folded plate slab = 100 mm

Assuming effective cover = 25 mm

Effective depth $d = 75$ mm

Maximum moment = 6 kN·m

$$\therefore \qquad A_{st} = \left(\frac{6 \times 10^6}{230 \times 0.9 \times 75}\right) = 386 \text{ mm}^2$$

Provide 10 mm diameter bars at 200 mm centres.

At joint 2, the transverse moment causes tension at bottom of slab, i.e.

$$\therefore \qquad A_{st} = \left(\frac{1.5 \times 10^6}{230 \times 0.9 \times 75}\right) = 97 \text{ mm}^2$$

But minimum percentage of steel is:

$$\therefore \qquad 0.3\% = \left(\frac{0.3}{100} \times 1000 \times 100\right) = 300 \text{ m}^2$$

Provide 10 mm diameter bars at 200 mm centers.

b. Longitudinal reinforcement:

The maximum tensile stresses develop at joints 1 and 3, hence considering plate 1 is shown in Fig. 15.14.

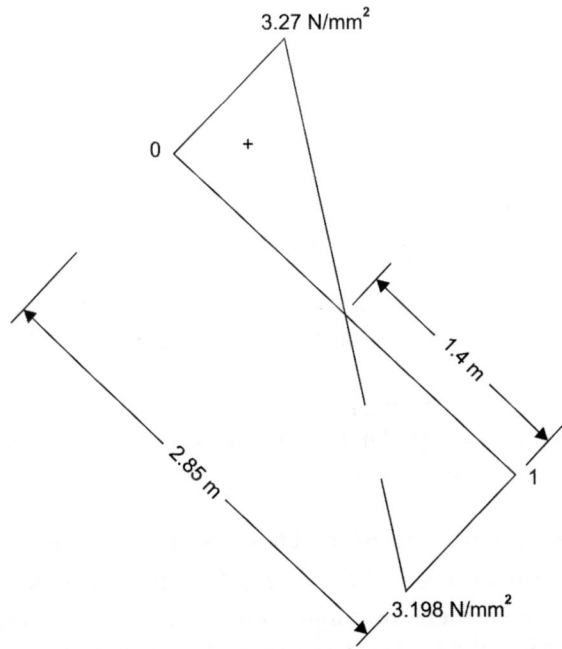

Fig. 15.14: Stress distribution in folded plate

Reinforcement in the tension zone is

$$A_{st} = \left[\frac{(1.4 \times 10^3 \times 100) \times \left(\dfrac{3.198}{2} \right)}{230} \right] = 973 \text{ mm}^2$$

Provide 5 bars of 16 mm diameter in the tension zone as detailed in Fig. 15.15. For the compression zone portion, provide 10 mm diameter bars at 200 mm centers.

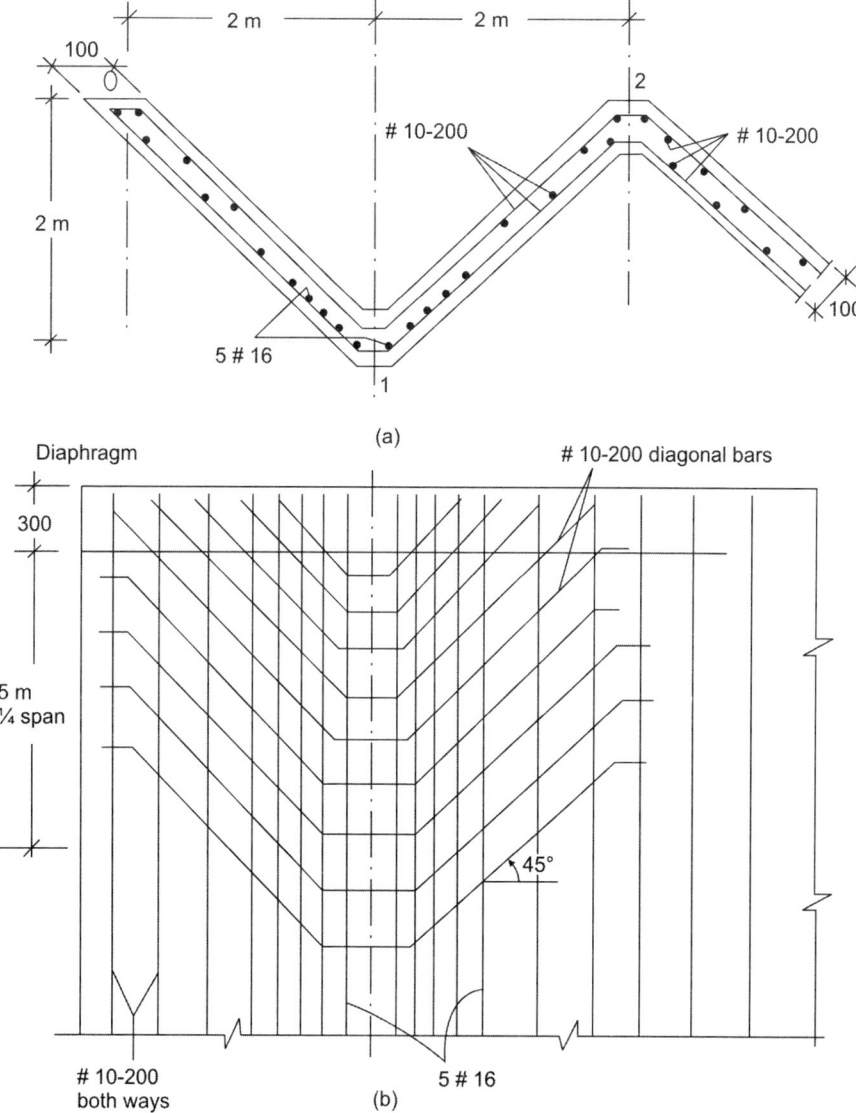

Fig. 15.15: Reinforcement details in folded plate

4. Analyse the symmetrical V-shaped folded plate of 20 m span given in design Example 15.13 by the beam method and compare the results of stresses obtained by the iteration and beam methods.

i. *Data*:

Span L = 20 m   Depth of plate = $2\sqrt{2}$ m

Thickness t = 100   Live load = 0.6 kN/m^2

ii. *Section properties*:

For the cross-section of the V-shaped folded plate shown in Fig. 15.16, the second moment of area is calculated as shown below.

For each plate inclined at an angle θ to axis XX

$$I_{xx} = \left[\frac{2t\sin^2\theta(L/2)^2}{3}\right]$$

where,

t = thickness of the plate

L = length of the plate

θ = angle made by the plate with the horizontal axis XX

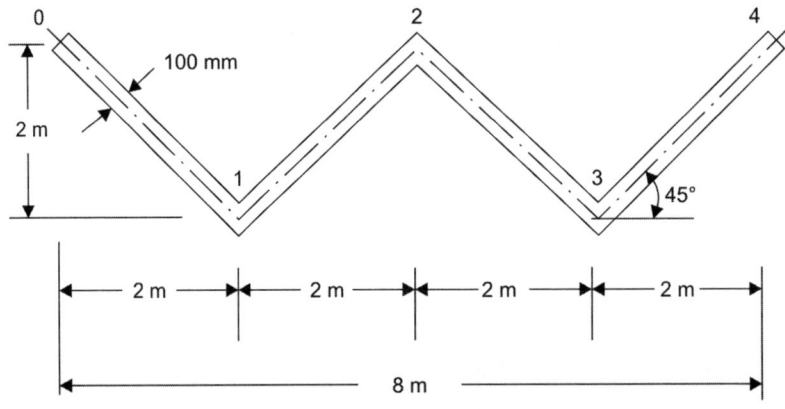

Fig. 15.16: Cross-section of folded plate

The second moment of area of the cross-section with four plates is given by

$$I_{xx} = 4\left[\frac{2\times0.1\times\sin^2 45°\times\dfrac{(2\sqrt{2})^3}{2}}{3}\right] = 0.38 \text{ m}^4$$

iii. *Loads*:

Dead load = $(0.1\times24\times4\times2\sqrt{2})$ = 27.2

Live load = $(4 \times 2 \times 0.6)$ = 4.8

Total load w = 32.0 kN/m

iv. *Bending moment*:

$$M_{max} = \left(\frac{wL^2}{8}\right) = \left(\frac{32 \times 20^2}{8}\right) = 1600 \text{ kN·m}$$

v. *Longitudinal stresses*:

Bending stress

$$\sigma = \left(\frac{My}{I}\right) = \left(\frac{1600 \times 10^6 \times 1000}{0.38 \times 10^{12}}\right) = 4.2 \text{ N/mm}^2$$

The maximum stresses developed in the plate are 4.2 N/mm^2 (compression at points 0, 2 and 4 and tension at points 1 and 3).

vi. *Comparision of stresses*:

Joint	Iteration method	Beam method
0	+3.277	+4.20
1	−3.198	−4.20
2	+3.119	+4.20
3	−3.198	−4.20
4	+3.277	+4.20

(Stresses in N/mm^2)   +Compression   −Tension

The beam method results in prediction of stress values nearly 30% higher than those obtained by the iteration method.

References

1. ASCE Task Committee, Report on folded plate construction, *Journal of the Structural Division*, Vol. 89, No. 6, Nov/Dec. 1963, pp. 365–460.

2. Winter G and Pei M, Hipped plate construction, *ACI Journal*, Vol. 43, No. 1, 1947, pp. 505–532.

3. Gaafar I, Hipped plate analysis considering joint displacements, Paper 2696, *Transactions of the ASCE*, Vol. 119, 1954, pp. 743–784.

4. Simpson H, Design of folded plates roofs, *ASCE Journal of Structural Division, Proceedings of ASCE*, Vol. 84, Jan. 1958.

5. Whitney CS and Anderson BG, Reinforced concrete folded plate structures, *Journal of the Structural Division, ASCE*, 1959.

6. Ramaswamy GS, *Design and Construction of Concrete Shell Roofs*, McGraw-Hill, New York, 1968.

7. Evans HR and Rockey KC, A critical review of the methods of analysis of folded plates, *Proceedings of the Institution of Civil Engineers*, London, Vol. 51, No. 3, March 1972, pp. 581–589.

8. Rao PS, Continuous folded plates, *IASS Bulletin*, Vol. 67, 1970, pp. 37–44.

9. Aroni S, Folded plate roofs, *Architectural Science Review*, Vol. 7, No. 4, 1964, pp. 146–150.

10. Cheung YK, Folded plate structures by finite strip method, *Journal of the Structural Division, ASCE*, Vol. 95, No. 12, Dec. 1969, pp. 2963–2982.

11. Guralnick SA, Swartz SE and Loginow A, Buckling of reinforced concrete folded plates, *ACI-Special Publication SP-67*, 1981, pp. 111–134.

12. Guralnick SA, Swartz SE, Approximate analysis of the stability of single cell folded plate structure, *IABSE Publications*, Vol. 29, 1979, pp. 217–236.

13. Brielamaier, Prismatic folded plates, *ACI Journal Proceedings*, 1962, Vol. 59, No. 32 pp. 407–426.

14. Indian standard IS:2210-1988 (reaffirmed 2003), *Criteria for the Design of Reinforced Concrete Shell Structures and Folded Plates*, Bureau of Indian Standards, New Delhi, 1988, pp. 1–21.

15. Krishna Raju N, *Prestressed Concrete*, 5 edn, McGraw-Hill Education (India), New Delhi, 2002, pp. 549–573.

Assignment

1. Analyse the folded plate ABCDEF using the theorem of the three edge shears. The plates AB and EF are vertical, while the plate CD is horizontal. The plates BC and DE are inclined at 45° to the horizontal. All the plates are 12 cm thick and their widths are as follows:

    AB and EF = 1.25 m

    BC and DE = 3.90 m

    CD = 3.00 m

    Live load = 0.6 kN/m^2 of covered area

    Distance between the traverses = 10 m.

    Draw the stress distribution diagram for the plates. In what way the stresses differ if the hipped plate structure is analysed as a beam spanning horizontally between the traverses.

2. A V-shaped folded plate ABCDE spans over 12 m between the traverses and the folds have a uniform thickness of 100 mm. The plates are inclined at 45° to the horizontal and vertical and the horizontal projections of each of the plates is 3 m. Edges B and D are the ridges and edge C form the valley portion. The folded plate is to be analysed for incidental live loads of 0.75 kN/m^2 of covered area. Analyse the stress distribution for the section at centre of span using the following methods:
    a. theorem of three edge shears
    b. stress distribution procedure (no rotation solution)
    Draw the stress diagrams.

3. Design a folded plate roof with V-shaped units to cover a storehouse 9 m wide × 30 m long. Six plates, each having a thickness of 120 mm may be used with plates inclined at 45° to the horizontal. The vertical and horizontal projections of the plates = 1.5 m. Live load = 0.75 kN/m^2. Analyse the folded plate using the iteration method and design the longitudinal and transverse reinforcements in the plate. Adopt M20 grade concrete and tor steel reinforcements. Compare the stresses developed at the centre of span with those resulting from the beam method.

4. Analyse the hipped plate roof of a small single storey industrial building shown in Fig. 15.17, using the theorem of three edge shears.

    Thickness of plates = 125 mm

    Live load = 0.75 kN/m^2

    Design suitable reinforcements and sketch the details of transverse and longitudinal reinforcements with folded plate roof. Adopt M20 grade concrete and Fe 415 tor steel reinforcements.

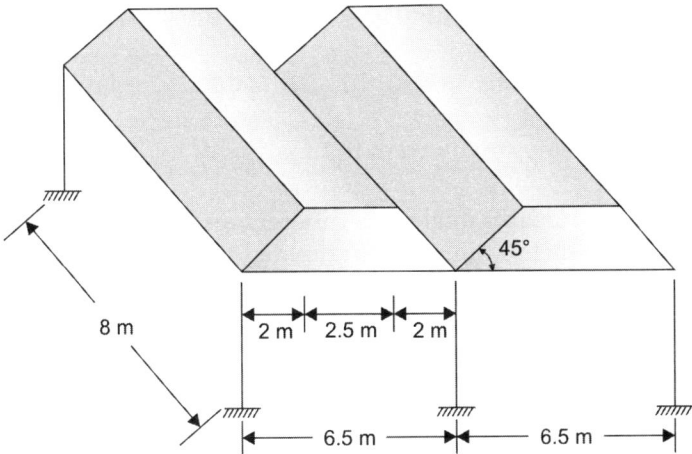

Fig. 15.17: Hipped plate roof

5. A V-shaped folded plate roof has to be designed for a warehouse to cover a span of 18 m. The width of the warehouse is 40 m. Thickness of the plates may be assumed as 100 mm. Live load is 0.75 kN/m². Adopt M25 grade concrete and Fe 500 HYSD bars. Design the folded plate and sketch the details of reinforcements.

6. Design the folded plate roof shown in Fig. 15.18, by Simpson's method using the following data.

Span = 20 m

Live load = 0.75 kN/m²

Adopt M20 grade concrete and Fe 415 grade tor steel. Sketch the details of reinforcements in the folded plate.

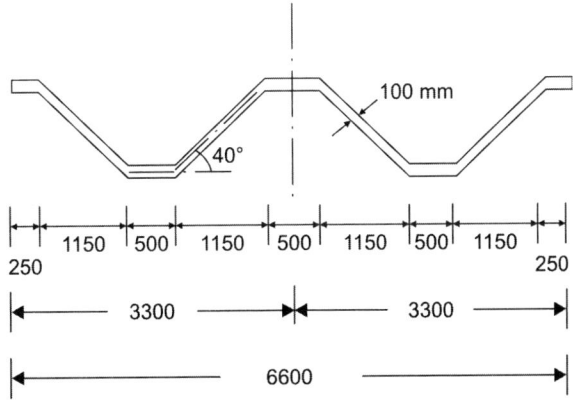

Fig. 15.18: Trough shaped folded plate

Review Questions

1. Briefly outline the specific advantages of folded plate roofs for covering large areas in comparison with the shell roofs.

2. Explain with sketches the different types of structural configurations adopted in folded plate roofs.

3. What are the various methods used to analyse the forces developed in folded plates subjected to imposed loads?

4. Explain the terms (a) plate action (b) slab action, in relation to the structural behaviour of folded plates.

5. What are the salient steps involved in the analysis of folded plates using Whitney's method?

6. In what way the Simpson's method is different and advantageous in comparison with the Whitney's method in analysing folded plates?

7. What are the limitations of Winter and Pei's method of analysis of folded plates? Specify the simplifying assumptions made in this method?

8. Briefly explain the equation of three shears used in the analysis of folded plates with the help of sketches.

9. What type of analysis you would resort to, for evaluating the joint loads and forces in the plane of the plates?

10. Sketch the typical reinforcements used in a V-shaped folded plate showing their arrangement in the cross-section and plane of the folded plate.

Objective Type Questions

1. Folded plate structures are generally made up of
   a. curved elements
   b. plane elements
   c. hyperboloid elements

2. Folded plate structural elements are ideally suited for use in
   a. water tanks
   b. warehouse roofs
   c. transportation structures

3. The thickness of reinforced concrete folded plates are generally of the order of
   a. 30 mm to 50 mm
   b. 150 mm to 250 mm
   c. 50 mm to 120 mm

4. In the case of reinforced concrete folded plate roof structures, compressive stresses develop at the
   a. valley junctions
   b. ridge junctions
   c. centre of the plates

5. The force components developed in a folded plate structure is determined by the
   a. theorem of three moments
   b. moment distribution
   c. equation of three shears

6. The beam method of analysis is applicable for folded plates with span to depth ratio
   a. less than 5
   b. exceeding 10
   c. equal

7. In the analysis of forces in the folded plates, joint displacements can be ignored for
   a. long span folded plates
   b. medium span folded plates
   c. short span folded plates

8. In case of folded plates used as a roof structure, each plate bends in the longitudinal direction in its own plane between the traverses due to
   a. slab action
   b. plate action
   c. shear action

9. In case of long V-shaped folded plates supported on diaphragms at the ends, the main longitudinal reinforcements to resist bending is provided in the vicinity of
   a. ridge junctions
   b. valley junctions
   c. support junctions

10. Shear reinforcements in folded plates are generally provided near the
    a. centre of span
    b. supports
    c. ridge junctions

16

Grid or Coffered Floors

16.1 GENERAL FEATURES

Reinforced concrete grid or waffle or coffered slab systems[1,2] consists of beams spaced at regular intervals in orthogonal directions with an integral slab at the top. They are generally used to cover large floor spaces of parking garages, commercial and industrial buildings, airports and residential structures. The grid floors are often preferred for marriage halls where column free space is required over a large area. Grid floors[3,4] are preferred by architects since the rectangular or square voids between the ribs can be advantageously utilised for concealed architectural lighting.

The beams or ribs in a coffered slab may be arranged in a perpendicular direction to the supporting edges or at 45° to the supports. Grid slab with inclined beams to the edge supports is termed diagrid. Figure 16.1 shows the different types of grid floors. The ceiling of a typical waffle slab having square grids is shown in Fig. 16.2.

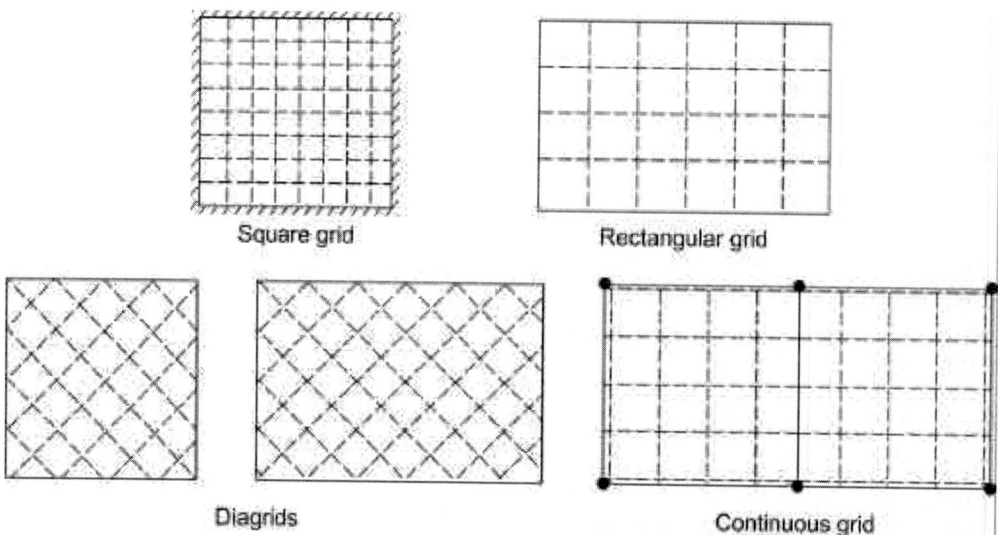

Square grid   Rectangular grid

Diagrids   Continuous grid

Fig. 16.1: Types of grid floors

Fig. 16.2: Typical ceiling of a reinforced concrete grid floor

16.2 ANALYSIS OF GRID FLOORS

Approximate Methods

IS:456 Code Method

Grid floors can be analysed for moments and design forces developed under the action of dead and imposed loads by using approximate methods specified in the Indian standard code and adopting certain simplifying assumptions. According to the Indian standard code IS:456-2000[5], the ribbed slab system can be analysed as a solid slab, if the structure satisfies the following requirements regarding spacing of beams, thickness of slab and edge beams:

   i. The width of the *in situ* ribs should be not less than 65 mm.

   ii. The spacing of the ribs should not exceed 1.5 m.

   iii. The depth of the ribs, excluding any topping, should not exceed 4 times their width.

   iv. The width of the ribs along the edges should be as wide as the width of the bearing.

   The moments and shear forces per unit width of grid slab are determined using Table 26 of IS:456-2000 code and suitable reinforcements are designed to resist the design force components. The slab reinforcement generally consists of a mesh or fabric. The detailing of reinforcements in the grid floor should conform to the clause 36.7 of IS:456 code.

Rankine–Grashoff Method

The moment coefficients prescribed in the Annexure D of the IS:456 code (clause D 2, Table 26) for two way slabs are based on the Rankine–Grashoff theory[6]. According to this theory, the slab can be divided into a system of orthogonal unit beam strips and the load can be apportioned to the short and long span strips in such a way that the deflections of the two strips along the two centre lines should be the same at the junction. By equating the deflection component of the strips expressed as a function of the unknown unit loads, the magnitude of loads shared in the orthogonal directions can be determined.

Consider a grid floor shown in Fig. 16.3 in which the spacings of the ribs are a_1 and b_1 in the x- and y-directions respectively.

Let $\qquad q$ = total load per unit area

q_1 and q_2 = the loads shared in the x- and y-directions

a = shorter dimension of grid

b = longer dimension of grid

The deflections of the ribs AB and CD at the junction O must be the same and by equating the deflections, we have

$$\delta = \left(\frac{5q_1 a^4}{384EI}\right) = \left(\frac{5q_2 b^4}{384EI}\right)$$

$\therefore \qquad\qquad q_1 a^4 = q_2 b^4 \qquad\qquad\qquad\qquad\qquad\qquad ...(16.1)$

Also $\qquad\qquad\quad q = q_1 + q_2 \qquad\qquad\qquad\qquad\qquad\qquad ...(16.2)$

Solving Eqs (16.1) and (16.2), we have

$$q_1 = q\left(\frac{b^4}{a^4 + b^4}\right) \text{ and } q_2 = q\left(\frac{a^4}{a^4 + b^4}\right)$$

The bending moments for the central ribs are given by

$$M_{AB} = \left(\frac{q_1 b_1 a^2}{8}\right) \qquad M_{BD} = \left(\frac{q_2 a_1 b^2}{8}\right)$$

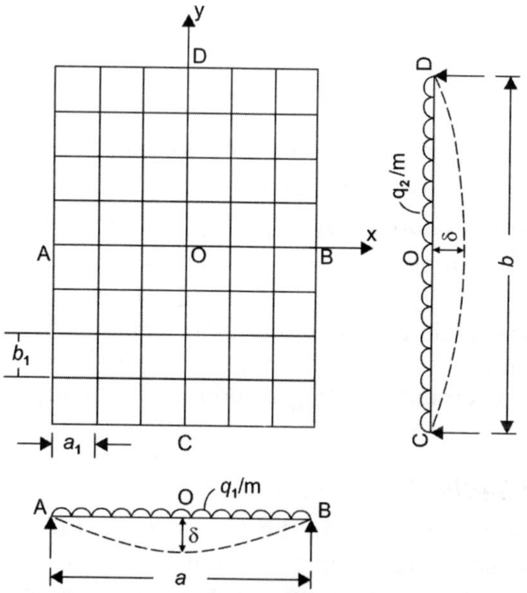

Fig. 16.3: Deflections of ribbed slabs

The bending moments in the other ribs can also be determined in direct proportion to their distances from the centre. The ribs are designed as flanged sections to resist the moments and shears. However, the approximate methods do not yield the twisting moments in the beams. For small span grids with spacings of ribs not exceeding 1.5 m, approximate methods can be used, but for grids of larger spans with spacings of ribs exceeding 1.5 m, a rigorous analysis based on orthotropic plate theory is generally used.

Rigorous Methods

Plate Theory

The theory of plates developed by Timoshenko *et al*[7] can be advantageously used for the analysis of grid floors. The main advantage of this method is that we can incorporate the influence of ribs in the orthogonal directions by considering the grib slab with ribs as an orthotropic plate. By using the rigorous analysis of orthotropic plates applied to a grid slab subjected to transverse loads, it is possible to compute the moment and shear force components developed in the grid slab at critical locations.

The vertical deflection a at the point of the grid shown in Fig. 16.4 is expressed as

$$a = \frac{16}{\pi^6}\left(\frac{\sin\left(\dfrac{\pi x}{a_x}\right)\sin\left(\dfrac{\pi y}{b_y}\right)}{\dfrac{D_x}{a_x^4} + \dfrac{2H}{a_x^2 b_x^2} + \dfrac{D_y}{b_y^4}}\right)$$

where,

q = total uniformly distributed load per unit area

a_x, b_y = length of plate along x- and y-directions respectively

D_x, D_y = flexural rigidity per unit length of plate along x- and y-directions respectively

C_x, C_y = torsional rigidity per unit length of plate along x- and y-directions respectively.

If a_1 and b_1 are the spacings of the ribs along x- and y-directions respectively, then we have the relations

$$D_x = (EI_1/b_1) \qquad\qquad C_x = (C_1/b_1)$$
$$D_y = (EI_2/a_1) \qquad\qquad C_y = (C_2/a_1)$$

where, EI_1, EI_2, C_1 and C_2 are the flexural and torsional rigidities of the effective section in x- and y-directions. The moments and shears are computed using the following expressions:

$$M_x = -D_x\left(\frac{\partial^2 a}{\partial x^2}\right) \qquad M_y = -D_y\left(\frac{\partial^2 a}{\partial y^2}\right)$$

$$T_{xy} = -\frac{C_1}{b_1}\left(\frac{\partial^2 a}{\partial x \partial y}\right) \qquad T_{xy} = -\frac{C_2}{a_1}\left(\frac{\partial^2 a}{\partial x \partial y}\right)$$

$$Q_x = -\frac{\partial}{\partial x}\left[D_x\left(\frac{\partial^2 a}{\partial x^2}\right) + \frac{C_2}{a_1}\left(\frac{\partial^2 a}{\partial x \partial y}\right)\right]$$

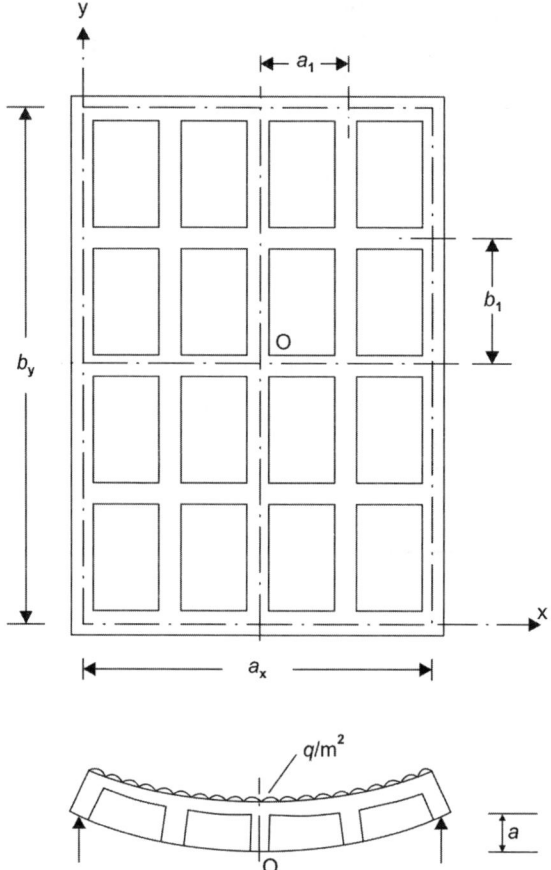

Fig. 16.4: Deflection characteristics of grid floors

$$Q_y = -\frac{\partial}{\partial x}\left[D_y\left(\frac{\partial^2 a}{\partial y^2}\right) + \frac{C_1}{b_1}\left(\frac{\partial^2 a}{\partial x \partial y}\right)\right]$$

Maximum bending moments develop at the centre of span while maximum torsional moments are generated at the corners of the grid and maximum shear forces develop at mid points of longer side supports.

Stiffness Method

In stiffness method using matrix analysis and computers, the grid slab floor can be accurately analysed for force components in all types of support conditions. Grid floors supported on columns are often cast integral with continuous beams running along the columns at the edges of the grid slab. In such cases, the grid floor support conditions should be treated as built-in or rigid at the edge supports since this will influence the moment and shear force components developed in the grid floor.

Halkude and Mahamuni[8] have reported a study on the comparison of various methods of analysis of grid floor frames in which they have compared the results of force components developed in grid floors by using Rankine–Grashoff method, rigorous plate theory and stiffness method. Sathawane *et al*[9] have also reported on the comparative analysis of both flat and grid slabs. Numerical and experimental studies have been conducted by Schwetz *et al*[10] regarding the analysis and design of waffle slabs. Mallick and Bhushan[11] have also presented an exhaustive report on the various methods of analysis of grid floors. Chowdhury and Singh[12] have investigated the analysis and design of waffle slabs with different boundary conditions.

In the case of grid floors[13] having large spans in the orthogonal directions, the depth of ribs to cover the large areas will be more and to limit the overall depth of the grid slab, it will be necessary to resort to the use of prestressing the ribs with high tensile cables housed in the ribs in the orthogonal directions. The reader may refer elsewhere to the complete design of a prestressed concrete grid floor by the author[14].

16.3 DESIGN EXAMPLE

A reinforced concrete grid floor is to be designed to cover a floor area of size 12 m × 16 m. The spacings of the ribs in mutually perpendicular directions is 2 m c/c and live load is 1.5 kN/m². Adopt M20 grade concrete and Fe 415 HYSD bars. Analyse the grid floor for moments and shears using the following methods:

a. Approximate method (Rankine–Grashoff theory)

b. Rigorous analysis (plate theory)

Design suitable reinforcements at critical sections.

i. *Data*:

Size of grid = 12 m × 16 m

Spacings of ribs = 2 m c/c

Concrete = M20 grade

Steel = Ribbed Fe 415 HYSD bars

ii. *Dimensions of slab and beams*:

Adopt thickness of slab = 100 mm

Depth of ribs based on $\left(\dfrac{\text{span}}{\text{depth}}\right)$ = 20 is obtained at depth = $\left(\dfrac{12 \times 10^3}{20}\right)$ = 600 mm

Width of rib = 200 mm

Adopt overall depth of ribs = 600 mm

Figure 16.5 shows the details of the grid and the section of the ribs in both directions.

iii. *Loads*:

Weight of slab = (0.1 × 24) = 2.4 kN/m²

Total load of slab = (2.4 × 16 × 12) = 460.8 kN

Weight of ribs = (0.2 × 0.5 × 24) = 2.4 kN/m

Total weight of beams (x-direction) = (7 × 2.4 × 12) = 201.6 kN

Total weight of beams (y-direction) = (5 × 2.4 × 14.6) = 175.2 kN

Total weight of floor finish = (0.6 × 12 × 16) = 115.2 kN

Total live load = (1.5 × 12 × 16) = 288 kN

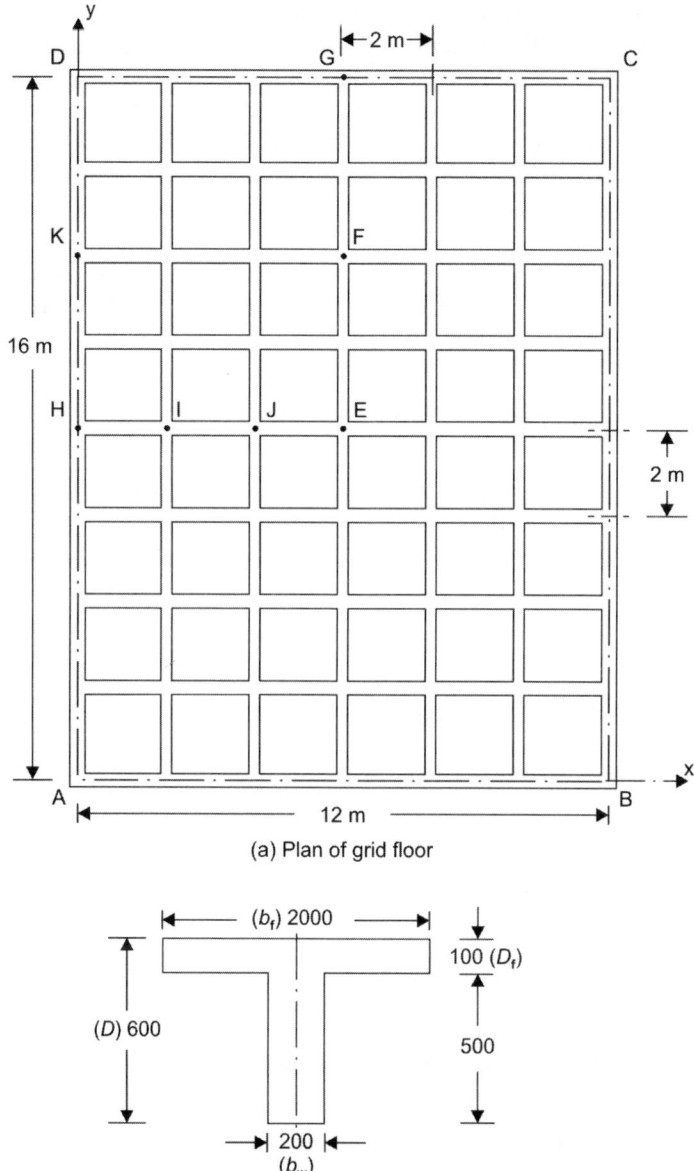

(a) Plan of grid floor

(b) Section of ribs in x- and y-directions

Fig. 16.5: Reinforced concrete grid floor

∴   Total dead and live loads on grid floor

$$= (460.8 + 201.6 + 175.2 + 115.2 + 288.0) = 1240.8 \text{ kN}$$

Load per m² $= q = \left(\dfrac{1240.8}{16 \times 12}\right) = 6.5 \text{ kN/m}^2$

iv. *Approximate method (moments):*

If q_1 and q_2 are the loads shared in the x- and y-directions, then

$$q_1 = q\left(\frac{b_y^4}{a_x^4 + b_y^4}\right) = 6.5\left(\frac{16^4}{12^4 + 16^4}\right) = 5 \text{ kN/m}^2$$

$$q_2 = q\left(\frac{a_x^4}{a_x^4 + b_y^4}\right) = 6.5\left(\frac{12^4}{12^4 + 16^4}\right) = 1.5 \text{ kN/m}^2$$

Moments in x- and y-directions at the centre of grid for 2 m width is obtained as

$$M_x = \left(\frac{q_1 b_1 a^2}{8}\right) = \left(\frac{5 \times 2 \times 12^2}{8}\right) = 180 \text{ kN·m}$$

$$M_y = \left(\frac{q_2 a_1 b^2}{8}\right) = \left(\frac{1.5 \times 2 \times 16^2}{8}\right) = 96 \text{ kN·m}$$

$$Q_x = \left(\frac{5 \times 2 \times 12}{2}\right) = 60 \text{ kN}$$

$$Q_y = \left(\frac{1.5 \times 2 \times 16}{2}\right) = 24 \text{ kN}$$

The approximate methods neglect the twisting moments in the grid.

v. *Rigorous method (plate theory)*:

  a. Section properties:

$$\left(\frac{D_f}{D}\right) = \left(\frac{100}{600}\right) = 0.166$$

$$\left(\frac{b_w}{b_f}\right) = \left(\frac{200}{2000}\right) = 0.10$$

From design table (Reynolds' RC designer's handbook),

$$I = C b_w D^3$$

The value of constant $C = 0.191$

∴ $I = (0.191 \times 200 \times 600^3) = 83 \times 10^8 \text{ mm}^4$

If I_1 and I_2 are the second moment of area of the tee-sections about their centroidal axis in the x- and y-directions respectively, then

$$D_1 = EI_1 \text{ and } D_2 = EI_2$$

In the present case $I_1 = I_2 = I$ and $a_1 = b_1 = 2$ m

∴ $$D_x = D_y = \left(\frac{83 \times 10^8 \times E}{2 \times 10^{12}}\right) = 0.00415E$$

The torsional rigidity in the x- and y-directions are given by

$$C_1 = C_2 = k_1 G(2a)^3 2b$$

where,

$2a = 200$ mm (width of beam)

$2b = 600$ mm (depth of beam)

For $\left(\dfrac{b}{a}\right) = \left(\dfrac{300}{100}\right) = 3, \quad k_1 = 0.263$ (from Reynolds' Designer's Handbook)

From tables in Timoshenko, theory of elasticity

$\therefore \qquad C_1 = C_2 = k_1 \left[\dfrac{E}{2(1+\mu)}\right](2a)^3 2b$

$$= \left[\dfrac{0.263 \times E \times (200)^3 (600)}{2(1+0.15)}\right] = 5.488 \times 10^8 E \text{ mm}^4$$

$$= 0.0005488E \text{ m}^4$$

$$2H\left[\dfrac{C_1}{b_1} + \dfrac{C_2}{a_1}\right] = E\left[\dfrac{0.0005488}{2} + \dfrac{0.0005488}{2}\right] = 0.0005488E$$

b. Deflections at centre of span:

$$q = 6.5 \text{ kN/m}^2, \qquad E = 5700\sqrt{f_{ck}} = 25.49 \times 10^6 \text{ kN/m}^2$$

$$\left(\dfrac{D_x}{a_x^4}\right) = \left(\dfrac{0.00415E}{12^4}\right) = \left(\dfrac{0.00415 \times 25.49 \times 10^6}{12^4}\right) = 5.089$$

$$\left(\dfrac{D_y}{b_y^4}\right) = \left(\dfrac{0.00415E}{16^4}\right) = \left(\dfrac{0.00415 \times 25.49 \times 10^6}{16^4}\right) = 1.61$$

$$\left(\dfrac{2H}{a_x^2 b_y^2}\right) = \left(\dfrac{0.0005488E}{12^2 \times 16^2}\right) = \left(\dfrac{0.0005488 \times 25.49 \times 10^6}{12^2 \times 16^2}\right) = 0.3794$$

The deflection at the centre of the plate is given by the equation

$$a = \dfrac{16q}{\pi^6}\left[\dfrac{\sin\dfrac{\pi x}{a}\sin\dfrac{\pi y}{b}}{\dfrac{D_x}{a_x^4} + \dfrac{2H}{a_x^2 b_y^2} + \dfrac{D_y}{b_y^4}}\right]$$

$$= \left(\dfrac{16 \times 6.5}{963}\right)\left(\dfrac{1}{5.089 + 0.3794 + 1.61}\right) = 0.015 \text{ m}$$

Assume creep coefficient $\theta = 2$

The modified modulus of elasticity $E_{ce} = \left(\dfrac{E_c}{1+\theta}\right)$

$$= \left(\dfrac{E_c}{1+2}\right) = \left(\dfrac{E_c}{3}\right)$$

$\therefore$ Long term deflection = $(0.015 \times 3) = 0.045$ m

According to IS:456-2000, the long term deflection should not exceed $\left(\dfrac{\text{span}}{250}\right)$.

Hence, maximum permissible long term deflection = $\left(\dfrac{12}{250}\right) = 0.048$ m > 0.045 m.

The maximum deflection including long term effects is within permissible limits.

c. Design moments and shears:

The bending moments, torsional moments and shears at various salient points E, F, G, H, I, J, K and D marked in Fig. 16.5 are computed using the equations given below:

$$M_x = -D_x\left(\frac{\partial^2 a}{\partial x^2}\right) = D_x\left(\frac{\pi}{a_x}\right)^2 \frac{16q}{\pi^6}\left[\frac{\sin\dfrac{\pi x}{a_x}\sin\dfrac{\pi y}{b_y}}{\dfrac{D_x}{a_x^4}+\dfrac{2H}{a_x^2 b_y^2}+\dfrac{D_y}{b_y^4}}\right]$$

$$= 0.00415 \times 25.49 \times 10^6\left(\frac{\pi}{12}\right)^2(0.015)\sin\frac{\pi x}{a_x}\sin\frac{\pi y}{b_y}$$

$$= 108\left(\sin\frac{\pi x}{a_x}\right)\left(\sin\frac{\pi y}{b_y}\right)$$

$$M_y = -D_y\left(\frac{\partial^2 a}{\partial y^2}\right) = D_y\left(\frac{\pi}{b_y}\right)^2 \frac{16q}{\pi^6}\left[\frac{\sin\dfrac{\pi x}{a_x}\sin\dfrac{\pi y}{b_y}}{\dfrac{D_x}{a_x^4}+\dfrac{2H}{a_x^2 b_y^2}+\dfrac{D_y}{b_y^4}}\right]$$

$$= 0.00415 \times 25.49 \times 10^6\left(\frac{\pi}{16}\right)^2(0.015)\left(\sin\frac{\pi x}{a_x}\right)\left(\sin\frac{\pi y}{b_y}\right)$$

$$= 61\left(\sin\frac{\pi x}{a_x}\right)\left(\sin\frac{\pi y}{b_y}\right)$$

$$M_{xy} = -\left(\frac{C_1}{b_1}\right)\left(\frac{\partial^2 a}{\partial x \partial y}\right) = -\frac{C_1}{b_1}\left(\frac{\pi^2}{ab}\right)\frac{16q}{\pi^6}\left[\frac{\cos\dfrac{\pi x}{a_x}\cos\dfrac{\pi y}{b_y}}{\dfrac{D_x}{a_x^4}+\dfrac{2H}{a_x^2 b_y^4}+\dfrac{D_y}{b_y^4}}\right]$$

$$= -0.0002744 \times 25.49 \times \frac{10^6 \times \pi^2}{12 \times 16} \times 0.015 \left(\cos \frac{\pi x}{a_x} \right) \left(\cos \frac{\pi y}{b_y} \right)$$

$$= -5.3 \left(\cos \frac{\pi x}{a_x} \right) \left(\cos \frac{\pi y}{b_y} \right)$$

$$M_{yx} = -\left(\frac{C_2}{a_1} \right) \left(\frac{\partial^2 a}{\partial x \partial y} \right)$$

$$= -\frac{C_2}{a_1} \left(\frac{\pi^2}{ab} \right) \frac{16q}{\pi^6} \left[\frac{\cos \dfrac{\pi x}{a_x} \cos \dfrac{\pi y}{b_y}}{\dfrac{D_x}{a_x^4} + \dfrac{2H}{a_x^2 b_y^4} + \dfrac{D_y}{b_y^4}} \right]$$

$$= -0.000274 \times 25.49 \times 10^6 \times \frac{\pi^2}{12 \times 16} \times (0.015) \left(\cos \frac{\pi x}{a_x} \right) \left(\cos \frac{\pi y}{b_y} \right)$$

$$= -5.3 \left(\cos \frac{\pi x}{a_x} \right) \left(\cos \frac{\pi y}{b_y} \right)$$

$$Q_x = -\frac{\partial}{\partial x} \left[D_x \frac{\partial^2 a}{\partial x^2} + \frac{C_2}{a_1} \frac{\partial^2 a}{\partial y^2} \right]$$

$$= -\frac{16q}{\pi^6} \left[\frac{\cos \dfrac{\pi x}{a_x} \sin \dfrac{\pi y}{b_y}}{\dfrac{D_x}{a_x^4} + \dfrac{2H}{a_x^2 b_y^2} + \dfrac{D_y}{b_y^4}} \right] \left[D_x \frac{\pi^3}{a_x^3} + \frac{C_2}{a_1} \frac{\pi^3}{a_x b_y^2} \right]$$

$$= -0.015 \left[0.00415 \times 25.49 \times 10^6 \times \frac{\pi^3}{12^3} \right.$$

$$\left. + 0.0002744 \times 25.49 \times 10^6 \left(\frac{\pi^3}{12 \times 16^2} \right) \right] \left(\cos \frac{\pi x}{a_x} \right) \left(\sin \frac{\pi y}{b_y} \right)$$

$$= -10.107 \left(\cos \frac{\pi x}{a_x} \right) \left(\sin \frac{\pi y}{b_y} \right)$$

$$Q_y = -\frac{\partial}{\partial y} \left[D_y \frac{\partial^2 a}{\partial y^2} + \frac{C_1}{b_1} \frac{\partial^2 a}{\partial x^2} \right]$$

$$= -\frac{16q}{\pi^6}\left[\frac{\sin\dfrac{\pi x}{a_x}\cos\dfrac{\pi y}{b_y}}{\dfrac{D_x}{a_x^4}+\dfrac{2H}{a_x^2 b_y^2}+\dfrac{D_y}{b_y^4}}\right]\left[D_y\frac{\pi^3}{b_y^3}+\frac{C_1}{b_1}\frac{\pi^3}{b_y a_x^2}\right]$$

$$= -0.015\left[0.00415\times 25.49\times 10^6\times\frac{\pi^3}{16^3}\right.$$

$$\left.+\,0.0002744\times 25.49\times 10^6\left(\frac{\pi^3}{16\times 12^2}\right)\right]\left(\sin\frac{\pi x}{a_x}\right)\left(\cos\frac{\pi y}{b_y}\right)$$

$$= -13.387\left(\cos\frac{\pi x}{a_x}\right)\left(\sin\frac{\pi y}{b_y}\right)$$

Using these moment and shear equations, the values of bending moment, twisting moment and shear forces are computed at the various salient points of the grid and tabulated as shown in Table 16.1.

Table 16.1: Moments and shear forces per metre width of grid

Points	x (m)	y (m)	M_x (kN·m)	M_y (kN·m)	M_{xy} (kN·m)	M_{yx} (kN·m)	Q_x (kN)	Q_z (kN)
E	6	8	108	61	0	0	0	0
F	6	12	77	43	0	0	0	9.4
G	6	16	0	0	0	0	0	13.4
H	0	8	0	0	0	0	10.1	0
I	2	8	54	30.5	0	0	8.74	0
J	4	8	94	53	0	0	5.05	0
K	0	12	0	0	3.74	3.74	7.14	0
D	0	16	0	0	5.30	5.30	0	0

vi. *Comparison of moments*:

The moments per metre width computed by the rigorous and approximate methods are shown in Table 16.2.

The approximate method underestimates the bending moments developed in x- and y-directions, to the extent of 17% and 21% respectively. The moments are very much underestimated in the long span direction.

Table 16.2: Comparison of maximum moments in grid

Method	Moments per metre width (kN·m)	
	M_x	M_y
Approximate method (Rankine–Grashoff theory)	90	48
Rigorous analysis (plate theory)	108	61

vii. *Design of reinforcements*:

Maximum working moment M_w = 108 kN·m/m

Moment resisted by central rib in x-direction over 2 m width = (2 × 108) = 216 kN·m

Ultimate moment M_u = (1.5 × 216) = 324 kN·m

Moment capacity of flange section M_{uf} = $0.36 f_{ck} b_f D_f (d - 0.42 D_f)$

$$= (0.36 \times 20 \times 2000 \times 100) \times (550 - 0.42 \times 100)$$

$$= 731 \times 10^6 \text{ N·mm} = 7731 \text{ kN·m}$$

Since $M_u < M_{uf}$ neutral axis falls within the flange

$$M_u = 0.87 f_y A_{st} d \left[1 - \frac{A_{st} f_y}{b d f_{ck}} \right]$$

$$320 \times 10^6 = 0.87 \times 415 \times A_{st} \times 550 \left[1 - \frac{A_{st} \times 415}{2000 \times 550 \times 20} \right]$$

Solving, A_{st} = 1700 mm²

Provide 4 bars of 25 mm diameter (A_{st} = 1964 mm²)

Maximum ultimate shear = (1.5 × 2 × 13.4) = 40.2 kN

$$\therefore \qquad \tau_v = \left(\frac{V_u}{bd} \right) = \left(\frac{40.2 \times 10^3}{200 \times 550} \right) = 0.379 \text{ N/mm}^2$$

Assume two bars to be bent up near supports.

A_{st} at support = 982 mm² (Refer to Table 19 of IS:456-2000)

$$\left[\frac{100 A_{st}}{bd} \right] = \left[\frac{100 \times 982}{200 \times 550} \right] = 0.89 \text{ and } \tau_c = 0.57 \text{ N/mm}^2$$

Since $\tau_v < \tau_c$ provide nominal shear reinforcements using 6 mm diameter two-legged stirrups spacing as

$$S_v = \left(\frac{A_{sv} 0.87 f_y}{0.4b} \right) = \left(\frac{2 \times 28 \times 0.87 \times 415}{0.4 \times 200} \right) = 252 \text{ mm}$$

Provide 6 mm diameter two-legged stirrups at 250 mm c/c at supports and the spacing gradually increased to 400 mm towards the centre of span. Maximum moment in the central rib in the y-direction (long span) is

$$M_u = (2 \times 77 \times 1.5) = 231 \text{ kN·m}$$

Provide 4 bars of 20 mm diameter (A_{st} = 1256 mm²) and 6 mm diameter two-legged stirrups at 250 mm centres near the supports. The beam section at D is checked for the torsion and suitably reinforced. The moments in the slab being small, mesh reinforcement consisting of 6 mm diameter at 200 mm centres provided both ways for positive and negative moments in the slab. The reinforcement details in the typical sections of the grid is shown in Fig. 16.6.

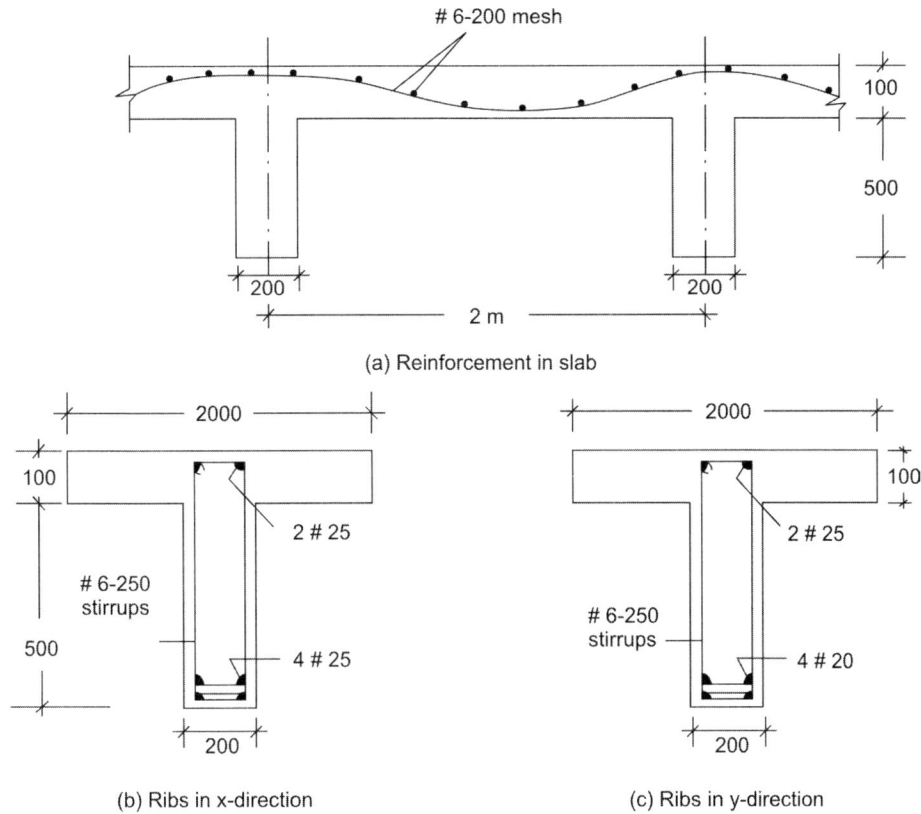

(a) Reinforcement in slab

(b) Ribs in x-direction

(c) Ribs in y-direction

Fig. 16.6: Reinforcement details in grid floor

References

1. Gambhir ML, *Design of Reinforced Concrete Structures*, PHI, 2008, pp. 308–330.

2. Puroshothaman P, *Reinforced Concrete Structural Elements, Behaviour, Analysis and Design*, Tata McGraw-Hill, New Delhi, 1984.

3. Park R and Paulay T, *Reinforced Concrete Structures*, John Wiley & Sons Inc, New York, 1975.

4. Dunham CW, *Advanced Reinforced Concrete*, McGraw-Hill Book Co, New York, 1964.

5. IS:456-2000, Indian Standard Code of Practice for Plain and Reinforced Concrete, Bureau of Indian Standards, New Delhi, 2000.

6. Unnikrishna Pillai S and Menon D, *Reinforced Concrete Design*, 3 edn, Tata McGraw-Hill Education Pvt. Ltd, 2009, pp. 427–430.

7. Timoshenko S and Woinowsky Krieger S, *Theory of Plates and Shells*, McGraw-Hill Book Co, New York, 1959.

8. Halkude SA and Mahamuni SV, Comparison of various methods of analysis of grid floor frames, *International Journal of Engineering Science Inventions*, Vol. 3, No. 2, Feb. 2014, pp. 1–7.

9. Sathawane A and Deotale RS, Analysis and design of flat slab and grid slab and their comparison, *International Journal of Engineering Research and Applications*, Vol. 1, No. 3, 2002, pp. 837–848.

10. Schwetz PF, Gastal FPSL and Silva LCP, Numerical and experimental study of real scale waffle slab, *IBRACON Structures and Materials Journal*, Vol. 2, No. 4, 2009, pp. 380–403.

11. Mallick SK and Bhushan N, Methods of analysis of reinforced concrete grids for roofs and floors, *Indian Concrete Journal*, Vol. 57, No. 9, 1983, pp. 241–246.
12. Chowdhury I and Singh PJ, Analysis and design of waffle slab with different boundary conditions, *Indian Concrete Journal*, Vol. 84, 2010.
13. Galeb AO and Atiyah ZF, Optimum design of reinforced concrete waffle slabs, *International Journal of Civil and Structural Engineering*, Vol. 1, No. 4, 2011.
14. Krishna Raju N, The design of prestressed concrete grid floors, *Indian Concrete Journal*, Vol. 39, No. 1, Jan. 1965, pp. 33–35.

Assignment

1. A two way slab 5 m × 5 m size with ribs at 1 m intervals is to be designed to support a live load of 4 kN/m². Adopting M20 grade concrete and Fe 415 HYSD bars, design a suitable grid floor and sketch details of reinforcements in ribs.

2. A reinforced concrete grid floor of size 9 m × 12 m is required for an assembly hall. Assuming rib spacing of 1.5 m in the short span direction and 2 m in the long span direction, design the grid floor. Adopt M20 grade concrete and Fe 415 HYSD bars. Live load may be assumed as 4 kN/m².

3. An orthotropic reinforced concrete grid 16 m × 20 m is required for the roof of an auditorium. The ribs are spaced at 2 m intervals. Live load on roof is 1.5 kN/m². Adopt M20 grade concrete and Fe 415 HYSD bars. Design suitable reinforcements in the grid beams and sketch the details of reinforcements.

4. A coffered reinforced concrete grid of overall size 15 m × 18 m covers the roof of a conference hall. The ribs are spaced at 1.5 m intervals. The live load on the roof is 2 kN/m². Adopting M25 grade concrete and Fe 415 HYSD bars, design the grid beams and slab, and sketch the details of the reinforcements in the grid floor.

5. It is proposed to cover a marriage hall of size 12 m × 12 m by a waffle slab floor having beams in the orthogonal directions at intervals of 1m. The live load on the roof may assumed to be 2 kN/m². Design a suitable reinforced concrete waffle slab using M25 grade concrete and Fe 500 grade HYSD bars. Sketch the details of the waffle slab showing the details of reinforcements in the ribs and slab.

6. A reinforced concrete grid floor of size 14 m × 16 m is to be covered by a coffered slab roof with beams spaced at intervals of 1.6 m in the perpendicular directions. Assuming the grid to be freely supported at the edges, design the coffered slab using M25 grade concrete and Fe 500 HYSD reinforcements. Sketch the details of reinforcements in the coffered slab.

Review Questions

1. Under what situations would you recommend the use of coffered slab roofs in preference to other traditional types of roofing solutions?

2. What are the structural advantages of grid slab floors?

3. Explain with sketches the different types of reinforced concrete grid floors used in the construction industry.

4. How do you analyse a grid floor for moment and force components developed when it is subjected to imposed loads?

5. List the various approximate and rigorous methods of analysing the force components in a waffle slab mentioning their merits and demerits. How do you check

the ribbed slab for the limit state of collapse according to the principles of limit state design?

6. How do you satisfy the limit state of deflection in a coffered slab subjected to imposed loads? Specify the limits permitted in the Indian standard code.

7. What structural techniques would you adopt to reduce the depth of beams in a waffle slab floor with larger spans in the orthogonal directions?

8. What are the various reinforcements used in the ribs of coffered slab floor?

9. Explain with sketches the reinforcement details in the slab and beams of a grid floor.

Objective Type Questions

1. Reinforced concrete grid floors are generally used in structures like
   a. water tanks
   b. retaining walls
   c. conference halls

2. Waffle slab floors comprise structural elements like
   a. thick uniform slab
   b. beams in the shorter span direction
   c. beams in the orthogonal directions

3. The spacings of ribs in a typical reinforced concrete waffle slab floor is of the order of
   a. 3 to 4 m
   b. 0.5 to 0.75 m
   c. 1 to 2 m

4. In case of diagrid floors, the ribs are arranged
   a. perpendicular to the supporting edges
   b. inclined to the supporting edges
   c. perpendicular along short span and inclined along long span directions

5. In case of a rectangular coffered floor slab supporting uniformly distributed loads, the load shared by the strips per unit width is
   a. the same in both long and short span directions
   b. higher in the long span direction
   c. higher in the short span direction

6. In a simply supported grid slab supporting transverse imposed loads, the maximum bending moments develop at
   a. quarter span points
   b. corners
   c. centre of span points

7. In a coffered floor slab, the maximum shear forces develop at
   a. centre of slab
   b. mid points of shorter supports
   c. mid points of longer supports

8. The maximum torsional moments in a coffered slab develop at
   a. centre of span
   b. mid points of supports
   c. corners

9. The ribs of a waffle slab are generally designed as a
   a. rectangular section
   b. tee beam section
   c. doubly reinforced section

10. In case of coffered slab floors covering larger spans, the depth of the ribs can be reduced by using
    a. doubly reinforced sections
    b. larger width of section
    c. prestressing techniques

17

Vierendeel Girders

17.1 INTRODUCTION

The Belgian professor and innovator Arthur Vierendeel (1852–1940)[1,2] developed a rigid frame with open web girders having rigid joints for the use of industrial structures in the year 1896. Basically a Vierendeel girder comprises a top and bottom chord with vertical members connecting them without any inclined members. During this period, steel was the ruling structural material and concrete was in its infancy. Steel Vierendeel girders[3] were used extensively for industrial, commercial and residential buildings. During the first half of 19th century, reinforced concrete Vierendeel girders more or less replaced the steel structures mainly due to the enhanced durability and significant reduction in maintenance costs in comparison with the steel girders.

The salient feature of the Vierendeel girder being the absence of diagonal members and the integrity and stability of the frame entirely depends upon the rigidity of the joints. The use of Vierendeel girders is very popular in Europe and especially Russia[4,5] and Belgium where pioneering work on the development and use of this type of girders was done. In industrial structures like workshop floors, warehouse sheds and factories where large column free space is required, Vierendeel girders are generally used. Also the free unobstructed space between the top and bottom chords is advantageously used for clerestory lighting in churches. The various types of Vierendeel girders used in practice are compiled in Fig. 17.1.

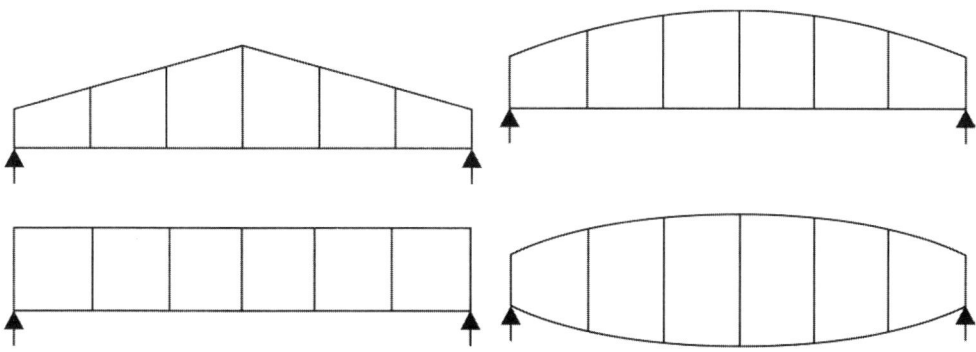

Fig. 17.1: Types of Vierendeel girders

343

17.2 ANALYSIS OF VIERENDEEL GIRDERS

Reinforced concrete Vierendeel girder with top and bottom chords connected by vertical stringers is analysed as a statically indeterminate structure using the classical structural methods[6,7,8] generally used for indeterminate structures. Several methods have been developed for the analysis of Vierendeel girders such as:

a. Statically determinate analysis

b. Naylor's moment distribution

c. Modified moment distribution

d. Computer analysis based on generated slope deflection equations

Two simplified methods of analysing the forces developed in the Vierendeel girder subjected to loads based on the first two methods (a) and (b) are presented in the following subsections.

a. Statically Determinate Analysis

In this method, hinges are assumed at the mid span of the chord and vertical members of the Vierendeel girder as shown in Fig. 17.2.

Let W be the load acting on the joints of the top boom of the Vierendeel girder (loads on roof). The vertical shear forces acting at the mid sections of each of the panels of the Vierendeel girder are determined and equally divided between the top and bottom chords.

Accordingly, the shear force at section XX = $V = W$

Shear force at each of the hinges = $(W/2)$

The bending moment in the chords are obtained by multiplying the shears at the hinge by half the length of the panel. The shear in the verticals is obtained by dividing the moment in the verticals at the joints by half the height. The axial force in the chords

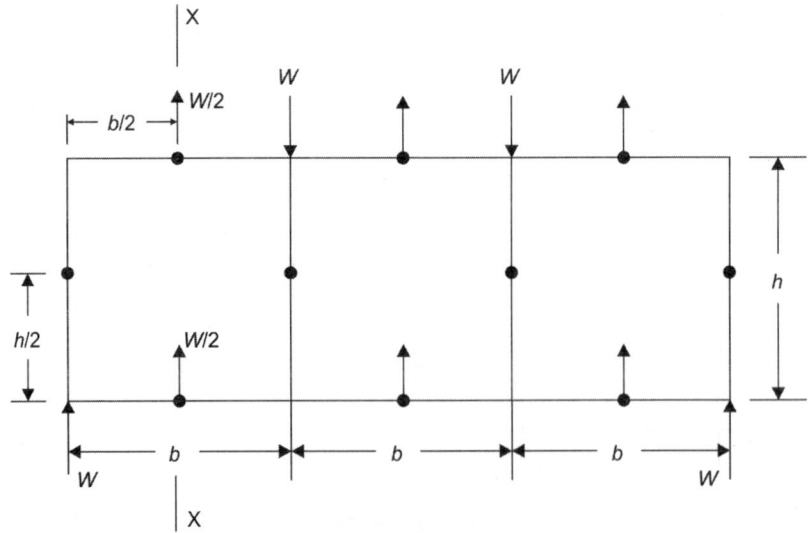

Fig. 17.2: Analysis of Vierendeel girder

is obtained by summing up the horizontal shears. The simplified method of statically determinate analysis is applied to the following numerical example.

b. Naylor's Moment Distribution Method

In case of Vierendeel girders having the same section for the top and bottom chords, Naylor's modified moment distribution procedure can be advantageously used for the evaluation of forces in the members. In the following example, the stiffness values k of the chords and verticals are assumed to be the same. However, in the Naylor's method only half the frame is considered with hinges assumed at mid heights of vertical members.

The following salient features of Naylor's method are noteworthy:

  i. Since only half the length of the vertical members are considered with hinges at mid height, the modified stiffness of verticals are obtained as $6k$, where k is the original stiffness of the vertical members.

 ii. The carry over factor is –1.

iii. The fixed end moments in each panel is obtained from the shears in the panels.

The distribution factors at each of the joints are computed using the modified stiffness values of the members and the moment distribution is carried out to determine the final moments and shear forces.

17.3 ANALYSIS EXAMPLE

A Vierendeel girder spanning over 12 m has six bays of 2 m each. The vertical struts connecting the top and bottom chords are 2 m in height. The girder supports loads of 10 kN at each of the nodes. Analyse the forces in the members of the Vierendeel girder using (a) statically determinate analysis and (b) Naylor's moment distribution procedure. Compare the results obtained by both these methods.

a. Statically determinate analysis:

The Vierendeel girder supporting the load at the nodes is shown in Fig. 17.3. Compute the shear forces in the panels by considering half the frame and assuming hinges at centre of panels.

Shear in panel 1-2 = 25 kN

Shear at hinges of top and bottom chords = (0.5 × 25) = 12.5 kN

Shear in panel 2-3 = (25 – 10) = 15 kN

Shear at hinges = (0.5 × 15) = 7.5 kN

Shear in panel 3-4 = (25 – 10 – 10) = 5 kN

Shear at hinges = (0.5 × 5) = 2.5 kN

The chord moments are obtained by multiplying the pin shears by half the length of the panel.

$$M_{1-2} = M_{8-7} = (12.5 \times 1) = 12.5 \text{ kN·m}$$
$$M_{2-3} = M_{7-6} = (7.5 \times 1) = 7.5 \text{ kN·m}$$
$$M_{3-4} = M_{6-5} = (2.5 \times 1) = 2.5 \text{ kN·m}$$

Considering the equilibrium of joints, we have

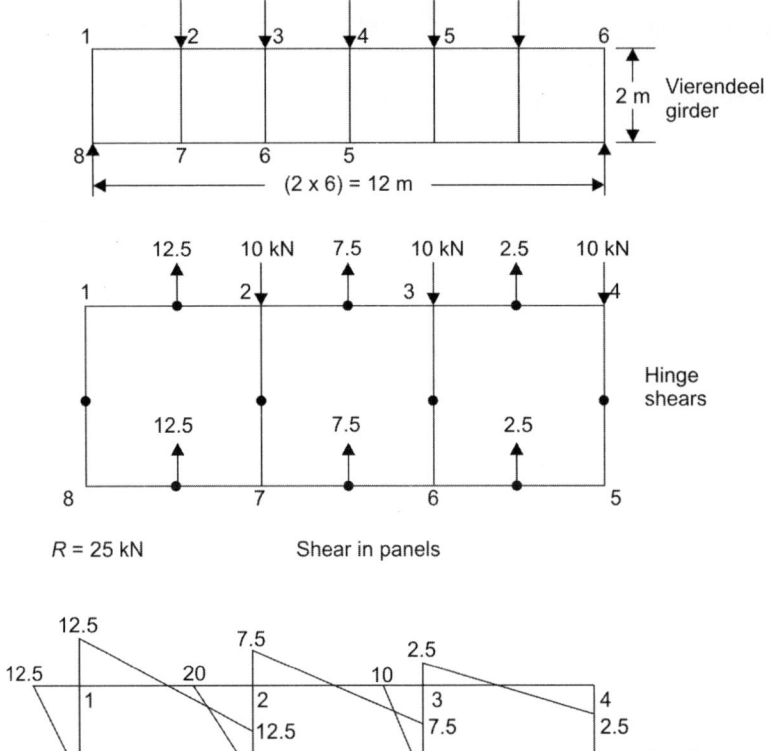

Fig. 17.3: Analysis of forces in Vierendeel girder

$$M_{1\text{-}8} = 12.5 \text{ kN·m}$$
$$M_{2\text{-}7} = (12.5 + 7.5) = 20 \text{ kN·m}$$
$$M_{3\text{-}6} = (7.5 + 2.5) = 10 \text{ kN·m}$$
$$M_{4\text{-}5} = 0$$

The shear forces in the vertical members are obtained by dividing the moment in the verticals at the joints by half their height.

For member 1-8, shear force $V_{1\text{-}8} = \left(\dfrac{12.5 \times 2}{2}\right) = 12.5 \text{ kN}$

For member 2-7, shear force $V_{2\text{-}7} = \left(\dfrac{20 \times 2}{2}\right) = 20 \text{ kN}$

For member 3-6, shear force $V_{3-6} = \left(\dfrac{10 \times 2}{2}\right) = 10$ kN

The bending moment diagram for the Vierendeel girder is shown in Fig. 17.3.

The axial forces in the chords are evaluated by summing up the horizontal shear in the vertical members.

Axial forces in the members are computed as below:

$$H_{1-2} = 12.5 \text{ kN (compression)}$$
$$H_{2-3} = (12.5 + 20) = 32.5 \text{ kN (compression)}$$
$$H_{3-4} = (12.5 + 20 + 10) = 42.5 \text{ kN (compression)}$$

The axial forces in the bottom boom members have the same magnitude as their top counterparts, but they will be in tension.

b. Naylor's moment distribution method:

In Naylor's method, only half the frame is considered with hinges assumed at mid heights of verticals of the Vierendeel girder as shown in Fig. 17.4. The distribution factors at each of the joints are computed using the modified stiffness values of the members and the moment distribution is carried out as shown in Table 17.1. The shears in each panel, fixed end moment and bending moment diagram for the Vierendeel girder is shown in Fig. 17.4.

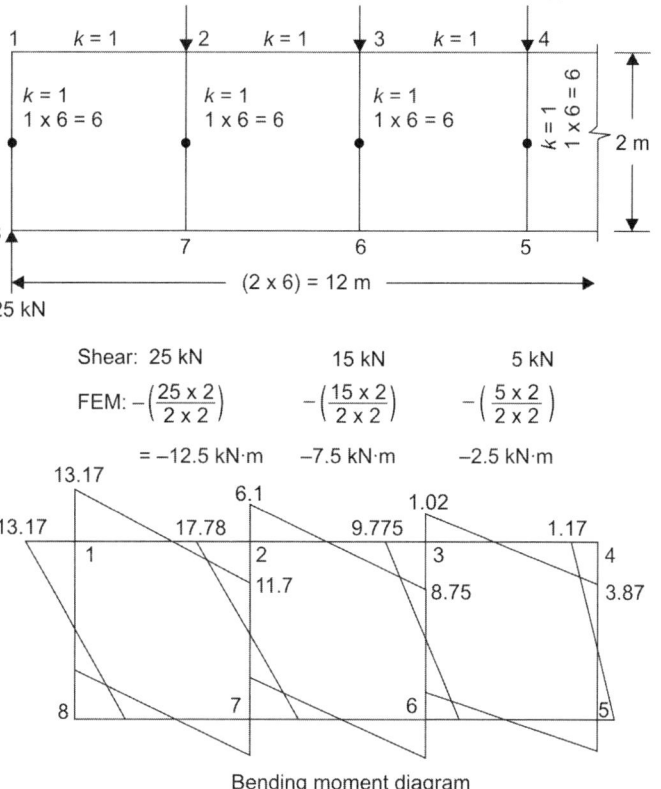

Bending moment diagram

Fig. 17.4: Naylor's moment distribution for Vierendeel girder

Table 17.1: Naylor's moment distribution for Vierendeel girder

DF	$\dfrac{6}{7}$	$\dfrac{1}{7}$	$\dfrac{1}{8}$	$\dfrac{6}{8}$	$\dfrac{1}{8}$	$\dfrac{1}{8}$	$\dfrac{6}{8}$	$\dfrac{1}{8}$	$\dfrac{1}{8}$	$\dfrac{6}{8}$	$\dfrac{1}{8}$
FEM		−12.50	−12.50		−7.50	−7.50		−2.50	−2.50	−	+2.50
Dist	+10.70	+1.80	+2.50	+15.0	+2.50	+1.25	+7.50	+1.25	−	−	−
CO		−2.50	−1.80		−1.25	−2.50			+1.25		
Dist	+2.14	+0.36	+0.38	+2.28	+0.38	+0.31	+1.875	+0.31	+0.15	+0.94	+0.15
		−0.38	−0.36		−0.31	−0.38		−0.15	−0.31		
Dist	+0.33	+0.05	+0.08	+0.50	+0.08	+0.07	+0.40	+0.07	+0.04	+0.23	+0.04
Final moments (kN·m)	+13.17	− 13.17	−11.70	+17.18	−6.10	−8.75	+9.775	−1.02	−3.87	+1.17	+2.69

A comparative analysis of the moments obtained by statically determinate analysis and Naylor's moment distribution procedure indicates that the latter method results in higher moments in end verticals and lower moments in interior vertical members.

17.4 DESIGN OF MEMBERS OF VIERENDEEL GIRDER

The analysis indicates that the top boom members of Vierendeel girder are subjected to axial compression and bending moments while the bottom boom members are subjected to axial tension and bending moment. The vertical members are under axial thrust and bending moment. In addition, all the members are subjected to shear forces to a varying degree.

The top boom and the vertical members are designed for combined bending and compression using interaction curves[9,10] while the bottom boom members are designed for axial tension and associated moment in a manner similar to that of rectangular tank walls. The design of various members of a typical Vierendeel girder[11,12] is illustrated in the following example.

17.5 DESIGN EXAMPLE

A Vierendeel frame spanning over 12 m is made up of six bays of 2 m × 2 m. All the members of the Vierendeel frame have the same length of 2 m. The frame supports concentrated loads of 10 kN each at the top boom junction node points. The analysis indicates the magnitude of maximum forces, moments and shear forces developed in the members of the frame as shown in Table 17.2. Adopting M20 grade concrete and Fe 415 HYSD reinforcements, design suitable size and reinforcements in the various members.

Table 17.2: Design force components in Vierendeel frame

Member type	Maximum design working force components			
	Axial compression (kN)	Axial tension (kN)	Bending moment (kN·m)	Shear force (kN)
Top boom	12.5	–	12.5	12.5
Bottom boom	–	12.5	12.5	12.5
Verticals	10.0	–	20.0	20.0

i. *Top boom member*:

The top chord member is subjected to combined compression and bending moment. Restricting the ratio of effective length to least lateral dimension to 12, we have

$$\left(\frac{L_e}{D}\right) = 12 \quad \Rightarrow \quad D = \left(\frac{L_e}{12}\right) = \left(\frac{2000}{12}\right) = 166 \text{ mm}$$

Adopt a section 200 mm × 200 mm with effective depth of 170 mm.

Using a load factor of 1.5, we have the ultimate moment, thrust and shear as

$$M_u = (12.5 \times 1.5) = 18.75 \text{ kN·m}$$
$$P_u = (12.5 \times 1.5) = 18.75 \text{ kN}$$
$$V_u = (12.5 \times 1.5) = 18.75 \text{ kN}$$

$$\left[\frac{M_u}{f_{ck}bD^2}\right] = \left[\frac{18.75 \times 10^6}{20 \times 200 \times 200^2}\right] = 0.117$$

$$\left[\frac{P_u}{f_{ck}bD}\right] = \left[\frac{18.75 \times 10^3}{20 \times 200 \times 200}\right] = 0.023$$

Refer to interaction curves in Chart 33 of SP:16 (Fig. 17.5) and use the parameters for grade of steel used as $f_y = 415 \text{ N/mm}^2$ and the ratio as

$$\left(\frac{d'}{D}\right) = \left(\frac{30}{200}\right) = 0.15$$

Readout from the chart, $\left(\dfrac{p}{f_{ck}}\right) = 0.09$

$$p = (0.09 \times 20) = 1.8$$

$$A_s = \left(\frac{pbD}{100}\right) = \left(\frac{1.8 \times 200 \times 200}{100}\right) = 720 \text{ mm}^2$$

Using 4 bars of 16 mm diameter, $A_s = 804 \text{ mm}^2$

$$\tau_v = \left(\frac{V_u}{bd}\right) = \left(\frac{12.5 \times 10^3}{200 \times 170}\right) = 0.36 \text{ N/mm}^2$$

$$\left(\frac{100 A_s}{bd}\right) = \left(\frac{100 \times 402}{200 \times 170}\right) = 1.18$$

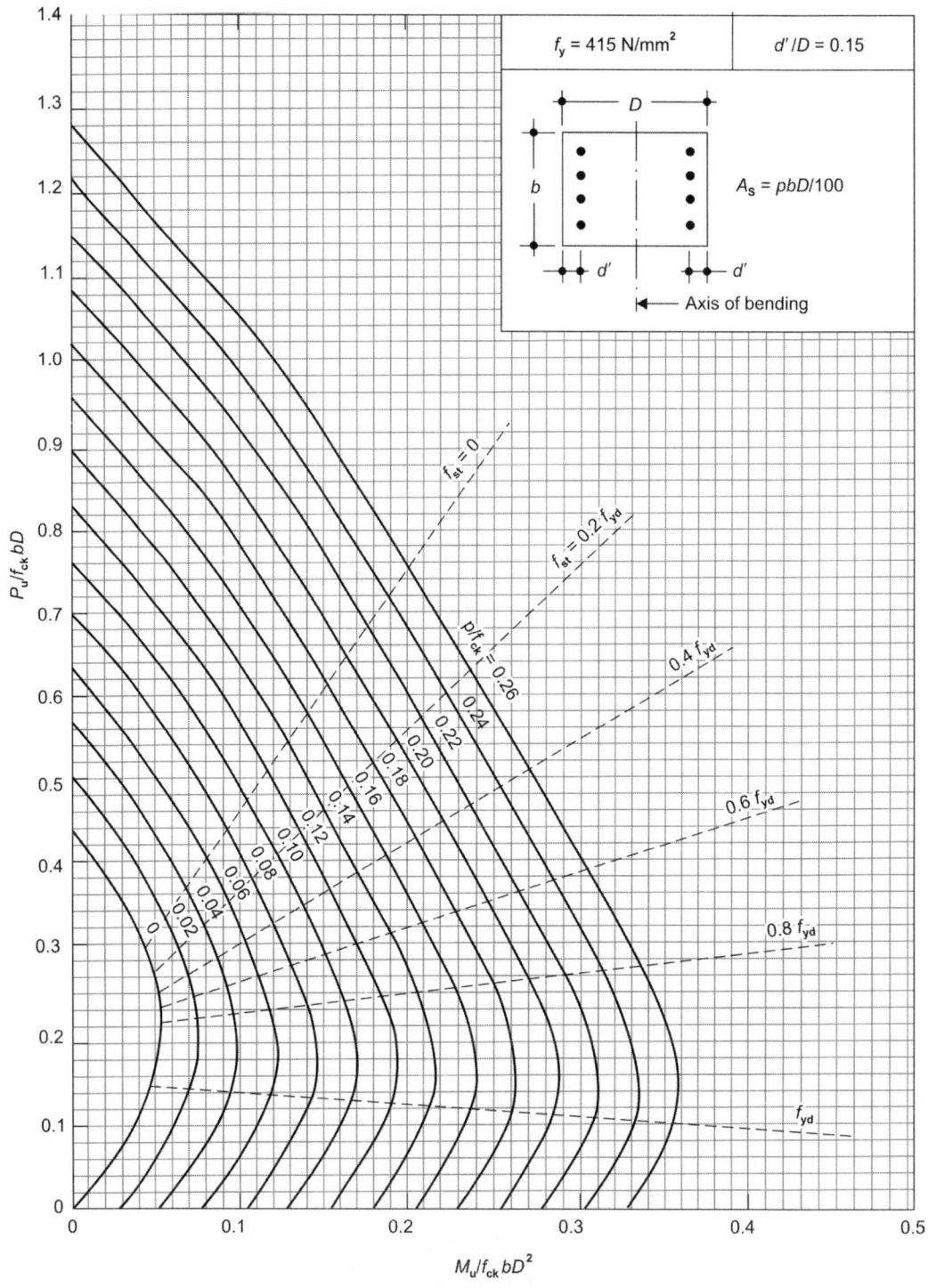

Fig. 17.5: Compression with bending rectangular section reinforcement distributed equally on two sides (SP:16, Chart 33)

From Table 19 of IS:456-2000, readout the permissible shear stress as $\tau_c = 0.65 \text{ N/mm}^2$.
Since $\tau_v < \tau_c$, nominal shear reinforcements are provided.
Using 6 mm diameter, two-legged stirrups, spacing is given by

$$S_v = \left(\frac{A_{sv} \, 0.87 \, f_y}{0.4b}\right) = \left(\frac{2 \times 28 \times 0.87 \times 415}{0.4 \times 200}\right) = 252 \text{ mm}$$

The spacing should not be greater than $0.75d = (0.75 \times 170) = 127.5$ mm
Adopt 6 mm two-legged stirrups at 120 mm centres.

ii. *Bottom boom member*:

The bottom boom member is subjected to axial tension and bending moment.

$$M_u = (12.5 \times 1.5) = 18.75 \text{ kN·m}$$
$$P_u = (12.5 \times 1.5) = 18.75 \text{ kN}$$
$$V_u = (12.5 \times 1.5) = 18.75 \text{ kN}$$

For the section adopted being 200 mm × 200 mm, $d' = 30$ mm and $D = b = 200$ mm

The ratio $\left(\dfrac{d'}{D}\right) = \left(\dfrac{30}{200}\right) = 0.15$

Referring to Chart 80 of SP:16 (Fig. 17.6) with reinforcement distributed equally on four sides, we have the parameters

$$\left[\frac{M_u}{f_{ck}bD^2}\right] = \left[\frac{18.75 \times 10^6}{20 \times 200 \times 200^2}\right] = 0.117$$

$$\left[\frac{P_u}{f_{ck}bD}\right] = \left[\frac{18.75 \times 10^3}{20 \times 200 \times 200}\right] = 0.023$$

Corresponding to the above two values and extrapolating, we have

$$\left(\frac{p}{f_{ck}}\right) = 0.11$$

$$p = (0.11 \times 20) = 2.2$$

$$A_s = \left(\frac{pbD}{100}\right) = \left(\frac{2.2 \times 200 \times 200}{100}\right) = 880 \text{ mm}^2$$

Alternatively using the working stress method, effective depth $d = 170$ mm
Distance of tensile steel from centre of section $x = (100 - 30) = 70$ mm

$$A_{st} = \left[\frac{M - Px}{\sigma_{st} jd}\right] + \left[\frac{P}{\sigma_{st}}\right]$$

$$= \left[\frac{12.5 \times 10^6 - (12.5 \times 10^3 \times 70)}{230 \times 0.9 \times 170}\right] + \left[\frac{12.5 \times 10^3}{230}\right] = 385 \text{ mm}^2$$

Using the higher value of steel, $A_{st} = 880 \text{ mm}^2$, provide 8 bars of 16 mm diameter equally distributed on all sides ($A_{st} = 1005 \text{ mm}^2$). Also adopt 6 mm two-legged stirrups at 120 mm centres.

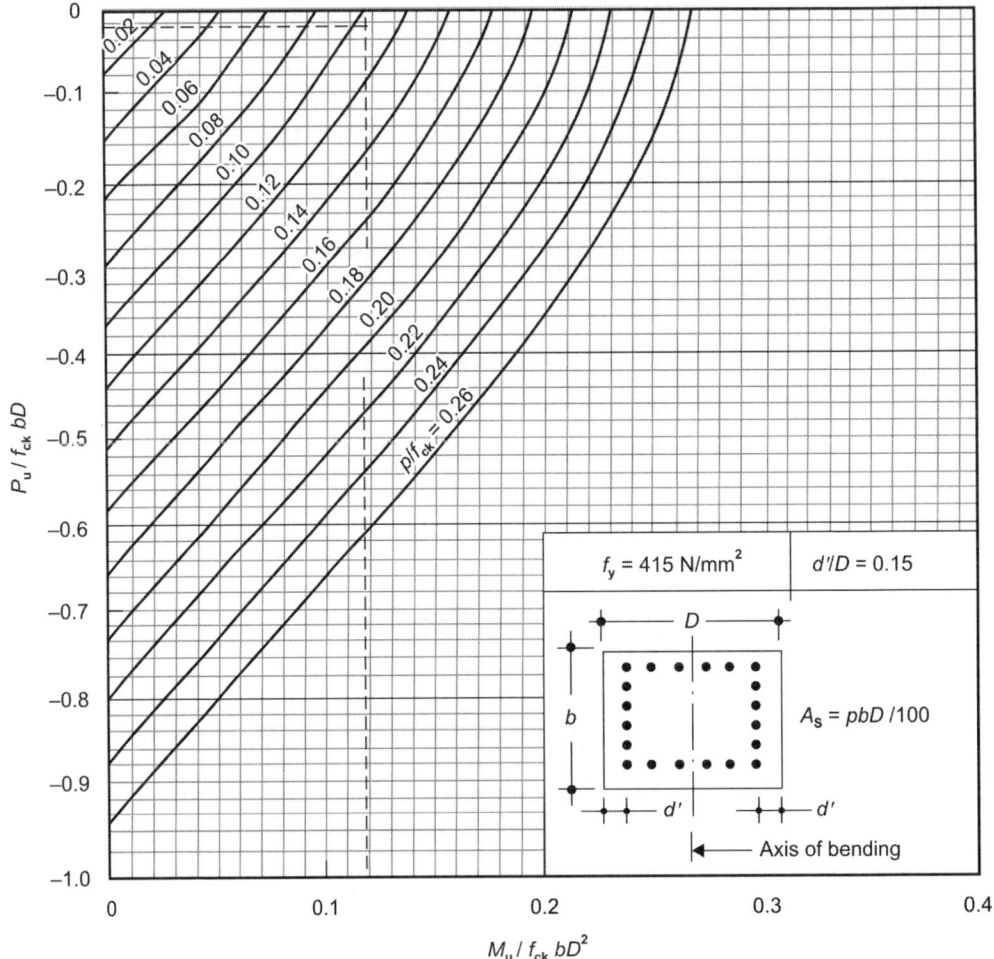

Fig. 17.6: Tension with bending rectangular section reinforcement distributed equally on four sides
(SP:16, Chart 80)

iii. *Vertical members*

$$M_u = (20 \times 1.5) = 30 \text{ kN·m}$$
$$P_u = (10 \times 1.5) = 15 \text{ kN}$$
$$V_u = (20 \times 1.5) = 30 \text{ kN}$$

Using a section 200 mm × 200 mm, compute the parameters as

$$\left[\frac{M_u}{f_{ck}bD^2}\right] = \left[\frac{20 \times 10^6}{20 \times 200 \times 200^2}\right] = 0.125$$

$$\left[\frac{P_u}{f_{ck}bD}\right] = \left[\frac{15 \times 10^3}{20 \times 200 \times 200}\right] = 0.018$$

Compute the ratio $\left(\dfrac{d'}{D}\right) = \left(\dfrac{30}{200}\right) = 0.15$

From Chart 33 of SP:16 (Fig. 17.5), extrapolate $\left(\dfrac{p}{f_{ck}}\right) = 0.10$

$$p = (0.10 \times 20) = 2.0$$

$$A_s = \left(\dfrac{pbD}{100}\right) = \left(\dfrac{2 \times 200 \times 200}{100}\right) = 800 \text{ mm}^2$$

Provide 4 bars of 16 mm diameter ($A_{st} = 804$ mm^2) with nominal shear reinforcements of 6 mm two-legged stirrups at 120 mm centres.

References

1. Wickershcimer DJ, The Vierendeel, *Journal of the Society of Architectural Historians*, 1976.
2. Verswijver K, Meyer RD, Denys R, The Writings of Belgian Engineer Arthur Vierendeel (1852–1940), Homo Universales or Contemporary propagandist. Proceedings of the Third International Congress on Construction Industry, University of Ghent 2009, pp. 1463–1470.
3. Schierle, Vierendeel Structures–Tata Steel Construction Reports-Wikipedia.
4. Sigalove E and Strongin S, *Reinforced Concrete*, Foreign Languages Publishing House, Moscow, 1962.
5. Murashev V, Sigalove E, Baikov V, *Design of Reinforced Concrete Structures*, Mir Publishers, Moscow, 1968.
6. Reddy CS, *Basic Structural Analysis*, Tata McGraw-Hill, New Delhi, 1981, p. 435.
7. Junarkar SB and Shah HJ, *Mechanics of Structures*, Vol. II, 13 edn, Charotar Publishing House, Anand, 1989, p. 724.
8. Wang CK, *Matrix Methods of Structural Analysis*, 2 edn, International Scranton, Pennsylvania, 1970.
9. SP:16-1980, Design Aids for Reinforced Concrete to IS:456, Bureau of Indian Standards, New Delhi, 1980.
10. Ghanekar VK, Chandra R and Sarkar S, Handbook for Ultimate Strength Design of Reinforced Concrete members, SERC Roorkee, 1970.
11. Johnson RP, *Structural Concrete*, Tata McGraw-Hill, New Delhi, 1967.
12. Dunham CW, *Advanced Reinforced Concrete*, McGraw-Hill, New York, 1964.

Assignment

1. A Vierendeel girder spanning over 9 m has three bays of 3 m each. The verticals have the same dimensions as the top and bottom boom members. The girder supports concentrated loads of 20 kN at the interior node points of the top boom. Assuming constant stiffness for all the members, evaluate the force components developed in the various members of the frame using the following methods:

   a. Statically determinate analysis

   b. Naylor's moment distribution

2. A Vierendeel girder having 10 bays of 2 m each has verticals of the same length as the other members. The frame supports vertical loads of 15 kN at each of the top

node points of the top boom. Design the typical members of the frame assuming M25 grade concrete and Fe 500 grade HYSD bars.

3. Design the typical members of a Vierendeel girder spanning over 20 m and having 10 bays of 2 m each. All the members of the frame have the same length and cross sectional dimensions.The girder supports point loads of 10 kN each at the node points of the top beam. Assume M30 grade concrete and Fe 500 grade HYSD bars.

4. The top chord member of a Vierendeel girder is subjected to a design axial compression of 20 kN and bending moment of 20 kN·m. The bottom chord member is also subjected to a design axial tensile force and moment of magnitude 18 kN and 18 kN·m respectively. Using M25 grade concrete and Fe 500 grade HYSD bars, design the top and bottom chord using a load factor of 1.5 for the working forces.

5. A reinforced concrete Vierendeel girder is to be designed for a workshop floor. The girder of spanning over 18 m is divided into bays of 2 m each and the depth of the frame is also 2 m. The loads at the joints may be assumed as 12 kN. Using M25 grade concrete and Fe 415 HYSD reinforcements, design the typical members of the frame and sketch the details.

6. A Vierendeel frame of 15 m span is required for an industrial workshop. The frame has five bays of 3 m each. The depth of the frame is 2 m. Design typical members of the frame using M25 grade concrete and Fe 500 HYSD reinforcements. Sketch the typical details of steel in the top and bottom boom and vertical members of the frame.

Review Questions

1. Specify the situations under which you will recommend the use of Vierendeel girders for construction.

2. Specify the difference between the structural members of a Vierendeel girder and truss.

3. What methods would you adopt for the analysis of forces in Vierendeel girders?

4. What are the economical span ranges of reinforced concrete Vierendeel girders?

5. Briefly outline the method of statically determinate analysis used for computing the forces in Vierendeel girders subjected to external loads.

6. Briefly explain the Naylor's moment distribution method used for analysing the forces in Vierendeel girders.

7. Compare the methods used for the analysis of Vierendeel girders and comment on the various force components resulting from these methods.

8. Explain the types of force components developed in the various members of Vierendeel girders subjected to loads at the node points.

9. How would you design the various members of Vierendeel girders?

10. Explain with typical sketches the reinforcements used in the various structural members of Vierendeel trusses.

Objective Type Questions

1. In the case of Vierendeel girders, there is no room for
   a. tension members
   b. compression members
   c. diagonal members

2. The joints in a Vierendeel girder are generally treated as
   a. rigid
   b. hinged
   c. moveable

3. Vierendeel girders find extensive applications in
   a. domestic structures
   b. liquid retaining structures
   c. industrial structures

4. Vierendeel girder can be classified as a
   a. determinate structure
   b. indeterminate structure
   c. complex structure

5. The top chord members of a Vierendeel girder are generally subjected to
   a. compression
   b. tension
   c. combined compression and bending

6. The bottom chord members of a Vierendeel girder are generally subjected to
   a. tension
   b. compression
   c. combined tension and bending

7. The vertical members of a Vierendeel girder are normally subjected to
   a. compression, bending and shear force
   b. tension
   c. torsion

8. In the design of structural members of a Vierendeel girder, it is advantageous to select
   a. variable size depending upon the nature of forces
   b. depth greater than width
   c. uniform size for all members

9. The philosophy of structural design of Vierendeel girders is simplified by using
   a. working stress method
   b. ultimate load method
   c. interaction charts

10. Vierendeel girders are generally reinforced with
    a. compression steel only
    b. tension steel only
    c. symmetrically reinforced with compression, tension and shear reinforcements

18

Trusses

18.1 GENERAL FEATURES

Reinforced concrete trusses were first used for covering the large span workshop floors in Russia during the middle of 19th century. In comparison with solid panel structures like grids, flat slabs and shells, trusses are skeletal structures and offer several advantages over the traditional types of reinforced concrete roofing solutions. Roof systems used for industrial structures in the coastal regions are normally subjected to adverse environmental conditions. In such situations reinforced concrete truss systems are preferred to steel trusses due to their superior durability characteristics. Many countries in Europe[1,2] and especially USSR have developed and extensively used the reinforced concrete trussed roofs of different configuration to suit the spans varying from 10 m to 30 m. The most favourable configuration of the top chord is obtained in the bow string type truss having top chord members of varying slopes from the centre of span and towards the supports. Various types of truss configurations used for spans in the range of 18 m to 30 m is shown in Fig. 18.1.

18.2 DIMENSIONS OF TRUSSES

The height of a reinforced concrete truss at mid span is in the range of 1/7 to 1/9 of its span length. The span of the trusses ranges from 18 m to 30 m and more. The width of various compression and tension members is kept constant at 200 mm to 350 mm depending upon the span of the truss. The depth of top boom members which are in compression generally is in the range of 200 mm to 300 mm. The bottom tie member should be of sufficient size to house the pretensioned wires or post tensioned cables. The depth is around 200 mm for spans of 15 m increasing to 300 mm for spans of 30 m. The depth of diagonal web members which are in compression and tension generally vary in the narrow range from 100 mm to 150 mm.

The use of concrete trusses with modular coordination for spans of 6, 9, 12, 15, 18, 24, 30 and 36 m and with a base module of 3 m is most common for industrial buildings of East European countries such as USSR, Poland, Yugoslavia and East Germany. Branko Zezelj has reported the construction of reinforced concrete trusses with prestressed tie members for spans up to 60 m in Yugoslavia.

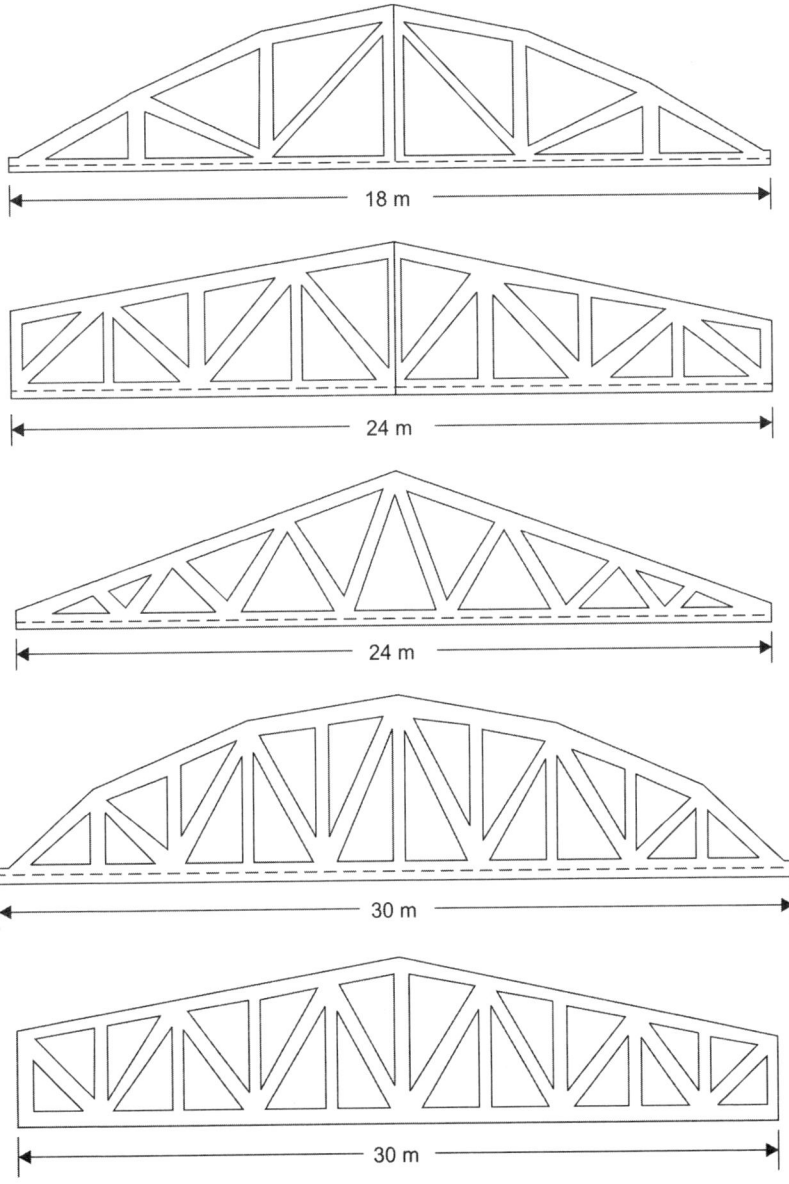

Fig. 18.1: Types of trusses

18.3 ▌ MATERIAL REQUIREMENTS

Concrete used in trusses is normally of grade ranging from M30 to M50 which can be considered as high strength concrete. The reinforcement consists of the mild steel or deformed tor steel together with high tensile steel wires used in the bottom tie member. The material requirement per truss varies with span and spacing of trusses. Table 18.1 shows the requirements of materials for precast reinforced concrete trusses designed for roof loads of 3.5 kN/m^2 to 5.3 kN/m^2.

Table 18.1: Material requirements for precast roof trusses

Types of truss	Weight of truss (kN)	Grade of concrete (N/mm²)	Material requirement per truss Steel (kg)	Concrete (m³)
1. Truss spacing = 6 m Prestressed bow string truss with cable reinforcement span				
a. 18 m	4.3–4.8	30	338–433	1.72–1.9
b. 24 m	8.8–10.0	30–40	621–689	3.50–4.0
c. 30 m	15.2–17.0	30–40	1041–1219	6.08–6.8
Polygonal, built up from blocks with prestressed bottom chord having wire cable reinforcement span				
a. 18 m	6.58	40	514–529	2.63
b. 24 m	9.60	40	744–765	3.85
c. 30 m	13.20	40	1135–1186	5.28
2. Truss spacing = 12 m Prestressed bow string of linear element with wire reinforcement span				
a. 18 m	7.6–9.1	30–40	491–759	3.06–3.63
b. 24 m	14.9–17.4	30–50	1018–1367	5.95–6.96
c. 30 m	25.5–29.8	30–50	1422–2213	10.20–11.90
Prestressed bow string with bar reinforcement span				
a. 18 m	7.6–9.1	30–40	563–962	3.06–3.63
b. 24 m	14.9–17.4	30–50	1238–1822	5.95–6.96
c. 30 m	25.5–29.8	30–50	1778–2981	10.20–11.90

18.4 CONSTRUCTIONAL FEATURES

The chords and struts of trusses are designed to have the same width for convenience in fabricating in a horizontal position. If precast roof slabs are used for roof covering, the upper chord panels are made equal to the width of precast slabs which is usually about 3 m. The lower tension chord is prestressed with the use of bunched high strength wires or cables house in preformed ducts. For spans in the range of 18 m to 24 m, the trusses are made in one piece but when spans run from 24 m to 30 m, they are made in two pieces with the joint in mid span.

Polygonal trusses with inclined top chords are generally made of 6 m blocks are half trusses with 3 m panels. Due to higher tensions developed in the diagonal members of large span trusses, prestressing them becomes inevitable. In general, polygonal trusses are less economical than the bow type with regard to material and labour costs. At the ends of trusses near the supports, 10 mm to 12 mm thick steel bearing plates are anchored and embedded while casting and these serve as bearing pads for fixing of trusses to the columns.

The use of reinforced concrete trusses with modular coordination for spans of 6, 9, 12, 15, 18, 24, 30 and 36 m having a base module of 3 m is most common for industrial buildings of East European countries such as USSR[3], Poland, Slovakia and other erstwhile communist countries. Branko Zezelj[4] has reported the construction of a reinforced concrete truss with prestressed tie for spans in the range of 60 m in erstwhile Yugoslavia.

18.5 ANALYSIS OF FORCES IN TRUSSES

The analysis of force components in reinforced concrete trusses subjected to external loads is more or less similar to the procedure adopted for steel trusses. It is well established that the assumption of hinged joints holds good for concrete trusses also. The rigidity of the joints does not significantly affect the forces developed in the members of the truss. Consequently, the analysis is generally carried out on the assumption of hinged joints.

The trusses are analysed for dead, wind and snow loads applied to the joints of the top chord as specified in the Indian standard codes[5,6]. The loads due to the suspended mechanical handling is applied at the panel joints of the bottom chord. When loads are applied to the chords of a truss between the panel points, the bending moments developed are determined by assuming the chord as a continuous beam with spans equal to the distance between the joints. In the design of trusses, the forces developed during fabrication and erection should also be analysed along with that of the initial stresses developed due to prestressing of the bottom chord member.

The bottom chord members of the trusses are generally subjected to tension and in the case of larger span ranges, the large magnitude of tension developed is generally resisted by concentric prestressing of the tie member using a high tensile cable. The reader may refer to a separate monograph by the author[7] for limit state design[8,9] of concrete members subjected to axial tension.

If Δ = contraction of the bottom chord due to precompression

P = prestressing force

A = cross-sectional area of bottom chord

L = length of the bottom chord

E_c = modulus of elasticity of concrete

Then, we have the relation

$$\Delta = \left(\frac{PL}{AE_c} \right)$$

The displacements at the ends of truss members in a direction perpendicular to their longitudinal axis caused by the elastic contraction, is determined with the aid of Williot Mohr diagram.

If δ = displacement at the ends of members

M = bending moment developed in the members

then the moments in the members are obtained by the relation,

$$M = \left(\frac{6EI\delta}{L^2} \right)$$

In case of large span trusses with larger magnitudes of prestressing force in the bottom chord, the secondary moments developed in various members due to contraction of the bottom chord should be investigated and considered in the design of the members.

18.6 DESIGN EXAMPLE

A reinforced concrete truss is to be designed for a workshop floor to suit the following data:

Effective span of the truss between the bearings = 25 m

Spacings of trusses = 5 m

Central rise of truss = 4.13 m

The fink type truss shown in Fig. 18.2 supports reinforced concrete purlins at intervals of 1.35 m and the roof is covered by coated metallic sheets. The bottom chord member is to be prestressed. The analysis of force components of the truss due to dead, live and wind loads are compiled in Table 18.2.

Design suitable reinforcements in the truss members. Adopt M35 grade concrete and Fe 415 grade HYSD bars as reinforcements. Freyssinet system high tensile cables comprising 12 wires of 7 mm diameter are available for prestressing the tie member. The design should conform to the specifications of the Indian standard codes, IS:456-2000[10] and IS:1343-2012[11]. Sketch the details of reinforcements in the members of the truss.

1. Design of compression members

    i. *Member AB*:

      $p_u = (1.5 \times 395) = 592.5$ kN $b = 250$ mm

      $M_u = (1.5 \times 4.3) = 6.45$ kN·m $D = 200$ mm

      $L = 2.11$ m $f_{ck} = 35$ N/mm^2

      $f_y = 415$ N/mm^2

      Effective length $L_e = (0.65 \times 2.11) = 1.37$ m

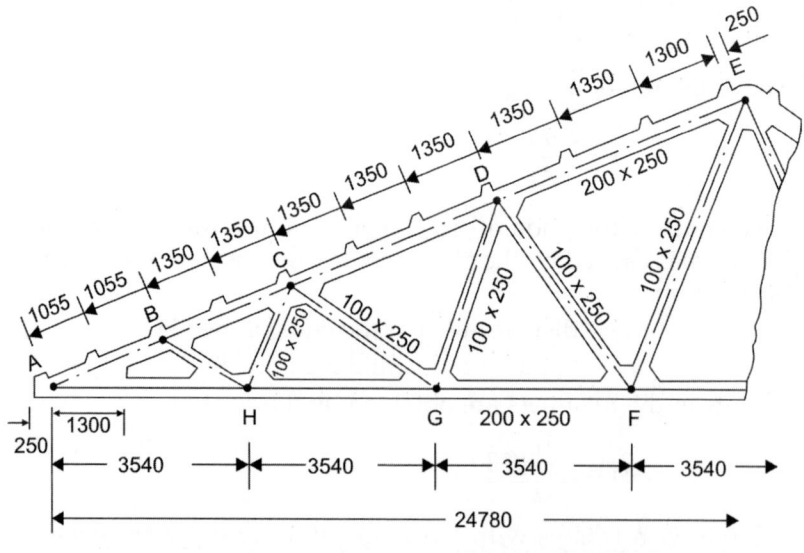

Fig. 18.2: Precast reinforced concrete truss for 25 m span

Table 18.2: Forces in truss members

Member	Cross-sectional dimensions (mm × mm)	Direct force Compression (kN)	Tension (kN)	Bending moment (kN·m)
AB	200 × 250	395	–	4.3
BC	200 × 250	364	–	12.0
CD	200 × 250	297	–	16.0
DE	200 × 250	236	–	16.0
BH	100 × 250	36	–	–
CG	100 × 250	62	–	–
DF	100 × 250	72	–	–
AH	200 × 250	–	377	4.0
HG	200 × 250	–	335	–
GF	200 × 250	–	268	–
FG	200 × 250	–	225	–
CH	100 × 250	–	18	–
DG	100 × 250	–	35	–
EF	100 × 250	–	59	–

$$\left(\frac{L_e}{D}\right) = \left(\frac{1.37}{0.2}\right) = 6.85 < 12$$

$$\left[\frac{P_u}{f_{ck}bD}\right] = \left[\frac{592.5 \times 10^3}{35 \times 250 \times 200}\right] = 0.338$$

$$\left[\frac{M_u}{f_{ck}bD^2}\right] = \left[\frac{6.45 \times 10^6}{35 \times 250 \times 200^2}\right] = 0.023$$

Adopting cover of 40 mm, $\left(\frac{d'}{D}\right) = \left(\frac{40}{200}\right) = 0.20$

Referring to Chart 34 of SP:16[12] (Fig. 18.3), readout the ratio $\left(\frac{p}{f_{ck}}\right) = 0$

Hence, provide minimum reinforcement of 0.8% of the cross-section

$$A_s = \left(\frac{0.8 \times 250 \times 200}{100}\right) = 400 \text{ mm}^2$$

Provide 4 bars of 12 mm diameter (A_s = 452 mm²) and 6 mm diameter ties at 120 mm centres.

ii. *Member BC:*

Length $L = 2.70$ m

Effective length $L_e = (0.65 \times 2.70) = 1.755$ m

Ultimate load $p_u = (1.5 \times 364) = 546$ kN

Ultimate moment $M_u = (1.5 \times 12) = 18.0$ kN·m

$$\left(\frac{L_e}{D}\right) = \left(\frac{1.755}{0.2}\right) = 8.77 < 12$$

$$\left[\frac{p_u}{f_{ck}bD}\right] = \left[\frac{546 \times 10^3}{35 \times 250 \times 200}\right] = 0.312$$

$$\left[\frac{M_u}{f_{ck}bD^2}\right] = \left[\frac{18 \times 10^6}{35 \times 250 \times 200^2}\right] = 0.051$$

Referring to Chart 34 of SP:16 (Fig. 18.3), readout $\left(\dfrac{p}{f_{ck}}\right) = 0.01$

∴ $p = (35 \times 0.01) = 0.35$

$$A_s = \left(\frac{0.35 \times 250 \times 200}{100}\right) = 175 \text{ mm}^2 < A_{smin}$$

Provide minimum reinforcement of 0.8% of cross-section

$$A_s = \left(\frac{0.8 \times 250 \times 200}{100}\right) = 400 \text{ mm}^2$$

Provide 4 bars of 12 mm diameter ($A_s = 452$ mm^2) and 6 mm diameter ties at 120 mm centres.

iii. *Member CD:*

Length $L = 4.05$ m

Effective length $L_e = (0.65 \times 4.05) = 2.632$ m

Ultimate load $p_u = (1.5 \times 297) = 445.5$ kN

Ultimate moment $M_u = (1.5 \times 16) = 24.0$ kN·m

$$\left(\frac{L_e}{D}\right) = \left(\frac{2.632}{0.2}\right) = 13.16 > 12$$

Hence, the member is to be designed as a long column and slenderness effects have to be considered in the design.

The additional moments to be considered in design is computed as,

$$M_{ux} = \left(\frac{p_u D}{2000}\right)\left(\frac{L_e}{D}\right)^2 = \left(\frac{445.5 \times 0.2}{2000}\right) \times (13.16)^2 = 7.71 \text{ kN·m}$$

∴   Total moment $M_u = (24 + 7.71) = 31.71$ kN.m and $p_u = 445.5$ kN

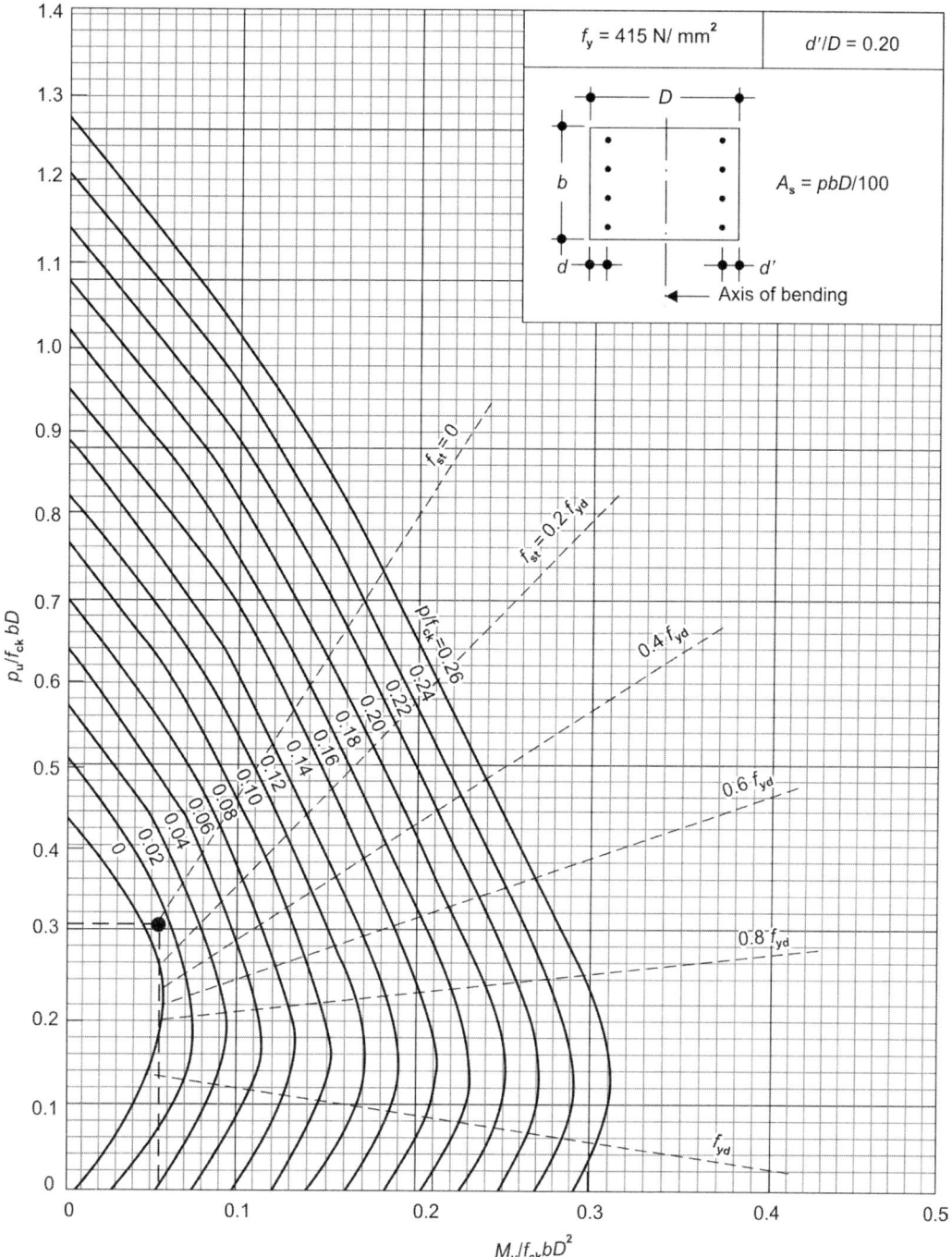

Fig. 18.3: Compression with bending rectangular section reinforcement distributed equally on two sides (SP:16, Chart 34)

$$\left[\frac{p_u}{f_{ck}bD^2}\right] = \left[\frac{445.5 \times 10^3}{35 \times 250 \times 200}\right] = 0.254$$

$$\left[\frac{M_u}{f_{ck}bD^2}\right] = \left[\frac{31.71 \times 10^6}{35 \times 250 \times 200^2}\right] = 0.09$$

Referring to Chart 46 of SP:16[12] (Fig. 18.4), readout $\left(\dfrac{p}{f_{ck}}\right) = 0.06$

$\therefore \quad p = (35 \times 0.06) = 2.10$

$$A_s = \left(\frac{2.10 \times 250 \times 200}{100}\right) = 1050 \text{ mm}^2$$

Provide 4 bars of 16 mm and 4 bars of 10 mm diameter ($A_s = 1120$ mm^2) and 6 mm diameter ties at 120 mm centres.

iv. *Member DE*:

Length L = 4.00 m

Effective length L_e = (0.65 × 4.00) = 2.6 m

Ultimate load p_u = (1.5 × 236) = 354 kN

Ultimate moment M_u = (1.5 × 16) = 24.0 kN·m

$$\left(\frac{L_e}{D}\right) = \left(\frac{2.6}{0.2}\right) = 13 > 12$$

Hence, the member is to be designed as a long column and slenderness effects have to be considered in the design.

The additional moments to be considered in design is computed as,

$$M_{ux} = \left(\frac{p_u D}{2000}\right)\left(\frac{L_e}{D}\right)^2 = \left(\frac{354 \times 0.2}{2000}\right) \times (13)^2 = 5.98 \text{ kN·m}$$

$\therefore \quad$ Total moment M_u = (24 + 5.98) = 29.98 kN·m and p_u = 445.5 kN

$$\left[\frac{p_u}{f_{ck}bD}\right] = \left[\frac{354 \times 10^3}{35 \times 250 \times 200}\right] = 0.202$$

$$\left[\frac{M_u}{f_{ck}bD^2}\right] = \left[\frac{29.98 \times 10^6}{35 \times 250 \times 200^2}\right] = 0.085$$

Referring to Chart 46 of SP: 16 (Fig. 18.4), readout $\left(\dfrac{p}{f_{ck}}\right) = 0.055$

$\therefore \quad p = (35 \times 0.055) = 1.925$

$$A_s = \left(\frac{1.925 \times 250 \times 200}{100}\right) = 963 \text{ mm}^2$$

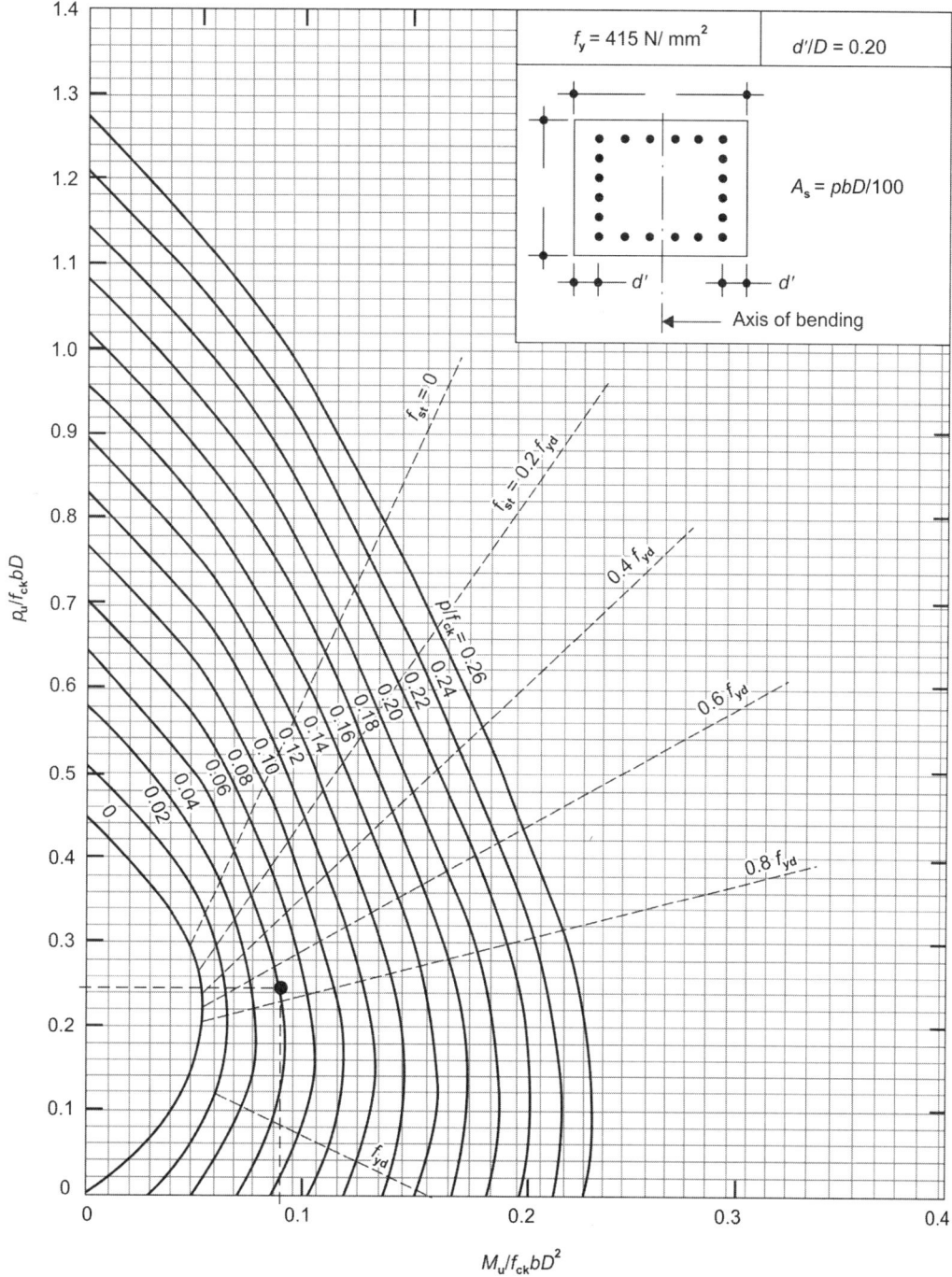

Fig. 18.4: Compression with bending rectangular section reinforcement distributed equally on four sides (SP:16, Chart 46)

Provide 4 bars of 16 mm and 4 bars of 10 mm diameter ($A_s = 1120$ mm^2) and 6 mm diameter ties at 120 mm centres.

v. *Member DF*:

$p_u = (1.5 \times 72) = 108$ kN

Section adopted is (100×250) mm

$$\left(\frac{L_e}{D}\right) = \left(\frac{0.65 \times 4.1}{0.1}\right) = 26.65 > 12$$

Additional moments to be considered is

$$M_{ux} = \left(\frac{p_u D}{2000}\right)\left(\frac{L_e}{D}\right)^2 = \left(\frac{108 \times 0.1}{2000}\right) \times (26.65)^2 = 3.83 \text{ kN·m}$$

Since, the forces and moments are of small magnitude, provide minimum reinforcements of 0.8% of the cross-section.

$$A_s = \left(\frac{0.8 \times 100 \times 250}{100}\right) = 200 \text{ mm}^2$$

Provide 2 bars of 12 mm diameter ($A_s = 226$ mm^2) with 6 mm ties at 120 mm centres.

2. Design of tension members

i. *Member AH*:

The tie member is designed as class 1 type structure without any cracks at working loads.

Precompression is provided by a prestreeing cable in the tie member.

Tensile force in the member $N_d = 377$ kN

Bending moment $M = 4.0$ kN·m

Permissible comspressive stress in concrete of M35 grade at transfer $\sigma_{ct} = 15$ N/mm^2

Permissible tensile stress at working loads $\sigma_{tw} = 0$ and loss ratio $\eta = 0.8$

∴ Area of concrete section required $\left(\dfrac{N_d}{\eta \sigma_{ct}}\right) = \left(\dfrac{377 \times 10^3}{0.8 \times 15}\right) = 31333$ mm^2

Section adopted is $(200 \times 250) = 50000$ mm^2

∴ Compressive prestress $= \left(\dfrac{377 \times 10^3}{0.8 \times 50000}\right) = 9.43$ N/mm^2

Prestressing force $P = \left(\dfrac{9.43 \times 50000}{1000}\right) = 471.5$ kN

Using 7 mm diameter high tensile wires initially stressed to 1100 N/mm^2 and having an ultimate tensile strength of 1500 N/mm^2, the number of wires required is given by

$$n = \left(\frac{471.5 \times 10^3}{38.5 \times 1100}\right) = 11.13$$

Use one Freyssinet cable containing 12 wires of 7 mm diameter with suitable end anchorages.

$$\text{Ultimate tensile strength of tie} = \left[\frac{12\times38.5\times0.87\times1500}{1000}\right] = 603 \text{ kN}$$

$$\text{Load factor against collapse} = \left(\frac{603}{377}\right) = 1.6 > 1.5$$

Assuming direct tensile strength of concrete as 4 N/mm²

$$\text{Cracking load} = \left[\frac{50000(0.8\times9.43)+4.0}{1000}\right] = 577.2 \text{ kN}$$

$$\therefore \quad \text{Load factor against cracking} = \left(\frac{577.2}{377}\right) = 1.53 > 1.25$$

Provide 4 bars of 12 mm and 4 bars of 10 mm diameter as untensioned reinforcement with 6 mm diameter ties at 200 mm centres.

The tie member AH can also be designed as a reinforced member without prestressing in which case more reinforcements are required to resist tension with bending moment.

$P_u = (1.5 \times 377) = 565.5$ kN

$M_u = (1.5 \times 4) = 6.0$ kN·m

$$\left[\frac{P_u}{f_{ck}bD}\right] = \left[\frac{565.5\times10^3}{35\times250\times200}\right] = 0.323$$

$$\left[\frac{M_u}{f_{ck}bD^2}\right] = \left[\frac{6\times10^6}{35\times250\times200^2}\right] = 0.017$$

Referring to Chart 80 of SP:16 (Fig. 18.5), readout $\left(\dfrac{p}{f_{ck}}\right) = 0.10$

$\therefore \quad p = (35 \times 0.10) = 3.5$

$$A_s = \left(\frac{3.5\times250\times200}{100}\right) = 1750 \text{ mm}^2$$

Provide 8 bars of 18 mm diameter distributed equally on all sides (A_{st} provided = 2035 mm²) with 10 mm stirrups at 120 mm centres.

ii. *Member EF*:

Adopting a section of size 100 mm × 250 mm, maximum design tensile force $N_d = 59$ kN

$$\therefore \quad \text{Area of steel } A_s = \left(\frac{59\times10^3}{230}\right) = 257 \text{ mm}^2$$

Modular ratio of M35 grade concrete $m = \left(\dfrac{280}{3 \times 11.5}\right) = 8.1$

Using 2 bars of 16 mm diameter ($A_s = 404$ mm²)

$$\text{Tensile stress} = \left[\frac{N_d}{A_c + (m-1)A_s}\right]$$

$$= \left[\frac{59 \times 10^3}{(250 \times 100) + (8.1 - 1) \times 404}\right] = 2.11 \text{ N/mm}^2 < 4 \text{ N/mm}^2$$

Provide single-legged ties of 6 mm diameter at 200 mm centres.
Typical reinforcement details are shown in Fig. 18.6.

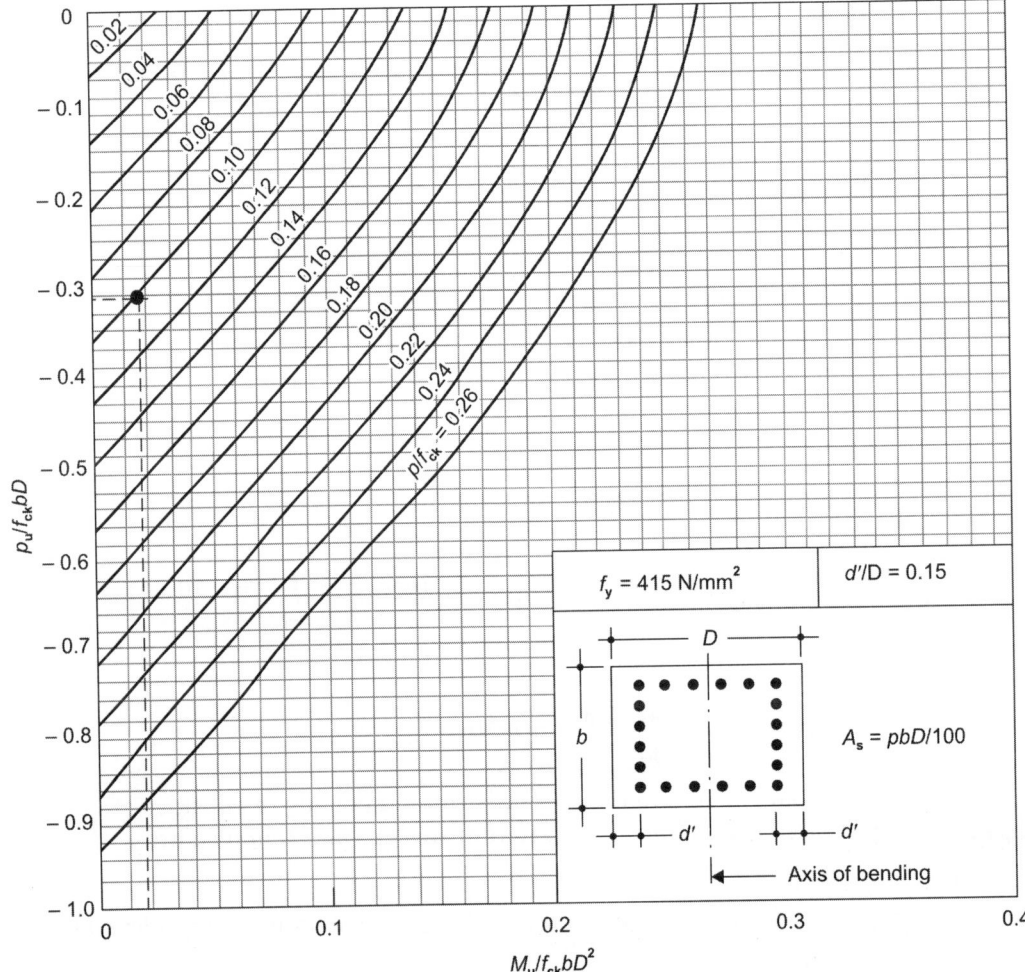

Fig. 18.5: Tension with bending rectangular section reinforcement distributed equally on four sides
(SP:16, Chart 80)

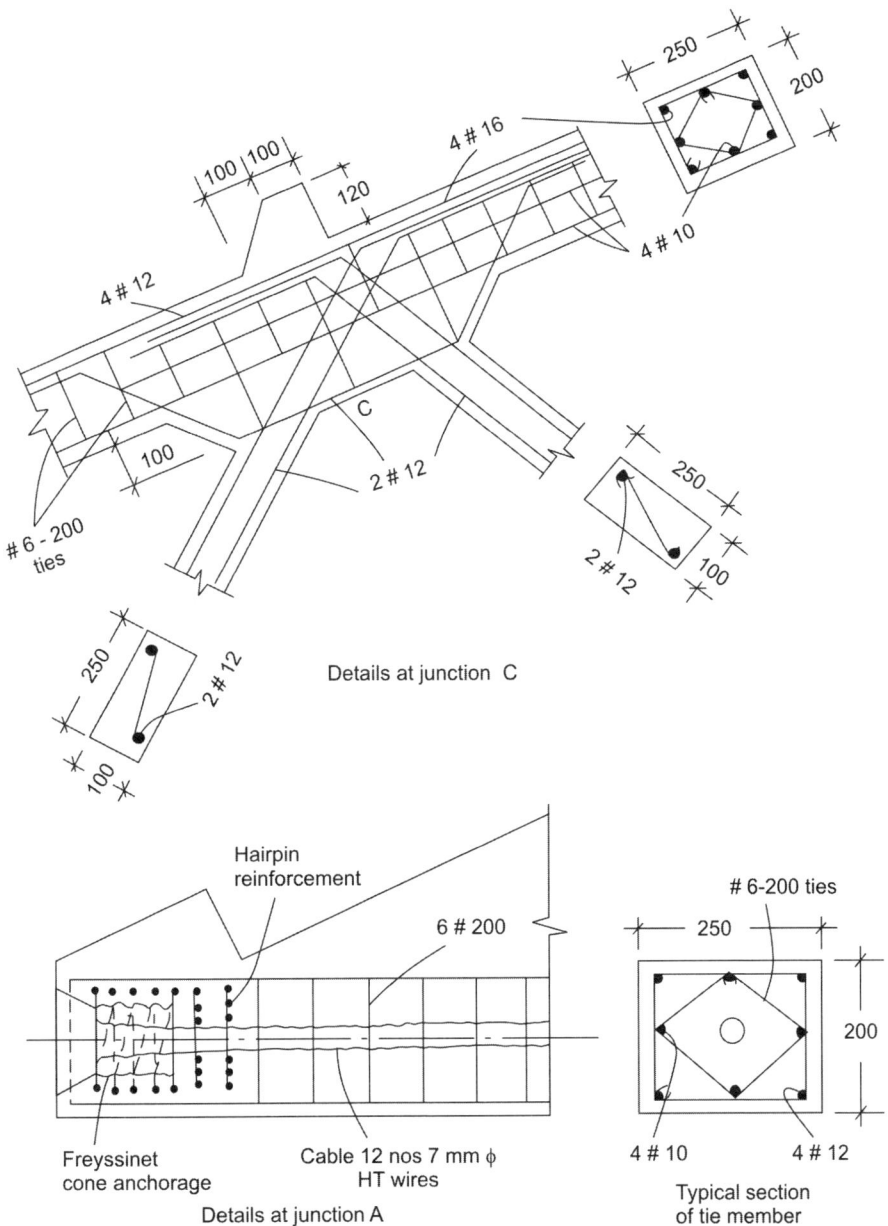

Fig. 18.6: Reinforcement details in truss

References

1. Murashev V, Sigalov E, Baikov V, *Design of Reinforced Concrete Structures*, Mir Publishers, Moscow, 1968.

2. Sigalov E, Strongin S, *Reinforced Concrete*, Foreign Languages Publishing House, Moscow, 1962.

3. Dunham CW, *Advanced Reinforced Concrete*, McGraw-Hill, New York, 1964.

4. Graduck H, *Prestressed Concrete*, Oxford and IBH, New Delhi, 1968.

5. IS:3201-1965, Indian Standard Specification: Criteria for Design and Construction of Precast Concrete Trusses, Bureau of Indian Standards, New Delhi, 1965.

6. Reynolds C and Steedman J, *Reinforced Concrete Designer's Handbook*, 8 edn, Cement and Concrete Association, London, 1976.

7. Krishna Raju N, *Prestressed Concrete*, 5 edn, McGraw-Hill Education (India) Pvt Ltd, New Delhi, 2002, pp. 337–379.

8. Rowe RE, Cranston WB, and Best BC, New concepts in the design of concrete, *Structural Engineer*, Vol. 43, 1965, pp. 339–403.

9. Bate SCC, Why limit state design, *Concrete*, March 1968, pp. 103–108.

10. IS:456-2000, Indian Standard Code of Practice for Reinforced Concrete (Fourth Revision) Bureau of Indian Standards, New Delhi, 2000, pp. 1–100.

11. IS:1343-2012, Indian Standard for Prestressed Concrete (Second Revision), Bureau of Indian Standards, New Delhi, 2012, pp.1–56.

12. SP:16-1980, Design Aids for Reinforced Concrete to IS:456, Bureau of Indian Standards, New Delhi, 1980.

Assignment

1. A reinforced concrete truss of polygonal parallel chord type is proposed for the roof system of an industrial structure for a span of 24 m, comprising 8 bays of 3 m each. The truss height is 2.5 m. Spacings of trusses is 6 m. The roof is covered by precast ribbed slabs of size 3 m × 6 m each weighing 24 kN. Live load on roof is 1.5 kN/m^2. The top and bottom boom members are 250 mm wide by 250 mm deep, while the diagonal and vertical members are of cross-section 250 mm × 150 mm. Analyse the frame for dead and live loads only and design the typical members of the truss using M40 grade concrete and Fe 415 HYSD bars. Also design the bottom tie as a prestressed member using Freyssinet cables comprising 12 wires of 7 or 8 mm diameter with an ultimate strength of 1500 N/mm^2.

2. The compression and tension members of an 18 m span reinforced concrete truss are subjected to the forces shown in the table given below:

Member	Cross-sectional dimensions (mm × mm)	Length of member (m)	Forces (kN)	Moment (kN·m)
AB	200 × 240	1.8	300(compression)	3.5
BC	200 × 240	3.6	240(compression)	12.0
AH	200 × 240	3.0	320(tension)	–

Using M40 grade concrete and Fe 415 HYSD bars, design suitable reinforcements in the members. Also design the tie AH as a prestressed member using suitable number of 7 mm diameter high tensile wires initially stressed to 1000 N/mm^2. Sketch the details of reinforcements in the cross-section of the members.

3. The top chord member of a reinforced concrete truss is subjected to a compressive force of 240 kN together with a bending moment of 3.0 kN·m. The bottom chord tie member is subjected to a tensile force of 220 kN. The members have a breadth of 200 mm and a depth of 180 mm. The length of compression member is 1.6 m while that of the tie is 2.5 m. Using M35 grade concrete and Fe 415 HYSD bars, design the

reinforcements in the members. Also design the tension member with the required magnitude of prestress using high tensile wire of 7 mm diameter having an ultimate tensile strength of 1500 N/mm^2 and initially stressed to 1000 N/mm^2.

4. A fink type reinforced concrete truss is to be designed for warehouse to suit the following data:

Effective span of the truss = 18 m

Type of truss = fink type

Roof cladding = asbestos sheets supported on precast pretensioned purlins

Spacings of trusses = 6 m

Wind pressure = 1.5 kN/m^2

Central rise = one fourth span

Grade of concrete = M40

Reinforcements = Fe 415 HYSD bars and high tensile wires of 7 mm diameter with an ultimate tensile strength of 1500 N/mm^2

Design the reinforcements in the typical compression and tie members.

Review Questions

1. In what situations you would recommend the use of reinforced concrete trusses for the roofing system of structures? Mention their advantages.

2. Explain with sketches the different types of truss configurations used for various spans in the range of 10 m to 30 m.

3. Explain briefly the constructional features of reinforced concrete trusses indicating the material requirements.

4. Outline the reasons for prestressing the bottom boom tie member of the concrete truss.

5. Sketch a typical tie member of a reinforced concrete truss showing the details of high tensile wires and the end anchorages.

6. Write a brief note on the nature of forces developed in various members of concrete trusses subjected to dead, live and wind loads.

7. What grade of concrete is generally used in the construction of reinforced concrete trusses? Explain with reasons the necessity for using higher grade concretes.

8. What are the forces to be considered in the design of a typical compression member of a reinforced concrete truss?

9. What are the simplifying assumptions made in the analysis of reinforced concrete trusses? How do you justify these assumptions?

10. What happens if the tie member is not prestressed? Is it advantageous to design the bottom boom tie as a reinforced member?

Objective Type Questions

1. Reinforced concrete trusses are preferred in structures located in
   a. arid zones
   b. hilly areas
   c. coastal areas with aggressive environmental conditions

2. Reinforced concrete trusses were first developed and widely used for structural roofs in
   a. Australia
   b. America
   c. Europe

3. The diagonal members of a reinforced concrete truss are generally subjected to
   a. bending
   b. torsion
   c. compression or tension

4. The bottom boom member of concrete trusses are primarily subjected to
   a. compression
   b. tension
   c. flexure

5. The type of reinforcements used for large span reinforced concrete trusses is
   a. mild steel
   b. high tensile steel
   c. HYSD bars

6. The critical forces in various members of the concrete truss in analysed for
   a. dead load
   b. live load
   c. dead load + live load + wind load

7. The diagonal members of a reinforced concrete truss are generally
   a. prestressed
   b. reinforced and prestressed
   c. reinforced

8. The tie member of a reinforced concrete truss is generally designed as
   a. axially prestressed
   b. transversely prestressed
   c. eccentrically prestressed

9. In the design of top boom members of a concrete truss loaded between the joints, the forces to be considered in design are
   a. axial forces
   b. eccentric forces
   c. axial force and bending moment

10. Reinforced concrete trusses are preferably
    a. cast at site in the vertical position
    b. cast at site in the horizontal position
    c. cast at site over the supporting columns

19

Poles

19.1 GENERAL FEATURES

Reinforced concrete poles have almost replaced the traditional wood and steel poles due to the inherent advantage of concrete to resist deterioration under adverse environmental conditions. Reinforced concrete poles[1,2] are mass produced and extensively used in most countries for railway, power and signal lines, lighting poles, antenna masts, telephone transmission, low and high voltage electric power transmission and substation towers. Mainly square, rectangular and circular poles have been used during the middle of 19th century. The main advantages of reinforced concrete poles are:

- very good fire resistance, particularly to grass and bush fired near the ground level
- superior freeze-thaw resistance essential for use in cold regions
- economical when mass produced using precast techniques
- durability from corrosion in humid and temperate climates
- resistance to erosion in desert areas
- resistance to termite attack when embedded in earth at the base
- excellent mould ability to the desired shapes
- easy installation in ground
- negligible maintenance
- progressive cracking and warning of impending failure
- comparatively cheaper than steel and other types of poles
- aesthetically pleasing with smooth surfaces.

The various advantages have resulted in the rapid development and use of different types of reinforced concrete poles in the developed and developing countries. In view of the crash programme of rural electrification in India, it is estimated that there will be a demand for two to three million poles annually which will further increase due to the electrification of railways and installation of new lines during the next decade.

19.2 CLASSIFICATION OF POLES

According to the Indian standard code IS:785-1964, the reinforced concrete poles were earlier classified under 11 categories depending upon their overall length

(7.5 m to 17 m) and transverse load carrying capacity (2 kN to 30 kN). However, in the second revised version of the Indian code[3] in the year 1998, all these categories are removed and only two length ranges covering the minimum and maximum are prescribed along with the minimum depth of planting in ground as shown in Table 19.1.

Table 19.1: Minimum depth of planting of reinforced concrete poles in ground (IS:785-1998, clause 6.3)

Length of pole (m)	Minimum depth of planting in ground (m)
6 to 7	1.20
7.5 to 9	1.50

The Indian standard code also prescribes that the minimum overall length of reinforced concrete poles should be not less than 6 m and the maximum length should not exceed 9 m. The length of the pole should be in steps of 0.5 m.

The Sri Lanka Electricity Board (Ceylon Electricity Board) specifications[4] for reinforced concrete poles permit four categories of reinforced concrete poles in the range of 6 m/50 kg, 8.3 m/100 kg, 9 m/115 kg and 10 m/300 kg for use in their country. Poles exceeding 9 m have been eliminated from the present Indian code mainly due to the introduction of prestressed concrete poles[5,6,7] for use on a large scale due to their slenderness and superior load resisting characteristics.

19.3 ▊ DESIGN FEATURES

Reinforced concrete poles are generally designed to resist bending moments of equal magnitude in opposite directions. The following critical loading conditions are to be considered in the design of reinforced concrete poles:

   i. Bending due to wind load on cable and exposed face of pole
  ii. Combined bending and torsion due to eccentric snapping of wires
 iii. Maximum torsion due to skew snapping of wires
  iv. Bending due to failure of all the wires on one side of the pole
   v. Handling and erection stresses.

The pole should be designed such that when it is vertical, its bending strength in the transverse direction should be sufficient to take an ultimate load equal to the horizontal wind load on wires and pole multiplied by the desired load factor. The load factor specified in the Indian code is that it should be not less than 2. The design should also ensure the limit state of cracking as specified in IS:456-2000 code[8]. In case of poles used for power transmission lines, the strength of the pole in the direction of lines should be not less than one fourth of the strength required in the transverse direction. The reinforcements in poles should conform to the specifications and norms prescribed in SP:34[9] and detailing practices outlined in the paper by Kalgal and Jayasimha[10].

The Indian standard code also specifies the details of holes to be provided to house the bracket fixtures and the eye holes required for handling of poles during transport and erection. The typical details of holes and the depth of planting the pole below the ground are shown in Fig. 19.1.

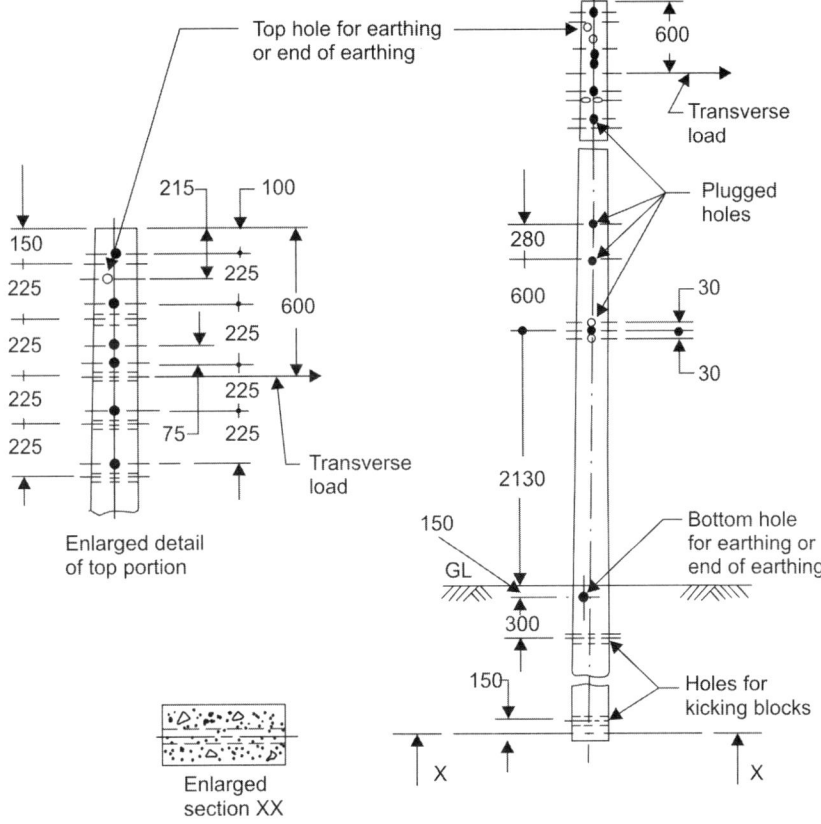

Fig. 19.1: Typical reinforced concrete pole (IS:1678-1998)

19.4 DESIGN EXAMPLE

A reinforced concrete pole is to be designed to suit the following data:

Line voltage: 3 phase, 400 V

Number of circuits: One

Number of conductors (including neutral and street lighting wire): Five

Size and material of conductor: 7.1 mm diameter copper

Size and material of neutral wire: 5 mm diameter copper

Size and material of street lighting wire: 4 mm diameter copper

Normal span: 50 m

Wind pressure: 1 kN/m²

Load factor: 2.5

Overall length of pole: 7 m

Material: M20 grade concrete and Fe 415 HYSD bars

Arrangement of conductors: As shown in Fig. 19.2

Tension in conductors: 3 kN

Design and detail the reinforcements in the pole.

i. *Wind load calculations*:

Wind load on three conductors/span = $\left(1.00 \times \dfrac{2}{3} \times \dfrac{7.1}{1000} \times 3 \times 50\right)$ = 0.71 kN

Wind load on neutral wire/span = $\left(1.00 \times \dfrac{2}{3} \times \dfrac{5}{1000} \times 50\right)$ = 0.167 kN

Wind load on street lighting wire = $\left(1.00 \times \dfrac{2}{3} \times \dfrac{4}{1000} \times 50\right)$ = 0.133 kN

ii. *Bending moments*:

Total bending moment at ground level due to windage on all the wires
$$= (0.71 \times 5.6) + (0.167 \times 5.6) + (0.133 \times 5.95) = 5.70 \text{ kN·m}$$

Equivalent load at 0.6 m from top corresponding to windage on all the wires

$$= \frac{1}{5.20}[(0.71 \times 5.6) + (0.167 \times 5.6) + (0.133 \times 5.95)] = 1.097 \text{ kN}$$

Assume the pole to have uniformly tapering square section having dimensions as shown in Fig. 19.2.

Wind load on pole above ground level = $1.00 \times 5.80 \left(\dfrac{0.125 + 0.2286}{2}\right) = 1.025$ kN

Bending moment at GL assuming the load to be acting at centre of gravity of pole

$$= 1.025 \times \frac{1}{3} \times 5.80 \left[\frac{0.2286 + 2 \times 0.125}{0.2286 + 0.125}\right] = 2.682 \text{ kN·m}$$

Equivalent load at 0.6 m from top corresponding to windage on pole

$$= \left(\frac{2.682}{5.2}\right) = 0.515 \text{ kN}$$

Total transverse load at 0.6 m from top corresponding to windage on the conductors and pole = (1.097 + 0.515) = 1.612 kg

Ultimate load with load factor = (1.612 × 2.5) = 4.03 kN

From Table 19.1, a reinforced concrete pole belonging to class 9 can be chosen.

iii. *Design of reinforcements*:

Using M20 grade concrete and Fe 415 grade tor steel, the ultimate moment capacity of the pole section at ground level is given by
$$M_{u \text{ (lim)}} = 0.138 f_{ck} bd^2$$
Assume an effective cover of 30 mm.

Effective depth at ground level = (228.6 − 30.0) = 198.6 mm

∴ $M_{u \text{ (lim)}} = (0.138 \times 20 \times 50 \times 198.6^2) = 272 \times 10^6$ N·mm
$$= 27.2 \text{ kN·m}$$

Total service load bending moment at GL due to windage on wires and pole = (5.70 + 2.682) = 8.382 kN·m

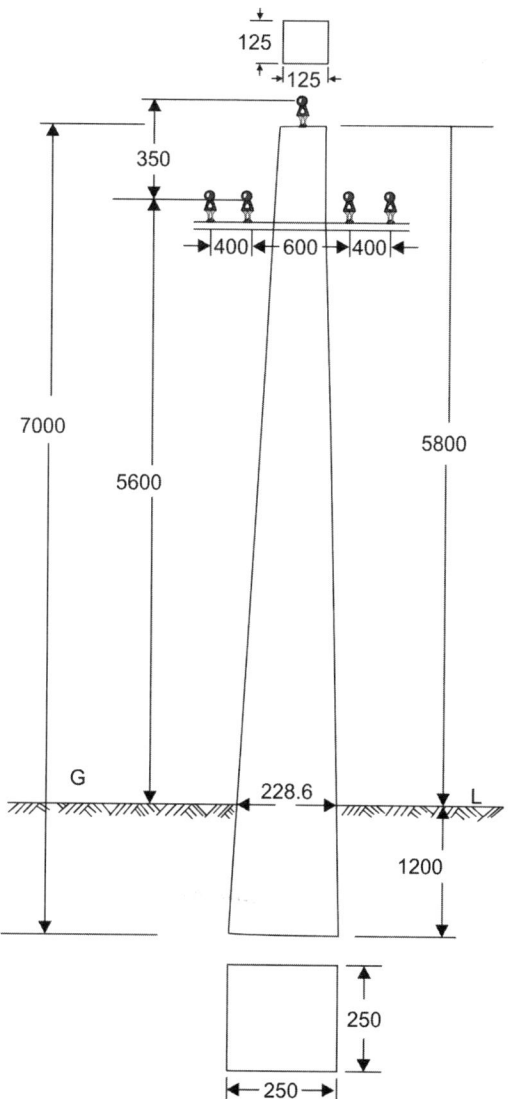

Fig. 19.2: RC transmission pole

Ultimate moment with a load factor of 2.5 M_u = (2.5 × 8.382)

$$= 20.955 \text{ kN·m}$$

Since $M_u < M_{u \text{ (lim)}}$, the section is under-reinforced.

$$\therefore \qquad M_u = 0.87 f_y A_{st} d \left[1 - \frac{A_{st} f_y}{b d f_{ck}} \right]$$

$$20.955 \times 10^6 = 0.87 \times 415 \times A_{st} \times 198.6 \left[1 - \frac{A_{st} \times 415}{250 \times 198.6 \times 20} \right]$$

Solving, A_{st} = 342 mm²

iv. *Torsional moments:*

Due to skew snapping of the wires torsional moments develop in the pole.

Torsional moment $T = 3(1.4 + 0.6) = 6$ kN·m

$$T_u = (2.5 \times 6) = 15 \text{ kN·m}$$

Bending moment at base $M_u = 20.955$ kN·m

$$V_u = 4.03 \text{ kN·m}$$

Equivalent bending moment $M_{el} = (M_u + M_t)$

where, $M_t = T_u \left[\dfrac{1 + (D/b)}{1.7} \right] = 15 \left[\dfrac{1 + (250/250)}{1.7} \right] = 17.6$ kN·m

∴ $M_{el} = (20.955 + 17.6) = 38.6$ kN·m

The longitudinal steel should be designed for this bending moment.

For the assumed section at ground level, $M_{u\ (lim)} = 27.2$ kN·m

$M_{el} - M_{u\ (lim)} = (38.6 - 27.2) = 11.4$ kN·m

$11.4 \times 10^6 = f_{sc} A_{sc}(d - d')$

Assuming cover of 30 mm, $d = 198.6$ mm, $d' = 30$ m

$$f_{sc} > 0.87 f_y = (0.87 \times 415) = 361 \text{ N/mm}^2$$

∴ $A_{sc} = \left(\dfrac{11.4 \times 10^6}{361 \times 168.6} \right) = 187.3 \text{ mm}^2$

$A_{st_1} = \left[\dfrac{0.36 f_{ck} b(0.48d)}{0.87 f_y} \right] = \left[\dfrac{0.36 \times 20 \times 250(0.48 \times 198.6)}{0.87 \times 415} \right] = 475 \text{ mm}^2$

$A_{st_2} = \left(\dfrac{A_{sc} f_{sc}}{0.87 f_y} \right) = \left(\dfrac{187.3 \times 361}{0.87 \times 451} \right) = 187.3 \text{ mm}^2$

∴ $A_{st} = \left(A_{st_1} + A_{st_2} \right) = (475 + 187.3) = 662.3 \text{ mm}^2$

Provide 4 bars of 22 m diameter, one each at the corners ($A_{st} = 720 \text{ mm}^2$).

v. *Transverse reinforcements:*

$$V_e = V_u + 1.6 \left(\dfrac{T_u}{b} \right) = 4.03 + 1.6 \left(\dfrac{15}{0.25} \right) = 100 \text{ kN}$$

$$\tau_{ve} = \left(\dfrac{V_e}{bd} \right) = \left(\dfrac{100 \times 10^3}{250 \times 198.6} \right) = 2 \text{ N/mm}^2$$

$$\left(\dfrac{100 A_{st}}{bd} \right) = \left(\dfrac{100 \times 720}{250 \times 198.6} \right) = 1.45$$

∴ $\tau_c = 0.70 \text{ N/mm}^2$

Balance shear $V_s = \left[100 - \dfrac{0.70 \times 250 \times 198.6}{1000} \right] = 65.3 \text{ kN}$

Using 10 mm diameter two-legged stirrups, spacing is

$$S_v = \left(\frac{0.87 \times 415 \times 2 \times 79 \times 198.6}{65.3 \times 10^3} \right) = 173 \text{ mm}$$

Adopt 10 mm diameter two-legged stirrups at 170 mm centres spacing through-out the length of the pole.

The reinforcement details are as shown in Fig. 19.3.

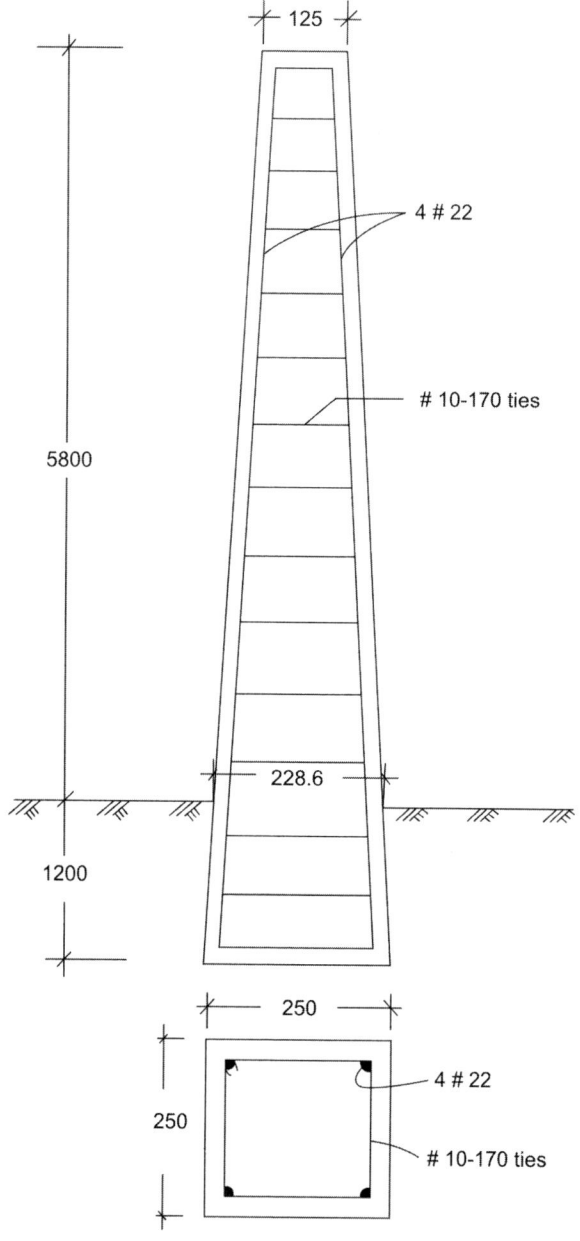

Fig. 19.3: Reinforcement details in RC pole

References

1. Report of the Task Committee of the ASCE, Guide for the Design and Use of Concrete Poles, ASCE, April 1987, pp. 1–49.
2. Wolman E, Utility of poles of reinforced and prestressed concrete pipe, *Journal of the American Concrete Institute*, Vol. 31, No. 10, April 1980.
3. IS:785-1998 (Second Revision), Reinforced Concrete Poles for Overhead Power and Telecommunication Lines, Bureau of Indian Standards, New Delhi, March 1998, pp. 1–8.
4. Ceylon Electricity Board Specifications for Reinforced Concrete Poles, Sri Lanka, 1996, pp. 1–24.
5. IS:1678-1960, Specification for Prestressed Concrete Poles for Overhead Power Traction Telecommunication Lines (Second Revision), Bureau of Indian Standards, New Delhi, 1998 pp. 1–10.
6. Krishna Raju N, *Prestressed Concrete*, 5 edn, McGraw-Hill Education Pvt. Ltd, New Delhi, 2002, pp. 574–616.
7. Krishna Raju N and Ramamurthy LN, Limit State Design of Pretensioned Partially Prestressed Concrete Poles, Proceedings of the International Symposium on Prestressed Concrete Pipes, Poles, Pressure Vessels and Sleepers, Madras 1972, Paper PL/6, pp. 1–15.
8. IS:456-2000, Indian Standard Code of Practice for Plain and Reinforced Concrete (Fourth Revision), Bureau of Indian Standards, New Delhi, July 2000.
9. SP:34, Handbook for Concrete Reinforcement and Detailing, Special Publication Bureau of Indian Standards, New Delhi, 1987.
10. Kalgal MR and Jayasimha KS, Detailing of Reinforcements, Proceedings of the Workshop on Reinforcements in Concrete, Tor Steel Research Foundation, Bangalore, September 1994, Paper No. 3.

Assignment

1. A reinforced concrete pole 10 m long is required to carry four conductors of 7 mm diameter each spaced at 500 mm intervals in a cross arm fixed at 600 mm from the top. The depth of embedment is 1.8 m below ground level.

   Spacings of poles = 50 m

   Wind pressure = 1.5 kN/m^2

   Load factor = 2.5

   Tension in conductors = 3 kN

   Adopting M20 grade concrete and Fe 415 HYSD bars, design a suitable pole for the transmission line.

2. A reinforced concrete pole 12 m long is required to carry four conductors of 7 mm diameter each spaced at 600 mm intervals in a cross arm fixed at 500 mm from the top. The pole is embedded to a depth of 2 m below ground level.

   Spacings of poles = 60 m

   Wind pressure = 1 kN/m^2

   Load factor = 2.5

   Tension in conductors = 4 kN

   Adopting M20 grade concrete and Fe 415 grade HYSD bars, design a suitable pole for the transmission line.

3. A reinforced concrete pole 9 m long is required for power lines of a small city. The pole should accommodate four conductors of 7 mm diameter each using a bracket 1.5 m in length and fixed at 600 mm from the top of the mast. Two wires are

500 mm apart on either side of the pole. The wind pressure in the locality is 1.6 kN/m². Tension in conductors is 2 kN. The depth of embedment of pole is 1.5 m. Adopt M25 grade concrete and Fe 500 HYSD reinforcements. Assume the spacings of the poles as 40 m. Design the pole to conform to IS code specifications.

4. A reinforced concrete pole is to be designed to suit the following data:

Overall length of the pole = 10 m

Length of embedment in ground = 2 m

Spacings of poles = 15 m

Number of conductors = 4

Length of bracket = 1.5 m

Spacings of conductors arranged symmetrically on either side of the bracket = 500 mm. Tension in conductors = 3 kN

Adopt M30 grade concrete and Fe 500 grade HYSD reinforcements to design a rectangular pole conforming to Indian standard code provisions.

Review Questions

1. In what ways are reinforced concrete poles more superior in comparison with steel, wooden and GI poles?

2. Specify the advantages of reinforced concrete poles with particular reference to durability and fire resistance.

3. What are the lengths of reinforced concrete poles specified in the Indian standard code IS:785-1998?

4. Specify the main reasons for limiting the lengths of reinforced concrete poles in the Indian standard code.

5. What are the main forces to be considered in the design of reinforced concrete poles?

6. How do you consider the wind loads acting on a reinforced concrete pole?

7. What is skew snapping of the conductor wires? How do you consider skew snapping in the design of poles?

8. What are the Indian code provisions regarding the embedment of reinforced concrete poles in ground?

9. Which is the most critical section to be designed in reinforced concrete poles? How do you design this section for flexure and torsion?

10. Explain with sketches the details of holes to be provided in an electric pole and the typical reinforcement details.

Objective Type Questions

1. An ideal type of electric pole to be used in a desert area with severe gusts of winds is
   a. wooden pole
   b. steel pole
   c. concrete pole

2. Reinforced concrete pole is an ideal choice for rural electrification projects mainly due to
   a. lighter self weight of the poles
   b. can be cast at site
   c. precasting techniques

3. The maximum bending moment developed in an electric pole is at
   a. the level of conductor wires
   b. the mid height of the pole
   c. the base of the pole

4. Torsional moments develop in the concrete poles mainly due to
   a. the wind pressure
   b. snapping of the conductors on one side of the pole
   c. skew snapping of the conductor wires

5. In the design of reinforced concrete poles, the critical direction of wind to be considered is
   a. along the direction of the conductors
   b. vertical to the direction of the pole
   c. perpendicular to that of the wires

6. In the limit state design of reinforced concrete poles, the load factor against collapse to be considered according to the Indian standard code is
   a. 1.2
   b. 2.5
   c. 2.0

7. In the case of reinforced concrete poles, the holes for earthing should be provided at
   a. the top
   b. the bottom
   c. both at the top and the bottom

8. The critical section of a reinforced concrete pole should be designed as
   a. over-reinforced section
   b. balanced section
   c. under-reinforced section

9. The Indian standard code restricts the overall length of a reinforced concrete pole to a maximum of 9 m mainly to reduce
   a. the cracks under imposed loads
   b. overall weight of the pole
   c. the possibility of sudden collapse

10. According to the Indian standard code, poles longer than 9 m should preferably be designed as
    a. over-reinforced poles
    b. under-reinforced poles
    c. prestressed poles

20

Deep Beams

20.1 GENERAL FEATURES

Deep beams are structural elements loaded as traditional simple beams in which a significant amount of the load is carried to the supports by a compression force combining the load and the reaction. A beam is considered as deep if the depth is large in relation to the span of the beam. According to the clause 29.1 of the Indian standard code IS:456-2000[1], a beam is classified as deep when the ratio of effective span to overall depth (L/D) is less than 2 for simply supported members and 2.5 for continuous members.

In deep beams, the bending stress distribution across any transverse section deviates from the straight line distribution assumed in the elementary beam theory[2,3]. Consequently, a transverse section which is plane before bending does not remain approximately plane after bending and the neutral axis does not usually lie at the mid depth. In such beams, shear-flexure[4] and shear modes[5] dominated by tensile cleavage failure are common. The ultimate failure due to shear[6] is generally brittle in nature in contrast to the ductile behaviour and progressive flexural failure with large number of cracks prevalent in normal beams.

In many cases of industrial and multistoried structures, deep beams are invariably used due to structural requirements. Floor slabs under horizontal load, short span beams carrying heavy loads, and transfer girders are examples of deep beams. The side walls of coal bunkers also behave as deep beams. The method of design of deep beams is different from that of the simple traditional beams having large span to depth ratios since several influencing parameters have to be considered in the design of deep beams.

20.2 MAJOR DIFFERENCES IN BEHAVIOUR OF DEEP BEAMS

The behaviour of deep beams is significantly different in comparison with the simple beams, and many assumptions made for the design of simple beams are not valid for deep beams since the structural behaviour of deep beams under transverse loads is complex. The major differences are listed below:

i. The stress and strain distribution is not linear since plane sections do not remain plane after bending.

ii. The structural behaviour is influenced by both support to depth and span to depth ratios.

iii. Shear deformation cannot be neglected as in simple beams.

iv. At ultimate limit stage, the stress block in concrete compression zone cannot assumed to be parabolic in shape as in simple beams.

v. Conventional shear investigations are not applicable for deep beams.

20.3 PARAMETERS INFLUENCING DESIGN OF DEEP BEAMS

During last sixty years, extensive research by several investigators on the analysis and behaviour of deep beams has helped to formulate design procedures for deep beams. The analysis and design of deep beams have been presented by Kong[7], Rogowsky et al[8]. The investigations on the design of reinforced concrete slender deep beams have been reported by the ACI committee 533 in 1971. The ultimate strength aspects of deep beams in shear has been studied by Ramakrishnan and Ananthanarayana[9]. The investigations on the shear strength of deep beams have also been reported by Smith and Vantsiotis[10]. Shear strength model and design formulas have been proposed by Russu and Venir[11] and these are helpful in the design of deep beams. Tan et al[12] have developed a direct design method for deep beams with web reinforcement.

The prominent parameters influencing the design of deep beams are:

C = length of support

D = overall depth of beam

L = effective span

b = width of beam

w = uniform load of the beam

The characteristic ratios used as parameters in design calculations and chart are

$$\varepsilon = \left(\frac{C}{L} \right) = \text{support to span ratio}$$

$$\beta = \left(\frac{D}{L} \right) = \text{depth to span ratio}$$

The design charts and the procedure of design outlined here are based on the reports published by the Portland Cement Association, USA and the Concrete Association of India. These report are based on the original work reported by Dischinger in 1932. The parameters used in design are illustrated in Fig. 20.1.

20.4 FLEXURAL BENDING STRESSES

Figure 20.2 shows the flexural bending stress at mid span of a continuous deep beam for ratio β having values 0.5, 0.67 and 1.0 and $\varepsilon = 0.1$. As β increases from 0.5 to 1.0, the compressive stress decreases rapidly at the top and the neutral axis moves towards the soffit of the beam. The tensile stress decreases gradually from $1.31(w/b)$ to (w/b).

Increasing the value of the depth D beyond the value of the span $L(D/L > 1)$, does not significantly influence the magnitude of the stresses within the depth equal to L. The curves are nearly the same for values of $\varepsilon = 0.05$ to 0.2.

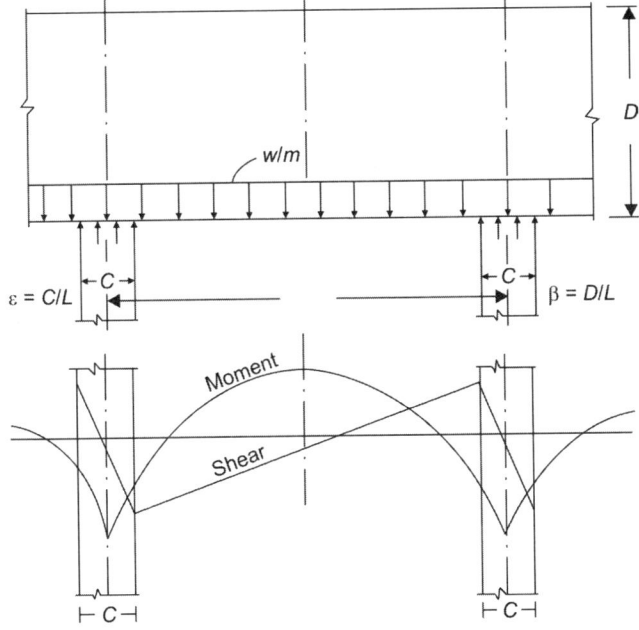

Fig. 20.1: Typical deep beam under uniform loading

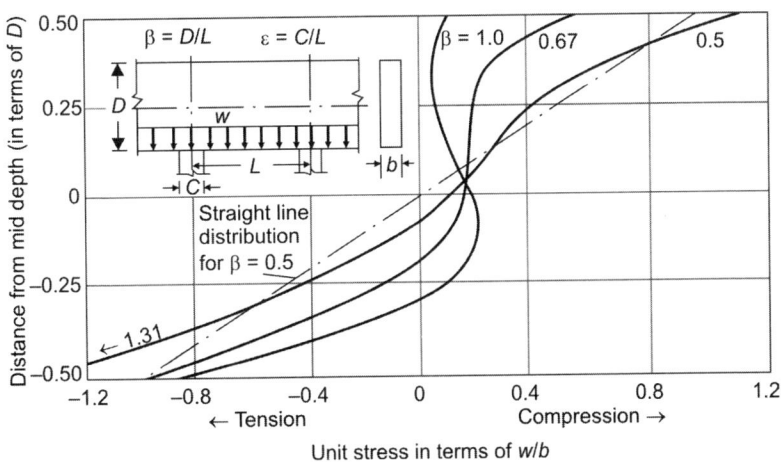

Fig. 20.2: Flexural stresses at mid span of deep beams with uniform loading

The variation of stresses at support sections of a continuous deep beam for different values of β varying from 0.5 to 1.0 and $\varepsilon = 0.1, 0.2$ and 0.05 are shown in Figs 20.3–20.5.

Assuming the uniformly distributed load shown in Figs 20.2–20.5, the mid span and support moments in a typical interior span is given by

$$M_{(span)} = \frac{wL^2}{24}(1-\varepsilon^2)$$

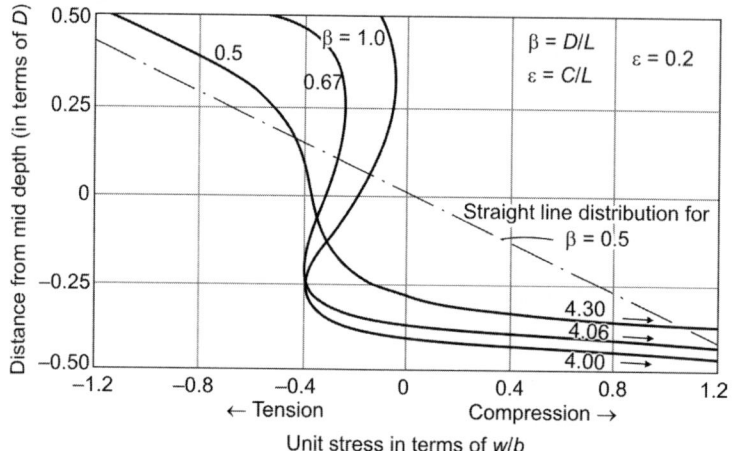

Fig. 20.3: Flexural stresses at support of deep beams with uniform loading

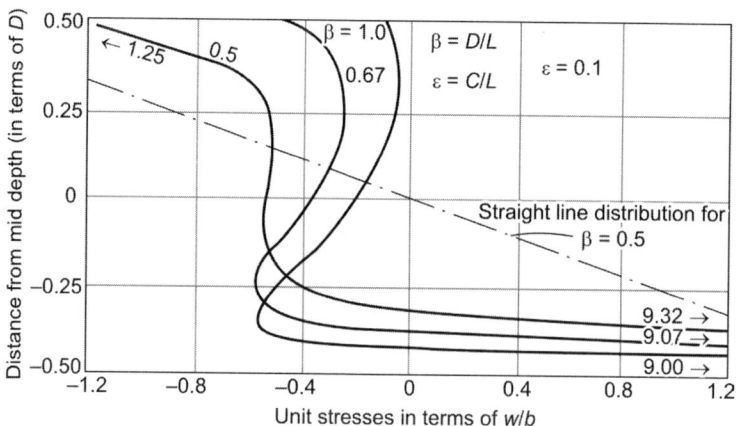

Fig. 20.4: Flexural stresses at support of deep beams with uniform loading

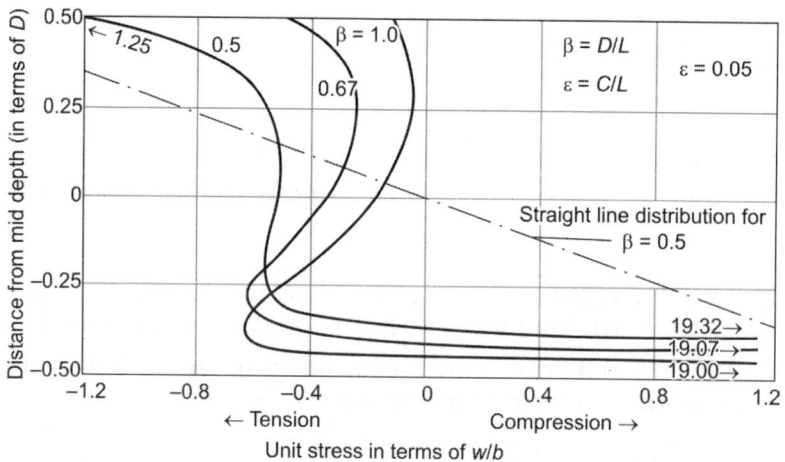

Fig. 20.5: Flexural stresses at support of deep beams with uniform loading

$$M_{(support)} = \frac{wL^2}{24}(1 - \varepsilon)(2 - \varepsilon)$$

The reinforcements are designed for tension T, expressed as the product of a coefficient and wL for mid span and support sections. The coefficient for tension computation is shown in Fig. 20.6. The flexural stresses developed in simply supported single span girders is shown in Fig. 20.7.

It is assumed that the stress distribution curves are similar at mid span for continuous girder with a length equal to $0.5L$ between the points of inflexion and that of a simply supported girder of span L. Hence, the curves for the continuous girder shown in Fig. 20.7 may be adopted for single span girders when $\beta = 0.5D/L$. The design data is, therefore, generally selected for single span girders for $\beta = 0.5D/L$ and $\varepsilon = 0.5$ by using the Figs 20.6 and 20.7.

The flexural stresses developed in continuous deep girder having concentrated loading at bottom and top are shown in Figs 20.8 and 20.9 respectively.

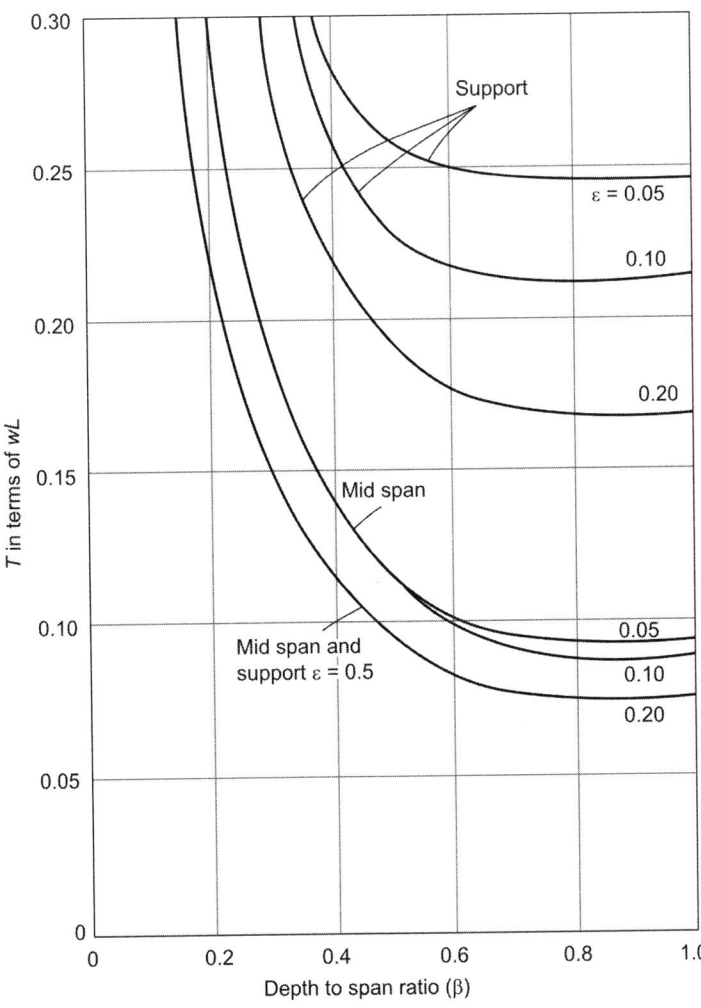

Fig. 20.6: Resultant tension in deep beams having uniform loading

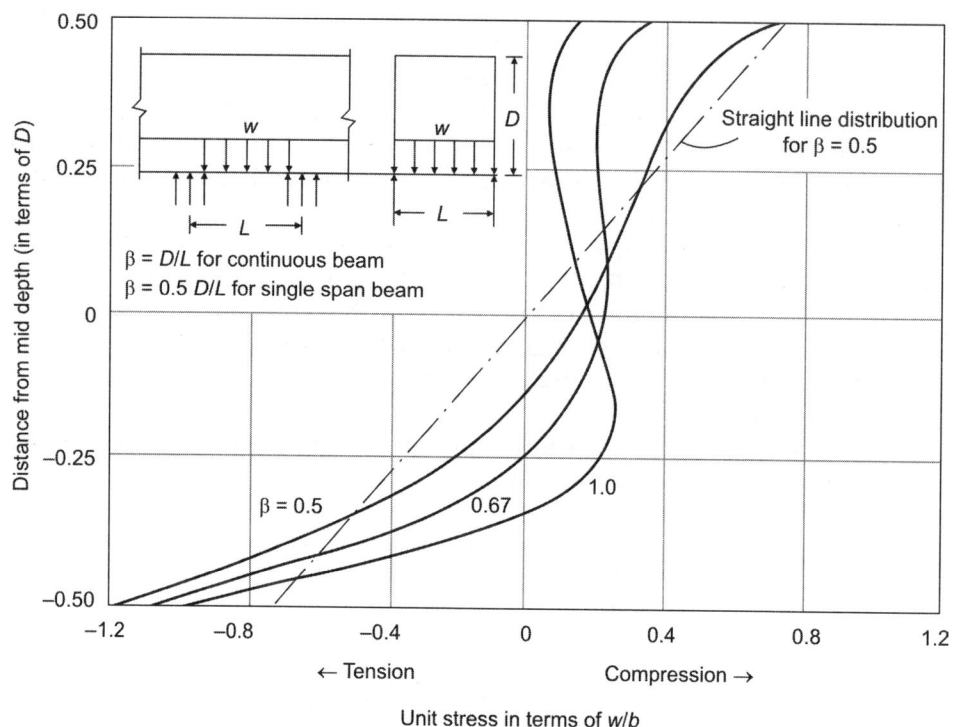

Fig. 20.7: Flexural stresses in simply supported single span deep girders

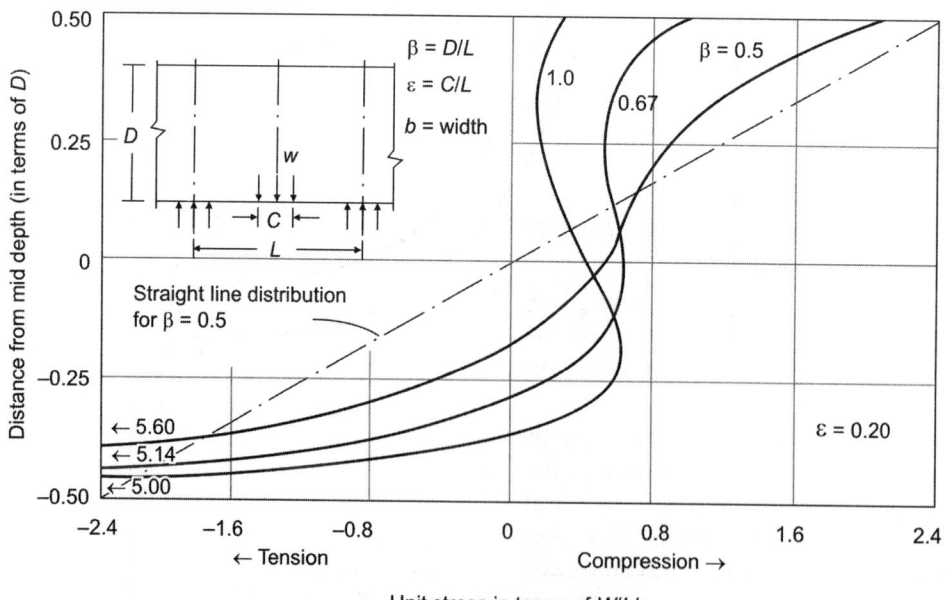

Fig. 20.8: Flexural stresses at mid span of deep beams with concentrated load at bottom edge

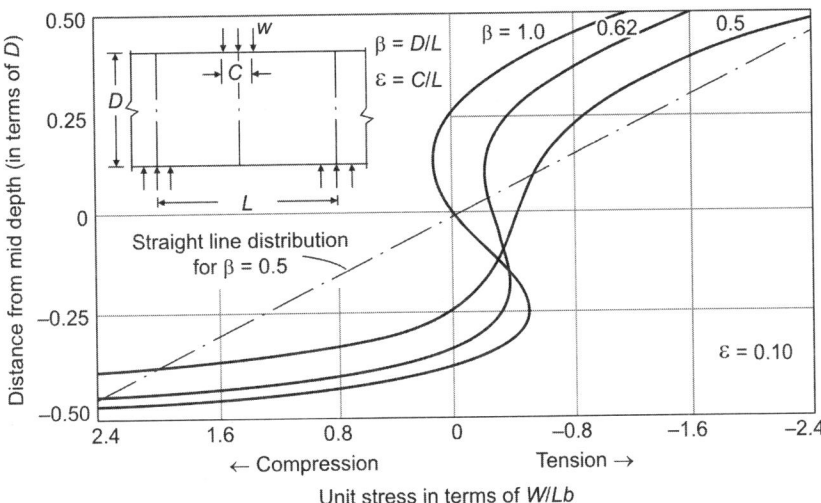

Fig. 20.9: Flexural stresses at centre line of support of deep beams with concentrated loading at top edge.

A comparison of the curves indicates that they are nearly alike for $\beta = 0.5$. But as β increases there is increasing divergence of stresses for the two cases of loading. For $\beta = 1$, when the girder is loaded at the top, the stress changes direction thrice and there are three neutral axis positions across the depth of the girder.

The resultant tension values as a function of depth to span ratio β and the parameter ε, are shown in Fig. 20.10. According to Dischinger, selection of T-values based on loads applied at the bottom gives conservative results for continuous beams with loads applied on the top edge. For single span girder of span L and depth D with a concentrated load W, the value of parameter β is taken as equal to $0.5D/L$ and $\varepsilon = 0.5C/L$. The tension is computed as the product of the coefficient from Fig. 20.10 and the concentrated force W. The reinforcements are provided as near the soffit of the girder as possible.

20.5 SHEAR STRESSES IN DEEP BEAMS

In the case of deep beams with relatively small percentage of reinforcement, the cracks develop vertically from the soffit and remain practically vertical in comparison with the diagonal tension cracks observed in conventional shallow beams as shown in Fig. 20.11.

It is clear from the figures that the diagonal tension which is characteristic of a shallow beam changes gradually into plain horizontal tension as the beam becomes a deep girder. Hence, the conventional shear investigations are not strictly applicable to deep beams. Experimental investigations by V Ramakrishnan and Y Ananthanarayan have revealed that deep and shallow beams exhibit practically similar modes of shear failure for low shear span to depth ratios of less than 2. In deep beams, the shear failure is always initiated by splitting action similar to that in a cylinder under diametrical compression as in the Brazilian splitting test.

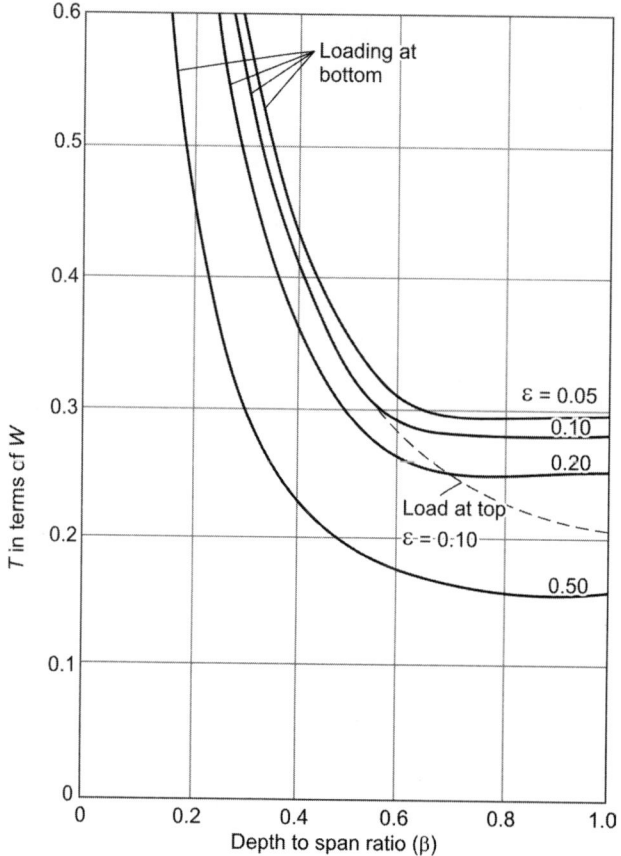

Fig. 20.10: Resultant tension in deep beams with concentrated load at mid span

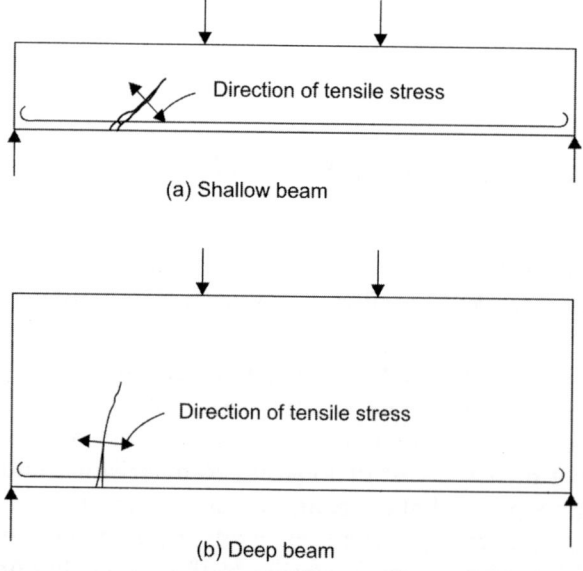

Fig. 20.11: Diagonal tension cracks in shallow and deep beams

20.6 ■ IS CODE PROVISIONS FOR DESIGN OF DEEP BEAMS

According to the Indian standard code IS:456-2000 provisions, a deep beam is a beam having a ratio of effective span to overall depth (L/D) less than 2 for simply supported and 2.5 for continuous members.

The reinforcements are provided for positive bending moment at mid span and negative bending moment at supports of continuous members computed on the basis of the lever arm, which is determined using the following relations:

a. For simply supported beams

$$Z = 0.2(L + 2D)$$

when $1 \leq L/D \leq 2$

or $Z = 0.6L$

when $L/D < 1$

b. For continuous beams

$$Z = 0.2 (L + 1.5D)$$

when $1 \leq L/D \leq 2.5$

or $Z = 0.5L$

when $L/D < 1$

where,

     Z = lever arm

     L = effective span taken as centre to centre distance between supports or 1.15 times the clear span, whichever is smaller

     D = overall depth

The positive reinforcements extending over the full span length without curtailment, should be placed within a zone of depth equal to ($0.25D - 0.05L$) adjacent to the tension face of the beam.

The negative reinforcement over supports of a deep beam with L/D ratio in the range of 1 to 2.5 should be distributed in two zones as detailed below:

   i. A zone of depth $0.2D$ adjacent to the tension face should contain a proportion of the tension steel given by $0.5[L/D - 0.5]$

  ii. A zone measuring $0.3D$ on either side of the mid depth of the beam, containing the remainder of the tension steel evenly distributed.

For beams having a span to depth ratio less than 1, the steel is evenly distributed over a depth of $0.8D$ measured from the tension face.

In addition, side face reinforcement to the extent of 0.12% of gross concrete area, when deformed bears are used and 0.15% when mild steel bars are used should be provided as vertical reinforcements. Also the minimum percentage of longitudinal reinforcement is fixed at 0.2% for deformed bars and 0.25% for other types of bars. The spacing of the horizontal reinforcement should be not greater than three times the wall thickness or 450 mm. A deep beam complying with these reinforcement details is deemed to satisfy the provisions for shear according to the IS code. The application of these principles are illustrated in the following design examples.

20.7 ☐ DESIGN EXAMPLES

1. Design a typical interior span of a continuous deep beam using the following data:

Span of beam (L) = 9 m
Overall depth (D) = 4.5 m
Width of supports (C) = 0.9 m
Width of beam (b) = 0.4 m
Uniformly distributed load (including self weight) w = 200 kN/m
Concrete = M20 grade
Reinforcements = Fe 415 HYSD bars

Sketch the details of reinforcements at centre of span and support sections.

i. *Design parameters*:

$$\varepsilon = \left(\frac{C}{L}\right) = \left(\frac{0.9}{9}\right) = 0.1$$

$$\beta = \left(\frac{D}{L}\right) = \left(\frac{4.5}{9}\right) = 0.5$$

$$L = 9 \text{ m} \qquad D = 4.5 \qquad w = 200 \text{ kN/m}$$

ii. *Moments*:

$$\text{Moment at mid span} = \frac{wL^2}{24}(1-\varepsilon^2)$$

$$= \left(\frac{200\times 9^2}{24}\right)(1-0.01) = 675 \text{ kN·m}$$

$$\text{Moment at support} = \left(\frac{wL^2}{24}\right)(1-\varepsilon)(2-\varepsilon)$$

$$= \frac{wL^2}{24}(1-0.1)(2-0.1)$$

$$= \left(\frac{wL^2}{14}\right) = \left(\frac{200\times 9^2}{14}\right) = 1150 \text{ kN·m}$$

iii. *Tensile reinforcements (using design graphs)*:

The resultant T of all tensile stresses in concrete is obtained by interpolating coefficients from Fig. 20.6.

For the given parameters, ε = 0.1 and β = 0.5

Coefficient for mid span = 0.12

Coefficient for interior support = 0.23

∴   Tension = (coefficient) × wL

Tension at mid span = $(0.12 \times 200 \times 9) = 216$ kN

Tension at support = $(0.23 \times 200 \times 9) = 414$ kN

Mid span $A_s = \left(\dfrac{216 \times 10^3}{230}\right)$

Support $A_s = \left(\dfrac{414 \times 10^3}{230}\right) = 1800 \text{ mm}^2$

iv. *Steel reinforcements according to IS:456 code procedure*:

Lever arm for continuous beam,

$$Z = 0.2(L + 1.5D)$$

when $\qquad 1 \le L/D \le 2.5$

In the present case, $\left(\dfrac{L}{D}\right) = \left(\dfrac{9}{4.5}\right) = 2.0$

∴ $\qquad Z = 0.2(9 + 1.5 \times 4.5) = 3.15 \text{ m}$

∴ Tension at mid span $= \left(\dfrac{M}{Z}\right) = \left(\dfrac{675}{3.15}\right) = 214 \text{ kN}$

Tension at support $= \left(\dfrac{M}{Z}\right) = \left(\dfrac{1150}{3.15}\right) = 365 \text{ kN}$

Mid span $A_s = \left(\dfrac{214 \times 10^3}{230}\right) = 930 \text{ mm}^2$

Support $A_s = \left(\dfrac{365 \times 10^3}{230}\right) = 1586 \text{ mm}^2$

v. *Minimum reinforcements*:

According to IS:456 code, minimum percentage of horizontal reinforcements is given by

$$A_s = (0.002 \times 400 \times 4500) = 3600 \text{ mm}^2$$

Vertical reinforcements are given by

$$A_{sv} = (0.0012 \times 400 \times 1000) = 480 \text{ mm}^2/\text{m}$$

vi. *Arrangement of reinforcements*:

According to IS:456 code, the reinforcements for positive and negative bending moments should be arranged as detailed below.

For positive BM (centre of span):

Zone of depth $= (0.25D \times 0.05L) = (0.25 \times 4.5 \times 0.05 \times 1) = 0.675 \text{ m}$

$A_s = 3600 \text{ mm}^2$, to be distributed over a depth of 0.675 m from the tension face. Use 20 bars of 16 mm diameter ($A_s = 4020 \text{ mm}^2$) arranged in 5 rows of 4 bars each.

For negative BM (support section):

Zone of depth $= 0.2D = (0.2 \times 4.5) = 0.9 \text{ m}$

Proportion of steel $= 0.5\left(\dfrac{L}{D} - 0.5\right) = 0.5\left(\dfrac{9}{4.5} - 0.5\right) = 0.75$

∴ $A_s = (0.75 \times 3600) = 2700 \text{ mm}^2$

Use 16 bars of 16 mm diameter over a depth of 900 mm from tension face (4 rows of 4 bars each).

Remaining steel = (3600 – 2700) = 900 mm²

Zone = 0.3D = (0.3 × 4.5) = 1.35 m on either side of the mid depth of beam.

Use 20 bars of 8 mm (5 rows of 4 bars). For vertical reinforcements at support and span sections, adopt four-legged stirrups of 6 mm diameter at 230 centres. The details of reinforcements are shown in Fig. 20.12.

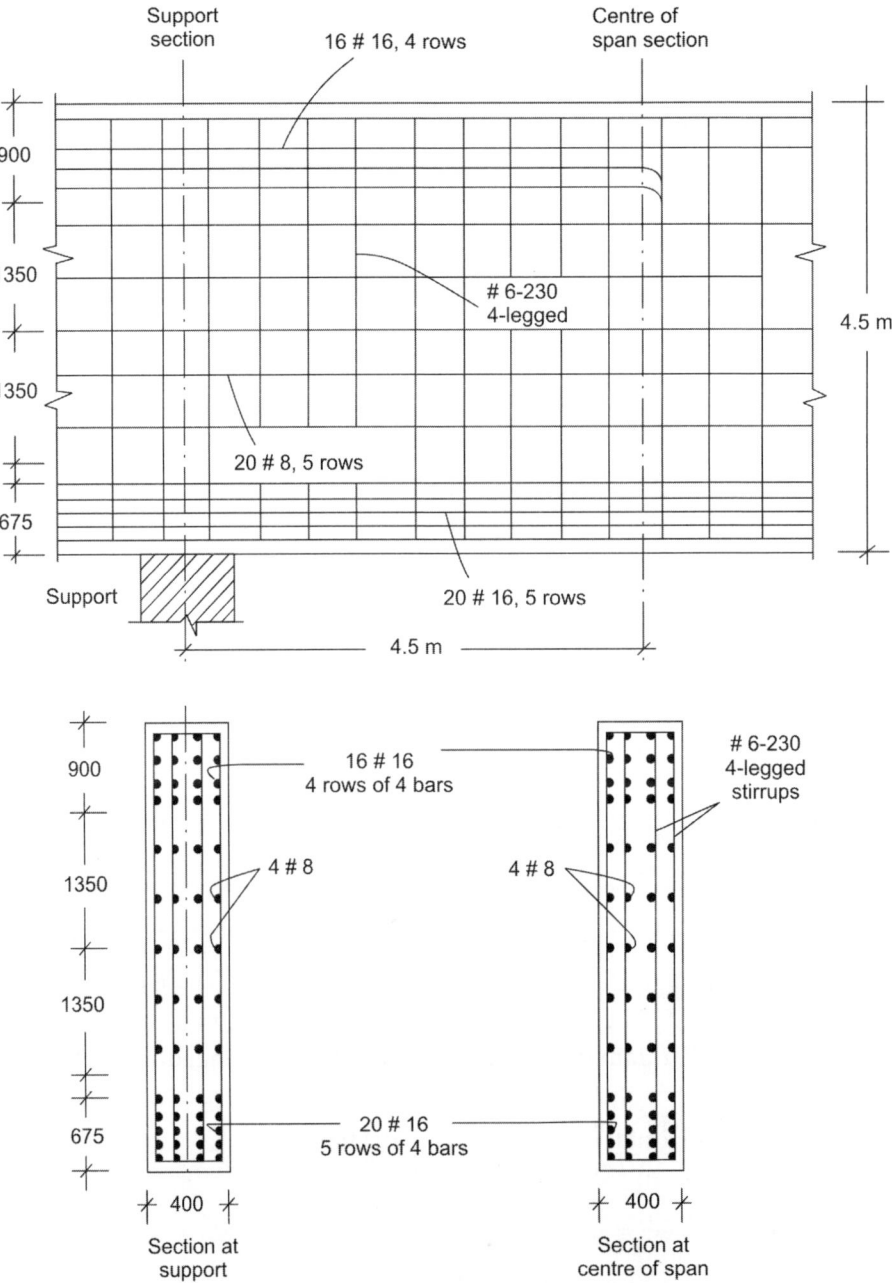

Fig. 20.12: Reinforcement details in deep beam

2. Design a single span deep beam to suit the following data:

Effective span = 6 m

Overall depth = 6 m

Width of support = 0.6 m

Width of beam = 0.4 m

Total load on beam including self weight = 400 kN/m

Concrete = M20 grade

Steel = Fe 415 HYSD bars

i. *Design parameters*:

$L = 6$ m, $C = 0.6$ m

$D = 6$ m, $w = 400$ kN/m

For single span deep beams, the length of the support is disregarded and reaction assumed to act at the centre line of the support. Using the parameters of the continuous girder, we have

$$\varepsilon = 0.5 \text{ and } \beta = \left(\frac{D}{2L}\right) = \left(\frac{6}{2 \times 6}\right) = 0.5$$

The total tension obtained from Fig. 20.6 is

$$T = 0.095 \, w(2L) = (0.095 \times 400 \times 2 \times 6) = 456 \text{ kN}$$

$\therefore$ $A_s = \left(\dfrac{456 \times 10^3}{230}\right) = 1983 \text{ mm}^3$

Using the IS:456 code procedure,

$$\text{maximum moment} = \left(\frac{wL^2}{8}\right) = \left(\frac{400 \times 6^2}{8}\right) = 1800 \text{ kN·m}$$

For simply supported beams,

$$Z = 0.2(L + 2D) \text{ for } 1 \leq L/D \leq 2$$

In this case $(L/D) = 1$

$\therefore$ $Z = 0.2(6 + 2 \times 6) = 3.6$

ii. *Reinforcements*:

$$\text{Tension} = \left(\frac{M}{Z}\right) = \left(\frac{1800}{3.6}\right) = 500 \text{ kN}$$

$\therefore$ $A_s = \left(\dfrac{500 \times 10^3}{230}\right) = 2174 \text{ mm}^2$

Minimum horizontal reinforcement = $(0.002 \times 400 \times 6) = 4800 \text{ mm}^2$

Zone of depth = $(0.25D - 0.05L) = (0.25 \times 6 - 0.05 \times 6) = 1.2$ m

The tension reinforcement of 4800 mm², is arranged within a depth of 1.2 m from the tension face.

Adopt 24 bars of 16 mm diameter in 6 rows of 4 bars each.

Vertical reinforcement area = $(0.0012 \times 400 \times 1000) = 480 \text{ mm}^2$

Use four-legged stirrups of 6 mm diameter at 230 mm centres.

The reinforcement details are as shown in Fig. 20.13.

3. A coal bunker 3 m × 3 m square with a height of 3 m stores 300 kN of coal. The side walls are 180 mm thick. Depth of hopper bottom is 1.2 m with a central opening of 0.5 m × 0.5 m. The bunker is supported on four RC columns at the corners. If the total weight of hopper bottom is 60 kN, design the side walls of the bunker as a deep beam using M20 grade concrete and Fe 415 HYSD bars.

i. *Data*:

Weight of coal = 300 kN

Weight of hopper bottom = 60 kN

Size of bunker = 3 m × 3 m × 3 m

Thickness of side walls = 180 mm

ii. *Loads*:

Weight of one side wall = $(3 \times 3 \times 0.18 \times 24) = 40$ kN

Total load on one wall acting as deep beam = $\left(\dfrac{360}{4} + 40\right) = 130$ kN

iii. *Moments*:

Span of deep beam = 3 m

$$M = \left(\frac{WL}{8}\right) = \left(\frac{130 \times 3}{8}\right) = 48.75 \text{ kN·m}$$

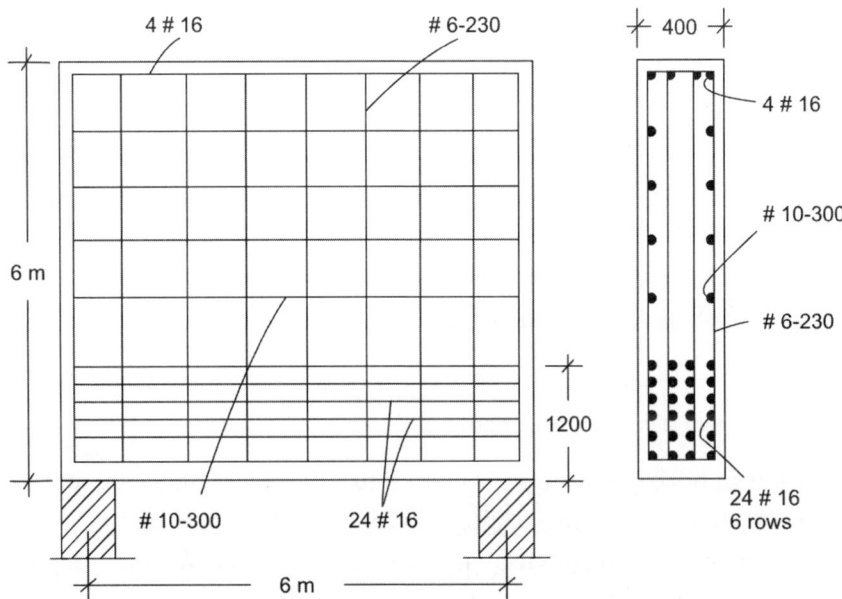

Fig. 20.13: Reinforcement details in single deep beam

iv. *Reinforcements*:

$$\left(\frac{L}{D}\right) = \left(\frac{3}{3}\right) = 1$$

$$Z = 0.2(L + 2D) = 0.2(3 + 2 \times 3) = 1.8 \text{ m}$$

Tension $\quad T = \left(\frac{M}{Z}\right) = \left(\frac{48.75}{1.8}\right) = 27 \text{ kN}$

$$\therefore \qquad A_s = \left(\frac{27 \times 10^3}{230}\right) = 118 \text{ mm}^2$$

Minimum horizontal reinforcement = $(0.002 \times 180 \times 3000) = 1080 \text{ mm}^2$

Use 10 bars of 12 mm diameter in a zone of depth

$$= (0.25D \times 0.05L) = (0.25 \times 3000 \times 0.05 \times 3000) = 900 \text{ mm}$$

Vertical reinforcement = $(0.0012 \times 180 \times 1000) = 216 \text{ mm}^2$

Use 6 mm diameter bars at 240 mm centres in vertical direction in both faces.

References

1. IS:456-2000, Indian Standard Code of Practice for Plain and Reinforced Concrete (Fourth Revision), Bureau of Indian Standards, New Delhi, July 2000.

2. Taylor FW and Thompson SE, *A Treatise on Concrete, Plain and Reinforced*, Bureau of Indian Standards, 1 edn, 1905; 2 edn 1912; 3 edn, 1916.

3. Faber Oscar and Bowie PG, *Reinforced Concrete–Theory and Practice*, Vol 1 and 2, 1912–1920.

4. Bresler BE and Macgregor JG, Review of concrete beams failing in shear, *ASCE Journal (Structural Division)*, Vol. 93, Feb. 1967, pp. 343–372.

5. ACI-ASCE Committee 326, Report on shear and diagonal tension, *Journal of the American Concrete Institute*, Vol. 59, Jan, Feb and March 1962, pp. 1–30, 277–334 and 352–396.

6. ACI-ASCE Committee 426, The shear strength of reinforced concrete members, *ASCE Journal Structural Division*, Vol. 99, June 1973, pp. 1091–1187.

7. Kong FK, *Reinforced Concrete Deep Beams*, Van Nostrand Reinhold, New York, 2002.

8. Rogowsky DM and Macgregor JG, Design of reinforced concrete deep beams, *Concrete International*, Vol. 8, No. 8, August 1986, pp. 46–58.

9. Ramakrishnan V and Ananthanarayana Y, Ultimate strength of reinforced concrete deep beams in shear, *Proceedings of the American Concrete Institute Journal*, Vol. 65, No. 2, 1968, pp. 87–98.

10. Smith KN and Vantsiotis AS, Shear strength of deep beams, *ACI Journal Proceedings*, Vol. 79, No. 3, May 1982, pp. 201–213.

11. Russu G, Venir R and Pauletta M, Reinforced concrete deep beams shear strength model and design formula, *ACI Journal Proceedings*, Vol. 102, No. 3, May-June 2005, pp. 429–437.

12. Tan KH, Tang CY and Tong K, A direct design method for deep beams with web reinforcements, *Magazine of Concrete Research*, London, Vol. 55, No. 1, February 2003, pp. 53–63.

Assignment

1. A single span deep beam has an overall depth of 4 m and an effective span of 6 m. The width of the beam is 400 mm. The beam supports a uniformly distributed live load of 300 kN/m, over the entire span. Using M20 grade concrete and Fe 415 HYSD bars, design suitable reinforcements for the beam and sketch the details.

2. A continuous deep beam spanning over three equal spans of 8 m each has an overall depth of 4 m. The width of support is 0.8 m and the width of beam is 0.4 m. The beam supports a uniformly distributed live load of 160 kN/m. Using M20 grade concrete and Fe 250 grade mild steel, design suitable reinforcements for the central span of the continuous deep beam. Sketch the details of reinforcements.

3. A continuous deep beam with an interior span of 9 m, has an overall depth of 4.5 m. The width of support is 0.9 m and the width of beam is 400 mm. The beam supports a uniformly distributed live load of 100 kN/m together with concentrated loads of 200 kN at centre of span points. Using M20 grade concrete and Fe 415 HYSD bars, design suitable reinforcements for the beam.

4. The total weight of contents and the hopper bottom of a bunker having four walls of size 4 m × 4.7 m is 840 kN. The thickness of walls is 150 mm. The bunker is supported on four columns located at the corners. Adopting M20 grade concrete and Fe 415 HYSD grade bars, design the walls of the bunker assuming them as deep beams.

5. A transfer girder has to be designed as a deep beam. Using the following data, design the deep beam and sketch the details of reinforcements.

   Effective span = 9 m

   Overall depth of beam = 5 m

   Width of beam = 500 mm

   Uniformly distributed live load = 200 kN/m

   Grade of concrete = M25

   Type of reinforcements = Fe 500 HYSD bars

6. Design a deep continuous beam using the following data:

   Number of spans = 2

   Effective length of each span = 10 m

   Overall depth of beam = 4.5 m

   Width of beam = 500 mm

   Uniformly distributed load = 200 kN/m

   Concentrated load at centre of spans = 200 kN

   Grade of concrete = M25

   Type of reinforcements = Fe 500 HYSD bars

Review Questions

1. In what way the design of a deep beam different from that of a simple beam?

2. How do you classify a beam as a deep beam? What are the main parameters used in designating the beam as deep?

3. What are the types of failures encountered in the case of reinforced concrete deep beams?

4. Explain with sketches the variation of flexural stresses at mid span of deep beams supporting uniformly distributed load.

5. How does the flexural stresses vary at the support of deep beams supporting uniformly distributed load?

6. What are the main parameters to be considered in the design of reinforced concrete deep beams?

7. Explain with sketches the types of diagonal tension cracks developed in reinforced concrete deep beams. In what way they are different in comparison with that of ordinary beams?

8. What are the specifications incorporated in the Indian standard code for the design of reinforcements in deep beams to resist the positive and negative moments?

9. Explain with sketches the resultant tension developed in deep beams subjected to concentrated loads.

10. How do you arrange the reinforcements in a deep beam? Explain with sketches the typical arrangement of reinforcements in the longitudinal cross-sections of a deep beam.

Objective Type Questions

1. In the case of simply supported deep beams, the ratio of effective span to overall depth is
   a. greater than 2
   b. less than 2
   c. in the range of 3 to 4

2. In deep beams, the bending stress distribution across a transverse section
   a. is the same as that of an ordinary beam
   b. deviates significantly from the straight line distribution
   c. can be evaluated using simple beam theory

3. The flexural stresses developed at mid span of a deep beam supporting uniformly distributed load depends upon
   a. the grade of concrete in the beam
   b. the span length
   c. the depth to span ratio

4. In deep beams, the resultant tension developed due to loading will be maximum at
   a. quarter span
   b. mid span
   c. supports

5. A continuous beam is designated as a deep beam when the span/depth ratio is
   a. greater than 3
   b. less than 2.5
   c. less than 2.0

6. In continuous deep beams supporting imposed loads, the maximum moment develops at
   a. mid span
   b. quarter span
   c. supports

7. According to the Indian standard code, the minimum percentage of horizontal reinforcement to be provided in deep beams is
   a. 0.3%
   b. 1.0%
   c. 0.2%

8. The Indian standard code specifies that the minimum vertical reinforcements in deep beams should be not less than
   a. 0.5%
   b. 1.0%
   c. 0.12%

9. The minimum percentage of side face reinforcements to be used in deep beam is
   a. 1.00%
   b. 0.12%
   c. 0.30%

10. In the case of deep beams having a span to depth ratio less than 1, the IS code specifies that the reinforcements should be evenly distributed over a depth expressed as a function of overall depth equal to
    a. 0.5
    b. 0.75
    c. 0.80

21

Pipes

21.1   INTRODUCTION

Reinforced concrete pipes[1,2] are widely used for various types of applications like water supply and sanitary systems, bridge structures, gravity mains for carrying water under hydrostatic pressure. The main advantage of reinforced concrete pipes is its negligible maintenance costs in comparison with the traditional steel pipes which are costly and prone to corrosion damage. In coastal regions where the structure has to withstand severe environmental conditions, reinforced concrete pipes are invariably used due to its superior durability[3,4] characteristics. Rapid improvements in the manufacturing process of precast pipes by spinning process[5] has reduced the costs and also significantly improved the structural quality of reinforced concrete pipes. For the last few decades, concrete pipes have more or less replaced steel pipes due to their superior durability and reduction in costs.

The application of prestressing techniques[6,7] has further improved the structural performance of concrete pipes for large scale water supply systems using large diameter pipes to convey water under hydrostatic pressure. Prestressed concrete pipes have been used for the water supply systems of metropolitan areas in many countries. Reinforced concrete hume pipes have been used for culverts of bridges and for drainage of storm water at city road crossings. The reader may refer to investigative research reports of Rao[8] and Frank[9] and other prominent codes[10] for design aspects of reinforced concrete pipes. The various tests to be conducted for certification of RCC pipes can be obtained from Indian standard code IS:3597-1966[11]. Specifications for laying of concrete pipes is codified in IS:783-1985[12].

21.2   CLASSIFICATION OF PIPES

According to the Indian standard code IS:458-1971, reinforced cement concrete pipes are classified as non-pressure and pressure pipes with their applications as detailed below in Table 21.1.

Table 21.1: Classification of pipes

	Pipe designation	Conditions where used
NP-1	Unreinforced concrete non-pressure pipe	For drainage and irrigation use above ground or in shallow trenches
NP-2	Reinforced concrete light duty non-pressure pipe	For drainage and irrigation use for culverts carrying light traffic
NP-3	RC heavy duty non-pressure pipe	For drainage and irrigation use and for culverts carrying heavy traffic
NP-4	RC heavy duty non-pressure pipe	For drainage and irrigation use and for culverts carrying very heavy traffic such as railway loadings
P-1	RC pressure pipes tested to a hydrostatic pressure of 0.2 N/mm^2 (20 m head)	For use in gravity mains the design pressure not exceeding 2/3 of test pressure
P-2	RC pressure pipes tested to a hydrostatic pressure of 0.4 N/mm^2 (40 m head)	For use in pumping mains, the design pressure not exceeding half the test pressure
P-3	RC pressure pipes tested to a hydrostatic pressure of 0.6 N/mm^2 (60 m head)	For use in pumping mains, the design pressure not exceeding half the test pressure

21.3 DESIGN PRINCIPLES

Reinforced concrete pipes either spun or cast are designed to withstand the internal hydrostatic pressure without exceeding the permissible stresses of 126.5 N/mm^2, the mild steel and 140 N/mm^2 in case of cold drawn steel wires. The thickness of concrete pipe is designed in such a way that under specified test pressure, the maximum tensile stress in concrete, when considered as effective to take stress along with the tensile reinforcement, should not exceed 2 N/mm^2. The minimum thickness of pipe varies with the internal diameter and classification of pipes. For pressure pipes the thickness varies from 25 mm for diameter of 80 mm to 65 mm for a diameter of 1200 mm. The spigot dimensions of NP-1 class pipes are designated as shown in Fig. 21.1. The structural design of pipes should be in accordance with the Indian standard code IS:783-1985[12].

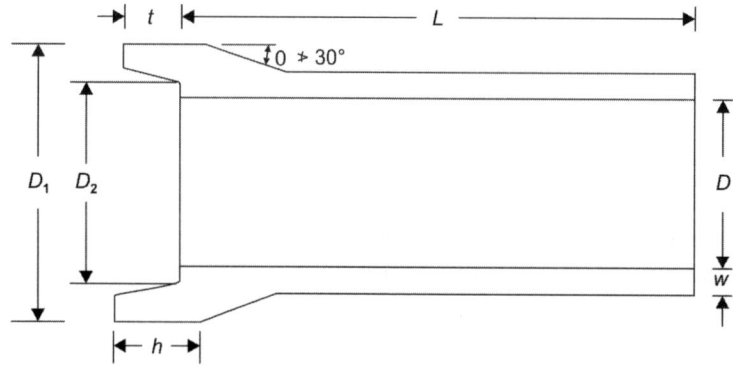

Fig. 21.1: Spigot dimensions of NP-1 class RCC pipes

The longitudinal reinforcement is designed to support the RCC pipe as a circular beam loaded with twice the self weight of the pipe and twice the weight of water to fill the pipe across a span equal to the length of the pipe. Under these loading conditions, the stresses in the reinforcement should not exceed the permissible stresses.

21.4 REINFORCEMENTS IN PIPES

The circumferential and longitudinal reinforcements are designed for the loads but minimum quantity of steel reinforcement are specified for different classes of pipes in IS:458-1971. The typical reinforcement requirements for pipes of class P-1 are shown in Table 21.2.

Table 21.2: Reinforcement requirements in pipes of class P-1

Internal diameter (mm)	Reinforcements	
	Longitudinal mild steel at permissible stress of 126.5 N/mm² (kg/m)	Spiral hard drawn steel wire at premissible stress of 140 N/mm² (kg/m)
100	0.863	0.327
200	0.863	0.575
400	1.00	3.800
600	1.25	8.150
800	1.78	14.500
1000	2.50	22.50
1200	3.36	32.50

The pitch of spirals should neither be more than 100 mm or 4 times the thickness of the barrel, whichever is less, nor less than the maximum size of aggregate plus the diameter of the bar used.

The minimum clear cover for concrete pipes specified in the IS code for different types of pipes are as shown in Table 21.3.

Table 21.3: Cover requirements

Barrel thickness	Four spun pipes (mm)	For pipes other than spun pipes (mm)
25 mm and below	8.5	12.0
Over 25 and including 30 mm	9.0	12.0
Over 30 and below 75	12.0	16.0
75 mm and above	18.0	18.0

21.5 ░ TESTS ON PIPES

The RCC pipes should conform to the following tests specified in Indian standard code IS:3597-1966:

a. Hydrostatic test

b. Three edge bearing test or sand bearing test

c. Absorption test

d. Bursting test

The requirements of load to produce a crack of size 0.25 mm and the ultimate load to be sustained before failure for different classes of pipes are specified in the relevant IS code.

21.6 ░ DESIGN EXAMPLE

A reinforced concrete pressure pipe is to be designed to withstand a working pressure of $0.2\ N/mm^2$. The internal diameter of the pipe is 1000 mm and the length of the pipe is 3 m. Design the pipe and sketch the details of reinforcements. Adopt M20 grade concrete and hard drawn steel wire conforming to IS:432.

i. *Data*:

Intensity of pressure = $0.2\ N/mm^2 = 200\ kN/m^2$

Internal diameter = 1000 mm = 1 m

Length of pipe = 3 m, m = 13 m

ii. *Permissible stresses*:

According to IS:458-1971

$$\sigma_{st} = 140\ N/mm^2$$
$$\sigma_{ct} = 2\ N/mm^2$$

iii. *Hoop tension and reinforcement*:

Maximum hoop tension = $\left(\dfrac{pD}{2}\right) = \left(\dfrac{200 \times 1}{2}\right) = 100\ kN/m$

$$\therefore \qquad A_{st} = \left(\dfrac{100 \times 10^3}{140}\right) = 714\ mm^2/m\ (length)$$

Provide 10 mm diameter bars at 70 mm c/c as spiral reinforcement ($A_{st} = 1122\ mm^2$)

Minimum quantity of steel as per IS:458 is 22.5 kg/m length of pipe

Assuming 80 mm as thickness of pipe,

weight of one spiral of 10 mm diameter = $(\pi \times 1.06 \times 0.612) = 2.03\ kg$

Number of spirals in 1 m = $\left(\dfrac{1000}{70}\right) = 14.3$

$\therefore$ Weight of spiral reinforcement/metre length of pipe

$$= (2.03 \times 14.3) = 29.02\ kg/m$$

The quantity of steel reinforcement provided is greater than the minimum of 22.5 kg/m specified in the code.

iv. *Thickness of pipe*:

Let t be thickness of pipe from cracking considerations:

$$\left[\frac{(pD/2)}{1000t + (m-1)A_{st}}\right] = \sigma_{ct}$$

$$\left[\frac{100 \times 10^3}{1000t + (13-1) \times 1122}\right] = 2$$

Solving, $t = 36.53$ mm

But minimum thickness is not less than 55 mm. Adopt $t = 60$ mm.

v. *Longitudinal reinforcement*:

Assuming the pipe to span over a length of 3 m,

self weight of pipe $= (\pi \times 1.06 \times 0.06 \times 24) = 4.79$ kN/m

Weight of water $= \left(\frac{\pi \times 1^2}{4} \times 10\right) = 7.85$ kN/m

∴ Total design load $= 2(4.79 + 7.85) = 25.28$ kN/m

$$I = \frac{\pi}{64}\left[D^4 - d^4\right] = \frac{\pi}{64}(1.12^4 - 1^4) = 0.028 \text{ m}^4$$

$$M_{max} = \left(\frac{25.28 \times 3^2}{8}\right) = 28.44 \text{ kN·m}$$

$$\text{Stress } \sigma = \left(\frac{28.44 \times 10^6 \times 560}{0.028 \times 10^{12}}\right) = 0.568 \text{ N/mm}^2$$

Stresses are negligibly small. Provide minimum longitudinal reinforcement of 2.5 kg/m using 5 mm diameter bars,

weight of each bar $= \left(\frac{\pi \times 0.005^2}{4} \times 7800\right) = 0.153$ kg/m

∴   Number of bars required $= \left(\frac{2.5}{0.153}\right) = 16.33$

∴   Provide 20 bars of 5 mm diameter spaced 166 mm along the circumference as longitudinal reinforcement.

21.7 ▌ DESIGN OF NON-PRESSURE RCC PIPES FOR CULVERTS

RCC pipes are commonly used as a cross-drainage work for a road or railway embankment. The hydraulic design consists of computing the area of the pipe required to pass a given discharge. If Q is the discharge, A the cross-sectional area of the pipe, v the velocity of flow, and d the diameter of the pipe, then:

$$A = \left(\frac{\pi d^2}{4} \right) = \left(\frac{Q}{v} \right) \qquad \therefore \quad d = \sqrt{\frac{4Q}{\pi v}}$$

The structural design of the pipe involves the computations of the three edge bearing strength of pipe, the weight of earthfill over the fill and the load on pipe due to a surface concentrated live load, each associated with a strength factor generally taken as 1.5. The type of non-pressure pipe and bedding are so chosen that under the worst combination of field loading, a factor of safety of 1.5 is available as given by the equation:

$$\left[\frac{\text{Three edge bearing strength (kN/m)}}{\text{Factor of safety (1.5)}} \right] = \left[\frac{W \text{ due to filling material (kN/m)}}{\text{Corresponding strength factor}} \right]$$

$$+ \left[\frac{W \text{ due to surface load (kN/m)}}{\text{Strength factor (1.5)}} \right]$$

The load enforced on the pipe due to the soil in embankment is computed from the equation:

$$W = C_e w D^2$$

where,

W = vertical external load in kN/m of pipe due to embankment material as shown in Table 21.4

C_e = coefficient depending on the ratio of height of embankment H to the external diameter of the pipe and condition of laying as specified in IS:783-1959

w = density of the embankment material in kN/m³

D = external diameter of pipe (m)

Table 21.4: Load on pipe due to earth fill

Pipe size d (mm)	Outer dia. D (mm)	Embankment loading on pipe in kN/m for various depths H in metre									
		1	2	3	4	5	6	7	8	9	10
NP-3											
500	650	16.8	34.9	54.0	65.3	99.0	118.5	150.0	171.0	191.0	212.0
600	760	18.7	42.6	62.4	75.5	103.0	133.0	160.5	186.0	197.0	230.0
700	860	20.0	48.0	68.0	86.5	118.5	154.5	174.0	209.0	228.0	246.0
800	980	24.2	55.3	79.5	107.0	136.5	163.0	202.0	230.0	260.0	288.0
900	1100	28.3	58.7	91.5	122.0	150.0	185.0	209.0	257.0	297.0	338.0
1100	1200	28.6	59.5	101.0	132.0	163.0	202.0	228.0	259.0	324.0	357.0
1200	1430	33.4	74.0	115.0	159.0	200.0	226.0	274.0	315.0	352.0	427.5

The load on the pipe due to a concentrated highway wheel load P is obtained from the equation:

$$W = 4C_s IP$$

where,

W = vertical external load in kN/m due to concentrated surface load

C_s = influence coefficient depending upon D and H as compiled in Table 21.5.

H = vertical depth of top of pipe below the surface (m)

D = external diameter of pipe (m)

P = concentrated wheel load (kN)

I = impact factor (1.5 for highways)

Table 21.5: Influence coefficient C_s for concentrated surface load for highways

Pipe size d (mm)	Outer dia. D (mm)	C_s for various depth H in metre									
		0.1	0.2	0.3	0.4	0.6	0.8	1	2	3	4
NP-3											
500	650	0.246	0.228	0.198	0.169	0.117	0.083	0.060	0.017	0.008	0.005
600	760	0.247	0.234	0.210	0.182	0.131	0.094	0.068	0.022	0.010	0.006
700	860	0.247	0.236	0.215	0.186	0.140	0.102	0.075	0.024	0.010	0.006
800	980	0.249	0.240	0.220	0.196	0.149	0.110	0.083	0.027	0.013	0.007
900	1100	0.249	0.241	0.255	0.202	0.156	0.117	0.089	0.029	0.014	0.008
1100	1200	0.249	0.242	0.228	0.205	0.162	0.123	0.095	0.032	0.015	0.010
1200	1430	0.249	0.242	0.230	0.209	0.171	0.131	0.104	0.036	0.020	0.011

In case of railway loading, the load is uniformly distributed because of the sleepers and ballast. The load on a buried pipe in a railway embankment is given by the equation:

$$W = 4C_s UD$$

where,

W = load on pipe in kN/m

C_s = influence coefficient depending on the length of the sleeper, distance between two axles and depth of the top of pipe below surface

U = uniformly distributed load in kN/m², on the surface directly above the

$$\text{pipe} = \left[\frac{PI}{4AB} + 2W_t B \right]$$

P = axle load in kN (229 kN for BG)

A = half the length of the sleeper in metre (1.35 m for BG)

B = half the distance between the two driving axles in metre (0.92 m for BG)

W_t = weight of track structure in kN/m (generally 3 kN/m)

D = external diameter of the pipe in metre.

For broad gauge loading, the equation reduces to $W = 339C_sD$. The values of the coefficient C_s are compiled in Table 21.6.

Table 21.6: Influence coefficient C_s for broad gauge railway loading

H (m)	C_s	H (m)	C_s
0.1	0.250	1.0	0.183
0.2	0.249	2.0	0.094
0.3	0.245	3.0	0.052
0.4	0.240	4.0	0.032
0.5	0.233	5.0	0.021
0.6	0.224	6.0	0.015
0.7	0.218	7.0	0.011
0.8	0.205	8.0	0.009
0.9	0.193	9.0	0.007
		10.0	0.005

21.8 DESIGN EXAMPLE

Design a suitable pipe culvert to suit the following data:

Discharge through pipe culvert = 1.57 m³/s
Velocity of flow through pipe = 2 m/s
Width of road (two lane) = 7.5 m
Top width of embankment = 15 m
Side slopes of embankment = 1.5:1
Bed level = 100.00
Top of embankment = 103.00
Loading: IRC class AA
Draw the longitudinal section, and end view of the pipe culvert.

   i. *Diameter of pipe culvert*:

   Discharge $Q = Av$

$$\therefore \quad A = \left(\frac{Q}{v}\right) = \left(\frac{1.57}{2}\right) = 0.785 \text{ m}^2$$

$$\left(\frac{\pi d^2}{4}\right) = 0.785 \quad \therefore \quad d = 1.00$$

Adopt NP 3, RCC heavy duty non-pressure pipe for carrying heavy road traffic. From IS:458-1971, for a pipe of internal diameter 1 m, the external diameter $D = 1.2$ m.

ii. *Load due to earth fill*:

Height of embankment over pipe = 2 m

From Table 21.4, for d = 1000 mm and H = 2 m,

load due to earth fill = 59.5 kN/m

iii. *Load on pipe due to IRC class AA wheel loads*:

Assuming IRC class AA tracked vehicle

$$\text{Load/m (length)} = \left(\frac{350}{3.6}\right) = 97.22 \text{ kN/m}$$

Loading on pipe = $4C_sIP$

From Table 21.5, for d = 1000 mm and

$$H = 2 \text{ m}, \qquad C_s = 0.032, \qquad I = 1.5, \qquad P = 97.22$$

Loading on pipe = $(4 \times 0.032 \times 1.5 \times 97.22) = 18.66$ kN/m

iv. *Check for safety factor*:

Referring to IS:458-1971,

three edge bearing strength for NP-3 class pipe of 1000 mm internal diameter is 111 kN/m.

∴ The strength factor (SF) required for bending is obtained from the equation:

$$\left(\frac{111}{1.5}\right) = \left(\frac{59.5}{\text{SF}}\right) + \left(\frac{18.66}{1.5}\right) \qquad \therefore \quad \text{SF} = 0.96$$

The strength factor for first class bending is 2.3 and for concrete cradle bedding shown in Fig. 21.2 is 3.7. Any of these two beddings can be provided for the pipe culvert.

v. *Reinforcements*:

The minimum reinforcements in the pipe according to IS:458-1971 are as follows:

Spiral reinforcement of hard drawn steel wire with a permissible stress of 140 N/mm^2 = 44 kg/m.

Longitudinal reinforcement of mild steel with a permissible stress of 126.5 N/mm^2 = 5.80 kg/m.

Provide 12 mm diameter bars at 60 mm c/c as spiral reinforcement.

Average diameter of spiral = 1.1 m.

Weight of one spiral of 12 mm diameter = $(\pi \times 1.1 \times 0.88) = 3.045$ kg

$$\text{Number of spirals in 1 m} = \left(\frac{1000}{60}\right) = 16.66$$

∴ Weight of spiral reinforcement/metre length of pipe

$$= (3.045 \times 16.66) = 50.7 \text{ kg/m}$$

The quantity of spiral steel provided is greater than the minimum of 44 kg/m specified in the code. Provide 6 mm diameter mild steel bars as longitudinal reinforcement.

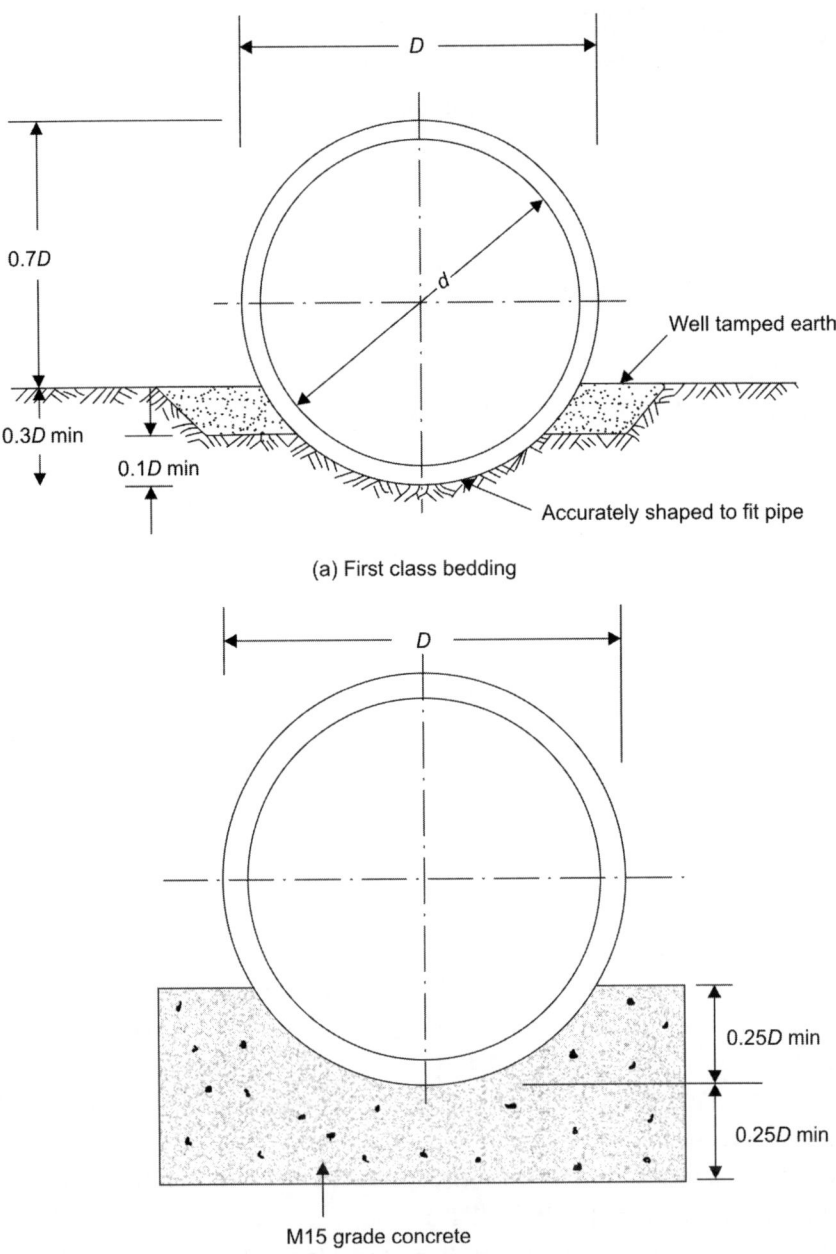

(a) First class bedding

(b) Concrete cradle bedding

Fig. 21.2: Beddings for concrete pipes in pipe culverts

$$\text{Weight of each bar} = \left(\frac{\pi \times 0.006^2}{4} \times 7800 \right) = 0.22 \text{ kg/m}$$

$$\therefore \quad \text{Number of bars required} = \left(\frac{5.80}{0.22} \right) = 26.36$$

$$\text{Spacing} = \left(\frac{\pi \times 1100}{26.36}\right) = 131 \text{ mm}$$

Adopt 130 mm spacing for the longitudinal reinforcement.
The details of the pipe culvert are shown in Fig. 21.3.

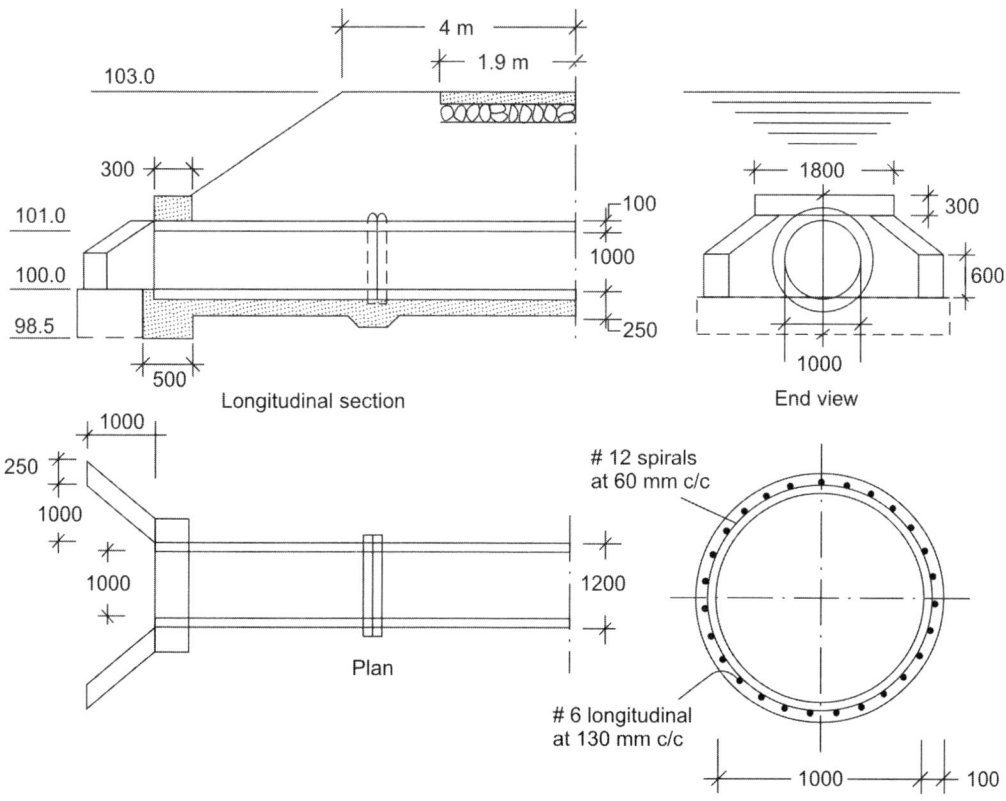

Fig. 21.3: RCC pipe culvert

References

1. IS:458-1971, Indian Standard Specification For Concrete Pipes (With and Without Reinforcement, Second Revision), V- Reprint, Bureau of Indian Standards, New Delhi, April 1983, pp. 1–25.

2. Bacher AE, Banke AN, and Kirkland DE, Reinforced Concrete Pipe Culverts, Design Summary and Implementation, 61st Annual Meeting of the Transportation Research Board, Washington, USA, 1982, pp. 83–92.

3. ACI Committee 201, Guide to durable concrete, *Journal of the American Concrete Institute*, Vol. 74, 1977, pp. 573–609.

4. Unnikrishna Pillai S and Menon D, *Reinforced Concrete Design*, 3 edn, McGraw-Hill Education Pvt Ltd, New Delhi, 2009, p. 62.

5. Reinforced Cement Concrete Spun/Hume Pipe Manufacturing Co, IS and ISO: 9001-2000 (Web Portal), Rajasthan, 2000.

6. Joshi NG, Prestressed Concrete Pipes, State of the Art, International Symposium on Prestressed Concrete Pipes, Poles, Pressure Vessels and Sleepers, Proceedings, Vol. 2, Madras, 1972, pp. p/3–1 to p/3–40.

7. Krishna Raju N, *Prestressed Concrete*, 5 edn, McGraw-Hill Education Pvt Ltd, New Delhi, 2012, pp. 488–521.

8. Rao VVS, Structural design of concrete pipes in accordance with IS:783-1959, *Journal of the Institution of Engineers (India)*, Vol. 50, No. 5 Part CI-3, Jan. 1970, pp. 99–107.

9. Frank HJ, Structural design method for precast RCC pipes, 61st Annual Meeting of the Transportation Research Board, Washington, USA, 1982, pp. 93–100.

10. A. 124-1962, Australian Standard for Concrete Pressure Pipes, Australian Water Resources Council Technical Paper No. 9, Canberra, 1975, pp. 105–110.

11. IS:3597-1966, Indian Standard Code of Practice for Test on RCC Pipes, Bureau of Indian Standards, New Delhi, 1966, pp. 1–5.

12. IS:783-1985 (First Revision), Indian Standard Code of Practice for Laying of Concrete Pipes, Bureau of Indian Standards, New Delhi, 1985, pp. 1–77.

Assignment

1. A reinforced concrete pressure pipe is to be designed to withstand a working pressure of 0.1 N/mm². The internal diameter of the pipe is 500 mm and the length of the pipe is 3 m. Design the pipe according to the IS code requirements and sketch the details of reinforcements. Adopt M20 grade concrete and hard drawn steel wire conforming to IS:432.

2. A RCC pipe is required to withstand 15 m head of water using M20 grade concrete and hard drawn steel wire, design the pipe and sketch the details of reinforcements.

3. Design a suitable pipe culvert to carry a discharge of 1 m³/s with a velocity of 2 m/s. The depth of earth filling over the pipe is 2 m. Adopt IRC class AA loading with M20 grade concrete and steel conforming to IS:432. Sketch the details of reinforcements and bedding for the pipe.

4. A reinforced concrete pipe is required to withstand a service pressure of 150 kN/m². The internal diameter of the pipe is 1200 mm and the length of the pipe is 3 m. Design the required thickness of concrete for the pipe and suitable longitudinal and hoop steel reinforcements according to the Indian standard code IS:432. Adopt M25 grade concrete and Fe 500 grade HYSD bars. Sketch the detail steel reinforcements in the cross and longitudinal sections of the pipe.

5. Design a suitable RCC pipe culvert for a highway crossing using the following data:

    Maximum likely discharge through pipe = 1.3 m³/s

    Velocity of flow through the pipe = 1.8 m/s

    Width of road (two lane traffic) = 7.5 m

    Top width of embankment = 18 m

    Side slopes of embankment = 2:1

    Bed level = 100.00

    Top of embankment = 104.00

    Loading: IRC class AA

    Sketch the details of the culvert using longitudinal and cross-sections.

6. A reinforced concrete pipe is to be designed for a factory in a coastal region. Design the pipe to conform to the specifications of the Indian standard codes using the following data:

Intensity of pressure inside the pipe = 0.3 N/mm^2

Internal diameter of the pipe = 750 mm

Length of pipe = 4 m

Grade of concrete = M25

Grade of steel = Fe 500 HYSD bars

Review Questions

1. Under what situations would you prefer to adopt reinforced concrete pipes in place of steel pipes?

2. What are the advantages of reinforced concrete pipes in comparison with steel and cast iron and GI pipes?

3. Mention the types of pipes you would select for carrying water under pressure from a dam site to a city in the coastal region. Specify the reasons for your selection.

4. Discuss briefly the classification of reinforced concrete pipes according to the Indian standard codes, mentioning the situations under which they are used.

5. Briefly outline the design principles to be followed, while designing reinforced concrete pressure pipes according to the specifications of the Indian standard codes.

6. What are the various types of reinforcements used in concrete pipes? Mention the structural purpose of using these reinforcements.

7. What are the minimum cover requirements adopted in reinforced concrete pipes designed according to the IS codes?

8. Briefly explain the method of considering the loads on pipe due to earth fill and concentrated loads due to vehicles on highway and loading due to railways.

9. Explain with sketches the various types of bedding used to lay the reinforced concrete pipes.

10. Explain with sketches a typical NP-1 class reinforced concrete pipe showing the details of spigot portion of the pipe.

Objective Type Questions

1. Reinforced concrete pipes are preferred to steel and other types of metallic pipes mainly due to
   a. lighter weight
   b. durability aspects
   c. faster construction

2. Reinforced concrete circular pipes are generally made
   a. at the site
   b. by casting using vertical forms
   c. by spinning process

3. Reinforced concrete pipes of different diameter are generally made in lengths of
   a. 1 to 2 m
   b. 5 to 8 m
   c. 3 to 4 m

4. The thickness of the concrete pipe is designed by the considerations of
   a. flexural stresses developed
   b. limiting the tensile stress from cracking considerations
   c. depth of soil above the pipe

5. The minimum clear concrete cover required for spun concrete pipes of thickness more than 75 mm according to the specifications of Indian standard code is
   a. 10 mm
   b. 30 mm
   c. 18 mm

6. The minimum quantity of spiral hard drawn wire at permissible stress of 140 N/mm² specified in the IS code for pipes of 1000 m diameter is
   a. 3.8 kg/m
   b. 50 kg/m
   c. 22.5 kg/m

7. In the structural design of non-pressure reinforced concrete pipes, used in embankments should be tested for safety using
   a. hydrostatic test
   b. absorption test
   c. three edge bearing test

8. In the case of beddings used to support the pipes, well tamped earth is used in
   a. concrete cradle bedding
   b. first class bedding
   c. continuous concrete raft bedding

9. In case of pipes required to be used as gravity mains, the class of pipe to be used according to the IS code is
   a. NP-1
   b. P-3
   c. P-1

10. The thickness required for a concrete pipe to resist hydrostatic pressure is determined using the limit state of
    a. collapse
    b. deflection
    c. cracking

22

Column Brackets
(Corbels)

22.1   INTRODUCTION

In machinery workshops and industrial structures, the reinforced concrete columns in the main shop floor are often provided with bracket projections to support the gantry girder crane. These brackets which are deep and transfer the crane loads to the main columns are referred to as corbels or nibs. A typical corbel subjected to loads at a short distance from the face of the column is shown in Fig. 22.1a. Corbels are also provided at the cantilever end of girders in double cantilever balanced reinforced concrete bridges[1,2] to support the end spans of the bridge.

Corbels are short cantilevers whose shear span/depth (a_v/d) ratio is less than 1.0 and the depth (D_f) at the end face is not less than one half of the depth D_s at support. In case of corbels, the load transfer at support is mainly by *strut* action than by simple flexure as shown in Fig. 22.1b. The revised IS:456-2000 code[3] does not specify any method for the design of corbels except prescribing the enhanced shear strength of concrete near the supports.

However, the revised British Euro code[4] and the American Concrete Institute code[5], based on several research investigations[6,7,8] have recommended some design principles which are outlined in the subsequent sections.

22.2   SHEAR SPAN/DEPTH RATIO AND SHEAR RESISTANCE

In case of corbels, heavy loads are transmitted very near to the supporting column and the shear resistance of reinforced concrete members is different from that in which the loads are applied far away from the supports. The shear resistance of concrete depends upon the shear span/depth (a_v/d) ratio and it varies as shown in Fig. 22.2. As the shear span/depth ratio increases from 0.5 to 2, the shear strength of concrete decreases rapidly as indicated by the ratio of enhanced shear strength to the normal shear strength (τ_m/τ_c) which decreases from 4 to 1.

The IS:456-2000 code considers this enhancement of shear strength in clause 40.5.1 and the reinforcements from sections close to the supports are designed from the balance shear as per clause 40.5.2. The maximum shear strength τ_m is limited to the values specified in Table 20 of IS:456 code.

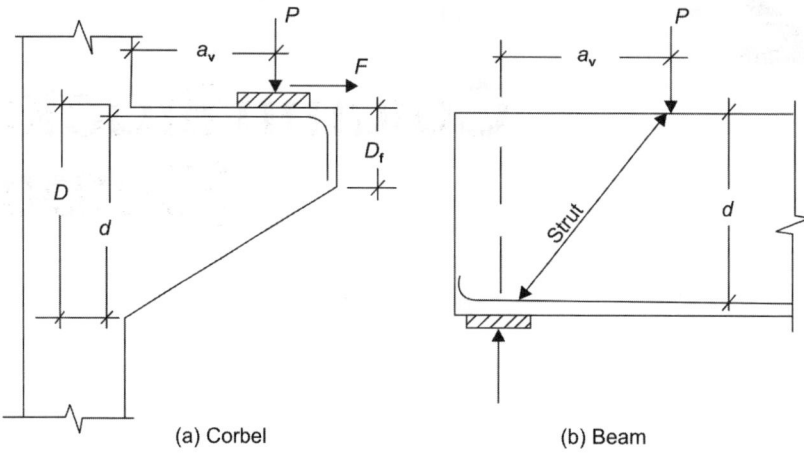

(a) Corbel   (b) Beam

Fig. 22.1: Shear span/depth ratio in corbels and beams

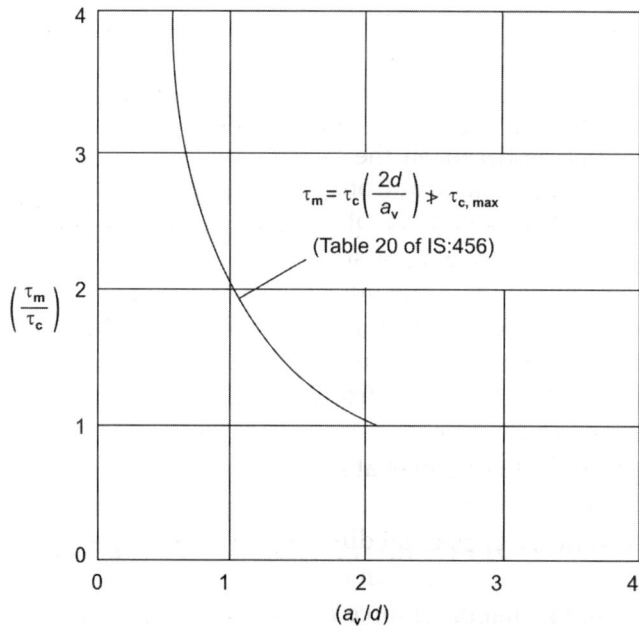

$$\tau_m = \tau_c \left(\frac{2d}{a_v} \right) \ngtr \tau_{c, max}$$

(Table 20 of IS:456)

Fig. 22.2: Influence of shear span/depth ratio on the ratio of enhanced to normal shear strength

22.3 ▌ DIMENSIONING CORBELS

The initial dimensions of a corbel are based on the permissible bearing stresses in concrete which are compiled by Mosley et al[9] based on the British Euro code recommendations. The design and detailing of a corbel should conform to the following requirements:

i. The safe bearing stress on the loaded area of concrete should not exceed a value expressed by the relation
$$f_{br} = 0.48[1 - f_{ck}/250] f_{ck}$$

ii. A horizontal force $H = 0.2 F_v$ must also be resisted. This force acts at the level of the top of the bearing plate at a distance a_H above the horizontal tie.

iii. The main tension steel A_{st} must be fully anchored into the column and the other end of these bars must be welded to an anchorage device or loops of reinforcing bars.

iv. The angle of inclination, θ of the compression strut must be within the limits expressed as $68° \geq \theta \geq 45°$ or $2.5 \geq \tan\theta \geq 1.0$.

v. The design stress, f_{cd} of the concrete strut model must not exceed $(\sigma_{cc} f_{ck}/\gamma_c)v_1$, where
$$v_1 = 0.6 [1 - (f_{ck}/250)]$$
$$\sigma_{cc} = 0.85$$
$$\gamma_c = 1.5, \text{ the partial safety factor for concrete in compression}$$
Therefore, f_{cd} must not exceed $0.34 f_{ck}[1 - (f_{ck}/250)]$.

vi. Closed horizontal links each of area $A_{s,link} \geq 0.5 A_s$ should be provided to confine the concrete in the compression strut.

22.4 ANALYSIS OF FORCES IN A CORBEL

The design of corbels is based on the assumption of strut action recommended by both the British and the American Concrete Institute codes. The procedure outlined below follows the recommendations of the British Euro code BS EN 1992-2004.

The forces acting on a corbel are shown in Fig. 22.3. The vertical force F_{ed} is in equilibrium under the action of the horizontal tensile force F_{td} in steel reinforcement and the inclined compressive force F_{cd} developed in concrete simulating strut action.

The following notations are used for the analysis of forces in corbels.

a_v = distance of vertical force from the face of the support
b = breadth of corbel
d = effective depth of corbel at support
F_v = applied vertical load
F_{td} = tension in the horizontal direction
F_{cd} = compression developed in concrete due to strut action
θ = angle of inclination of force F_{cd} to the horizontal
D_f = depth of corbel at free end
D_s = depth of corbel at support
t = thickness of the bearing plate over the corbel
a_{tt} = distance between the bearing plate and tension tie
a_t = distance from the face of column to the point B (Fig. 22.3)
   $= (a_v + 0.2a_{tt})$

The force in the concrete strut is F_{cd} and F_{td} in the horizontal ties respectively.

a. *Force in the concrete strut* (F_{cd})

The design stress for the concrete strut is $f_{cd} = 0.34 f_{ck}(1 - f_{ck}/250)$.

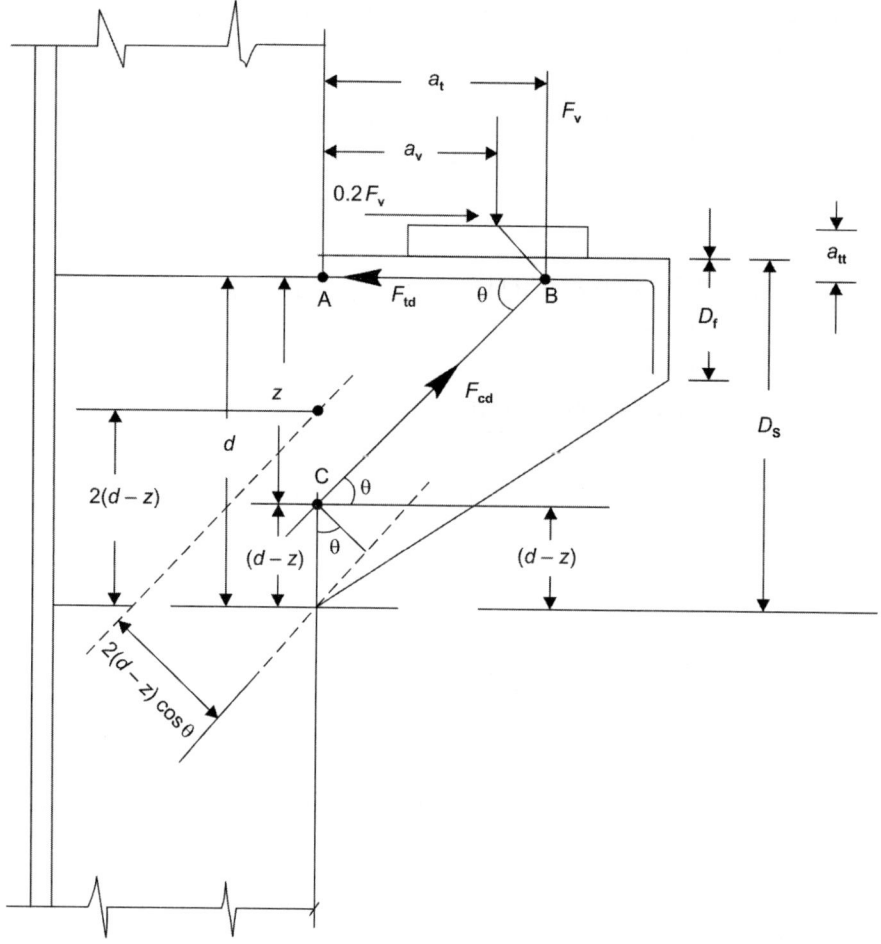

Fig. 22.3: Strut and tie system of forces in a corbel

From the geometry of Fig. 22.3, the width of the concrete strut measured vertically is $2(d-z)$. Hence, the width of the strut measured at right angles to its axis is given by $w_{strut} = 2(d-z)\cos\theta$.

Thus, the force F_{cd} in the concrete strut is expressed as

$$F_{cd} = (f_{cd} \times w_{strut} \times b\cos\theta) \qquad \qquad ...(22.1)$$

where, b = width of the corbel.

b. *Angle of inclination of the concrete strut*

Resolving the forces vertically at point B, we have

$$F_v = F_{cd}\sin\theta$$
$$= [f_{cd} \times 2(d-z) \times b\cos\theta]$$
$$= [f_{cd} \times (d - a_t\tan\theta)b\sin 2\theta]$$
$$= \left[f_{cd} \times d \times b\left\{1 - \left(\frac{a_t}{d}\right)\tan\theta\right\}\sin 2\theta \right]$$

On rearranging, we get

$$\left(\frac{F_v}{f_{cd}db}\right) = \left[1-\left(\frac{a_t}{d}\right)\tan\theta\right]\sin 2\theta \qquad\qquad ...(22.2)$$

This equation cannot be solved directly for θ. Also Table 22.1 which has been developed directly from this equation can be used in practical problems.

Table 22.1: Values of θ to satisfy relation (22.2)

θ (degrees)	$\left[\dfrac{F_v}{f_{cd}db}\right]$						
	$(a_t/d) = 0.9$	0.8	0.7	0.6	0.5	0.4	0.3
45	0.100	0.200	0.300	0.400	0.500	0.600	0.700
46	0.068	0.171	0.275	0.378	0.482	0.585	0.689
47	0.035	0.142	0.249	0.356	0.463	0.570	0.677
48	0.000	0.111	0.221	0.332	0.442	0.553	0.663
49		0.079	0.193	0.307	0.421	0.535	0.649
50		0.046	0.163	0.281	0.398	0.515	0.633
51		0.012	0.133	0.253	0.374	0.495	0.616
52			0.101	0.225	0.349	0.474	0.598
53			0.068	0.196	0.323	0.451	0.579
54			0.035	0.166	0.297	0.427	0.558
55			0.000	0.134	0.269	0.403	0.537
56				0.102	0.240	0.377	0.515
57				0.070	0.210	0.351	0.492
58				0.036	0.180	0.323	0.467
59				0.001	0.148	0.295	0.442
60					0.116	0.266	0.416
61					0.083	0.236	0.389
62					0.049	0.205	0.361
63					0.015	0.174	0.333
64						0.142	0.303
65						0.109	0.273
66						0.075	0.242
67						0.041	0.211
68						0.007	0.179

c. *Main tension steel* (A_{st})

Resolving the forces at B in the horizontal direction, the force F_t in steel is given by

$$F_t = F_{cd}\cos\theta = F_v\,(\cos\theta/\sin\theta) = F_v\cot\theta$$

The total force F_t' in the steel tie, including the effect of the horizontal force of $0.2F_v$ is given by

$$F_t' = (F_v \cot \theta + 0.2F_v) \qquad \qquad ...(22.3)$$
$$= F_v(\cot \theta + 0.2)$$

The area of main tension steel in the corbel (A_{st}) is given by

$$A_{st} = \left(\frac{F_t'}{0.87 f_{yk}} \right) = \left(\frac{F_t'}{f_s} \right) \qquad \qquad ...(22.4)$$

22.5 DESIGN PROCEDURE FOR CORBELS

When the vertical load has a stiff bearing, a_v may be measured to the edges of the bearing but where a flexible bearing is used, the distance is measured to the vertical force. A stepwise procedure is recommended for the design of corbels, where the steps i. to vi. are mentioned in Section 22.3 and steps vii. to xvi. are given below:

vii. *Breadth of bearing plate*: Based on design bearing pressure and the length of bearing plate being equal to the width of column, the width of bearing plate is determined.

viii. *Corbel depth at support*: Due to enhanced shear strength near supports permitted by the IS:456 code, assume a suitable value for τ_c, nearer to $\tau_{c,max}$ (Table 20 of IS:456) but not exceeding this value and compute the effective depth d at the support section using the relation,

$$d = \left[\frac{F_v}{\tau_c b} \right]$$

Hence, overall depth $D_s = (d + \text{effective cover})$.

ix. *Check for corbel dimensions*: The value of (a_v /d) should preferably be less than 0.6 but not greater than 1.0.

x. *Determination of lever arm depth* (z): Assume the thickness of bearing plate = t.
Point B is distant $a_t = (a_v + 0.2t)$ from the face of the column because of the effect of the horizontal force $H = 0.2F_v$.
From the geometry of the triangle ABC,
lever arm depth $z = a_t \tan \theta = (a_v + 0.2t) \tan \theta$
Compute $f_{cd} = 0.34 f_{ck}[1 - (f_{ck}/250)]$

Compute the ratios (a_t/d) and $\left(\dfrac{F_v}{f_{cd} db} \right)$

Corresponding to these two ratios, interpolate the value of θ from Table 22.1. Calculate z by using the value of the angle θ.

xi. *Computation of force* (F_t'):
$$F_t' = F_v (\cot \theta + 0.2)$$

xii. *Area of main reinforcement*:
$$A_{st} = \left(\frac{F_t'}{f_s} \right)$$

xiii. *Check for minimum and maximum percentages of steel*: The area of steel A_{st} should not exceed 1.3% and not less than 0.4% of the cross-section (*bd*). If it exceeds the maximum limit, increase the depth of the corbel and redesign.

xiv. *Area of horizontal shear reinforcement* (A_{sh}): The shear reinforcements are provided as closed loops in the upper two-third portion of the total depth of corbel at support.

xv. *Check for shear*: Knowing the percentage of steel ($100A_{st}/bd$), the exact value of the permissible shear stress is given by

$$\tau_m = \tau_c(2d/a_v)$$

The support section is checked for safety against shear.

xvi. *Reinforcement detailing*: The detailing of reinforcements in the corbel should conform to clause 7.7, Figs 7.18 and 7.19 of SP:34[10], which are reproduced as Figs 22.4 and 22.5 respectively in the text.

22.6 DESIGN OF NIBS (BEAM SHELVES)

In prefabricated structural systems, the reinforced concrete walls or columns are provided with nibs or beam shelves to support floor units comprising slabs and beams. Continuous nibs less than 300 mm in depth are designed as cantilever slabs with suitable reinforcements provided in the form of horizontal loops to resist the shear forces applied close to the supports similar to corbels. The following guidelines recommended by the Cement and Concrete Association, UK are useful in the design of nibs.

i. The bending moment and enhanced shear strength are computed by considering the distance a_v representing the line of action of the load as the distance from the centre line of the nearest vertical leg of the stirrup in the beam to the outer face of the main horizontal reinforcement of the nib as shown in Fig. 22.6.

ii. Additional ties or links are provided as hangers in the beam connected to the nib. The load on the nib has to be resisted by the compression zone of the supporting

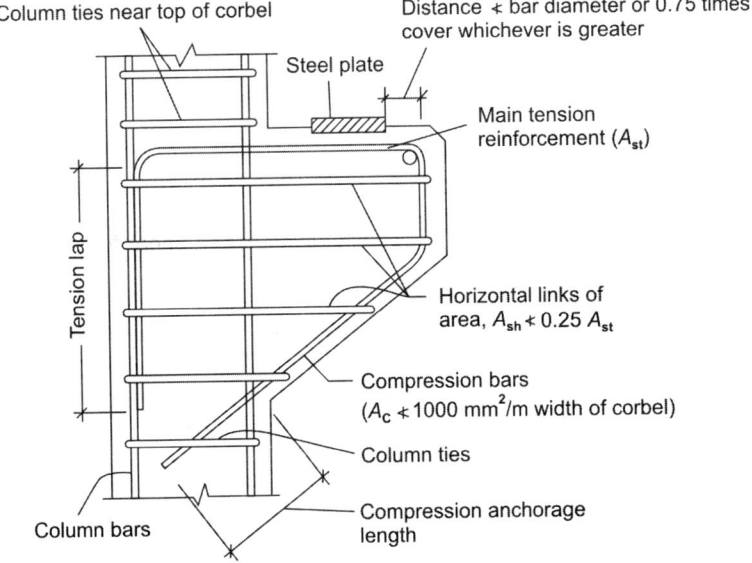

Fig. 22.4: Reinforcement details in corbel with main reinforcement of 16 mm diameter and less (SP:34)

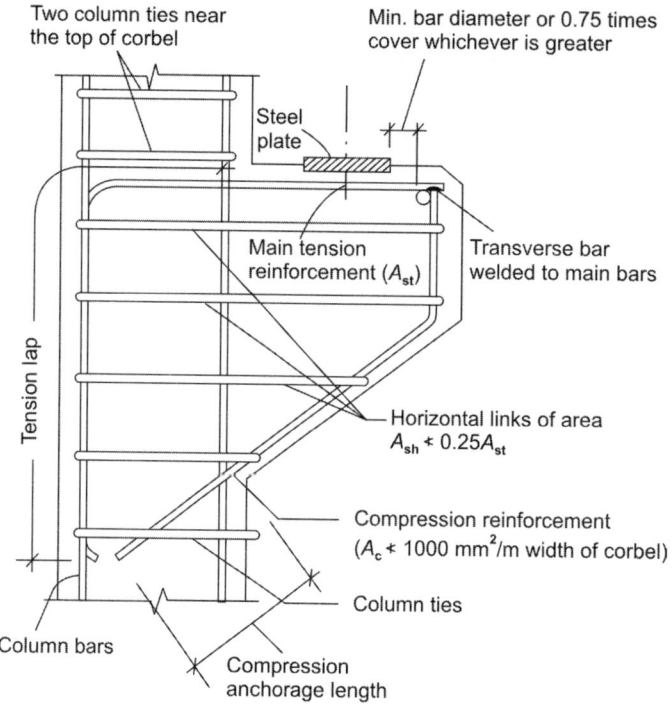

Fig. 22.5: Reinforcement details in corbel with main reinforcement of 18 mm diameter or more (SP:34)

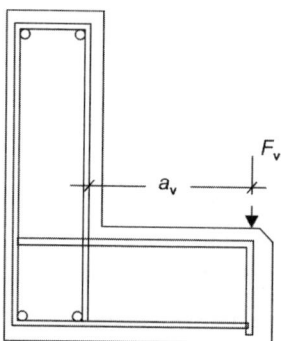

Fig. 22.6: Distance a_v for bending moment and shear force

beam. Hanger bars are used not only to resist shear in the beam but also to transfer the load from the nib to the compression side of the beam.

The additional reinforcement area (in addition to the area necessary to support the shear force) for the hangers is computed using the relation,

$$A_s = \left(\frac{F_v}{0.87 f_y} \right)$$

According to the recommendations of the Cement and Concrete Association, UK, the cantilever portion of the nibs should be reinforced with both horizontal and vertical

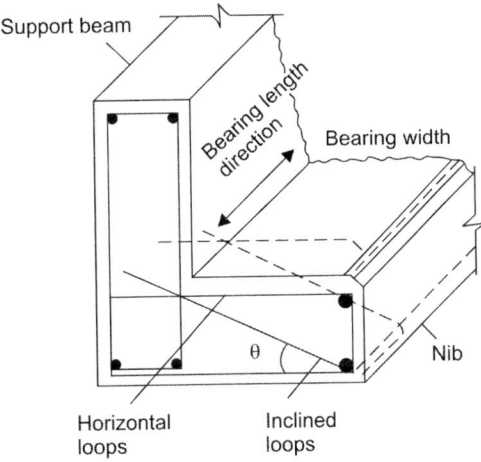

Fig. 22.7: Reinforcements in nibs with large loads

systems of reinforcements as shown in Fig. 22.7. The area of horizontal nib steel is given by

$$A_{sh} = \left(\frac{F_v a_v}{0.87 f_y z} \right)$$

where,

a_v = distance of the load F_v from the nearest hanger bar

z = lever arm

If inclined nibs are used, the area of the inclined nib reinforcement is given by

$$A_{si} = \left(\frac{F_v}{0.87 f_y \sin \theta} \right)$$

where, θ = angle of the inclined loop with the horizontal (Fig. 22.7).

These reinforcements are to be held securely in position by using additional fixing bars running parallel to the nib as shown in Fig. 22.7.

In the case of nibs supporting light loads, the horizontal reinforcement may be bent as a loop as shown in Fig. 22.8.

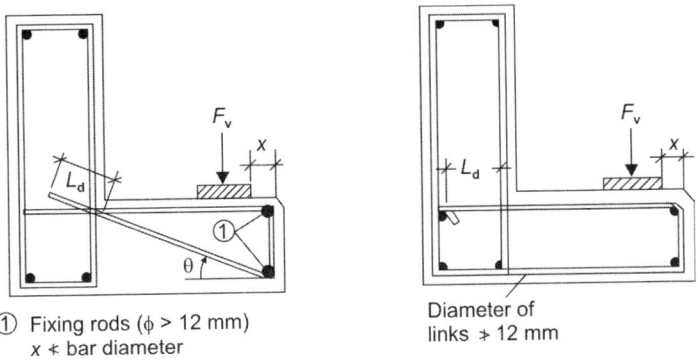

① Fixing rods (ϕ > 12 mm)
x ≮ bar diameter

Diameter of
links ≯ 12 mm

Fig. 22.8: Reinforcements in nibs with light loads

22.7 ▌ DESIGN EXAMPLES

1. Design a corbel to support a factored load of 400 kN at a distance of 200 mm from the face of a column of cross-section 300 mm × 400 mm. Adopt M25 grade concrete and Fe 415 grade HYSD bars. Sketch the details of reinforcements in the corbel.

   i. *Data*:

     Factored load F_v = 400 kN

     Width of column = length of corbel = 300 mm

     Distance between the bearing plate and tension tie (a_{tt}) = 75 mm

     Shear span length a_v = 200 mm

     Materials: M25 grade concrete (f_{ck} = 25 N/mm²)

                Fe 415 HYSD bars (f_y = 415 N/mm²)

  ii. *Dimensions of corbel*:

     Bearing length = width of column = 300 mm

     Adopt a bearing plate of 300 m width over corbel.

     Overall size of bearing plate = 300 mm × 300 mm

     The bearing is flexible and at a distance of 75 mm above the tension tie (a_{tt} = 75 mm)

     Bearing stress should not exceed

$$f_{br} = 0.48[1 - (f_{ck}/250)]f_{ck}$$
$$= 0.48[1 - (25/250)]25$$
$$= 10.8 \text{ N/mm}^2$$

     Providing corbel projection from column face = 400 mm

$$\text{Actual bearing stress} = \left[\frac{300 \times 10^3}{300 \times 300}\right] = 3.33 \text{ N/mm}^2 < 10.8 \text{ N/mm}^2$$

  iii. *Depth of corbel*:

     From Table 19 of IS:456, for M25 grade concrete $\tau_{c,max}$ = 3.1 N/mm²

$$\therefore \qquad d = \left(\frac{F_v}{\tau b}\right) = \left(\frac{400 \times 10^3}{3.1 \times 300}\right) = 430 \text{ mm}$$

     Adopt effective depth d = 450 mm

     Distance $a_t = (a_v + 0.2a_{tt}) = [200 + (0.2 \times 75)] = 215$ mm

     Also ratio $(a_t/d) = (215/450) = 0.47 < 0.6$

     Total depth at support is

$$D_s = (d + \text{cover} + \left(\frac{1}{2}\right) \text{ diameter of bar})$$

$$= (450 + 40 + 10) = 500 \text{ mm}$$

     Depth at face $D_f = (0.5D_s) = (0.5 \times 500) = 250$ mm

     The corbel dimensions are shown in Fig. 22.9.

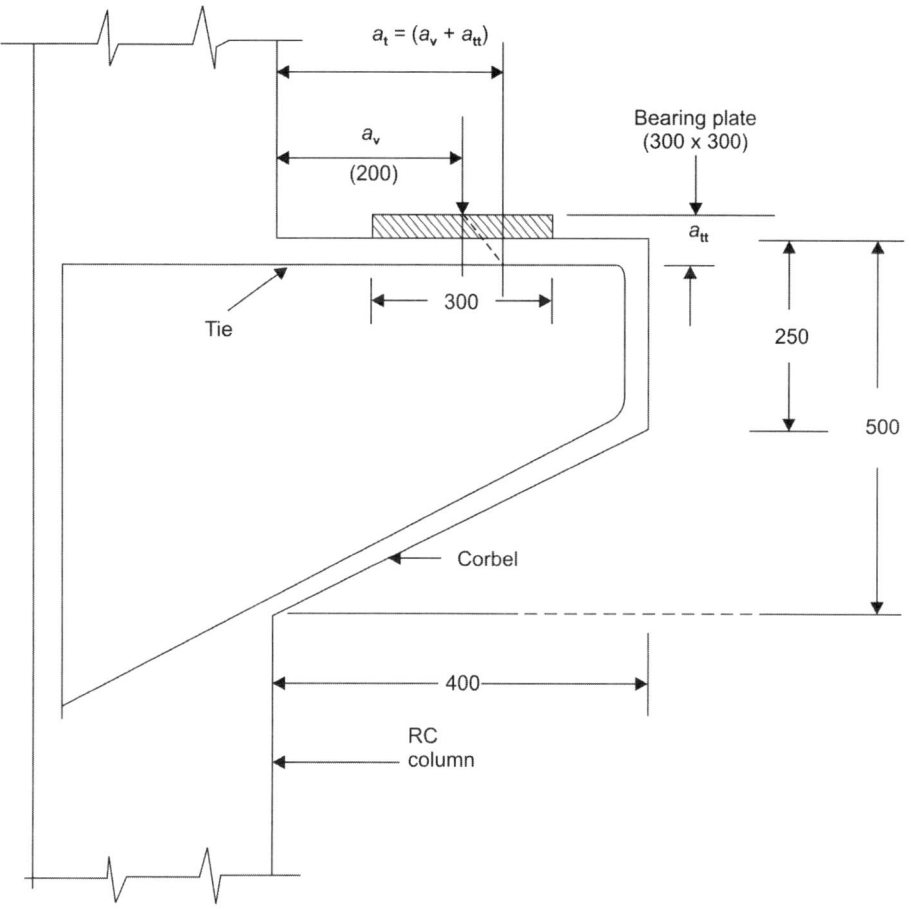

Fig. 22.9: Dimensions of corbel

iv. *Check for strut action*:

Distance of centre of the force $a_t = 215$ mm

Ratio $(a_t/d) = 0.47 < 0.6$, hence acts as a strut.

v. *Determination of lever arm depth (z)*:

$$f_{cd} = 0.34\,f_{ck}[1 - (f_{ck}/250)]$$
$$= (0.34 \times 25)[1 - (25/250)] = 7.65 \text{ N/mm}^2$$

Effective depth $d = 450$ mm

Ratio $(a_t/d) = (215/450) = 0.47$

and $\left(\dfrac{F_v}{f_{cd}db}\right) = \left(\dfrac{400 \times 10^3}{7.65 \times 450 \times 300}\right) = 0.38$

Hence from Table 22.1, value of θ corresponding to this ratio is

$$\theta = 55°$$

$\therefore$ $z = (a_v + 0.2t)\tan\theta$

$$= (200 + 0.2 \times 75)\tan 55° = 307 \text{ mm } (> a_v = 200 \text{ mm})$$

Force $F_t' = F_v(\cot\theta + 0.2) = 400(0.7 + 0.2) = 360$ kN

vi. *Area of tension reinforcement*:

$$A_{st} = \left(\frac{F'_t}{0.87 f_{yk}} \right) = \left(\frac{360 \times 10^3}{0.87 \times 415} \right) = 998 \text{ mm}^2$$

Provide 4 bars of 20 mm diameter (A_{st} = 1017 mm²).

vii. *Area of compression reinforcement*:

A_c not less than 1000 mm² per metre width of corbel.

Provide 4 bars of 16 mm diameter.

viii. *Area of shear reinforcement*:

$$A_{sv(min)} = \left(\frac{1}{2} A_{st} \right) = 508.5 \text{ mm}^2$$

Provide 4 numbers of 10 mm diameter two-legged horizontal links in the upper two third depth (A_{sv} = 628 mm²) at 75 mm spacing.

ix. *Check for shear*:

$(100 A_{st} / bd) = (100 \times 1017)/(300 \times 450) = 0.753$

Using Table 19 of IS:456, for M25 grade concrete and 0.753% steel,

$$\tau_c = 0.57 \text{ N/mm}^2 \text{ and } \left(\frac{a_v}{d} \right) = 0.444$$

∴ Enhanced shear strength = $0.57 \left(\frac{2 \times 450}{200} \right) = 2.565 \text{ N/mm}^2$

Hence, shear capacity of concrete is computed as

$$V_c = \left(\frac{2.565 \times 300 \times 450}{1000} \right) = 346 \text{ kN}$$

Shear capacity of steel = $\left(\frac{0.87 f_y A_s d}{S_v} \right) = \left(\frac{0.87 \times 415 \times 157 \times 450}{75 \times 1000} \right) = 340 \text{ kN}$

∴ Total shear capacity = (346 + 340) = 686 kN > 400 kN. Hence, the design is safe.

x. *Reinforcement details*:

The detailing of reinforcements in the corbel should conform to the specifications of SP:34[10], which are shown in Fig. 22.10.

2. Design a continuous nib (beam support) projecting from a RCC wall to support a prefabricated slab unit transmitting a service shear force of 15 kN/m, assuming the following data.

i. *Data*:

$F_v = (1.5 \times 15) = 22.5 \text{ kN/m}$

Projection of nib = 200 mm

$a_v = 100 \text{ mm}$

$f_{ck} = 30 \text{ N/mm}^2 \text{ and } f_y = 415 \text{ N/mm}^2$

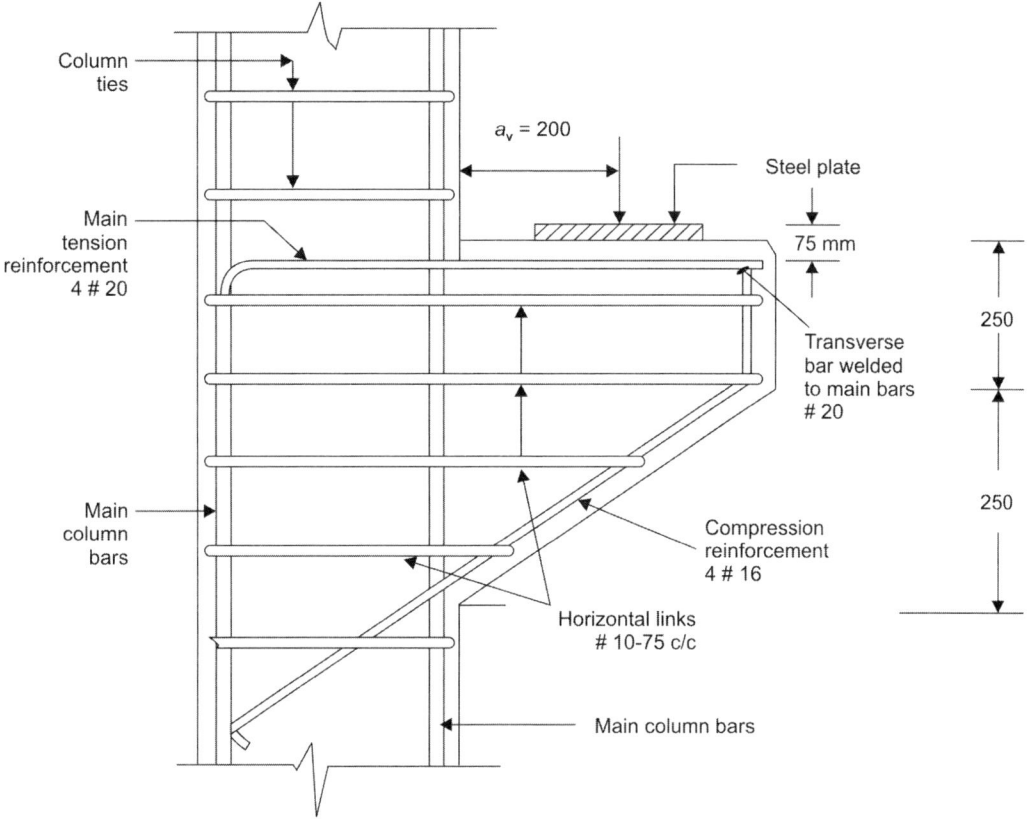

Fig. 22.10: Reinforcement details in corbel

ii. *Dimensions of nib*:

Since the shear force is small, adopt an overall depth of nib $D = 200$ mm and effective depth $d = 150$ mm

iii. *Bending moments and shear forces*:

Maximum bending moment at the face of RC wall is computed as

$$M = (F_v a_v) = (22.5 \times 10^3 \times 100) = (2.25 \times 10^6) \text{ N·mm}$$
$$V = F_v = 22.5 \text{ kN}$$

iv. *Reinforcements*:

Assuming a lever arm depth $z = 0.8d = (0.8 \times 150) = 120$ mm

$$A_{st} = \left(\frac{F_v a_v}{0.87 f_y z} \right) = \left(\frac{22.5 \times 10^3 \times 100}{0.87 \times 415 \times 120} \right) = 52 \text{ mm}^2/\text{m}$$

Provide minimum area of reinforcement of 0.4%.

$$\therefore \qquad A_{st} = (0.004 \times 1000 \times 150) = 600 \text{ mm}^2/\text{m}$$

Adopt 10 mm diameter bars at 130 mm centres ($A_{st} = 604$ mm^2) both at top and bottom of the section.

v. *Check for shear stress*:

$$V_u = 22.5 \text{ kN}$$

$$\left(\frac{100A_{st}}{bd}\right) = \left(\frac{100 \times 604}{1000 \times 150}\right) = 0.40$$

From Table 19 (IS:456) for M30 grade concrete,

$$\tau_c = 0.45 \text{ N/mm}^2$$

$$\text{But } \tau_v = \left(\frac{V_u}{bd}\right) = \left(\frac{22.5 \times 10^3}{1000 \times 150}\right) = 0.15 \text{ N/mm}^2$$

Since $\tau_c > \tau_v$, shear stresses are within safe permissible limits.

vi. *Details of reinforcements*:

The reinforcement details in the nib are shown in Fig. 22.11.

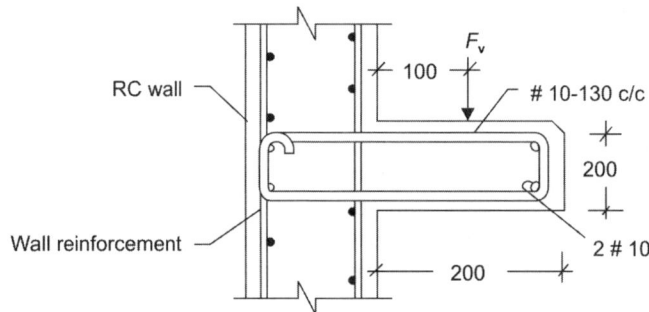

Fig. 22.11: Reinforcement details in nib

References

1. Raina VK, *Concrete Bridge Practice, Analysis, Design and Economics*, Tata McGraw-Hill, New Delhi, 1991, pp. 184–200.

2. Krishna Raju N, *Design of Bridges*, 4 edn, Oxford & IBH, New Delhi, 2009, pp. 315–362.

3. IS:456-2000, Indian Standard Code of Practice for Plain and Reinforced Concrete (Fourth Revision), Bureau of Indian Standards, New Delhi, 2000, p. 37.

4. BS EN:1992-1-1, Euro Code-2, Design of Concrete Structures, General Rules and Rules for Buildings, British Standards Institution, London, 2004.

5. ACI:318M-11, Building Code Requirements for Structural Concrete and Commentary, American Concrete Institute, Farmington Hills, Michigan, 2011.

6. Collins MP and Mitchell D, *Prestressed Concrete Structures*, Prentice Hall, Englewood Cliffs, NJ, 1991, pp. 338–411.

7. Vecchio FJ and Collins MP, The modified compression field theory for reinforced concrete elements subjected to shear, *Journal of the American Concrete Institute*, Vol. 83, March-April 1986, pp. 219–231.

8. Collins MP, Mitchell D, Adebar PE and Vecchio FJ, A general shear design method, *ACI Structural Journal*, Vol. 93, No.1, Jan-Feb, 1996, pp. 36–45.

9. Mosley B, Bunget J and Husle R, *Reinforced Concrete Design to Euro Code-2*, 7 edn, Palgrave MacMillan, London, 2012, pp. 204–210.

10. SP:34, Handbook of Concrete Reinforcement Detailing, Bureau of Indian Standards, New Delhi, 1987.

Assignment

1. Design a corbel for a factory shed column 500 mm × 300 mm to support a vertical ultimate load of 500 kN, with its line of action 200 mm from the face of the column. Assume M25 grade concrete and Fe 415 grade HYSD bars for the construction.

2. Design a corbel to support a reaction due to a characteristic dead load of 80 kN and live load of 120 kN. This reaction acts at 200 mm from the face of the column which is 350 mm^2 in section. There is also a horizontal reaction of 30 kN due to shrinkage restraint of beams, etc. Design the corbel and sketch the details of reinforcement. Assume $f_{ck} = 25$ N/mm^2 and $f_y = 415$ N/mm^2.

3. Design the reinforcement for a corbel projecting 400 mm from a reinforced concrete column of width 400 mm. The service load on the corbel is 400 kN with its line of action at a distance of 250 mm from the column face. Adopt M30 grade concrete and Fe 415 HYSD bars for the corbel. The design should conform to the standards of the British Euro code BS EN 1992-1-1-2004. Sketch the details of reinforcements in the corbel.

4. A continuous concrete nib is to be provided to a reinforced concrete beam cast *in situ*. The nib is to support a series of precast floor units 450 mm wide and 150 mm deep. These floor units have a clear span of 3.5 m and exert an ultimate total reaction of 25 kN per metre length on the nib. The dry bearing of the floor units on the beam can exert a pressure of 0.4 f_{ck}. Assuming that an allowance of 20 mm has to be provided for spilling and an allowance of 25 mm has to be made for the face of column for inaccurate dimension, design a suitable nib and sketch the details of reinforcements. Assume $f_{ck} = 30$ N/mm^2 and $f_y = 415$ N/mm^2.

Review Questions

1. Outline the salient differences between the structural behaviour of a corbel and a cantilever beam.

2. What is shear span? In what way the shear span ratio affects the structural behaviour of short cantilever beams?

3. Briefly explain the significance of enhanced shear strength. What are the various parameters which influence this shear strength according to various prominent codes?

4. List the significant parameters influencing the determination of dimensions of a corbel.

5. Explain with a sketch the various forces acting on a corbel.

6. What are the design considerations in the design of structures using the strut and tie action.

7. Explain the terms: (a) shear span ratio, (b) bearing stress, (c) effective depth and (d) lever arm depth with reference to the analysis of forces on a corbel.

8. In what type of structures would you adopt the use of nibs or beam shelves? Briefly mention the structural purpose of nibs.

9. What are the guidelines to be followed in the design of nibs according to the Cement and Concrete Association, UK?

10. Explain with sketches the detailing of reinforcements in a typical corbel and a nib.

Objective Type Questions

1. A corbel is a short cantilever subjected to forces like
   a. direct tension
   b. compression
   c. tension, compression and bearing

2. The primary forces in a corbel are
   a. torsion
   b. shear
   c. compression, tension and shear

3. The design of a corbel is based on
   a. simple bending theory
   b. combined bending and torsion
   c. strut and tie action

4. The depth of a corbel should be
   a. minimum at the support
   b. maximum at the support and minimum at the free end
   c. maximum at the free end

5. The main tension reinforcement in a corbel is provided
   a. horizontally at the top of the corbel
   b. inclined at an angle close to the free end face
   c. vertical in the form of stirrups

6. The depth of a corbel at the support is determined mainly by
   a. the shear at the support
   b. the flexure at the support
   c. compression at the support

7. The angle of inclination of the compression strut in a corbel should be
   a. less than 30°
   b. greater than 70°
   c. in the range between 45° and 68°

8. Closed horizontal links are used in a corbel mainly to
   a. resist tension
   b. to confine the concrete
   c. to resist compression

9. In the case of corbels, the minimum percentage of steel to be used is
   a. 0.15%
   b. 1.00%
   c. 0.40%

10. The maximum area of reinforcement provided in a corbel should not exceed
    a. 3.00%
    b. 1.30%
    c. 0.50%

23

Bridge Deck Systems

23.1 HISTORICAL REVIEW

The first reinforced concrete bridge was built by Adair in 1871 across the river Waveney at Homersfield, near Norfolk, England spanning across 15 m according to an historical survey by Shirley-Smith[1]. The adaptability of reinforced concrete in architectural form was demonstrated by Maillart in Switzerland in building arched bridges using reinforced concrete. An excellent historical survey of the world's greatest bridges is reported by Virola[2] in 1968. The first reinforced concrete railway bridge was the 28 ft span bridge built at Dundee, Scotland in 1903. Salginatobel and Schawanadbach bridges built by Maillart in 1930 and 1933 respectively are classical examples of aesthetically beautiful and efficient use of materials coupled with economy in bridge construction.

Reinforced concrete was preferred to steel as a suitable material for short and medium span bridges due to the added advantage of durability and minimum maintenance against aggressive environmental conditions in comparison with steel. The longest span concrete arch bridge is the Wnagxiang bridge in China built in 1996 with a span of 420 m. The development of prestressed concrete by Freyssinet[3] has paved the way for the construction of new types concrete bridges of longer spans.

23.2 TYPES OF REINFORCED CONCRETE BRIDGES

Reinforced concrete bridges can be of different types depending upon the span, type of crossing such as roadway, railway or pedestrian use and its location and density of traffic. Broadly, reinforced concrete bridges[4] can be grouped under the following types:

i. Slab type used for small spans in the range of 3 m to 5 m

ii. Simply supported tee beam and slab type generally adopted for medium spans in the range of 5 m to 10 m in most of the highway crossings in all types of roads (Refer to Fig. 23.1)

iii. Continuous beam and slab bridges[5] for road crossings of medium to long spans

iv. Arched bridges of open spandrel type for road and river crossings, mainly selected from aesthetic considerations

v. Bow string girder bridges[6] used for crossings in national highways for longer spans

vi. Reinforced concrete pipe culverts[7] mostly used for railway crossings of small streams

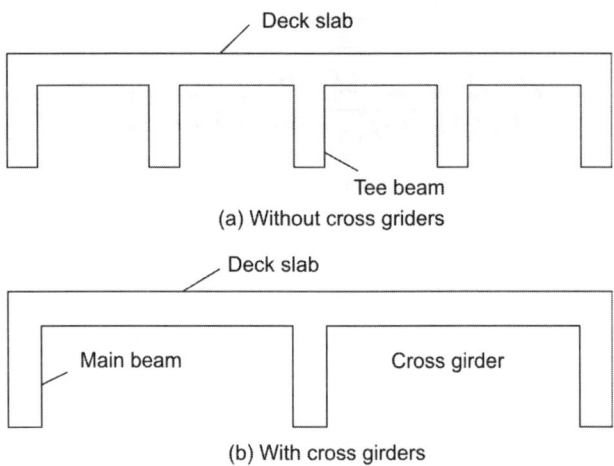

Fig. 23.1: Tee beam and slab bridge decks

vii. Rigid frame bridges[8] consisting of portal frames used in the span range of 10 m to 30 m for road crossings in embankments

viii. Balanced cantilever type bridges[9] with varying depth of beams to achieve economy preferred in situations where redundant structures are not desirable

ix. Box culverts[10] generally preferred for road or railway crossings with high embankments crossing a stream with a limited flow.

23.3 ANALYSIS AND DESIGN OF SLAB BRIDGE DECKS

a. Slabs Spanning in One Direction

For slabs spanning in one direction, the dead load moments can directly be computed assuming the slab to be simply supported between the supporters. Bridge deck slabs have to be designed for IRC loads, specified as class AA or class A depending on the importance of the bridge. The different classes of IRC loads specified in IRC bridge code are shown in Figs 23.2 and 23.3. For slabs supported on two opposite sides, the maximum bending moment caused by a wheel load may be assumed to be resisted by an effective width of slab measured parallel to the supporting edges. For a single concentrated load, the effective width may be calculated by the equation:

$$b_e = kx\left(1 - \frac{x}{L}\right) + b_w$$

where,

b_e = the effective width of slab in which the load acts
L = effective span
x = distance of centre of gravity of load from nearer support
b_w = breadth of concentration area of load
k = a constant depending on the ratio (B/L), where B is width of the slab.

The values of the constant k for different values of the ratio (B/L) are compiled in Table 23.1.

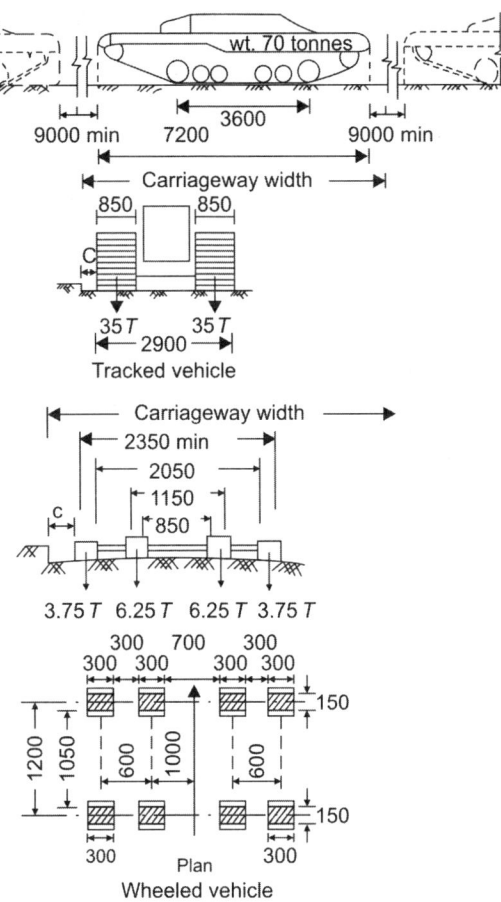

Fig. 23.2: IRC class AA loading

Notes for Fig. 23.2.

  i. The nose to tail spacing between two successive vehicles shall not be less than 90 mm.
 ii. For multilane bridges and culverts, one train of class AA tracked or wheeled vehicles whichever creates severe conditions shall be considered for every two traffic lane width.
iii. No other live load shall be considered on any part of the said two-lane width carriageway of the bridge when the above mentioned train of vehicles is crossing the bridge.
 iv. The maximum loads for the wheeled vehicles shall be 20 tonnes for a single axle or 40 tonnes for a bogie of two axles spaced not more than 1.2 m centres.
  v. The minimum clearance between the road face of the kerb and the outer edge of the wheel or track C, shall be as under:

Carriage width		Minimum value of C
3.8 m and above	Single lane bridges	0.3 m
Less than 0.5 m	Multi lane bridges	0.6 m
5.5 m or above		1.2 m

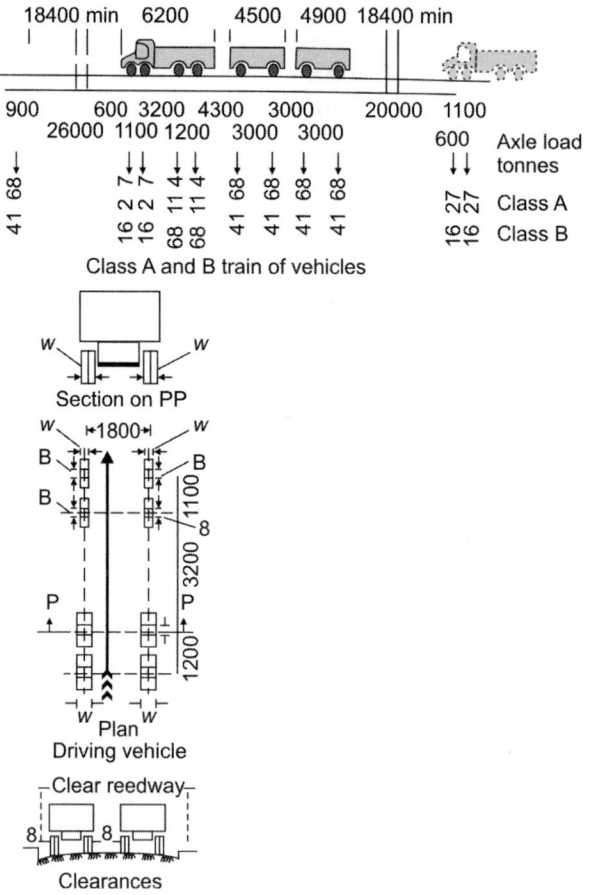

Fig. 23.3: IRC class A and B loadings

Notes for Fig. 23.3.

i. The nose to tail distance between successive trains shall not be less than 18.4 m.

ii. No other live load shall cover any part of the carriageway when a train of vehicles (or trains of vehicles in multi lane bridge) is crossing the bridge.

iii. The ground contact area of the wheels shall be as under:

Axle load tonnes	Ground contact area	
	B (mm)	W (mm)
11.4	250	500
6.8	200	380
4.1	150	300
2.7	150	200
1.6	125	175

iv. The minimum clearance f, between outer edges of the wheel and the roadways face of the kerb, and the minimum clearance g, between the outer edge of passing or crossing vehicles on multilane bridges shall be as given below:

Clear carriageway width	*g*	*f*
5.5 to 7.5 m	Uniformly increasing from 0.4 to 1.2 m	150 mm for all carriage-way widths
Above 7.5 m	1.2 m	

Table 23.1: Values of *k*

B/L	k (for simply supported slab)	k (for continuous slab)	B/L	k (for simply supported slab)	k (for continuous slab)
0.1	0.40	0.40	1.1	2.60	2.28
0.2	0.80	0.80	1.2	2.64	2.36
0.3	1.16	1.16	1.3	2.72	2.40
0.4	1.48	1.44	1.4	2.80	2.48
0.5	1.72	1.68	1.5	2.84	2.48
0.6	1.96	1.84	1.6	2.88	2.52
0.7	2.12	1.96	1.7	2.92	2.56
0.8	2.24	2.08	1.8	2.96	2.60
0.9	2.36	2.16	1.9	3.00	2.60
1.0	2.48	2.24	2 and above	3.00	2.60

The maximum bending moment in the slab develops for class AA wheeled vehicle for spans up to 4 m and for class AA tracked vehicle for spans exceeding 4 m. For two lane bridges, the shear due to class AA tracked vehicle controls the design for all spans from 1 m to 8 m. The distribution reinforcement is designed for 0.3 times the live load moment and 0.2 times the dead load moment.

b. Slabs Spanning in Two Directions

In the case of bridge decks with tee beam and cross-girders, the deck slab is supported on all the four sides and is spanning in two directions. The moments in the two directions can be computed by using the design curves developed by M Pigeaud. The method developed by Pigeaud is applicable to rectangular slabs supported freely on all the four sides and subjected to a symmetrically placed concentrated load as shown in Fig. 23.4.

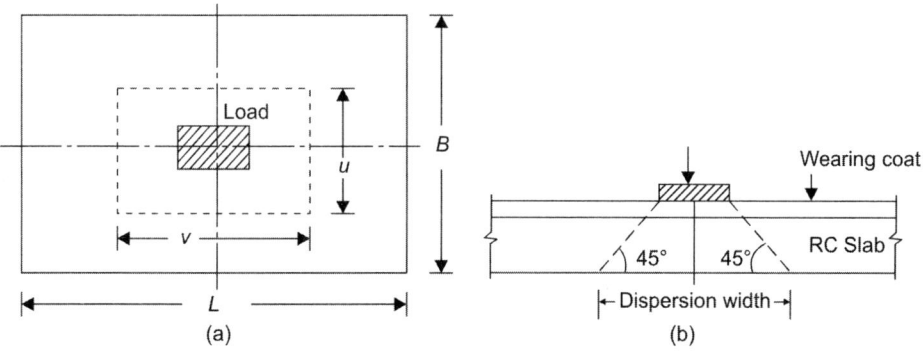

Fig. 23.4: Dispersion of live load through deck slab

The notations used are as follows:

L = long span length

B = short span length

u, v = dimensions of the load spread after allowing for dispersion through the deck slab

k = ratio of short to long span = (B/L)

M_1 = moment in the short span direction

M_2 = moment in the long span direction

m_1 and m_2 = coefficients for moments along the short and long spans

μ = Poisson's ratio for concrete generally assumed as 0.15

W = load from the wheel under consideration

The dispersion of the load may be assumed to be at 45° through the wearing coat and deck slab.

The bending moments are computed as:

$$M_1 = (m_1 + \mu m_2)W$$
$$M_2 = (m_2 + \mu m_1)W$$

The values of the moment coefficients m_1 and m_2, depend upon the parameters, (u/B), (v/L) and K.

Figures 23.5–23.11 are the Pigeaud's curves used for the estimation of moment coefficients m_1 and m_2 for values of k ranging from 0.4 to 1.0.

Moment coefficients for slabs completely loaded with uniformly distributed load for different values of k and $1/k$ are obtained from Fig. 23.12.

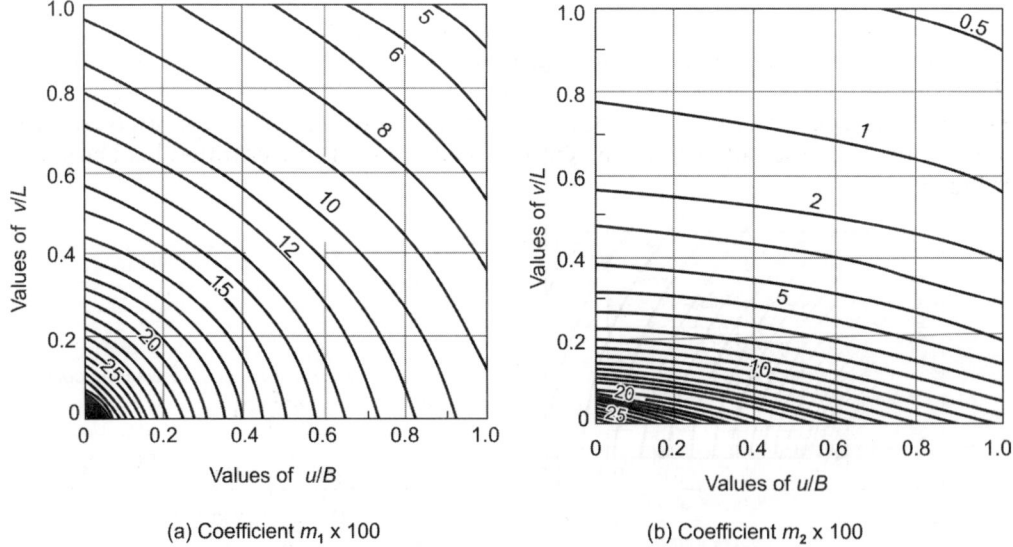

(a) Coefficient m_1 x 100

(b) Coefficient m_2 x 100

Fig. 23.5: Moment coefficients m_1 and m_2 for $k = 0.4$

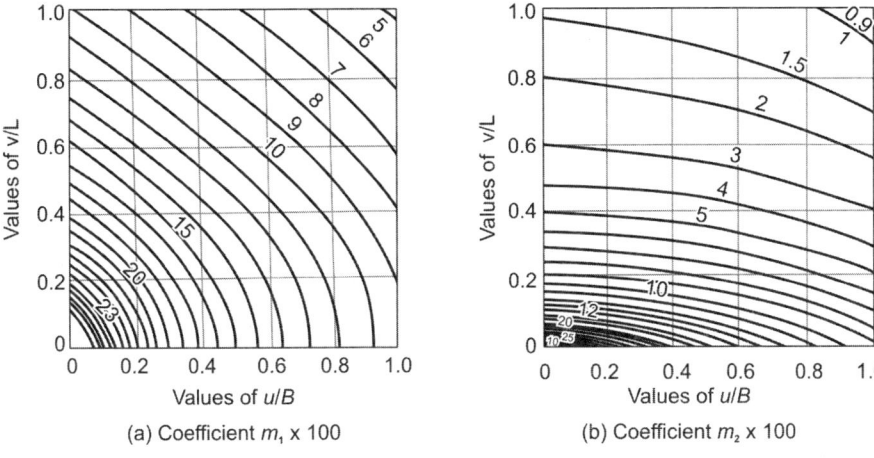

(a) Coefficient m_1 × 100

(b) Coefficient m_2 × 100

Fig. 23.6: Moment coefficients m_1 and m_2 for $k = 0.5$

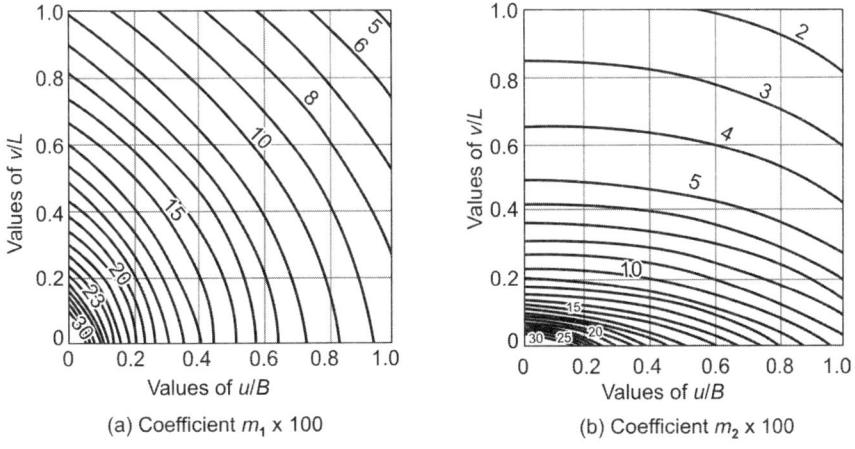

(a) Coefficient m_1 × 100

(b) Coefficient m_2 × 100

Fig. 23.7: Moment coefficients m_1 and m_2 for $k = 0.6$

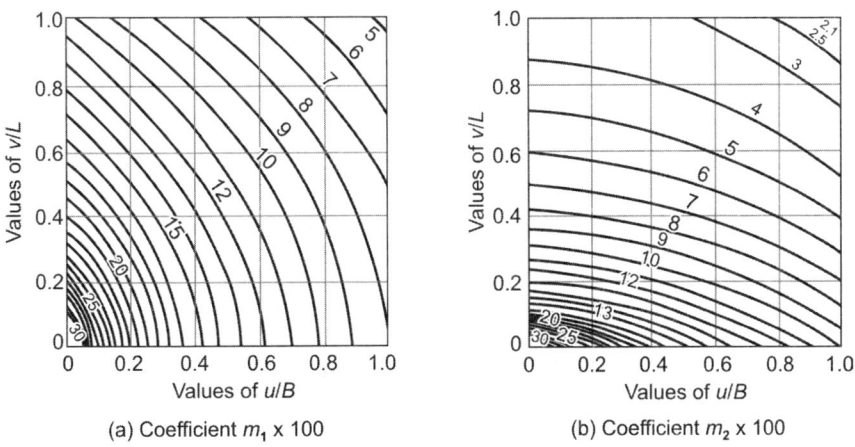

(a) Coefficient m_1 × 100

(b) Coefficient m_2 × 100

Fig. 23.8: Moment coefficients m_1 and m_2 for $k = 0.7$

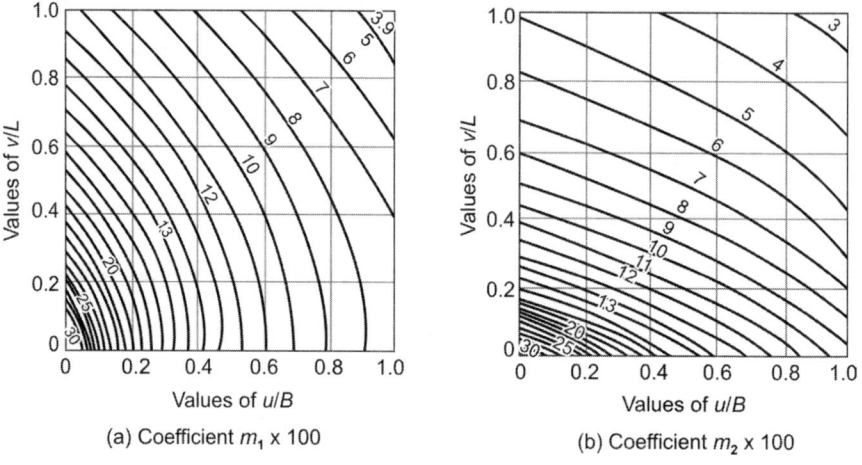

(a) Coefficient $m_1 \times 100$

(b) Coefficient $m_2 \times 100$

Fig. 23.9: Moment coefficients m_1 and m_2 for $k = 0.8$

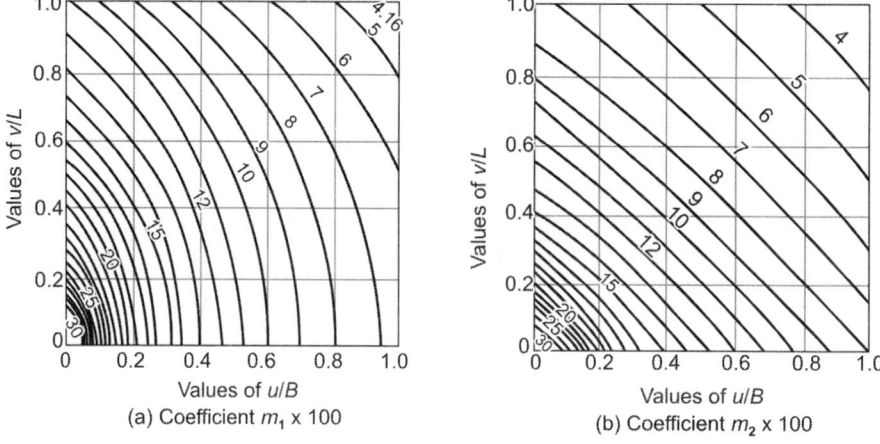

(a) Coefficient $m_1 \times 100$

(b) Coefficient $m_2 \times 100$

Fig. 23.10: Moment coefficients m_1 and m_2 for $k = 0.9$

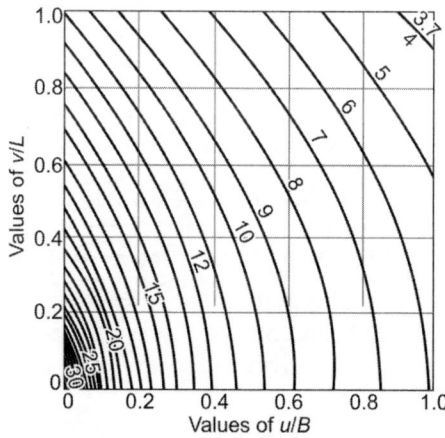

Fig. 23.11: Moment coefficient m_1 (or m_2) $\times 100$ for $k = 1.0$

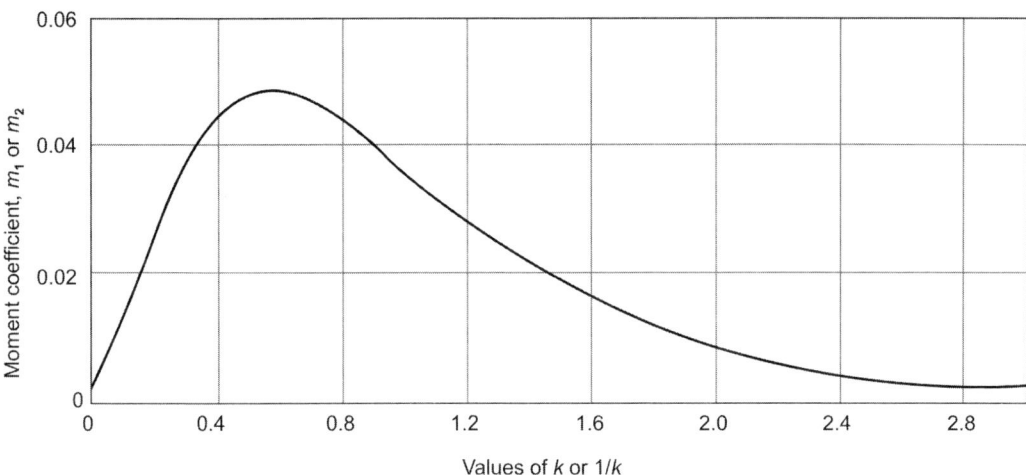

Fig. 23.12: Moment coefficient for slabs completely loaded with uniformly distributed load, coefficient is m_1 for k and m_2 for $1/k$

23.4 ANALYSIS AND DESIGN OF TEE BEAM AND SLAB DECKS

A typical tee beam deck slab generally comprises the longitudinal girders, continuous deck slab between the tee beams and cross girders to provide lateral rigidity to the bridge deck. The distribution of live loads among the longitudinal griders can be estimated by any of the following rational methods:

  i. Courbon's method

 ii. Hendry-Jaegar method

iii. Guyon-Massonnet method.

Among those methods, Courbon's method is the simplest and is applicable when the following conditions are satisfied:

   a. The ratio of span to width of deck is greater than 2 but less than 4.

   b. The longitudinal girders are interconnected by at least five symmetrically spaced cross girders.

   c. The cross girders extend to a depth of at least 0.75 of the depth of the longitudinal girder.

Courbon's method is popular due to the simplicity of computation as detailed below:

When live loads are positioned nearer to the kerb, the centre of gravity of live load acts eccentrically with the centre of gravity of the girder system. Due to this eccentricity, the loads are increased and decreased on each girder, depending upon the position of the girders. This is calculated by Courbon's theory by a reaction factor given by:

$$R_x = \frac{\Sigma W}{n}\left[1 + \frac{\Sigma I}{\Sigma d_x^2 I}d_x e\right]$$

where,

R_x = reaction factor for girder under consideration

I = moment of inertia of each longitudinal girder

d_x = distance of the girder under consideration from the central axis of the bridge

W = total concentrated live load

n = number of longitudinal girders

e = eccentricity of live load with respect to the axis of the bridge

The live load bending moments and shear forces are computed for each of the girders. The maximum design moments and shear forces are obtained by adding the live load and dead load bending moments. The reinforcements in the main longitudinal girders are designed for the maximum moments and shears developed in the girders.

An approximate method may be used for the computation of the bending moments and shear forces in cross girders. The cross girders are assumed to be rigid so that the reactions due to dead and live loads are assumed to be equally shared by the cross girders. This assumption will simplify the computation of bending moments and shear forces in the cross girders.

The complete design of slab deck and tee beam decks are illustrated by the following design examples.

23.5 DESIGN EXAMPLE OF SLAB BRIDGE DECK

Design a reinforced concrete slab culvert for a state highway to suit the following data:

Carriageway: Two lane 7.5 m wide

Materials: M25 grade concrete and Fe 415 HYSD bars

Kerbs: 600 mm wide

Clear span = 6 m, wearing coat = 80 mm

Width of bearing = 400 mm

Loading: IRC class A or AA, whichever gives the worst effect.

Design the reinforced concrete deck slab and sketch the details of reinforcements in the longitudinal and cross-sections of the slab. The design should conform to the specifications of the Bridge code IRC:21-1987.

    i. *Data*:

      Clear span = 6 m

      Width of bearing = 400 mm

      Kerbs = 600 mm

      Materials: M25 grade concrete and Fe 415 HYSD bars

    ii. *Allowable stress* (according to IRC:21-1987):

$$f_{ck} = 25 \text{ N/mm}^2 \qquad m = 10$$
$$f_y = 415 \text{ N/mm}^2 \qquad j = 0.90$$
$$\sigma_{cb} = 8.3 \text{ N/mm}^2 \qquad Q = 1.1$$
$$\sigma_{st} = 200 \text{ N/mm}^2$$

iii. *Depth of slab and effective span*:

Assume thickness of slab at 80 mm per metre of span for highway bridge decks.

∴   Overall slab thickness = (80 × 6) = 500 mm

Using 25 mm diameter bars with clear cover of 25 mm

Effective depth = [500 – (25 + 12.5)] = 462.5 mm

Width of bearing = 0.4 m

∴   Effective span is the least of:

1. Clear span + effective depth = (6 m + 0.4625 m) = 6.4625 m
2. Centre to centre of bearings = (6 m + 0.4 m) = 6.4 m

∴   Effective span L = 6.4 m

The cross-section of the deck slab is shown in Fig. 23.13.

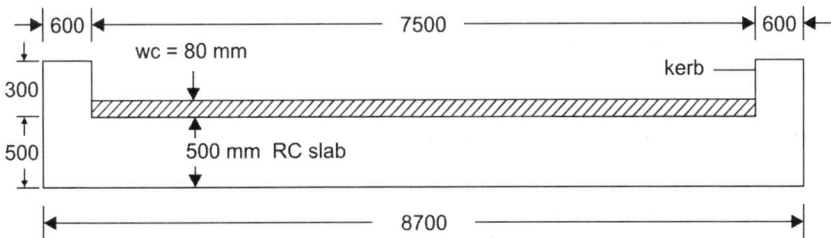

Fig. 23.13: Cross-section of deck slab

iv. *Dead load bending moments*:

Dead weight of slab = (0.5 × 24) = 12 kN/m²

Wearing coat = (0.08 × 22)   = 1.76

∴   Total load   = 13.76 kN/m²

∴   Dead load bending moment = $\left(\dfrac{13.76 \times 6.4^2}{8}\right)$ = 70.4 kN·m

v. *Live load bending moment*:

The bending moment will be maximum for IRC class AA tracked vehicles and hence computations will be made for this class of loading. Impact factor of class AA tracked vehicle is 25% for 5 m span decreasing linearly to 10% for 9 m span.

∴   For 6.4 m span:

Impact factor = $\left[25 - \dfrac{15}{4}(6.4 - 5)\right]$ = 19.7%

The tracked vehicle is placed symmetrically on the span,

effective length of load = [3.6 + 2(0.5 + 0.08)] = 4.76 m

Effective width of slab, perpendicular to span is expressed as:

$$b_e = kx(1 - x/L) + b_w$$

Referring to Fig. 23.14

$$x = 3.2 \text{ m}, \qquad L = 6.4 \text{ m}, \qquad B = 8.7 \text{ m}, \qquad \left(\frac{B}{L}\right) = 1.36$$

$$b_w = (0.85 + 2 \times 0.08) = 1.01 \text{ m}$$

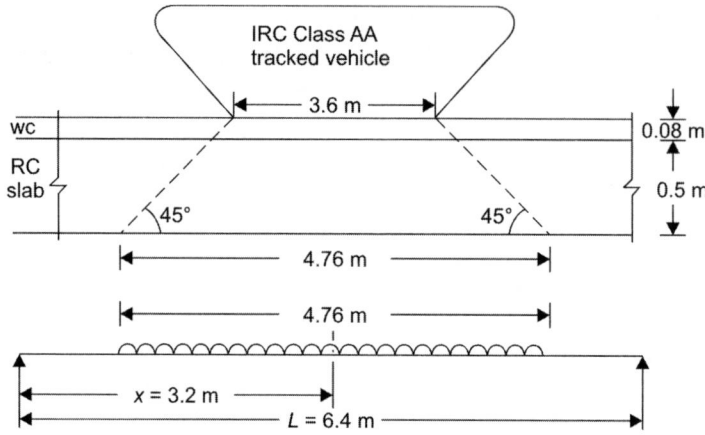

Fig. 23.14: Dispersion of loads due to IRC class AA vehicle

From Table 23.1, for $\left(\dfrac{B}{L}\right) = 1.36$, simply supported slabs, $k = 2.77$

$$\therefore \qquad b_e = 2.77 \times 3.2\left(1 - \frac{3.2}{6.4}\right) + 1.01 = 5.442 \text{ m}$$

The tracked vehicle is placed close to the kerb with the required minimum clearance as shown in Fig. 23.15.

Net effective width of dispersion = 6.996 m

Total load of two tracks with impact = $(700 \times 1.197) = 838$ kN

Average intensity of load = $\left(\dfrac{838}{4.76 \times 6.996}\right) = 25.17$ kN/m²

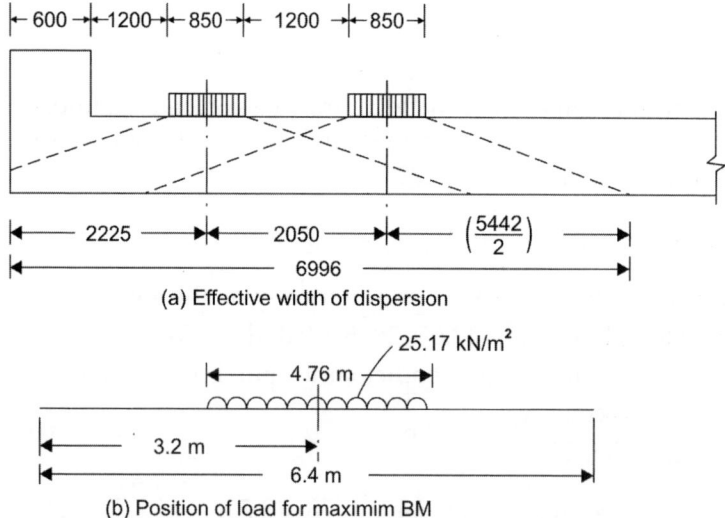

(a) Effective width of dispersion

(b) Position of load for maximim BM

Fig. 23.15: Effective width of dispersion of live loads

Maximum bending moment due to live load is given by:

$$M_{max} = \left(\frac{25.17 \times 4.75}{2} \times 3.2 \right) - \left(\frac{25.17 \times 4.76}{2} \times \frac{4.76}{4} \right) = 120.36 \text{ kN·m}$$

∴ Total design bending moment = (120.36 + 70.4) = 191 kN·m

vi. *Shear due to class AA tracked vehicle*:

For maximum shear at support, the IRC class AA tracked vehicle is arranged as shown in the Fig. 23.16.

Effective width of dispersion is given by:

$$b_e = kx(1 - x/L) + b_w$$

where,

$x = 2.38$ m $\qquad \left(\dfrac{B}{L} \right) = 1.36$

$B = 8.7$ m

$L = 6.4$ m $\qquad k = 2.77$

$b_w = 1.01$ m

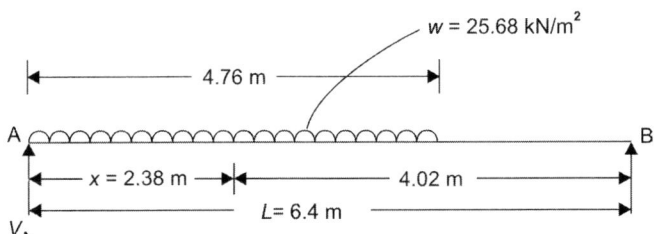

Fig. 23.16: Load position for maximum shear

∴ $\qquad b_e = 2.77 \times 2.38 \left(1 - \dfrac{2.38}{6.4} \right) + 1.01 = 5.16$ m

∴ Width of dispersion $= \left(2225 + 2050 + \dfrac{5160}{2} \right) = 6.855$ m

(Refer to Fig. 23.15)

∴ $\qquad w = \left(\dfrac{838}{4.76 \times 6.855} \right) = 25.68 \text{ kN/m}^2$

∴ Shear force $V_A = \left(\dfrac{25.68 \times 4.76 \times 4.02}{6.4} \right) = 76.80$ kN

Dead load shear $= \left(\dfrac{13.76 \times 6.4}{2} \right) = 43.75$ kN

∴ Total design shear = (76.80 + 43.75) = 121 kN

vii. *Design of deck slab*:

Effective depth required is computed as:

$$d = \sqrt{\frac{M}{Qb}} = \sqrt{\frac{191 \times 10^6}{1.1 \times 1000}} = 417 \text{ mm}$$

Effective depth provided = 462.5 mm

$$A_{st} = \left(\frac{M}{\sigma_{st} jd}\right) = \left(\frac{191 \times 10^6}{200 \times 0.90 \times 462.5}\right) = 2294 \text{ mm}^2$$

Spacing of 25 mm diameter bars = $\left(\dfrac{1000 \times 491}{2294}\right) = 214$ mm

Provide 25 mm diameter bars at 200 mm centres.

Bending moment for distribution reinforcement is:

$$= (0.3 M_L + 0.2 M_d) = [(0.3 \times 120.36) + (0.2 \times 70.4)] = 50.2 \text{ kN·m}$$

Using 12 mm diameter bars,

effective depth = $[462.5 - (12.5 + 5)] = 444$ mm

$$A_{st} = \left(\frac{50.2 \times 10^6}{200 \times 0.90 \times 444}\right) = 628 \text{ mm}^2$$

Spacing of 12 mm diameter bars = $\left(\dfrac{1000 \times 113}{628}\right) = 180$ mm

Provide 12 mm diameter bars at 150 mm centres.

viii. *Check for shear stress*:

As per IRC:21-1987, shear stress in the slab is checked as follows:

Design shear stress $\tau = \left[\dfrac{V}{bd}\right]$

where,

V = design shear force

b = width of section

d = effective depth

Permissible shear stress in slabs without shear reinforcement is computed as:

$$\tau_c = (k_1 k_2 \tau_{co})$$

where,

τ_c = the permissible shear stress

$k_1 = (1.14 - 0.7d) \geq 0.5$, where d is expressed in metres

$k_2 = (0.5 + 0.25\rho) \geq 1$

ρ = percentage of longitudinal reinforcement = $\left(\dfrac{100 A_s}{bd}\right)$

τ_{co} = basic values of shear stress for different grades of concrete as shown in Table 23.2 (Refer to IRC:21-1987)

A_s = area of longitudinal reinforcement which continues at least to a distance d beyond the section considered or fully anchored when support section is considered.

b = width of the section

Table 23.2: Basic values of permissible shear stress (τ_{co})

Grade of concrete	M15	M20	M25	M30	M35	M40
τ_{co}	0.28	0.34	0.40	0.45	0.50	0.50

Hence shear stress $\tau = \left[\dfrac{V}{bd}\right] = \left[\dfrac{121\times10^3}{1000\times462.5}\right] = 0.26\ \text{N/mm}^2$

$k_1 = (1.14 - 0.7 \times 0.4625) = 0.817 > 0.5$

$k_2 = (0.5 - 0.25\rho)$, where $\rho = [100A_s/bd]$

Assuming the longitudinal reinforcement of 25 mm diameter bars spaced at 400 mm centres at support section (alternate bars bent up), we have:

$$A_s = \left[\dfrac{1000\times491}{400}\right] = 1227.5\ \text{mm}^2$$

$$\rho = \left[\dfrac{100A_s}{bd}\right] = \left[\dfrac{100\times1227.5}{1000\times462.5}\right] = 0.265$$

$\therefore$ $k_2 = [0.5 - 0.25 \times 0.265] = 0.566 \geq 1$

$\therefore$ $k_2 = 1$

For M20 grade concrete, $\tau_{co} = 0.34\ \text{N/mm}^2$ (Refer to Table 23.2)

$\therefore$ $\tau_c = (k_1 k_2 \tau_{co})$

  $= (0.817 \times 1 \times 0.34)$

  $= 0.277\ \text{N/mm}^2 > \tau$

Since $\tau < \tau_c$, the shear stresses are within safe permissible limits.

The details of reinforcements in the slab are shown in Fig. 23.17

23.6 DESIGN EXAMPLE OF TEE BEAM AND SLAB BRIDGE DECK

Design a RCC tee beam girder bridge to suit the following data:

Clear width of roadway = 7.5 m

Span (centre to centre of bearings) = 16 m

Live load: IRC class AA or A whichever gives the worst effect

Average thickness of wearing coat = 80 mm

Materials: M25 grade concrete and Fe 415 HYSD bars

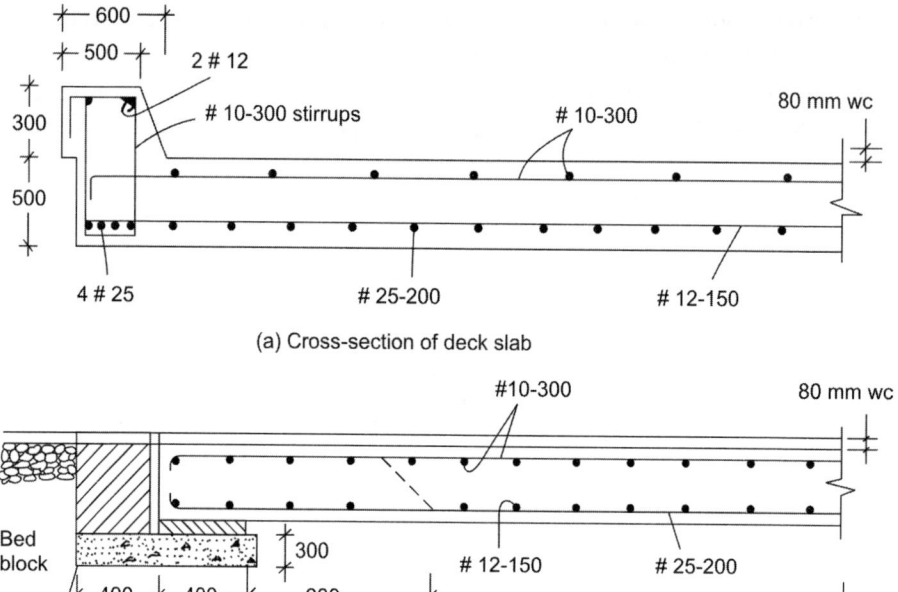

(a) Cross-section of deck slab

(b) Longitudinal section of deck slab

Fig. 23.17: Reinforcement details in deck slab

Design the deck slab, main girder and cross girder and sketch the typical details of reinforcements.

   i. *Data*:

     Effective span of beams = 16 m

     Road width = 7.5 m

     Thickness of wearing coat = 80 mm

     Materials: M25 grade concrete and Fe 415 HYSD bars.

  ii. *Permissible stresses* (according to IRC:21-1987)

     $\sigma_{cb} = 8.3\ \text{N/mm}^2$ $m = 10$

     $\sigma_{st} = 200\ \text{N/mm}^2$ $j = 0.90$

                            $Q = 1.1$

 iii. *Cross-section of deck*:

     Three main girders are provided at 2.5 m centres.

     Thickness of deck slab = 200 mm

     Wearing coat = 80 mm

     Width of main girders = 300 mm

     Kerbs 600 mm wide and 300 mm deep (cross girders are provided at every 4 m intervals)

     Breadth of cross girder = 300 mm

Depth of main girders = 1600 mm (at the rate of 100 mm per metre of span)
The depth of cross girder is taken as equal to the depth of main girder to simplify computations. The cross-section of the deck and the plan showing the spacing of cross-girders are shown in Fig. 23.18.

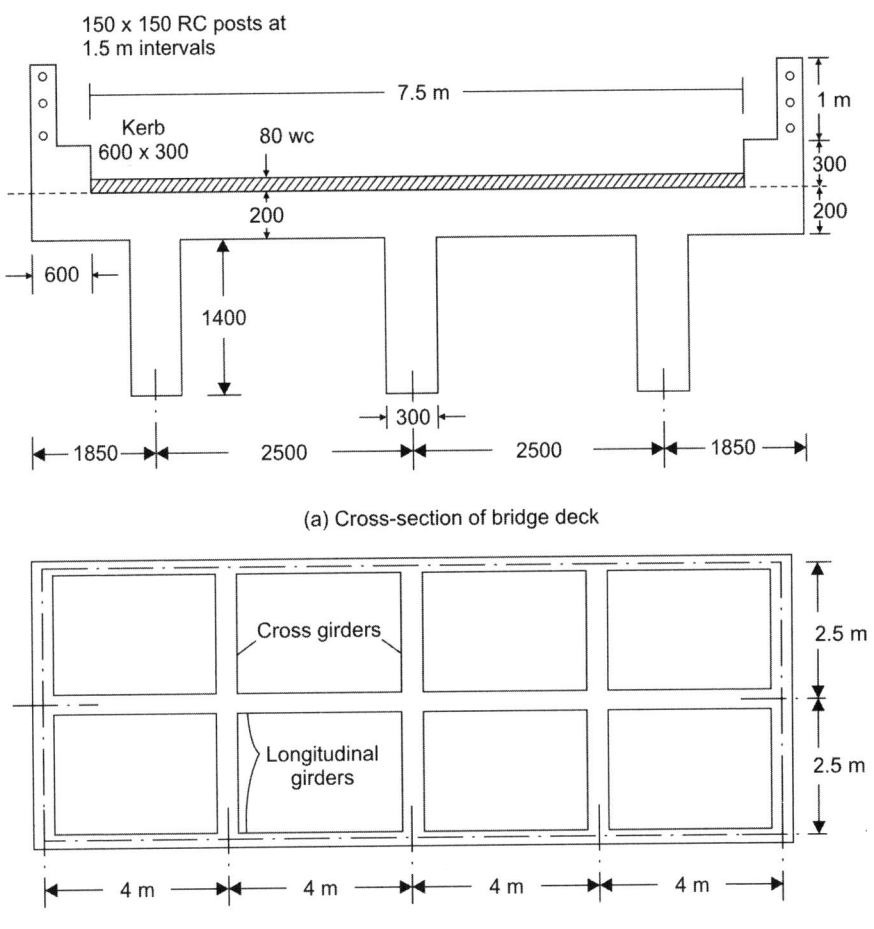

(a) Cross-section of bridge deck

(b) Plan of bridge deck

Fig. 23.18: Tee beam and slab bridge deck

iv. *Design of interior slab panels*:
   a. Bending moments
   Dead weight of slab = $(1 \times 1 \times 0.2 \times 24) = 4.80$
   Dead weight of wc = (0.08×22) $\quad = 1.76$
   Total dead load $\quad\quad\quad\quad\quad = 6.56 \text{ kN/m}^2$
   Live load: Class AA tracked vehicle. One wheel is placed at the centre of panel as shown in Fig. 23.19
   $$u = (0.85 + 2 \times 0.08) = 1.01 \text{ m}$$
   $$v = (3.60 + 2 \times 0.08) = 3.76 \text{ m}$$

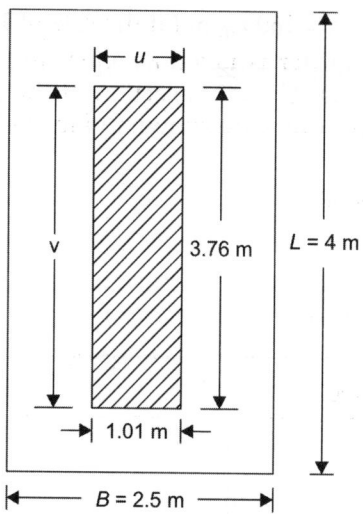

Fig. 23.19: Position of wheel load for maximum bending moment

$$\left(\frac{u}{B}\right) = \left(\frac{1.01}{2.5}\right) = 0.404 \qquad \left(\frac{v}{L}\right) = \left(\frac{3.76}{4.0}\right) = 0.94$$

$$k = \left(\frac{B}{L}\right) = \left(\frac{2.5}{4.0}\right) = 0.625$$

Referring to Pigeaud's curve (Fig. 23.7)

$$m_1 = 8.5 \times 10^{-2} \text{ and } m_2 = 2.4 \times 10^{-2}$$

$$\therefore \qquad M_B = W(m_1 + 0.15m_2)$$
$$= 350(8.5 \times 10^{-2} + 0.15 \times 2.4 \times 10^{-2}) = 31.01 \text{ kN·m}$$

As the slab is continuous,

design BM = $0.8M_B$

M_B (including impact and continuity factor)

$$= (1.25 \times 0.8 \times 31.01) = 31.01 \text{ kN·m}$$
$$M_L = 350(2.4 \times 10^{-2} + 0.15 \times 8.5 \times 10^{-2}) = 12.86 \text{ kN·m}$$

Design $\quad M_L = (1.25 \times 0.8 \times 12.86) = 12.86 \text{ kN·m}$

b. Shear forces

Dispersion in the direction of span = $[0.85 + 2(0.08 + 0.2)] = 1.41$ m. For maximum shear, load is kept such that the whole dispersion is in span, the load is kept at $\left(\frac{1.41}{2}\right) = 0.705$ m from the edge of beam as shown in Fig. 23.20.

Effective width of slab = $kx(1 - x/L) + b_w$

Breadth of cross girder = 30 cm

Clear length of panel $L = 3.7$ m

$$\therefore \qquad \left(\frac{B}{L}\right) = \left(\frac{3.7}{2.2}\right) = 1.68$$

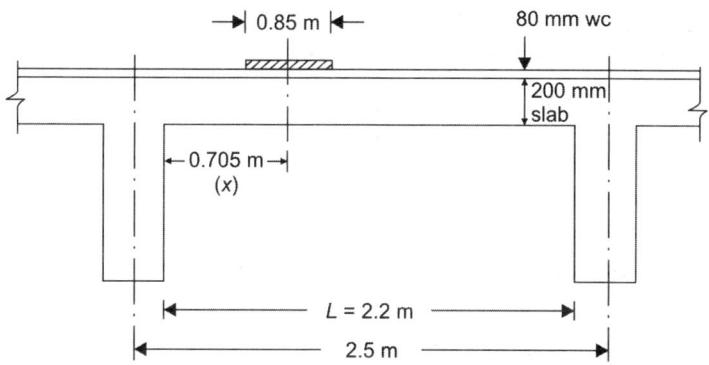

Fig. 23.20: Position of wheel load for maximum shear

From Table 23.1, k for continuous slab is obtained as $k = 2.52$

Effective width of slab $= \left[2.52 \times 0.705 \left(1 - \dfrac{0.705}{2.2} \right) + 3.6 + (2 \times 0.08) \right] = 5 \text{ m}$

$\therefore$ Load/m width $= \left(\dfrac{350}{5} \right) = 70 \text{ kN}$

$\therefore$ Shear force $= 70 \left[\dfrac{2.2 - 0.705}{2.2} \right] = 47.60 \text{ kN}$

Shear force with impact $= (1.25 \times 47.60) = 59.50 \text{ kN}$

c. Dead load bending moments and shear forces

Dead load $= 6.56 \text{ kN/m}^2$

Total load on panel $= (4 \times 2.5 \times 6.56) = 65.6 \text{ kN}$

$\left(\dfrac{u}{B} \right) = 1, \ \left(\dfrac{v}{L} \right) = 1$ as panel is loaded with uniformly distributed load

$$ k = \left(\dfrac{B}{L} \right) = \left(\dfrac{2.5}{4} \right) = 0.625 \qquad \text{and} \quad \dfrac{1}{k} = 1.6 $$

From Pigeaud's curve (Fig. 23.12)

$\qquad\qquad m_1 = 4.9 \times 10^{-2} \qquad\qquad \text{and } m_2 = 1.5 \times 10^{-2}$

$\therefore \qquad\qquad M_B = 65.6(4.9 \times 10^{-2} + 0.15 \times 1.5 \times 10^{-2}) = 3.36 \text{ kN·m}$

Taking continuity into effect:

$\qquad\qquad M_B = (0.8 \times 3.36) = 2.688 \text{ kN·m}$

$\qquad\qquad M_L = 65.6(1.5 \times 10^{-2} + 0.15 \times 4.9 \times 10^{-2}) = 1.468 \text{ kN·m}$

Taking continuity into effect:

$\qquad\qquad M_L = (0.8 \times 1.468) = 1.174 \text{ kN·m}$

Dead load shear force $= \left(\dfrac{6.56 \times 2.2}{2} \right) = 7.216 \text{ kN}$

d. Design moments and shears

Total M_B = (31.01 + 2.688) = 33.698 kN·m

Total M_L = (12.86 + 1.174) = 14.03 kN·m

Total shear force = (59.5 + 7.216) = 66.716 kN

e. Design of section

Effective depth $d = \sqrt{\dfrac{33.698 \times 10^6}{1.1 \times 1000}}$ = 175 mm

Adopt overall depth = 200 mm

$$A_{st} = \left(\frac{33.698 \times 10^6}{200 \times 0.90 \times 175} \right) = 1069 \text{ mm}^2$$

For short span use 16 mm diameter HYSD bars at 150 mm centres (A_{st} = 1341 mm²)

Effective depth for long span using 10 mm diameter bars is computed as:

$$d = (175 - 8 - 5) = 162 \text{ mm}$$

$$\therefore \qquad A_{st} = \left(\frac{M_u}{bd^2} \right) = 344 \text{ mm}^2$$

For long span adopt 10 mm diameter bars at 150 mm centres (A_{st} = 524 mm²)

f. Check for shear stress

$$\text{Nominal shear stress } \tau = \left(\frac{V}{bd} \right) = \left(\frac{66.716 \times 10^3}{10^3 \times 175} \right) = 0.381 \text{ N/mm}^2$$

$$k_1 = (1.14 - 0.7 \times 0.175) = 1.0175 \geq 0.5$$

$k_2 = (0.5 + 0.25\rho)$, where $\rho = (100 A_s / bd)$ and A_s = 1341 mm²

$$\rho = [(100 \times 1341)/(100 \times 175)] = 0.766$$

$\therefore \qquad k_2 = [0.5 + (0.25 \times 0.766)] = 0.69 \geq 1$

Hence, $\qquad k_2 = 1$

For M25 grade concrete, τ_{co} = 0.40 N/mm² (Refer to Table 23.2)

$\therefore \qquad \tau_c = (k_1 k_2 \tau_{co}) = [1.0175 \times 1 \times 0.40] = 0.407 \text{ N/mm}^2 > 0.381 \text{ N/mm}^2$

Since $\tau < \tau_c$, shear stresses are within safe permissible limits.

v. *Design of longitudinal girders:*

a. Reaction factors

Using Courbon's theory, the IRC class AA loads are arranged for maximum eccentricity as shown in Fig. 23.21.

Reaction factor for outer girder is:

$$R_A = \frac{2W_1}{3} \left(1 + \frac{3I \times 2.5 \times 1.1}{2I \times 2.5^2} \right) = 1.107 \, W_1$$

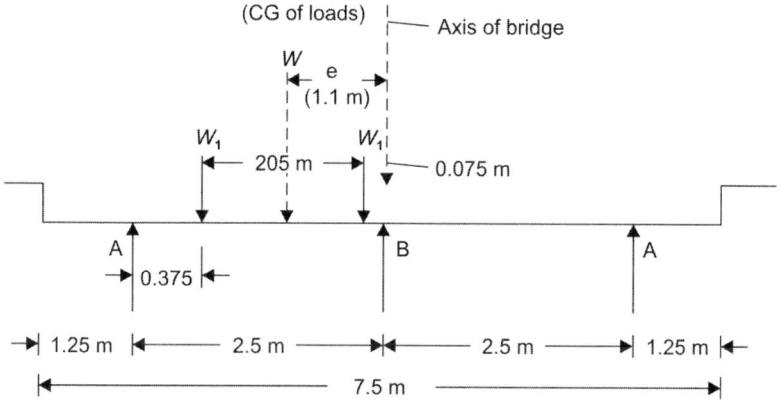

Fig. 23.21: Transverse disposition of IRC class AA tracked vehicle

Reaction factor for inner girder is:

$$R_B = \frac{2W_1}{3}[1+0] = \left(\frac{2W_1}{3}\right)$$

If W = axle load = 700 kN

$W_1 = 0.5\,W$

∴ $R_A = (1.107 \times 0.5\,W) = 0.5536\,W$

$R_B = (0.667 \times 0.5\,W) = 0.3333\,W$

b. Dead load from slab per girder

The dead load of deck slab is calculated with reference to Fig. 23.22

1. Parapet railing   = 0.700
2. Wearing coat = $(0.08 \times 1.10 \times 22)$ = 1.936
3. Deck slab = $(0.2 \times 1.10 \times 24)$ = 5.280
4. Kerb = $(0.5 \times 0.6 \times 1 \times 24)$ = 7.200
                                                      = 15.116 kN/m

Total dead load of deck = $(2 \times 15.116) + (6.56 \times 5.3) = 65$ kN/m

It is assumed that the dead load is sheared equally by all the girders.

∴ Dead load/girder = $\left(\frac{65}{3}\right)$ = 21.66 kN/m

c. Live load bending moments in girder

Span of girder = 16 m

Impact factor (for class AA) = 10%

The live load is placed centrally on the span as shown in Fig. 23.23.

Bending moment = $\left(\frac{4+3.1}{2}\right)700$ = 2485 kN·m

∴ Bending moment including impact and reaction factor for outer girder is:
= $(2485 \times 1.1 \times 0.5536) = 1513$ kN·m

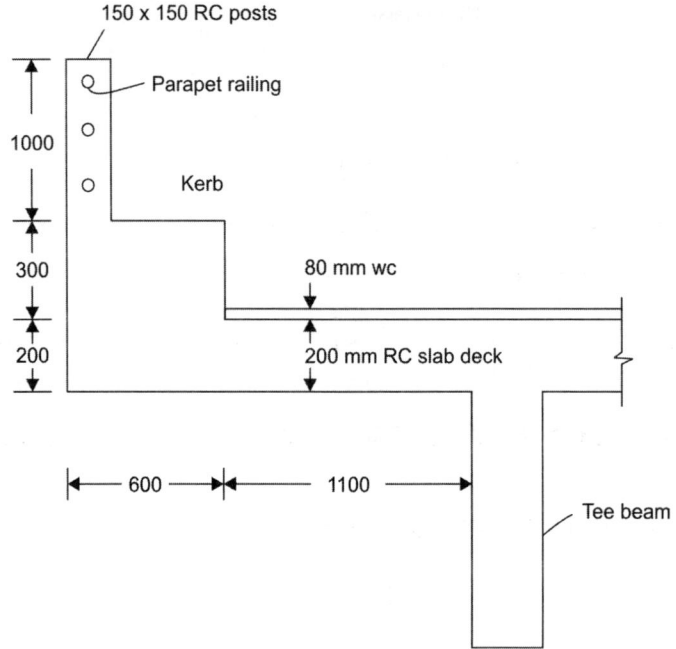

Fig. 23.22: Details of deck slab, kerb and parapet

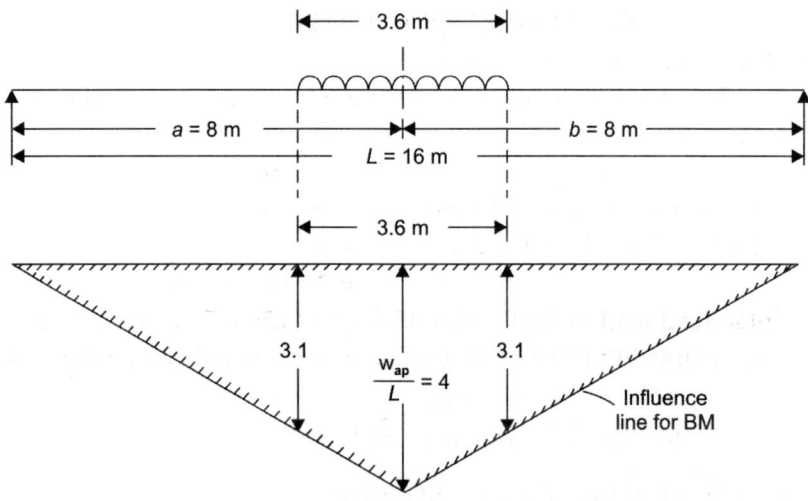

Fig. 23.23: Influence line for bending moments in girder

Bending moment including impact and reaction factor for inner girder is:
$$= (2485 \times 1.1 \times 0.3333) = 912 \text{ kN·m}$$

d. Live load shear

For estimating the maximum live load shear in the girders, the IRC class AA loads are placed as shown in Fig. 23.24.

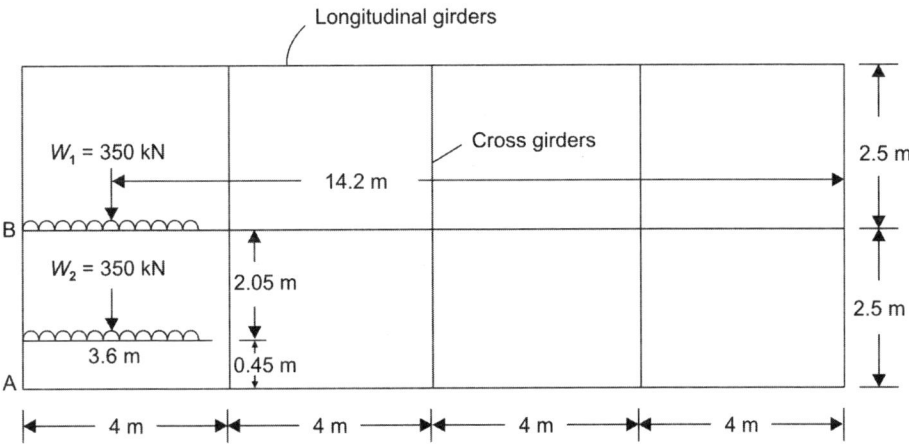

Fig. 23.24: Position of IRC class AA loads for maximum shear

$$\text{Reduction of } W_2 \text{ on girder B} = \left(\frac{350 \times 0.45}{2.5}\right) = 63 \text{ kN}$$

$$\text{Reaction of } W_2 \text{ on girder A} = \left(\frac{350 \times 2.05}{2.5}\right) = 287 \text{ kN}$$

$$\text{Total load on girder B} = (350 + 63) = 413 \text{ kN}$$

$$\text{Maximum reaction in girder B} = \left(\frac{413 \times 14.2}{16}\right) = 366 \text{ kN}$$

$$\text{Maximum reaction in girder A} = \left(\frac{287 \times 14.2}{16}\right) = 255 \text{ kN}$$

Maximum live load shears with impact factors in
inner girder = (366 × 1.1) = 402.6 kN and
outer girder = (255 × 1.1) = 280.5 kN

e. Dead load moments and shear force in main girder

The depth of girder is assumed as 1600 mm (100 mm for every metre of span)

Depth of rib = 1.4 m,   width = 0.3 m

Weight of rib/m = (1 × 0.3 × 1.4 × 24) = 10.08 kN/m

The cross girder is assumed to have the same cross-sectional dimensions of the main girder.

Weight of cross girder = 10.08 kN/m

Reaction on main girder = (10.08 × 2.5) = 25.2 kN

Reaction from deck slab on each girder = 21.66 kN/m

∴   Total dead load/m on girder = (21.66 + 10.08) = 31.74 kN/m

Referring to Fig. 23.25, the maximum bending moments are computed.

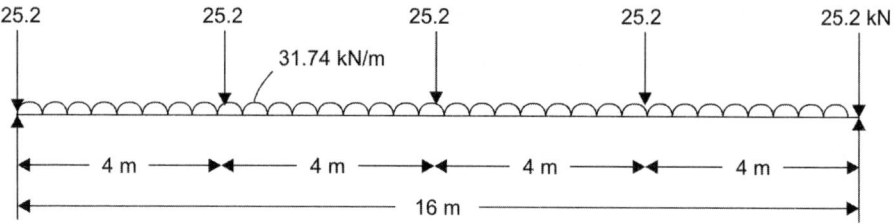

Fig. 23.25: Dead loads on main girders

$$M_{max} = \left(\frac{31.74 \times 16^2}{8}\right) + \left(\frac{25.2 \times 16}{4}\right) + \left(\frac{25.2 \times 16}{4}\right) \text{ (centre of span)}$$

$$= 1218 \text{ kN·m}$$

Dead load shear at support:

$$= \left(\frac{31.74 \times 16}{2}\right) + 25.2 + \left(\frac{25.2}{2}\right) = 292 \text{ kN}$$

f. *Design bending moments and shear forces*:

The design moments and shears are compiled in Table 23.3.

g. *Design of sections for maximum BM and SF*:

$$M_{max} = 2731 \text{ kN·m}$$
$$V_{max} = 694.6 \text{ kN}$$

The beam is designed as a Tee section

Assuming an effective depth d = 1450 mm

$$\text{Approximate lever arm} = \left(1450 - \frac{200}{2}\right) = 1350 \text{ mm}$$

$$A_{st} = \left(\frac{2731 \times 10^6}{200 \times 1350}\right) = 10114 \text{ mm}^2$$

Provide 16 bars of 32 mm diameter in four rows (A_{st} = 12864 mm²)

According to IRC:21-1987, the maximum size of reinforcements should not exceed 32 mm diameter.

Shear reinforcements are designed to resist the maximum shear at supports.

Table 23.3: Design moments and shear forces

BM	DL BM	LL BM	Total BM	Unit
Outer girder	1218	1513	2731	kN·m
Inner girder	1218	912	2130	kN·m
SF	DL SF	LL SF	Total SL	Unit
Outer girder	292	280.1	572.1	kN
Inner girder	292	402.6	694.6	kN

Nominal shear stress is computed as:

$$\tau_v = \left(\frac{V}{bd}\right) = \left(\frac{694.6 \times 1000}{300 \times 1450}\right) = 1.596 \text{ N/mm}^2 \text{ not greater than } 0.07 f_{ck}$$

which is equal to $(0.07 \times 25) = 1.75 \text{ N/mm}^2$, hence safe.

Assuming two bars of 32 mm diameter to be bent up at support section, shear resisted by the bent up bars is given by the relation:

$$V_s = (\sigma_{sv} A_{sv} \sin\alpha) = \left[(200 \times 2 \times 804 \times 1)/(1000 \times \sqrt{2})\right] = 227 \text{ kN}$$

Balance shear force $V_b = [V - V_s] = [694.6 - 227] = 467.6 \text{ kN}$

Using 10 mm diameter, 4 legged vertical stirrups, spacing is computed as

$$S_v = \left[\frac{\sigma_{sv} A_{sv} d}{V_b}\right] = \left[\frac{(200 \times 4 \times 79 \times 1450)}{(467.6 \times 1000)}\right] = 195 \text{ mm}$$

Provide 10 mm diameter 4 legged stirrups at 150 mm centres.

vi. *Design of cross girders*:

Self weight of cross girder = 10.08 kN/m

Referring to Fig. 23.26

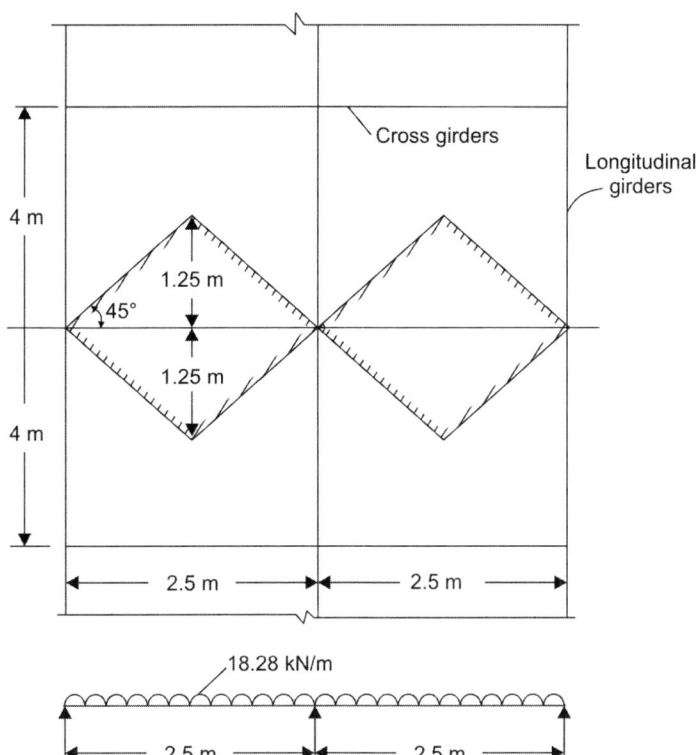

Fig. 23.26: Loads on cross girder

Dead load from slab $= \left(2 \times \dfrac{1}{2} \times 2.5 \times 1.25 \times 6.56 \right) = 20.5$ kN

Uniformly distributed load $= \left(\dfrac{20.5}{2.5} \right) = 8.2$ kN/m

Total load on cross girder $= (10.08 + 8.2) = 18.28$ kN/m

Assuming the cross girder to be rigid,

reaction on each cross girder $= \left(\dfrac{18.28 \times 5}{3} \right) = 30.47$ kN

For maximum bending moment in the cross girder, the loads of IRC class AA should be placed as shown in Fig. 23.27

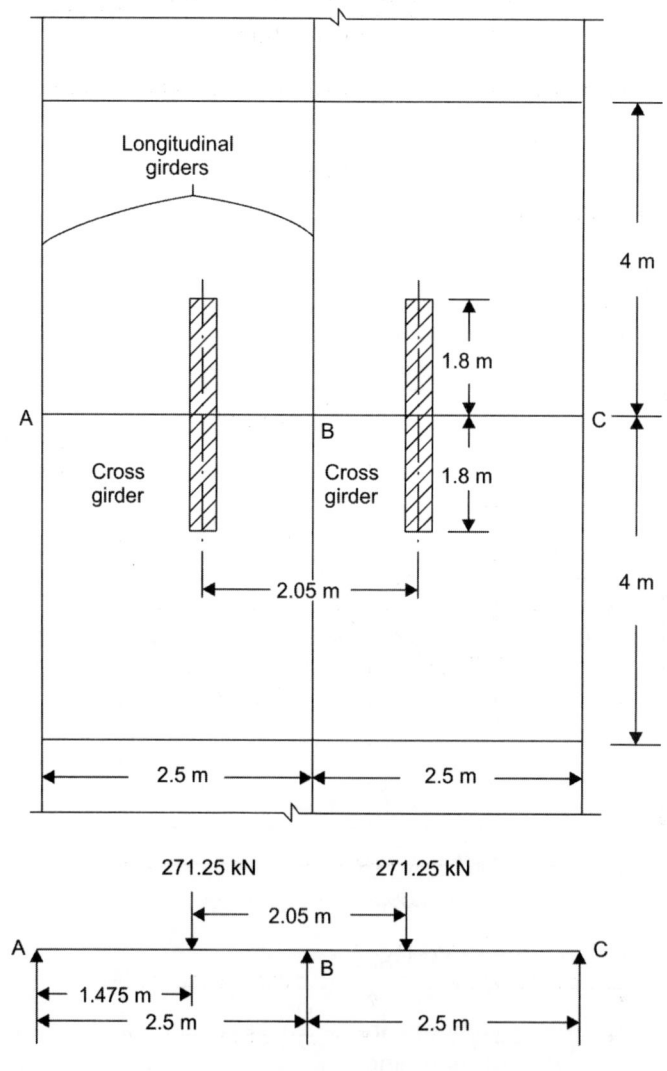

Fig. 23.27: Position of live loads for maximum BM in cross girder

$$\text{Load on cross girder} = \frac{350\left(4 - \dfrac{1.8}{2}\right)}{4} = 271.25 \text{ kN}$$

Assuming the cross girder as rigid, reaction on each longitudinal girder

$$= \left(\frac{2 \times 271.25}{3}\right)$$

$$= 180.83 \text{ kN}$$

Maximum BM in cross girder under the load = $(180.83 \times 1.475) = 266.7$ kN·m

LL BM including impact = $(1.1 \times 266.7) = 293.37$ kN·m

Dead load BM at 1.475 m from support:

$$= \left(30.47 \times 1.475 - 18.28 \times \frac{1.475}{2}\right)$$

$$= 25.10 \text{ kN·m}$$

∴   Total design BM = $(293.37 + 25.10) = 318.47$ kN·m

$$\text{Live load shear including impact} = \left(\frac{(2 \times 271.25)}{3} \times 1.1\right)$$

$$= 198.917 \text{ kN}$$

∴   Dead load shear = 30.47 kN

∴   Total design shear = $(198.917 + 30.47) = 229.39$ kN

Assuming an effective depth for cross girder as 1450 mm,

$$A_{st} = \left(\frac{318.47 \times 10^6}{200 \times 0.9 \times 1450}\right)$$

$$= 1220 \text{ mm}^2$$

Provide 4 bars of 20 mm diameter ($A_{st} = 1256$ mm²)

$$\text{Nominal shear stress } \tau_v = \left[\frac{(229.39 \times 1000)}{(300 \times 1450)}\right]$$

$$= 0.52 \text{ N/mm}^2 \text{ not greater than}$$

$(0.07 f_{ck}) = (0.07 \times 250) = 1.75$ mm², hence safe.

$$\text{Spacing } S_v = \frac{M_v}{bd^2} = \left[\frac{(200 \times 2 \times 79 \times 1450)}{(229.39 \times 1000)}\right] = 200 \text{ mm}$$

Adopt 10 mm diameter two-legged stirrups at 150 mm centres throughout the length of the cross girder.

The details of reinforcements are shown in the cross-section of deck slab and longitudinal section of main and cross girders in Figs 23.28 and 23.29.

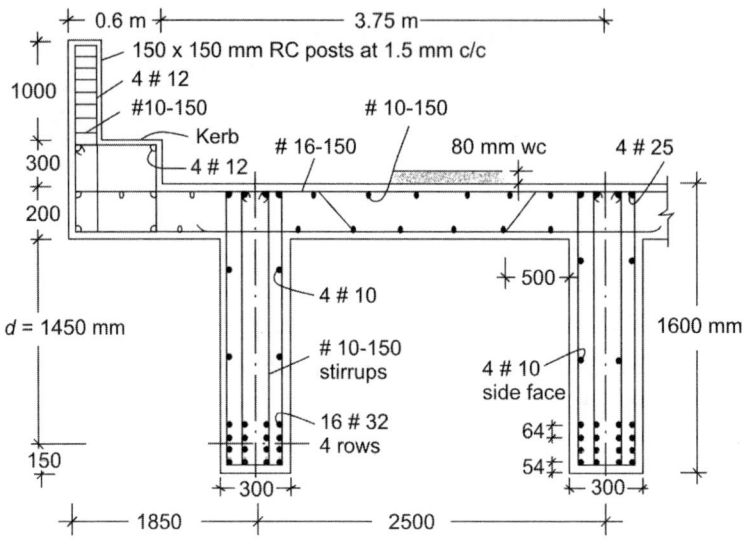

Fig. 23.28: Cross-section of tee beam and slab deck

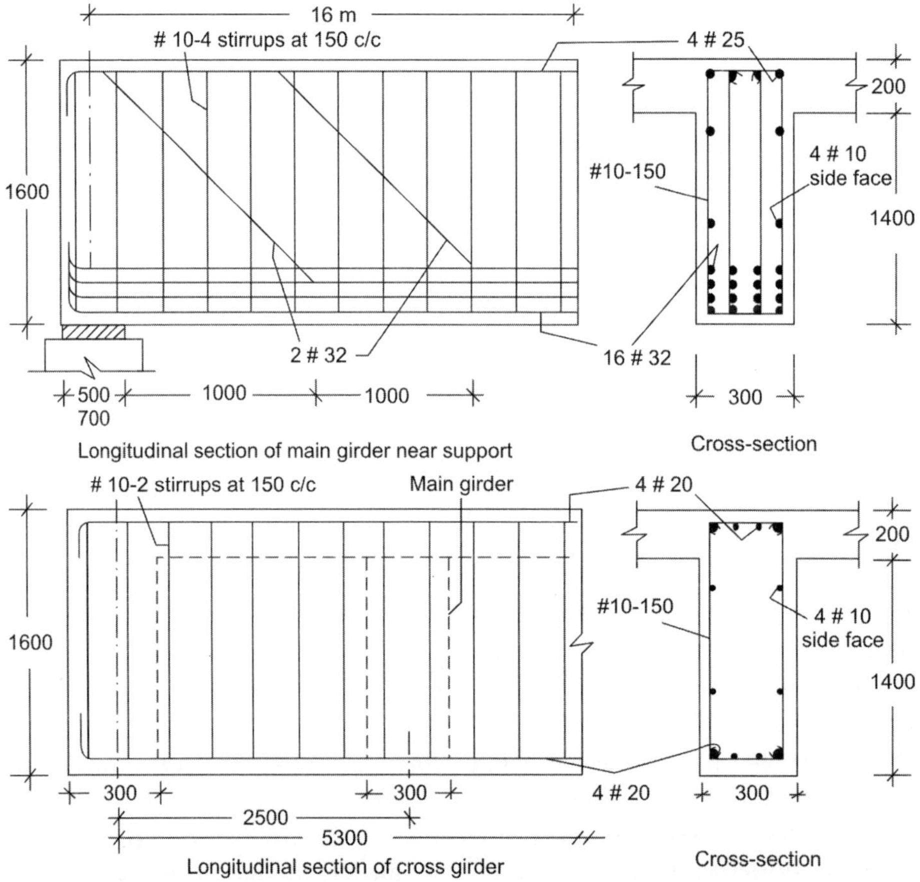

Longitudinal section of main girder near support

Cross-section

Longitudinal section of cross girder

Cross-section

Fig. 23.29: Reinforcement details in longitudinal and cross girder

References

1. Shirley-Smith DB, *World's Greatest Bridges*, English Language Book Society, London, 1964, pp. 1–250.

2. Virola J, The world's greatest bridges, *Civil Engineering, ASCE*, Vol. 38, No. 10, October 1968, pp. 52–55.

3. Freyssinet E, The Birth of Prestressing, Cement and Concrete Association, Translation No. CJ. 59, London, 1956, p. 44.

4. Johnson Vivtor, *Essentials of Bridge Engineering*, 5 edn, Oxford and IBH, New Delhi, 2009, pp. 1–9.

5. Rowe RE, *Concrete Bridge Design*, CR Books Ltd, London, 1962, pp. 1–336.

6. Taylor FW, Thompson SE and Smulski E, *Reinforced Concrete Bridges*, John Wiley & Sons, New York, 1955, pp. 466.

7. Guidelines for the Design of Small Bridges and Culverts, Special Publication, No. 13, Indian Roads Congress, New Delhi, pp 176.

8. Chettoe CS and Adams HC, *Reinforced Concrete Bridge Design*, Chapman & Hall, London, 1952, pp. 1–416.

9. Krishna Raju N, Advances in design and construction of concrete bridges, *Construction India*, Annual Publication, Bombay, 1992, pp. 50–53.

10. Reynolds CE and Steedman, *Reinforced Concrete Designer's Handbook*, Concrete Publications, London, 1974.

Assignment

1. A road bridge deck consists of a reinforced concrete slab continuous over the beams spaced at 2 m centres and cross girders spaced at 5 m centres. Thickness of the wearing coat is 100 mm. Type of loading: IRC class AA or A whichever gives the worst effect. Adopt M20 grade concrete and Fe 415 HYSD bars. Design the deck slab and sketch the details of reinforcements.

2. A reinforced concrete simply supported slab is required for the deck of a road bridge having the data given below:

   Width of carriageway = 7.5 m

   Kerbs = 600 mm wide

   Clear span = 5 m

   Type of loading = IRC class AA or A whichever gives the worst effect

   Materials: M20 grade concrete Fe 415 HYSD bars

   Design the deck slab and sketch the details of reinforcements in the longitudinal and cross-sections of the slab.

3. Design the centre girder of a three-girder RC tee beam and slab deck system using the following data:

   Clear width of roadway = 7.5 m

   Span (centre to centre of bearings) = 15 m

   Loading: IRC class AA and A

   Wearing coat = 100 mm thick

   Thickness of slab = 200 mm

   Spacing of cross girders = 3.75 m

Materials: Concrete M20 grade, tor steel of Fe 415 HYSD bars

Sketch the details of reinforcements in the girder.

4. An RCC tee beam and slab deck is to be designed for a major river crossing on a national highway. The following data is available:

Clear width of roadway = 7.5 m

Foot paths = 1.5 m on either side

Wearing coat = 100 mm thick

Number of main girders = 4

Span (centre to centre of bearings) = 20 m

Spacing of cross girders = 5 m c/c

Loading: IRC class AA or A whichever gives the worst effect

Materials: Concrete M20 grade tor steel of Fe 415 HYSD bars

Design the deck slab and the main girders of the deck and sketch the details of reinforcements.

5. The beam and slab deck of a highway bridge consists of two main longitudinal girders spaced at 4 m centres with cross girders at every 4 m intervals. Design the deck slab using the following data:

Thickness of wearing coat = 80 mm

Loading: IRC class AA or A whichever gives the worst effect

Materials: M20 grade concrete and Fe 415 HYSD bars

Design the deck slab and sketch the details of reinforcements in the slab.

6. A reinforced concrete simply supported slab deck is to be designed for a national highway road bridge having the following details:

Width of carriageway = 15 m

Kerbs = 600 mm wide

Clear span = 6 m

Type of loading: IRC class AA or A whichever gives the worst effect

Materials: M25 grade concrete and Fe 415 HYSD bars.

Design the deck slab and sketch the details of reinforcement in the longitudinal and cross-sections of the slab.

7. A RCC tee beam and slab deck is to be designed for a major river crossing across a national highway, using the following data:

Clear width of roadway = 15 m (6 lane traffic)

Span (centre to centre of bearings) = 16 m

Loading: IRC class AA or A whichever gives the worst effect

Wearing coat = 100 mm thick

Spacing of main girders = 2 m

Spacing of cross girder = 4 m

Materials: M25 grade concrete and Fe 415 HYSD bars.

Design the deck slab and one of the main beams of the bridge deck. Sketch the details of reinforcements in the various structural elements.

Review Questions

1. What are the advantages of reinforced concrete bridge decks over steel bridges? When do you prefer steel bridges in place of concrete bridge decks?

2. Briefly explain the types of IRC. Loading standards generally used for the design of bridges, mentioning the type of highways in which they are used.

3. What method do you adopt to determine the maximum design forces in slabs spanning in two directions?

4. Mention the IRC specifications used to fix the dimensions of (a) kerb (b) wearing coat foot paths in a reinforced concrete slab culvert.

5. Explain with sketches the disposition of IRC class AA tracked vehicle on a slab culvert to determine (a) the maximum bending moment (b) the maximum shear force.

6. What is effective width of dispersion? How do you calculate its value in the case of IRC class AA tracked vehicle?

7. What are the various methods of analysis used for the design of tee beam and slab decks? Explain the Courbon's method of analyzing the forces in the main girders of a bridge deck.

8. What are cross girders in a tee beam and slab bridge deck? What is the purpose of using the cross girders?

9. Explain the method of designing the cross girders in a tee beam and slab bridge desk.

10. Explain with sketches the typical reinforcement details in (a) slab culvert (b) tee beam and slab bridge deck.

Objective Type Questions

1. Reinforced concrete bridge decks are preferred to other types mainly due to
   a. faster construction
   b. superior durability
   c. initial cost considerations

2. The choice of an ideal concrete bridge type to cross a river in a deep valley is
   a. tee beam and slab deck
   b. continuous beam and slab bridge
   c. open spandrel arched type bridge

3. Reinforced concrete slab type bridge is suitable for spans in the range of
   a. 3 m to 4 m
   b. 10 m to 15 m
   c. 5 m to 10 m

4. In case of tee beam slab type bridge decks, the spacing of main girders is generally in the range of
   a. 1 m to 2 m
   b. 2 m to 3 m
   c. 4 m to 5 m

5. In tee beam and slab bridge decks, supporting IRC class AA tracked vehicle load, the maximum live load bending moment will develop in
   a. inner girder
   b. outer girder
   c. cross girder

6. The maximum shear force in the girders of a tee beam slab bridge deck carrying IRC class AA tracked vehicle load will develop in
   a. outer girder
   b. cross girder
   c. inner girder

7. Pigeaud's curves are generally used to determine the maximum moments in the
   a. main girders
   b. cross girders
   c. slab

8. In a reinforced concrete slab culvert, supporting highway loads, the slab is designed as
   a. two way supported slab
   b. continuous slab
   c. one way slab

9. According to IRC:21-1987, the maximum diameter of main reinforcements in main girders should not exceed
   a. 25 mm
   b. 40 mm
   c. 32 mm

10. The tee beam girders of a reinforced concrete bridge deck under live and dead loads should be checked for shear at
    a. centre of span section of inner girder
    b. near support section of outer girder
    c. near support section of inner girder

Appendix 1

Bending Moment and Shear Force Coefficients for Continuous Beams

<table>
<tr><td colspan="2" rowspan="2"></td><td colspan="2">All beams freely supported at end supports</td></tr>
<tr><td>All spans loaded</td><td>Incidental load</td></tr>
<tr>
<td rowspan="5">Max. bending moment coefficient</td>
<td>Uniformly distributed load</td>
<td>

0.125
↑ 0.071 ↑ 0.071 ↑
　0.100　0.100
↑ 0.080 ↑ 0.025 ↑ 0.080 ↑
　0.107　0.072　0.107
↑ 0.077 ↑ 0.036 ↑ 0.036 ↑ 0.077 ↑
　0.105　0.080　0.080　0.105
↑ 0.078 ↑ 0.003 ↑ 0.046 ↑ 0.033 ↑ 0.078 ↑

</td>
<td>

0.125
↑ 0.096 ↑ 0.096 ↑
　0.117　0.117
↑ 0.101 ↑ 0.075 ↑ 0.101 ↑
　0.121　0.107　0.121
↑ 0.099 ↑ 0.081 ↑ 0.081 ↑ 0.009 ↑
　0.120　0.111　0.111　0.120
↑ 0.100 ↑ 0.080 ↑ 0.086 ↑ 0.080 ↑ 0.100 ↑

</td>
</tr>
<tr>
<td>Central point load</td>
<td>

0.188
↑ 0.156 ↑ 0.156 ↑
　0.150　0.150
↑ 0.175 ↑ 0.100 ↑ 0.175 ↑
　0.161　0.107　0.161
↑ 0.169 ↑ 0.116 ↑ 0.116 ↑ 0.169 ↑
　0.158　0.119　0.119　0.158
↑ 0.171 ↑ 0.110 ↑ 0.130 ↑ 0.110 ↑ 0.171 ↑

</td>
<td>

0.188
↑ 0.203 ↑ 0.203 ↑
　0.175　0.175
↑ 0.213 ↑ 0.175 ↑ 0.213 ↑
　0.181　0.160　0.181
↑ 0.210 ↑ 0.183 ↑ 0.183 ↑ 0.210 ↑
　0.179　0.167　0.167　0.179
↑ 0.211 ↑ 0.181 ↑ 0.191 ↑ 0.181 ↑ 0.211 ↑

</td>
</tr>
<tr>
<td>Loads at 1/3 point</td>
<td>

0.167
↑ 0.111 ↑ 0.111 ↑
　0.133　0.133
↑ 0.123 ↑ 0.034 ↑ 0.123 ↑
　0.143　0.095　0.143
↑ 0.119 ↑ 0.056 ↑ 0.056 ↑ 0.119 ↑
　0.141　0.106　0.105　0.141
↑　　↑　　↑　　↑　　↑　　↑

</td>
<td>

0.167
↑ 0.139 ↑ 0.139 ↑
　0.157　0.157
↑ 0.145 ↑ 0.100 ↑ 0.145 ↑
　0.160　0.144　0.160
↑ 0.143 ↑ 0.111 ↑ 0.111 ↑ 0.143 ↑
　0.159　0.148　0.148　0.159
↑　　↑　　↑　　↑　　↑　　↑

</td>
</tr>
<tr>
<td rowspan="2">Max. forces (shear)
Uniformly distributed</td>
</tr>
<tr>
<td>

0.38　0.62
↑ 0.62 ↑ 0.38 ↑
0.40　0.50　0.60
↑ 0.60 ↑ 0.50 ↑ 0.40 ↑
0.39　0.54　0.46　0.61
↑ 0.61 ↑ 0.46 ↑ 0.54 ↑ 0.39 ↑
0.40　0.53　0.50　0.47　0.60
↑ 0.60 ↑ 0.47 ↑ 0.50 ↑ 0.53 ↑0.40 ↑

</td>
<td>

0.44　0.62
↑ 0.62 ↑ 0.44 ↑
0.45　0.58　0.62
↑ 0.62 ↑ 0.58 ↑ 0.45 ↑
0.45　0.60　0.57　0.62
↑ 0.52 ↑ 0.57 ↑ 0.60 ↑ 0.45 ↑
0.45　0.60　0.59　0.58　0.62
↑ 0.62 ↑ 0.58 ↑ 0.59 ↑ 0.60 ↑0.45 ↑

</td>
</tr>
</table>

BM coefficients: Multiply by (span x total load on span) coefficients above the line are for negative BM at supports those under the line are for positive mid-span BM
shear coefficients: multiply by (total load span)

Appendix 2

Values of Exponential Functions

x	e^{-x}	x	e^{-x}
0.1	0.9048	2.6	0.0743
0.2	0.8187	2.7	0.0872
0.3	0.7408	2.8	0.0608
0.4	0.6703	2.9	0.0550
0.5	0.6055	3.0	0.0498
0.6	0.5488	3.1	0.0450
0.7	0.4966	3.2	0.0408
0.8	0.4493	3.3	0.0369
0.9	0.4066	3.4	0.0334
1.0	0.3679	3.5	0.0302
1.1	0.3329	3.6	0.0273
1.2	0.3012	3.7	0.0247
1.3	0.2725	3.8	0.0224
1.4	0.2466	3.9	0.0202
1.5	0.2231	4.0	0.0183
1.6	0.2019	4.1	0.0166
1.7	0.1827	4.2	0.0150
1.8	0.1653	4.3	0.0136
1.9	0.1496	4.4	0.0123
2.0	0.1353	4.5	0.0111
2.1	0.1225	4.6	0.0101
2.2	0.1108	4.7	0.0091
2.3	0.1003	4.8	0.0082
2.4	0.0907	4.9	0.0074
2.5	0.0821	5.0	0.0067

Index

Reader's Notes

Reader's Notes